VOLUME XXI

THE CODE OF MAIMONIDES

(*MISHNEH TORAH*)

BOOK SEVEN

Moses ben Maimon.

The Code of Maimonides

BOOK SEVEN

THE
BOOK OF AGRICULTURE

TRANSLATED FROM THE HEBREW BY

ISAAC KLEIN

RABBI EMERITUS
CONGREGATION SHAAREY ZEDEK
BUFFALO, NEW YORK

NEW HAVEN AND LONDON, YALE UNIVERSITY PRESS

1979

*Published with assistance from
the National Endowment for the Humanities.*

*Designed by John O. C. McCrillis
and set in Granjon type.
Printed in the United States of America by
Vail-Ballou Press, Inc., Binghamton, N.Y.*

*Published in Great Britain, Europe, Africa, and
Asia (except Japan) by Yale University Press,
Ltd., London. Distributed in Australia and New Zealand
by Book & Film Services, Artarmon, N.S.W., Australia;
and in Japan by Harper & Row, Publishers, Tokyo Office.*

*The preparation of this volume was aided by a grant
from the National Endowment for the Humanities.
The findings, conclusions, etc. do not necessarily
represent the view of the Endowment.*

*Library of Congress Cataloging in Publication Data
Moses ben Maimon, 1135–1204.
The code of Maimonides.
(Yale Judaica series, v. 2–)
CONTENTS:
book 3. The book of seasons.
book 5. The book of holiness—book 6. The book of
asseverations.—book 7. The book of agriculture. [etc.]
1. Jewish law. I. Title. II. Series.
Law 296.1'8 49-9495
ISBN: 0-300-02223-9*

TO THE MEMORY OF

ROSE AND EPHRAYIM LEVINE

my parents-in-law

AND

ELKE AND SAMUEL KLEIN

my parents

CONTENTS

INTRODUCTION

The seventh Book of Maimonides' Code is entitled *Sefer Zĕra'im,* "Book of Seeds." In choosing this title Maimonides followed the Mishnah, where the title *Zĕra'im* is applied to the Order which covers roughly the same subject matter.

One of the well-known characteristics of Maimonides' Code, which differentiates it from the other codes of Jewish law, is that it is comprehensive and includes not only laws that were relevant to current Jewish practice, but also laws that were no longer in force in Maimonides' own time. He himself was well aware of this fact, as will be seen from his introductory statement cited below.

Why did Maimonides deviate from the practice of the other codifiers who preceded and followed him? A concomitant question is, what purpose did Maimonides wish to serve with his Code? The most obvious answer to the latter question is that the Code was meant to serve principally as a guide to practice, as an authoritative manual for those who had to render legal decisions. Is this not the purpose of all legal codes? But if this was what Maimonides had in mind, why did he include laws which no longer had any practical application in his own time?

The alternative answer is that Maimonides meant the Code to serve as an encyclopedia of Rabbinic law, in which he gathered all the relevant material scattered in the entire Rabbinic literature, from the Talmud down to his own time, and organized it into a logical system. We find some support for this answer in the words of Maimonides himself, for he says in his Introduction: "The main object of the matter is that no man shall have need for any other compilation whatsoever as regards any of the laws of Israel, so that this present compilation shall serve as a cyclopedia of the whole Oral Law, together with the enactments, customs, and edicts which were introduced from the days of our master Moses to the close of the Gemara . . . I have therefore named it *Mishneh Torah,* for if one studies the Torah first and then reads this work, he will therefrom obtain knowledge of the entire Oral Law, and will have no need to read any other book written in the interval between them."

The controversy about this point is an old one and continues down to our own day.[1] Some scholars tried to reconcile the two schools of thought by suggesting that Maimonides had both purposes in mind, i.e., his work was to serve as an authoritative practical code and as a textbook for study, for how else was one to know which laws were still applicable and which had become merely theoretical? To this other students of the Code reply that in his arrangement of the subject matter Maimonides made quite clear which laws were still in force and which were no longer so. They divide the fourteen Books of the Code into two major divisions: the first ten Books deal with subjects which may be subsumed under the heading of "matters between man and God"; the last four Books deal with civil law and may be subsumed under the heading of "matters between man and man." As to the former group, the first six Books of it deal with subjects that are still relevant, while the next four deal with laws that have become merely theoretical. In the case of the latter group, the first three Books deal with laws that are still applicable, while the fourth is only of academic interest.[2]

Solomon Zeitlin[3] suggests that Maimonides did not intend his work to be merely another code. Rather he fully expected the imminent return of the children of Israel to their ancestral homeland, and therefore prepared a work which was to serve as the constitution of the restored Jewish commonwealth. Chaim Tchernowitz[4] is even more explicit: he claims that a careful study of the Code cannot but lead to the conclusion that Maimonides intended it to become the constitution of the newly reestablished kingdom of Israel, for to him the "return to Zion" was a very real and active factor in the legal decisions which he rendered. In another earlier work,[5] however, Tchernowitz main-

1. Chaim Tchernowitz, *Tolĕḏoṯ hap-posĕḳim* (New York, 1946–47), *1*, 248; Bernhard Ziemlich, "Plan und Anlage der Mischneh Torah," *Moses ben Maimon, Sein Leben, seine Werke und sein Einfluss* (Leipzig, 1908–14), *1*, 249 ff.; Solomon Zeitlin, *Maimonides* (New York, 1935), p. 85.

2. Adolf Schwarz, *Der Mischneh Torah* (Vienna, 1905), p. 14.

3. Zeitlin, *op. cit.*, pp. 85 ff.

4. Tchernowitz, *op cit.*, *1*, 249 f.

5. *Miḳlaṭ*, *5* (1921), 361.

tained that Maimonides' purpose was to prepare an encyclopedic source-book of the Oral Law, and we may add that nowhere in the Code does Maimonides explicitly state that he had any such constitutional intention, except of course that like any other Jewish codifier he believed in the eventual restoration of Zion and the reestablishment of the Jewish people in their ancient homeland, at which time all the laws that could not be observed in exile will once more become applicable.

Whatever the true reason may have been for the inclusion of laws that were no longer in force, the seventh Book of the Code is an apt illustration of this practice of Maimonides. Most of the laws included in it were operative only in the days when there was a Jewish state with a basically agricultural economy, and when the laws set forth in this Book were of vital effect and interest.

In the present translation the term *Zĕra'im* is rendered not by its literal equivalent, "Seeds," but by the more appropriate modern term "Agriculture," which describes its contents just as, if not more, accurately.

CONTENTS

The Book of Agriculture includes laws concerning diverse kinds, gifts to the poor, heave offering, first tithe, second tithe, poor man's tithe, fourth year's fruit, *'orlah,* dough offering, firstlings of animals and humans, and Sabbatical and Jubilee years. A brief examination of these terms will be sufficient here.

Diverse kinds (Hebrew *ḳil'ayim*) deals with the laws prohibiting intermingled planting of heterogeneous seeds, the mating or yoking together of diverse species of animals, and the making and use of clothing material that is a composite of wool and linen. The Scriptural sources of these prohibitions are Lev. 19:19 and Deut. 22:9–13. Elsewhere [6] Maimonides gives the rationale for these laws as having to do with abominable pagan practices. Other commentators suggest that such mingling was regarded as a perversion of the process of creation: since God created all things

6. Maimonides' *Guide of the Perplexed,* III, 37, 39.

with an inherent potential for perfection, mingling of diverse kinds patently interferes with this potential and thus impugns the perfection of God's work.[7]

The Laws Concerning Gifts to the Poor form in fact a general disquisition on individual and public charity. Maimonides follows the Mishnah in including this subject in the Book of Agriculture. In the Mishnah it is covered by the Tractate Pe'ah (Corner Crop), which includes also the other gifts due to the poor out of the various crops, such as gleanings, defective clusters, forgotten sheaves, etc. These have their Scriptural source in Lev. 19:9–10, 23:22; Deut. 24:19–21. Maimonides, however, adds here his own essay on the virtue of charity, including the famous Eight Steps of Charity. Hence his choice of the more comprehensive title, Laws Concerning Gifts to the Poor.

The next two Treatises deal with specific gifts due to the priests and to the Levites. According to Scripture, the Tribe of Levi, which included the priests, was excluded from the distribution of land subsequent to the conquest of Canaan, because this Tribe was entrusted with the service of the Temple in Jerusalem (cf. Num. 18:8). In recompense they were granted the right to certain "gifts" that commoners were obligated to give them. The Levites were granted the tithe, and the priests the heave offering. Before these gifts were handed over to them, the produce liable to these gifts was forbidden for lay consumption, and even after they had been set aside, the heave offering retained a special sanctity that made its consumption restricted to the priests themselves and to their families.

The next two treatises deal with those portions of the husbandman's crops which he was not obligated to give to the priest or to the Levite, but was to consume himself, although subject to certain restrictions. The most important rule for these portions was that they, or their equivalents, were to be brought up to Jerusalem and consumed there. These portions were second tithe, set aside after the first tithe which was given to the Levite (Lev. 23:30; Deut. 14:22–26); first fruits, i.e., fruits which ripened

7. For other explanations see Yehudah Feliks, *Kil'e zĕra'im wĕ-harkabah* (Tel-Aviv, 1967), pp. 8 f.

first (Deut. 26:1–11); and fourth year's fruit, i.e., the fourth year's crop of a newly planted fruit tree, the crops of the preceding three years being forbidden for consumption altogether (Lev. 19:23–24).

The last Treatise deals with the Sabbatical [8] and Jubilee years.[9]

SEQUENCE OF SUBJECTS

In the sequence of subjects in this Book Maimonides deviates radically from the order of the corresponding Tractates in the Mishnah, and with good reason, The Mishnah follows not an order suggested by logic but rather a mechanical sequence governed by the respective length of the several Tractates, the longer preceding the shorter, or by the sequence in which these laws appear in Scripture. In the Mishnah the order of the Tractates is as follows:

1. *Bĕraḵoṯ* (Benedictions)
2. *Pe'ah* (Corner Crop)
3. *Dĕmay* (Doubtfully Tithed Produce)
4. *Kil'ayim* (Diverse Kinds)
5. *Šĕḇi'iṯ* (Sabbatical Year)
6. *Tĕrumoṯ* (Heave Offerings)
7. *Ma'ăśĕroṯ* (Tithes)
8. *Ma'ăśer Šeni* (Second Tithe)
9. *Ḥallah* (Dough Offering)
10. *'Orlah* (First Three Years' Crop of Fruit Tree)
11. *Bikkurim* (First Fruits)

In the Book of Agriculture the sequence is as follows:

1. *Kil'ayim* (Diverse Kinds)
2. *Mattĕnoṯ 'Ăniyyim* (Gifts to the Poor)
3. *Tĕrumoṯ* (Heave Offerings)
4. *Ma'ăśer* (Tithe)
5. *Ma'ăśer Šeni wĕ-Neṭa' Rĕḇa'i* (Second Tithe and Fourth Year's Fruit)

8. Exod. 23:10–11; Lev. 25:2–7, 20–22; Deut. 15:1–3.
9. Lev. 25:8–16.

6. *Bikkurim* (First Fruits [and Other Gifts to the Priests])
7. *Šĕmiṭṭah wĕ-Yobel* (Sabbatical and Jubilee Years)

It will be noted immediately that some Tractates of the Mishnah are missing from Maimonides's list, namely Bĕrakot, Pe'ah, Dĕmay, Ḥallah, and 'Orlah. The explanation varies: Maimonides either treated the subject matter of the particular Tractate of the Mishnah in another Book of the Code, where he felt it preferably belonged, or he regarded it as belonging in one of the other Treatises of this Book. Thus Bĕrakot, which deals with prayers and benedictions, is included where it properly belongs, in Book II of the Code, which sets forth the order of the liturgy. Pe'ah is included in the Treatise Concerning Gifts to the Poor, which is a more inclusive title and deals with all the gifts to the poor that are due from crops, and also with charity in general. Dĕmay is included in the Treatise on Tithes, since it covers a subject closely related to them. Ḥallah and 'Orlah are similarly included in Treatises which deal with closely related subjects.

The sequence of the Treatises in the Book of Agriculture is influenced by the chronology of the application of each type of law. Diverse Kinds deals with seeds at sowing time. Gifts to the Poor are due at harvest time. Heave Offerings and Tithes come into force after the produce has been harvested and threshed and is being made ready for consumption. Second Tithe, Fourth Year's Fruit, and First Fruits follow all these in time, and have in common the requirement that they, or their equivalent, must be brought up to Jerusalem and consumed there. Obviously Sabbatical and Jubilee years come last, because they occur at far longer intervals.

One of the major problems in the present translation is represented by the English renderings and the scientific equivalents of the names of plants and animals which abound in this Book. In the case of animals this problem is relatively minor, because the list is comparatively short, and most of those mentioned are familiar domesticated species. In the case of plants, however, the problem is substantial. To be sure, Maimonides usually simply quotes the Mishnah and leaves the problem of terminology to its interpreters, who were faced with the same difficulty of correct identification. But Maimonides himself wrote also a commentary on the Mishnah, where he could not escape this difficulty, since he had to translate these terms into Arabic. According to Immanuel Loew,[10] Maimonides made use of the traditional knowledge of these plants that was available in his time, and in addition possessed an extensive expertise in the medicinal herbs which he had to use in his daily work as a practicing physician. Loew made ample use of the Arabic botanical equivalents given in Maimonides' commentary on the Mishnah, and praises him highly for their relative accuracy. I have generally followed Loew's identifications, but I was also cognizant of the fact that a great deal of further study has been done recently in this field in Israel. I have therefore consulted the more recent Israeli works as well, especially the work of Dr. Yehudah Feliks,[11] as well as his lists of plants and animals in the Bible and the Mishnah printed in the *Encyclopaedia Judaica* (new English edition, *13,* 615–626; *3,* 9–16). In certain instances there is disagreement among the experts, but since the *Encyclopaedia Judaica* has adopted Dr. Feliks' lists, I followed them in such doubtful cases. In some instances the meaning of a term in the Mishnah and consequently in the Code varies from its meaning in modern spoken Hebrew; obviously I had to use the old meaning and not the one used in Israel today. Each such deviation has its own history, which does not concern us here. Glossaries of botanical and zoological terms (English—Latin—Hebrew) are appended here for the reader's convenience.

In addition to the special difficulties encountered in translating the terminology of flora and fauna which is peculiar to this Book, there were of course the usual general difficulties enumerated in my Introduction to the Book of Women (YJS, *19,* xxxii–xxxiv).

10. *Die Flora der Juden* (Vienna, 1924–34), 4 volumes, the basic work on this subject.

11. See above, n. 7.

SPECIAL TOPICS

Sacred Gifts

A large part of the Book of Agriculture is devoted to gifts due to the priests and Levites. The sixth Treatise enumerates the twenty-four gifts to the priests (*mattĕnoṯ ḵĕhunnah*), and these are divided according to the rules that govern them. As already mentioned above, these gifts were a special consideration given to the priests and Levites because they were required to render a lifetime of service at the Sanctuary in Jerusalem, and also because at the time when the Holy Land was divided among the Tribes, after its conquest by the children of Israel, the Tribe of Levi was not included in this division. Another source of livelihood had therefore to be provided for the members of that Tribe.[12]

As already stated, these gifts came out of field crops, domestic animals, and sacrificial offerings brought by laymen to the Temple. According to students of religion, all such gifts to priests were essentially sacrificial tributes to the deity itself and were not meant originally as revenue devoted to the physical maintenance of its priesthood.[13] While there is a trace of this even in the Talmud, where it is stated that "the priests enjoy this privilege from the table of the Most High,"[14] Scripture itself expressly designates it as statutory revenue for the support of priests[15] and Levites.[16] Certainly those gifts that are treated so exhaustively in the Book of Agriculture do fall into the category of revenue for the maintenance of priests and Levites, hence George Foot Moore treats this subject in the chapter entitled "Taxation."[17] But there were sacramental restrictions around these gifts that lent them a special aspect. First of all, any produce liable to these gifts had the status of *ṭeḇel,* i.e., it was forbidden for lay consumption, and only after the gifts had been taken out of it was this interdict lifted.[18]

12. Num. 18:23–24.

13. W. Robertson Smith, *Lectures on the Religion of the Semites,* 3rd ed. (New York, 1927), pp. 244 ff.

14. B. Ḥul. 120a. 15. Num. 18:20. 16. Num. 18:21–22.

17. *Judaism* (Cambridge, Mass., 1927–30), 2, 70 ff.

18. Code, V, II, x, 19–21 (YJS, *16,* 206 f.).

The first of these gifts was the great heave offering (*tĕrumah gĕdolah*). This was given to the priests, and only they and the members of their own households were permitted to consume it, and even so only when they were in a state of ritual cleanness. This gift alone did not lift the status of *ṭebel* from the produce involved: the husbandman had to proceed next with the tithing of his produce and give this first tithe to the Levite; only after this was done did the remainder of the produce cease to be ṭebel.

The Levite in his turn had to give to the priest one-tenth of his share of the first tithe, and this was termed heave offering of the tithe (*ma'ăśer min ham-ma'ăśer* or *tĕrumat ma'ăśer*). Once the Levite has given to the priest this tenth part, the remainder of the first tithe ceased to be sacred and might be consumed by any person.

The percentages to be given vary. In the case of tithes the amount is fixed at one-tenth, as the term itself indicates. In the case of the great heave offering, no amount is specified in Scripture, but the Rabbis suggested that a generous person should give one-fortieth of the total yield, the miserly one-sixtieth, and the middling one-fiftieth.

While all these gifts were mandatory and in the nature of taxes, they were not collected by any official administrative agency, and the owner of the produce was free to select any particular priest or Levite, respectively, whom he wished to benefit from his contribution. Hence the aura of private benevolence was preserved intact.

In addition to the tithe payable to the Levite, the Book of Agriculture deals with two more tithes, second tithe and poor man's tithe. The first tithe was set aside every year except the seventh, Sabbatical, year, when the husbandman was forbidden to cultivate his field or harvest it, and any crop that came up spontaneously was considered ownerless and was left to the poor and the stranger. The second tithe was set aside in the first, second, fourth, and fifth years of the septennial cycle, and this tithe, or its equivalent, had to be brought up to Jerusalem and consumed there.[19] The poor man's tithe was given to the poor in

19. Deut. 14:22–27.

the third and sixth years of the septennial cycle, and the needy recipient could consume it without any restriction whatsoever.

Akin to second tithe was fourth year's fruit.[20] The fruit of a newly planted tree was forbidden for consumption during the first three years; in the fourth year the fruit crop had to be brought up to Jerusalem and consumed there, subject to the same regulations as second tithe.

To return to priestly gifts out of the yield of the land, Maimonides, like the Mishnah, deals at length with the first ripened fruits of trees (*bikkurim*). These had to be brought up to the Temple in Jerusalem, where, after an elaborate ceremony, they were given to the particular division of priests whose turn it was to be on duty at that time. The first fruits of the womb of man or animal were also held holy unto the Lord. Animals permitted for consumption were given to the priests, who could consume their flesh after they had been brought to the altar. Animals unfit for consumption, such as unclean species—Scripture mentions specifically the ass—and human male first-born had to be redeemed.[21]

Another gift out of the produce of the field, but at a further stage of its development, was the dough offering (*ḥallah*), which the housewife or the professional baker had to set aside out of the dough and give to the priest.[22] Again Scripture does not specify the amount of it, but the Rabbis ruled that it must be one-twenty-fourth of the dough in the kneading trough in the case of the housewife, and one-forty-eighth in the case of the baker. Pious Jews still observe this rule, though not literally but symbolically, and their housewives, and the bakeries that cater to them, set aside a small piece of dough from each batch and incinerate it in the oven. This will explain to the uninitiated the statement "Ḥallah has been taken out" that appears on the wrapper of some packaged bread or other baked goods.

It should be noted that the previous statement that all these practices, although included in the Book of Agriculture, are only of academic interest today, needs some modification. Since the reestablishment of the State of Israel, some groups have made an effort to restore these practices as currently obligatory. To be

20. Lev. 19:23–25. 21. Exod. 13:2, 12–13. 22. Num. 15:20.

sure, these groups constitute only a small minority, but it is a vocal one. Some of the recently published traditional manuals which became popular as guides to Jewish religious practice, contain additions setting forth how these practices should be observed today.[23]

Charity

It is no accident that the subject of charity and benevolence is included in the Book of Agriculture which deals with obligations to priests and Levites. Gifts to the poor were also considered mandatory and were insisted upon as a matter of compulsory duty rather than of mere voluntary personal choice. Rabbinic literature has two terms for charity, *ṣĕḏaḵah* and *gĕmiluṯ ḥăsaḏim*. By the choice of these particular terms the Rabbis have indicated that charity is not a favor to the poor but something to which the poor have a valid and rightful claim, and the donor a binding obligation. Ṣĕḏaḵah means literally "righteousness, rectitude," and the Rabbis therefore rightly say, "The poor man does more for the householder (in accepting alms from him) than the householder does for the poor man (by giving him alms), for the latter gives to the former the opportunity to perform a religious duty (*miṣwah*)."[24]

The term *gĕmiluṯ ḥăsaḏim*, "act of loving-kindness," stresses the quality of the donor's act, the personal attention, sympathy, and service that go with it. It indicates the type of giving that not only helps the poor man physically but also does it in a way that preserves his human dignity and self-respect.[25] And it is significant that in the course of time this term came to refer specifically to interest-free loans granted to a person in order to enable him to avoid recourse to charity and to assist him in re-

23. Solomon Ganzfried, *Ḳiṣṣur Šulḥan 'aruḵ*, with a brief resumé of the laws valid only in the Land of Israel, by Rabbi J. M. Tykocinsky (Jerusalem, 5734 / 1974); *idem*, *Ḳiṣṣur Šulḥan 'aruḵ haš-šalem*, edited by Rabbi H. Y. hak-Kohen, with the laws valid only in the Land of Israel according to the *Ḥăzon 'iš* (by Rabbi A. Y. Karelitz), collected and edited by Rabbi K. Kahana (Jerusalem, 1954).

24. Lev. Rabbah 34:8. See also *Encyclopaedia Judaica* (English), *5*, 340.

25. Deut. 15:7–8. See also Hasting's *Encyclopaedia of Religion and Ethics*, *3*, 380.

establishing himself in some gainful employment. Practically every Jewish community today has such a Free Loan Society entitled with this term.

Solicitude for the poor and needy runs like a scarlet thread through the entire history of the Jewish people, from its very beginning in the Biblical period down to our own day. The tone is set already in Scripture, in the numerous passages that enjoin willing assistance to the needy and the helpless, for example:

> *If there be among you a needy man, one of thy brethren, within any of thy gates . . . thou shalt not harden thy heart, nor shut thy hand from thy needy brother; but thou shalt surely open thy hand unto him, and shalt surely lend him sufficient for his need in that which he wanteth . . . And thy heart shall not be grieved when thou givest unto him . . . For the poor shall never cease out of the land; therefore I command thee, saying: Thou shalt surely open thy hand unto thy poor and needy brother* (Deut. 15:7–11).

> *When thou reapest thy harvest in thy field, and hast forgot a sheaf in the field, thou shalt not go back to fetch it; it shall be for the stranger, for the fatherless, and for the widow . . . When thou beatest thine olive tree, thou shalt not go over the boughs again; it shall be for the stranger, for the fatherless, and for the widow. When thou gatherest the grapes of thy vineyard, thou shalt not glean it after thee; it shall be for the stranger, for the fatherless, and for the widow. And thou shalt remember that thou wast a bondman in the land of Egypt; therefore I command thee to do this thing* (Deut. 24:19–21).

Passages like these express the spirit in which charity should be dispensed. The substance of it is represented in the legislation which reflects the primitive agricultural economy of that day, and commands the husbandman to leave certain portions of the harvest to the poor. The last quoted Scriptural passage specifies some of these portions; a number of others will be cited by Maimonides further on.

In Mishnaic times we already have what may be regarded as organized charity. The community as a whole felt responsible for help to the needy, and special officials were charged with the task of collecting funds, victuals, and goods and dispensing them to the proper recipients. The most prestigious public office in the community was that of the *gabbay ṣĕḏaḵah,* the collector and dispenser of alms,[26] and the Talmud sets down specific regulations governing assessment for charity, persons subject to such assessment, and the method of alms distribution.

In the Middle Ages, when Jewish life was extremely precarious and constantly beset by oppression, persecution, and expulsion, the charitable institutions of the Jewish communities were tested to the utmost. Each community had to strain its resources to the limit and tax itself over and over again in order to meet the frequent emergencies as well as its ever present normal needs. A sign of the times was the prevalence of an organization for the ransoming of captives (*piḏyon šĕḇuyyim*). Pirates at sea and bands of official or unofficial brigands on land were common, and their captives, if not soon ransomed, were sold into slavery. Whenever humanly possible, such Jewish captives were redeemed by the nearest Jewish community.

Official public charities were supplemented by individuals or groups devoted to specific purposes. Active in the average Jewish community were voluntary associations bearing such names as *Biḳḳur Ḥolim,* "Visiting the Sick," which included helping the stricken family to take care of its sick by providing personal services and often also by extending financial assistance; *Leḥem 'Ăniyyim,* "Bread for the Needy"; *Malbiš 'Ărummim,* "Clothing the Naked"; *Niḥum 'Ăḇelim,* "Comforting the Bereaved"—obviously the comforting was not solely verbal whenever the loss incurred was that of the family's breadwinner; *Haḵnasaṯ Kallah,* "Endowing the Bride"—in an age when a bride was expected to bring in a dowry, families without means or of limited means often encountered great difficulty in marrying off their daughters; *Gĕmiluṯ Ḥăsaḏim,* "Free Loan"; and a host of others.[27]

The supreme test for Jewish charities came with the events of

26. Dem 3:1; B. Pes 49b.

27. Israel Abrahams, *Jewish Life in the Middle Ages* (London, 1932), Chapters 17–18.

our own time, during the two World Wars. After World War I Eastern European Jewry found itself prostrate. The ravages of the war itself—the fighting on the Eastern Front covered areas of heavy Jewish concentration—were followed by the massacres and despoliations of Jews during the Russian civil war and during the armed conflicts between the new independent Poland and Soviet Russia. When the tumult and the shouting died down, one hundred thousand Jews had been massacred and most of the affected communities were economically ruined. It was then that the American Joint Distribution Committee was born.[28] The resources of the American Jewish Community were quickly mobilized for rescue work, and the response showed that the great tradition of organized charity was the inheritance of the entire people, for the response was outstanding, not only in the size of the contributions but also in the number of individuals who responded to the call. World War II brought with it a tragedy that seemed unbelievable even to a people long used to tragedies. The holocaust, with its concentration—really extermination—camps and gas chambers again challenged American Jewry to come to the rescue of the survivors. Another challenge was posed by the establishment of the State of Israel. Every Jewish community was mobilized to cope both with its local needs and with the immense needs of the stricken Jewish communities overseas. The local Federations, as such organizations are called today, have become part of the familiar structure of Jewish communities, with professional staffs and hosts of volunteers. They embody the traditional Jewish idea of charity as a moral and social obligation as well as an expression of voluntary compassion, combining the mandatory aspect of the term ṣĕḏaḳah with the elective benevolent aspect of the term gĕmiluṯ ḥăsaḏim.

Throughout their long history charitable activities have been a necessity among the Jews, without which they could not have survived. It is, however, the sensitive manner of performing them and the profound respect for the feelings of the needy and for the inviolability of their human dignity that characterize the meaning of charity to the Jewish mind. Perhaps the matter-of-fact state-

28. Yehudah Bauer, *My Brother's Keeper* (Philadelphia, 1974).

ment, magnificent in its brevity and simplicity, found in the Mishnah, expresses this best: "There were two chambers in the Temple: one the Chamber of Secrets and the other the Chamber of Utensils. Into the Chamber of Secrets the devout used to put their gifts in secret, and the poor of good family received support therefrom in secret." [29] It was Maimonides, however, who formulated the noblest exposition of the Jewish idea of charity in his much quoted Eight Steps of Charity, in Chapter X of the second Treatise of this Book.[30]

Economic and Social Factors

One would assume that even, or rather especially, in a code of law the economic and social forces prevailing at the time would become apparent. When one reads the Book of Agriculture, however, one finds this assumption to be inapplicable. Maimonides seems to operate in a milieu far removed from his own time. How can we explain this fact?

As already stated in the discussion of Maimonides's purpose in writing his Code, the laws that are set forth in it constitute a restatement of the laws of the Mishnah, and therefore reflect the social and economic conditions of the Mishnaic period. As also stated before, Maimonides hoped for the restoration of a Jewish State which would be a replica of what had existed during the last days of Jewish national independence, coinciding roughly with the time of the composition of the Mishnah. Hence in perusing the Code it is necessary to be familiar with the agricultural and economic set-up of the Eastern dominions of the ancient Roman empire, rather than with that of these dominions under Islamic rule in Maimonides's own time. Several competent works that deal with the former subject are available.[31] The

29. Sheḳ 5:6.

30. Below, pp. 91–92.

31. G. Dalman, *Arbeit und Sitte in Palaestina* (Hildesheim, 1964; reprint of 1928–42 edition), 7 vols. in 8; Y. Feliks, *ha-Ḥaḳla'uṯ bĕ-'Ereṣ Yiśra'el bi-tĕḳufaṯ ham-Mišnah wĕhat-Talmuḏ* (Jerusalem, 1963); G. Glotz, *Ancient Greece at Work* (New York, 1926); J. Ph. Lévy, *The Economic Life of the Ancient World* (Chicago, 1967); M. I. Rostovtsev, *The Social and Economic History of the Hellenistic World* (Oxford, 1941), 3 vols.; *idem, The Social and Economic History of the Roman Empire,* 2d ed. (Oxford, 1957), 2 vols.; J. F. Toutain, *The Economic Life of the Ancient World* (New York, 1930).

only specific Jewish ingredient is that the Romans, in response to the Jewish revolts, confiscated much of the landed property in Palestine, whose original Jewish owners either became tenants of the new owners or left Palestine and emigrated to countries in which they believed they could live under more favorable regimes.

Many persons have been helpful to me in my work on this Book, and I am deeply indebted to them. I must mention especially Dr. Leon Nemoy, who as editor of the Yale Judaica Series has helped me far beyond the call of duty. Professor G. Evelyn Hutchinson and Dr. David Furth, both of Yale University, have very generously volunteered their expertise in zoology and botany and have checked the terminology of fauna and flora, a field entirely outside of my own and Dr. Nemoy's competence. To Henriette, my wife, who has been an unfailing source of inspiration and encouragement, I owe more than I can express in words.

I.K.

Illness prevented Dr. Klein from seeing the book through the press. I did the proofreading and compiled the indexes, and the responsibility for any inaccuracies is mine.

L.N.

THE BOOK OF
AGRICULTURE

COMPRISING SEVEN TREATISES IN
THE FOLLOWING ORDER

1. LAWS CONCERNING DIVERSE KINDS
2. LAWS CONCERNING GIFTS TO THE POOR
3. LAWS CONCERNING HEAVE OFFERINGS
4. LAWS CONCERNING TITHE
5. LAWS CONCERNING SECOND TITHE AND FOURTH YEAR'S FRUIT
6. LAWS CONCERNING FIRST FRUITS AND OTHER GIFTS TO THE PRIESTHOOD
7. LAWS CONCERNING THE SABBATICAL YEAR AND THE YEAR OF THE JUBILEE

TREATISE I

LAWS CONCERNING DIVERSE KINDS

Involving Five Negative Commandments
To Wit

1. Not to sow diverse kinds of seeds together;
2. Not to sow grain or vegetables in a vineyard;
3. Not to crossbreed cattle of diverse species;
4. Not to do work with cattle of diverse species joined together;
5. Not to wear garments woven of diverse sorts.

An exposition of these commandments is contained in the following chapters.

NOTE

In the list of 613 commandments prefixed to the Code, those dealt with in the present treatise appear in the following order:

Negative commandments:

[1] 215. Not to sow diverse kinds of seeds, as it is said: *Thou shalt not sow thy field with diverse seed* (Lev. 19:19);

[2] 216. Not to sow grain or vegetables in a vineyard, as it is said: *Thou shalt not sow thy vineyard with diverse seed* (Deut. 22:9);

[3] 217. Not to crossbreed cattle one species with another, as it is said: *Thou shalt not let thy cattle gender with a diverse kind* (Lev. 19:19);

[4] 218. Not to do work with beasts of two species joined together, as it is said: *Thou shalt not plow with an ox and with an ass together* (Deut. 22:10);

[5] 42. Not to wear garments of wool and linen mixed, such as idolatrous priests wear, as it is said: *Thou shalt not wear a garment of diverse sorts* (Deut. 22:11).

CHAPTER I

1. He who sows two species of seeds together in the Land of Israel is liable to a flogging, as it is said, *Thou shalt not sow thy field with diverse seed* (Lev. 19:19).

2. It matters not whether one sows, or weeds, or covers with soil; as, for instance, if a grain of wheat and a grain of barley, or a bean and a lentil, are lying on the ground, and he covers them with earth, whether with his hand, or with his foot, or with a tool, he is liable to a flogging.

It matters not whether one sows in the ground or in a perforated pot. If, however, he sows in an unperforated pot, he is liable to a flogging for disobedience.

3. It is forbidden to sow diverse kinds of seeds in behalf of a heathen, but it is permitted to tell him to sow them in his own behalf.

One is forbidden to retain diverse kinds of seeds in his field, and he must uproot them. But if he does retain them, he is not liable to a flogging.

An Israelite is permitted to sow diverse kinds of seeds with his own hand outside of the Land of Israel, and even to mix the seeds at the outset. The permissibility of sowing such mixed seeds outside the Land of Israel is a matter of tradition.

4. The prohibition of sowing diverse kinds of seeds applies only to seeds that are fit for human consumption. It does not apply to bitter grasses and similar roots that are suitable only for medical purposes and the like.

5. The planting of diverse kinds of trees is included in the general interdict *Thou shalt not sow thy field with diverse seeds* (Lev. 19:19). How so? If one grafts one tree onto another, as for example, if he grafts a shoot of an apple tree onto a citron (ethrog) tree, or vice versa, he is liable to a flogging on the authority of the Torah, in any place, whether he does it in the Land of Israel or outside of it.

Similarly he who grafts a vegetable onto a tree or a tree onto a vegetable is liable to a flogging in any place.

6. It is forbidden for an Israelite to allow a heathen to intergraft for himself diverse kinds of trees. It is permitted, however, to sow produce seeds and tree seeds together, or to mix seeds of various trees and plant them together, because the laws concerning diverse kinds do not apply to trees except only in the case of grafting.

7. In the case of him who sows diverse kinds of seeds, or intergrafts trees of diverse kinds, even though he is liable to a flogging, they are permitted to be eaten, even by the same person who had transgressed by sowing them, because the interdict covers only the sowing.

It is likewise permitted to plant a shoot of a tree which had been intergrafted with diverse kinds, or to sow the seeds of vegetables that had been sown with diverse kinds.

8. Seed plants are divided into three categories. The first is the one called "grain," consisting of five species, namely, wheat, emmer, barley, two-rowed barley, and spelt.

The second of them is the one called "pulse," consisting of all seeds edible by man other than grain, such as beans, chick-peas, lentils, durra, rice, sesame, millet, French vetch, and the like.

The third of them is the one called "garden seed plants," consisting of the rest of seed plants the seeds of which are not themselves edible by man, but the yield of which is so edible, such as the seed of onion and garlic, the seed of leek and fennel flower, the seed of turnip, and the like. The seed of flax is included among garden seed plants.

When all these species of seed plants are sown and sprout forth, the whole plant, as long as the seed is not visible, is called "grass," and is also called "vegetable."

9. Of the garden seed plants there are some that are customarily sown in fields, as for example, flax and mustard. These are called "species of seeds."

Other garden seed plants, however, are customarily sown only in small beds, as for example, turnip, radish, beet, spinach, onion, coriander, celery, horse-radish, and the like. These are called "species of vegetables."

CHAPTER II

1. If one kind of seed is intermixed with another kind, the rule is that if it amounts to one twenty-fourth of the total—as for example, if one sĕ'ah of wheat is intermixed with twenty-three sĕ'ah of barley—the mixture may not be sown until one reduces the amount of wheat or increases the amount of barley. If he sows the mixture nevertheless, he is liable to a flogging.

2. Whatever constitutes diverse kinds when mixed with another species of seed is joined together to make the one twenty-fourth. How so? If two ḳab̲ of barley, two ḳab̲ of lentils, and two ḳab̲ of beans are mixed with twenty-three sĕ'ah of wheat, one may not sow anything of the mixture until he reduces the one sĕ'ah of the admixture by removing a little of it, or else adds to the wheat, inasmuch as the barley, the lentils, and the beans are all diverse kinds vis-à-vis the wheat.

3. When does this apply? When species of grain or species of pulse are mixed with each other, or when grain is mixed with pulse, or vice-versa. If, however, one of the species of garden seeds is mixed with grain or pulse, the proportion is one twenty-fourth of the amount of such garden seeds that would be sown in a *bet̲ sĕ'ah*. If therefore such garden seeds are mixed with a sĕ'ah of grain or of pulse, one may not sow the mixture until he reduces the amount of the garden seeds or adds to the grain.

4. How so? In case mustard seed is mixed with grain, a ḳab̲ of mustard seed being proper for sowing a bet̲ sĕ'ah, if one twenty-fourth of a ḳab̲ is mixed with a sĕ'ah of grain or of pulse, one must reduce the amount of the mustard admixture.

Similarly, in case the particular species of garden seeds is sown two sĕ'ah to a bet̲ sĕ'ah, if half a ḳab̲ of it is mixed with each sĕ'ah of grain or of pulse, one must reduce the admixture.

5. Therefore, in case flaxseed is mixed with grain, if there are three-quarters of a ḳaḇ of flax with each sĕ'ah of grain, one must reduce the flaxseed; if less, no reduction is necessary, since three sĕ'ah of flaxseed are sown in a beṯ sĕ'ah. In the same manner one should calculate the amounts of all other admixed seeds.

6. When does all this apply? When one had no intention beforehand of mixing and sowing these two intermixed species. If, however, he intentionally mixed one species of seed with another, or sowed the two species together, the rule is that even if there is a single grain of wheat in a heap of barley, it is forbidden to sow it. The same rule applies to all similar cases.

7. If one sows one species of seed in his field, and after it sprouts forth notices diverse kinds therein, the rule is as follows: If the other species amounts to one twenty-fourth of the field, he must pluck some of it out until the admixture is reduced, for appearance's sake, lest people should say that he had intentionally sown diverse kinds.

It makes no difference whether the admixed species that has sprouted is grain among pulse, or pulse among grain, or garden seeds among grain, or pulse among garden seeds.

If the grown admixture is less than this, one need not reduce it.

8. When does this apply? When there is room for suspicion. When, however, it is evident that this was not the intention of the owner of the field, and that the admixture sprouted of itself, he need not be obligated to reduce it.

9. How so? When, for example, an aftergrowth of woad sprouts up among grain, or when fenugreek is planted for human consumption, and several species of grass sprout up among it, since such extraneous growth is injurious to the original crop. The same rule applies to all similar cases.

How can one tell that fenugreek was sown for human consumption? When it is sown in individual beds with a boundary around each.

10. The same applies to the ground of the threshing floor which sprouts many species of plants—the owner need not be

obligated to root them out, inasmuch as it is obvious that he does not wish any plants to sprout in the area of the threshing floor.

If, however, he removes some of them, he must be told, "Uproot them all but one species," inasmuch as he has revealed his mind to the effect that he wants the rest to remain.

11. One may not plant vegetables inside the stump of a sycamore or a similar tree.

If one hides a bunch of turnips, radishes, or the like under a tree or even under a vine, the rule is as follows: If some of the leaves are exposed to view, there is no cause for apprehension, inasmuch as it is not his intention that they take root. If they are not in a bunch, or if the leaves are not exposed to view, there is cause for apprehension of diverse kinds.

12. If in a sown field one harvests the seed crop, leaving the roots in the ground, the rule is that even if these roots bring forth no new sprouts until years later, he may not sow another species in that field until he pulls out these roots.

13. If one's field is sown with wheat, and he reconsiders and decides to sow it with barley before the wheat has sprouted, he must wait until the wheat seed has decomposed and rotted in the ground, or about three days if the field is well watered. Then he must turn it over with a plow, and may thereafter sow the other species. It is not necessary for him to turn over the entire field until no grain of wheat is left that is not uprooted; rather he may plow the field in the same manner as it is customarily plowed before the rain, in order to let it become well watered.

14. If the wheat has sprouted, and he then decides to sow the field with barley, he must first turn it over and then sow. But if he lets out his beast into the field and it crops the sprouts, he may sow another species therein.

15. On the first day of the month of Adar warning was given of diverse kinds, and each person then went into his garden and his field, and cleared it of diverse kinds. On the fifteenth of the month the court's emissaries set out on their rounds of inspection.

16. At first the emissaries would uproot the diverse kinds and throw them aside, and the owners would be pleased to have

them thus clear their fields. Subsequently it was ordained that all fields in which the emissaries found that diverse kinds had sprouted, should be declared ownerless, provided that the proportion of the admixed kind was one out of twenty-four; if it was less than this, the emissaries were not to touch it.

17. The court's emissaries returned during the intermediate days of Passover to look for late growths and diverse kinds that may have sprouted forth, in which case they did not wait, but rather proceeded against them immediately and declared the entire field ownerless, provided that the proportion was one of twenty-four.

CHAPTER III

1. There are individual species of seeds which appear in many different forms, due to variant location and variant manner of cultivation of the soil, so that one and the same species may appear as two different species. Even though these do not resemble one another, nevertheless they are not considered diverse kinds in relation to each other, inasmuch as they are in fact one and the same species.

2. Contrariwise, there are among the seeds cases of two species which resemble each other, their appearance being nearly the same, but which are nevertheless forbidden to be sown together, inasmuch as they are in fact two species.

3. How so? Garden lettuce planted together with wild lettuce, garden chicory with field chicory, garden leek with field leek, garden coriander with mountain coriander, common mustard with Egyptian mustard, Egyptian pumpkin with bitter gourd—all these are not accounted diverse kinds in relation to each other.

Similarly, wheat with darnel, barley with two-rowed barley, emmer with spelt, bean with French vetch, red grass pea with grass pea, white bean with Nile cowpea, chate melon with cucumber melon, kale cabbage with garden cabbage, and beet spinach with orach are not accounted diverse kinds in relation to each other.

Contrariwise, radish with rape, mustard with charlock, and Greek pumpkin with Egyptian pumpkin or bitter gourd, even though resembling each other, are accounted diverse kinds in relation to each other.

4. The same applies to trees which belong to two species, even though they resemble each other in their foliage or fruit—they are accounted diverse kinds, inasmuch as they are in fact two species.

How so? Apple with Syrian pear, peach with almond, and jujube with wild jujube are accounted diverse kinds in relation to one another, even though they resemble each other. Contrariwise, pear with crustuminian pear and quince with hawthorn berry are not accounted diverse kinds in relation to one another.

5. The same applies to any other seeds and trees—even though by their natural characteristics they are two species, so long as the leaves of the one closely resemble the leaves of the other, or the fruit of the one closely resembles the fruit of the other, so that they appear to be two kinds of the same species, there is no cause for apprehension of diverse kinds, because in matters of diverse kinds one must follow only the outward appearance.

6. How so? Turnip with radish are not diverse kinds in relation to each other because their fruits are alike, and turnip with rape are not diverse kinds in relation to each other because their leaves are alike. Radish with rape, however, are diverse kinds, even though both their leaves and their fruits resemble each other, because the taste of the one fruit is very far from the taste of the other. The same applies to all similar cases.

7. How far apart should two species that are diverse kinds be in relation to each other? Far enough to appear separated from one another. If they appear to be planted in a haphazard manner, it is forbidden.

8. There are numerous standard measurements of this distance depending upon the size of the sown field, the density of the foliage, and the reach of the young shoots.

9. How so? If one's field is sown with one species of grain, and he wishes to sow another species of grain in an adjacent field, he must leave between them the space of a *beṯ roḇa'*, which is approximately ten and one-fifth cubits square, either in the middle or on the side. If no such space is left between them, the sowing of the other species is forbidden, but no liability to flogging is incurred until the two species are closer to each other than the distance of six handbreadths.

10. If the field is sown with one species of vegetables, and one wishes to sow an adjacent field with another species of vegetables, be it even pumpkin, he must leave between the two fields a space six handbreadths square, either on the side or in the middle. If it is less than this, it is forbidden, but he is not liable to a flogging until the interval is less than one handbreadth.

11. If one of these two fields is sown with grain, while the other is sown with a vegetable or pumpkin, he must leave the space of a *beṯ roḇa'* between the two crops.

12. Where does the requirement of these various spaces apply? Between two fields. If, however, one's field is sown with one species of vegetable, and he wishes to plant alongside it one row of another species of vegetable, it is sufficient to make between the field and the row a furrow no more than six handbreadths long, its width equal to its depth.

13. If one's field is sown with grain, and he wishes to sow in its midst a row of vegetables, he must leave between them a space of six handbreadths, even if he plants pumpkins, which send forth long intertwining leaves. If these leaves spread out into the grain and become intertwined with it, he must uproot the grain that abuts on the pumpkins, so that the leaves will not intertwine. Needless to say, if he sows one row of one species and one row of another species, it is enough to have a furrow between them, as will be explained.

14. If one leaves the proper space between the two species, and one of the species grows far enough out to overhang the other, whether it is grain over grain, or vegetable over vegetable, or vegetable over grain, or grain over vegetable, all of it is permitted, inasmuch as he had left the proper space. Greek pumpkin alone is excepted, since it stretches exceedingly far; therefore, if it does overhang, he must uproot in front of it, as we have explained.

15. If between the two species there is fallow ground, newly plowed ground, a wall of loose stones, a road, a fence ten handbreadths high, a trench ten handbreadths deep and four handbreadths wide, a tree that overhangs the ground, or a rock ten handbreadths wide, it is permitted to plant the one species on one side of these barriers and the other species on the other side, for inasmuch as each such barrier separates them, they appear to be separated from one another.

16. When does the requirement of space or barrier apply? When one sows in his own field. If, however, A's field is sown with wheat, B may sow barley alongside it, as it is said, *Thou shalt not sow* THY *field with two kinds of seed* (Lev. 19:19)—the prohibition applies only to sowing diverse kinds in A's own field, for Scripture does not say, "Thou shalt not sow land with diverse kinds." Moreover, even if A sows barley in his own field next to wheat, but extends the barley so that it abuts on B's field which is sown with barley, it is permitted, because the barley in A's field appears to be the end of B's field.

17. If A's field is sown with wheat, and B's field adjacent to it is also sown with wheat, A is permitted to sow one furrow of flax alongside his own wheat and next to B's field, inasmuch as whosoever sees it would know that it is not common practice to sow just one furrow of flax, and that A merely wished to test his field as to whether it is or is not fit for the sowing of flax; hence it is the same as if he had sown it in order to waste it. Consequently it is forbidden to sow any other different species between two fields sown with the same species, unless one leaves an intervening space in his own field.

18. If A's own field and B's adjacent field are sown with two species of grain, A may not sow between them even one furrow of mustard or safflower, because people customarily sow only a single furrow of these plants.

If, however, these two fields are sown with two species of vegetables, it is permitted to sow between them mustard or safflower, because it is permitted to surround any species, with the exception of grain, with mustard or safflower, since they cause no harm to it.

Similarly, if a corner of this species of seed touches the side of the other species of seed within one's field, it is permitted, because they appear separated from each other. Needless to say, the same rule applies if the corner of one species of seed touches the corner of the other species of seed, because they appear to be the limit of one field touching the limit of the other, which is permitted without any intervening space or separating barrier, as we have explained.

CHAPTER IV

1. It is permitted to sow two rows side by side of chate melons, and next to them two rows of pumpkins and two rows of cowpeas, with a furrow between each species and the other. One may not, however, sow one row of chate melons, one row of pumpkins, and one row of cowpeas, even if there is a furrow separating each species from the other. For the leaves of these species are long, extend far out, and become intertwined; hence, if sown one row next to the other, they will all become intermixed and will appear as if sown in a haphazard manner.

2. If one's field is sown with a species of vegetables, and he wishes to sow within it several rows of pumpkins, he must uproot an area of vegetables sufficient for a row of pumpkins, with a furrow to separate pumpkin from vegetable. He must then skip an area of vegetables twelve cubits wide and sow a second row of pumpkins, again with a furrow separating gourd

from vegetable. He may continue in this manner as far as he wishes to go, so long as there is a distance of twelve cubits between one row of pumpkins and the other. If it is less than this, it is forbidden, because the leaves from either direction would become intertwined with the vegetables between them, and the whole would seem sown in a haphazard manner.

3. If there is a row sown with pumpkins, be it even a single pumpkin, and one wishes to sow grain alongside it, he must skip the space of a *beṯ roḇaʿ,* since pumpkin leaves spread out over a large area. Everything within the area of this *beṯ roḇaʿ* which separates the two species is included in the measurement, as for example, a grave, a rock, or the like.

4. A furrow or a water channel one handbreath deep may be sown with three species of seed, one on each side of the furrow, and one in the middle.

5. It is permitted to plant two species in the same hollow, be it even chate melon and pumpkin, provided that one species inclines in one direction over the edge of the hollow and the other species inclines in the other direction, so that they appear to be separate from one another.

Similarly, if one plants four species in the same hollow and turns them in four different directions, it is permitted.

6. If one wishes to sow his field in long beds, each with a different species, he must leave a space two cubits square between one bed and the other, and then progressively narrow it until at the end of the beds there is only a minimum of space between them, since this makes it evident that they were not sown in a haphazard manner.

7. If one wishes to lay out his field in patches, each sown with a different species, he may have no more than nine patches in each *beṯ sĕʾah,* each patch equal to a *beṯ roḇaʿ*. Hence the distance between one patch and the other will be approximately ten cubits less a quarter, because a *beṯ sĕʾah* is an area fifty cubits square.

8. What is the difference between a long bed and a patch? A long bed is oblong, while a patch is square.

9. In the case of species of vegetables that are customarily sown only a little at a time, as we have explained, one may sow even five such species in one bed six handbreadths square, provided that he sows four species in the four corners of the bed, and one in the center, leaving a space approximately one handbreadth and a half between one species and the other, so that the one could not draw sustenance from the other. One may not, however, sow more than five species, even if he leaves space to separate them, because the presence of so many species in the same bed makes them appear planted in a haphazard manner.

10. To what does this rule apply? To a bed laid out in a waste place with no other planting beside it. When the bed is situated among other beds, however, one may not sow five species therein, for if he should sow on each side of this bed, and thus on each side of the surrounding beds, the whole of it would appear planted in a haphazard manner. If therefore he inclines the leaves in one bed in one direction, and the leaves in the adjacent bed in another direction, so that they appear separated, it is permitted. It is likewise permitted if he makes a furrow between one bed and the other.

11. It is forbidden to sow on the outside of a bed without a furrow or without inclining the leaves, not even against the seedless corners of the bed. This is a precaution, lest one should sow four species in the four corners of the bed, and then sow other species outside of it against the corners, so that the whole would appear planted in a haphazard manner.

12. If the bed is six handbreadths square and has a boundary around it one handbreadth high and one handbreadth wide, it is permitted to sow in it even eighteen species, three on each side of the boundary and six in the middle, leaving a space of one handbreadth and a half between one species and the other.

One may not, however, plant a head of turnip inside the boundary, lest it should fill it.

More than that he may not sow.

13. It is forbidden to sow several species of seeds in this manner in a bed, because they would appear to be diverse kinds. Species of vegetables, however, customarily sown in small quantities, are permitted, as we have explained.

14. A boundary one handbreadth high, which, after the ground within it is sown with several species, as we have explained, is reduced to less than one handbreadth, remains permitted, inasmuch as it was permitted at the outset.

15. If one wishes to fill all of his garden with several species of vegetables without leaving space to separate them, he should divide the entire garden area into square beds, be they even no more than six handbreadths square. He should then make five circles in each bed, one in each corner of the bed, and one in the center. Thereupon he should sow one species in each circle and four other species in the four corners of the bed, making nine species in each bed, visibly separated from each other. He will thus waste only the space separating the circles, which he must leave fallow, so that the circles should appear separated from the corners as well as from each other.

If he wishes to waste no space at all, the rule is as follows: If the circles are sown lengthwise, he should sow the spaces separating them crosswise, and vice versa, so that they would appear separated.

16. From all this it should be clear to you that when two species are separated by a suitable space which prevents them from drawing sustenance from each other, there is no cause for apprehension about appearance, as we have explained. So also when they appear separated from each other, there is no cause for apprehension about their drawing sustenance from each other, even if they are next to each other, as has just been made clear.

CHAPTER V

1. He who sows two species of grain or two species of vegetables together with the seed of the vineyard is liable to a double

flogging, once for transgressing the commandment, *Thou shalt not sow thy field with two kinds of seed* (Lev. 19:19), and once for transgressing the commandment, *Thou shalt not sow thy vineyard with two kinds of seed* (Deut. 22:9).

2. He is not, however, liable to a flogging for sowing diverse kinds of the vineyard until he sows in the Land of Israel at least one grain of wheat, one grain of barley, and one grape seed at the same drop of the hand. Similarly, if he merely covers them with earth, he is liable to a flogging. Likewise, if he sows two species of vegetables with a grape seed, or one vegetable seed, one grain seed, and one grape seed at the same drop of the hand, he is liable to a flogging.

3. According to the Torah, liability is incurred only for hemp, arum, and similar seeds which become available simultaneously with the produce of the vineyard. Other seeds, however, are forbidden on Scribal authority only. Similarly it is forbidden on Scribal authority to sow diverse kinds in a vineyard outside the Land of Israel.

4. Why did the Scribal interdict extend only to diverse kinds of the vineyard outside the Land of Israel, and not also to diverse kinds of seeds? Because the law is more severe in the case of diverse kinds of the vineyard, for if they are sown in the Land of Israel, it is forbidden also to derive any benefit from them. Since it is forbidden also to derive any benefit from them when sown in the Land of Israel, it is forbidden to sow them outside the Land of Israel as well.

5. One may not hoe diverse kinds jointly with a heathen, but one may uproot with him in order to reduce what is improper.

6. The interdict of diverse kinds of the vineyard applies only to species of grain and species of vegetables. Other species of seeds, and needless to say, other trees, may be sown in a vineyard.

7. It is forbidden to sow vegetables or grain next to grapevines, or to plant a grapevine next to vegetables or grain. If one does

so nevertheless, even though he is not liable to a flogging, the mixture is forfeit, and both ingredients, the vegetable, or the grain, and the vines, are forbidden for benefit; both must be burned, as it is said, *lest the fullness of the seed . . . be forfeited* (Deut. 22:9). Even the straw of the grain and the wood of these grapevines are forbidden for benefit and must be burned. One may not use them to heat an oven or a stove, nor may one cook anything over their fire.

8. Whether one plants these diverse kinds or merely retains them, the rule is the same—once he notices diverse kinds growing in his vineyard and retains them, they become forfeit. A person may not, however, render forfeit a thing that is not his. Therefore, if A causes his grapevine to overhang B's grain, he forfeits his grapevine, but B's grain is not rendered forfeit. If A causes B's grapevine to overhang A's grain, the grain is rendered forfeit but B's grapevine is not. If A causes B's grapevine to overhang B's grain, neither is rendered forfeit.

Therefore, he who sows his vineyard during the Sabbatical year does not render it forfeit.

9. If A sees diverse kinds in B's vineyard and retains them, A, who has seen them, is forbidden to derive any benefit from them, but other people are permitted to do so; for had it been B, the owner of the vineyard, who had retained them, they would have become forfeit to everyone else, as we have explained.

10. If a usurper sows diverse kinds in another person's vineyard, the rule is as follows: If the original owner's name cannot be traced, even though he has not despaired of recovering the vineyard, the diverse kinds are rendered forfeit on the authority of the Torah. If the original owner's name is traceable, even though he has despaired of recovery, the diverse kinds are rendered forfeit by Scribal law only.

11. If the wind has uprooted some branches of the grapevine and has trained them over grain, one must cut them away immediately. But if he was accidentally prevented from removing them, they are permitted and do not become forfeit.

12. If a usurper had sown the vineyard, the owner must harvest the seed immediately upon the usurper's departure, even if this happens during the intermediate days of a festival. If he finds no laborers willing to do it, he should offer them up to one-third over their regular wages. If they demand more, or if he finds no laborers at all, he may proceed at leisure to harvest. If the seed remains until the time when it becomes forfeit, it does become forfeit and both become forbidden.

13. When does grain or vegetable become forfeit? When it strikes root. Grapes? When they reach the size of a hyacinth bean. For it is said, *The fullness of the seed which thou hast sown . . . together with the increase of the vineyard* (Deut. 22:9), implying that grain or vegetable must be rooted, and that grape must be grown.

Grain that has become fully dried and grapes that have fully ripened, however, do not become forfeit. How so? If, for example, one comes and plants a grapevine in the midst of fully dried grain, or similarly, if one sows grain or vegetables next to fully ripened grapes, even though this is not permitted, they do not become forfeit.

14. If the grapes in a vineyard have not yet reached the size of a hyacinth bean, and are still unripe, and one sows vegetables or grain in their midst, and they strike root, neither seed nor grape becomes forfeit. Nevertheless he must be fined, and the seed must be declared forbidden. The unripe grapes, however, are permitted.

If one uproots the seeds before the grapes have reached the size of a hyacinth bean, the grapes may be used for benefit. If some have reached the size of a hyacinth bean and some have not, those that have reached that size become forfeit, while those that have not are permitted.

15. If grapes have reached the size of a hyacinth bean, and one sows grain or some species of vegetables next to them, but gathers the seed before it strikes root, the grapes are permitted for benefit. If the seed has struck root, however, the grapes are forbidden.

16. If a grapevine's leaves have dried up and dropped off, as grapevines usually dry up in the cold season, one may not sow vegetables or grain next to it. If one has done so nevertheless, he has not rendered the vineyard forfeit.

Similarly, if one sows vegetables or grain in an unperforated pot placed in a vineyard, he does not render the grapes forfeit, but he is liable to a flogging for disobedience. A perforated pot, however, is deemed the same as the earth itself.

17. If, as one is passing through a vineyard, seeds happen to fall off of him, or if they come in with the manure or with the irrigation water; or if, as one is sowing or winnowing in a grain field, the wind drives some seeds behind him so that they fall into the vineyard and sprout, this does not render the vineyard forfeit, as it is said, *which thou hast sown* (Deut. 22:9), and this person did not actually sow.

He is, however, obligated to uproot these seeds once he sees them sprouting; if he retains them, he renders them forfeit. If the wind drives the seeds in front of him, and he observes them falling into the vineyard, it is as if he had sown them intentionally. What should he do if the stalks have already sprouted? He should turn them over with the plow, and this is enough. If he finds them grown into fresh ears, he must beat out those ears in order to destroy them, for all of it is forbidden for benefit. If he finds them matured into ripe corn, they must be burned. If he sees them and retains them, they must be burned together with the grapevines that are adjacent to them.

18. If one sees in the vineyard blades of such grass as is not customarily sown, they do not render the vineyard forfeit, even if he wishes to retain them for his cattle or for medicinal purposes, unless he retains something the like of which is retained by the majority of the people in that place. How so? If one retains thorns in a vineyard in Arabia, where people value thorns as feed for their camels, he renders them forfeit.

19. Iris, ivy, white lily, and other species of seeds are not accounted diverse kinds in a vineyard. Hemp, artichoke, and

cotton are like other species of vegetables, and render the plants in a vineyard forfeit.

Similarly, all kinds of grasses that grow spontaneously in the field render the plants in a vineyard forfeit. The cowpea, however, is regarded as a species of seed, and does not render forfeit. Reeds, roses, and boxthorn are considered as species of trees, and are not accounted diverse kinds in a vineyard.

20. The principle is this: that which brings forth leaves out of its root is a vegetable; that which does not bring forth leaves out of its root is a tree. The caper bush is accounted a tree in all respects.

21. If upon seeing vegetables in his vineyard one says, "When I come to them, I will pluck them," this is allowed. If upon coming to them he passes them by saying, "When I return, I will pluck them," the rule is that if he tarries until they ripen by one two-hundredth, they render the vineyard forfeit.

22. How is this amount to be estimated? One should observe how long it takes this vegetable or this species of grain, after it is cut off from the ground, to dry up. Suppose it takes one hundred hours for it to dry up until no sap at all is left in it. If therefore it remains attached to the ground for half an hour, it has ripened one two-hundredth and has become forbidden. If it has remained less than half an hour, it is allowed.

23. It is forbidden to pass through a vineyard carrying a perforated pot in which a vegetable has been sown. If one has placed the pot under a grapevine, and it remains there on the ground long enough to ripen by one two-hundredth, it renders the vineyard forfeit.

24. If one plants an onion in a vineyard, and if thereafter the vineyard is uprooted and onions sprout from the planted root, that root remains forbidden even if the sprouts represent a two-hundredfold increase over the root, because the permitted growth does not neutralize the forbidden root.

CHAPTER VI

1. If one sows vegetables or grain in a vineyard, or retains them until they ripen by one two-hundredth, he forfeits the grapevines around them to a distance of sixteen cubits in every direction, in a circle—not in a square. And the whole of this circle that is thirty-two cubits in diameter is regarded as if it were full of vegetables, so that every grapevine within this circle becomes forfeit together with the vegetables; whatever is outside the circle does not become forfeit.

2. When does this apply? When the distance between the perimeter of this circle and the rows of grapevines outside of it is more than four cubits. If, however, this distance is exactly four cubits or less, the circle is regarded as contiguous with the row nearest to it, just as if its diameter were forty cubits, so that every grapevine that falls within this circle of forty cubits is considered forfeit.

3. When does this apply? When one sows or retains these plants within the vineyard. If, however, he sows outside the vineyard but adjacent to it, he renders forfeit two rows of grapevines next to the seeds, to the entire length of the seeds, and four cubits beyond them, so that the width of four cubits, to the length of the entire row of the vineyard, is rendered forfeit by the seeds.

If he sows next to only one grapevine, he forfeits only six handbreadths of the seed in every direction.

4. A young vine shoot less than a handbreadth high does not render other seeds forfeit. When does this apply? When there are two vines parallel to two other vines, with one more projecting like a tail. If the whole vineyard, however, consists of such young shoots, it does cause forfeiture.

5. In the case of two gardens one above the other, the lower one made into a vineyard, one may sow the upper one until he reaches the airspace within ten handbreadths of the vineyard,

because it is forbidden to sow seeds within ten handbreadths of airspace next to a vineyard or to a single grapevine. If the upper garden has been made into a vineyard, one may sow the lower one up to three handbreadths below the roots of the vines.

6. If one has his field sown with vegetables or grain and decides to plant grapevines in it, he must first turn over the seeds with the plough and then plant the vines, not vice versa.

If his field is planted with grapevines, and he decides to sow it with seeds, he must first uproot the grapevines and then sow the seeds.

If he is willing to crop the grapevines first, so that less than a handbreadth of them remains close to the ground, he may then sow the seeds and thereafter go back and uproot what is left in the ground of the grapevines.

7. If one bends a grapevine into the soil or even leads it through a dried pumpkin that serves as a pipe, or through an earthenware pipe, the rule is as follows: If the soil above it is three handbreadths thick or more, one may sow on top of it. If the soil above it is less than this, one may not sow on top of it, but one is permitted to sow alongside it.

8. If one bends the vine into rocky soil, he may introduce seeds on top of it, even if the soil over it is no more than three fingers thick. When does this apply? When the root of the vine is not visible. If it is visible, one must leave a space of six handbreadths in every direction and then sow, the same as one must do in case of all single grapevines that have not been bent, as will be explained.

9. If one bends three vines, and their roots are visible, the rule is as follows: If the space between them measures from four to eight cubits, they may be combined with the other standing vines, as if they had not been bent. If the space measures less than this, they may not be combined.

10. If there are less than three bent vines, they may not be combined; rather one must leave six handbreadths' space in each direction and then sow.

11. If one sows under the branches or leaves that grow out of the vine, he renders the seeds forfeit, even if they are several cubits distant from the root of the vine.

12. If one trains the vine over part of a trellis, he may not sow under the rest of the same trellis, even if there are no vine leaves or branches on it. If he does so nevertheless, it is permitted, inasmuch as there are no seeds under the part covered by the vine.

The same rule applies if one trains the vine over some branches of a tree that bears no fruit, as for example, cedar or cypress.

If, however, one trains the vine over part of a fruit-bearing tree, he may sow under those branches of the tree which do not carry the trained vine shoots, inasmuch as no person would waste a fruit tree by making it a trellis for a vine. If the vine shoots spread after the sowing and overhang the seeds, one should turn them another way.

13. If one sows under the rest of the trellis or under the rest of a tree that bears no fruit, and the vine shoots spread and overhang the seeds, it is forbidden either to retain this or to turn back the shoots. What then should one do? He should uproot the seeds.

14. If poles protrude from an arbor, and one is reluctant to cut them down in order not to destroy the trellis, he may sow under them. If, however, he retains them in order to train the sprouting shoots and leaves of the vine over them, he is forbidden to sow under them.

15. If a blossom issues from the arbor or from the trained vine shoot, one must imagine that a plumb line is suspended from it down to the ground, and it is forbidden to sow seeds directly under it.

Similarly, if one stretches a vine shoot from one tree to another, it is forbidden to sow under it.

16. If one ties a rope or a band of reedgrass to a vine shoot and attaches the other end to a tree, he is permitted to sow seeds underneath the rope. If, however, he extends this rope in

order to train the shoots and leaves over it, it is considered the same as a trellis, and it is forbidden to sow under it.

CHAPTER VII

1. He who wishes to sow next to a vineyard must first leave a space of four cubits from the roots of the vines and then sow. If it is only a single vine, he must leave a space of six handbreadths, and then sow. If there is only a single row of vines arranged one alongside the other, be there even a hundred of them, it is accounted not a vineyard but rather the same as a single vine, and one need leave only a space of six handbreadths from the row, and may then sow. If there are two rows, they are accounted a vineyard, and one must leave a space of four cubits on each side, and may then sow.

2. How many vines constitute a row? Three or more. When does this apply? When there is an interval of four to eight cubits between each row of vines and the one next to it. If, however, there are eight cubits between the two rows, not counting the place occupied by the vines themselves, the rows are regarded as separated from each other and not forming one vineyard. One need therefore leave just six handbreadths' space for each row.

Similarly, if there is less than four cubits' space between them, they are regarded as one vine, and one is obligated to leave only six handbreadths in each direction.

3. If there are three rows, they are considered a vineyard, even if there is less than four cubits' space between one row and the next, and the middle row is regarded as if it did not exist.

Similarly, if there are three rows with an interval of eight or more cubits between one row and the next, one is permitted to sow between the rows.

4. Therefore, if one plants his vineyard at the outset with eight cubits' interval between each pair of rows, he is permitted to introduce seeds into it at only six handbreadths distance from each row.

If, however, he sows outside the vineyard, he must leave a space of four cubits from the outermost row, as in the case of other vineyards.

The rule governing the space between the rows of this vineyard is not the same as that governing a vineyard that has gone to waste in the middle, since one has originally planted the vines at the wider distance from each other.

5. If there is a row of vines in A's field, and another row facing it and close to it in B's field, the two rows combine together to form a vineyard, even if a private or public road or a barrier less than ten handbreadths high separates them, provided that the space between them is less than eight cubits wide.

6. If one plants one row on the ground and one on an embankment, the rule is as follows: If the embankment is ten handbreadths above the ground, the two rows do not combine together; if less, they do combine.

7. If one plants five vines, two and two opposite each other and one more projecting like a tail, this is called a small vineyard, and one must leave four cubits' space in every direction from them.

If, however, one plants them two and two opposite each other and one in the middle, or three in one row and two opposite in another row, this does not constitute a vineyard, and he need leave only six handbreadths' space in each direction.

8. If a vineyard has gone to waste, the rule is that if it is possible to gather in it ten vines for each *beṯ sĕ'ah,* planted in groups two opposite two and one more projecting like a tail, or if there are in it enough vines to align three opposite three, it is called a sparsely planted vineyard, and it is forbidden to sow in the whole of it.

9. In the case of a vineyard planted not in rows but in an irregular manner, the rule is as follows: If the vines in it can be aligned two opposite three, it is accounted a vineyard; if not, it is not a vineyard, and it is enough to leave a space six handbreadths wide from each vine, and then sow.

10. If the roots are in line but the shoots are not, it is accounted a vineyard. If the shoots are in line but the roots are not, it is not accounted a vineyard.

If when thin they are not in line, but as they thicken they come into line, it is accounted a vineyard.

How does one tell whether they are aligned? By fetching a measuring cord and stretching it from one to the other.

11. In the case of a vineyard that has gone to waste in its middle, but has remained intact all around it, the rule is as follows: If the bare patch in its middle measures sixteen cubits, one must leave a space of four cubits from the roots of the vines in each direction, and may then sow in the middle of the bare patch. If there are not sixteen cubits in the bare patch, he may not introduce any seeds there. If he does so nevertheless, he does not render it forfeit, so long as he has left a space of four cubits from the vines of the vineyard in each direction.

12. Similarly, in the case of the place that remains bare of vines, between the edge of the vineyard and its fence, called the perimeter of the vineyard, if it measures twelve cubits, one should leave four cubits' space from the vines, and may then sow the rest.

13. If the perimeter measures less than twelve cubits, one may not introduce seeds there. If he does so nevertheless, he does not render it forfeit, so long as he has left a space of four cubits.

Where does this apply? In a large vineyard. A small vineyard, however, has no perimeter, and one should merely leave a space of four cubits from where the vines end, and may then sow up to the fence.

Similarly, a large vineyard that has eight or more cubits' space between one row and the other, has no perimeter.

14. If the fence surrounding the vineyard is less than ten handbreadths high, or if it is ten handbreadths high but less than four handbreadths wide, the vineyard has no perimeter, and one must leave four cubits' space from the end of the vines, and may then sow up to the fence. Even if there is a space of

only four and one-half cubits between the vines and the partition, he may sow in that one-half cubit.

15. In the case of a fence ten handbreadths high, or a trench ten handbreadths deep, and four handbreaths wide, one is permitted to plant a vineyard on one side thereof and vegetables on the other side. Even a partition of reeds forms a valid separation between the vineyard and the vegetables, the same as a fence, provided that the space between one reed and the other is less than three handbreadths.

16. If a fence separating vineyard and vegetable is breached, the rule is as follows: Up to ten cubits, the breach is accounted an entrance, and is permitted. If the breach measures more than ten cubits, one is forbidden to sow opposite it until he leaves the required space from the vines.

If there are many breaches, the rule is as follows: If the intact part of the fence is equal to the breached part, it is permitted, the same as if there were no breach at all. If the breached part exceeds the part that remains standing, one may not sow opposite all the breached places unless he leaves the proper space.

17. If the partition of a vineyard is breached, the owner should be told, "Repair it." If he repairs it and it is breached again, he should be told again, "Repair it." If he despairs of it and fails to repair it, he renders the vines forfeit.

18. If a structure is half roofed and half unroofed, and vines are planted on one side of it, one may sow vegetables on the other side, because the edge of the roof is regarded as having extended downward and closed up the roofed half, thus forming a partition between the two halves.

If one completes the roof, the sowing is forbidden.

19. If a small courtyard is breached fully open onto a large one, and there are vines in the large one, it is forbidden to sow in the small one. If one does so nevertheless, the seeds are forbidden but the vines are permitted.

If the vines are in the small courtyard, it is permitted to sow in the large one, provided that the large one has posts on each side of the small one so that it appears separated from the small one, while the small courtyard does not appear separated from the large one.

20. If a trench ten handbreadths deep and four wide passes through a vineyard, the rule is as follows: If it cuts through all the way from one end of the vineyard to the other, it appears to lie between separate vineyards, and one is permitted to sow in it, provided that the vines do not overhang it, as we have explained.

If it does not cut through from one end to the other, it is the same as a wine press in the middle of a vineyard, which, even if ten handbreadths deep and four wide, or more, makes it forbidden for one to sow in it, unless this bare patch measures sixteen cubits.

21. A path between two vineyards is the same as a vineyard gone to waste in the middle. Therefore, if there are sixteen cubits between the two vineyards, one may leave four cubits space on each side, and sow in the rest. If the space is less than this, he may not introduce any seed there.

22. If a watchman's mound in a vineyard is ten handbreadths high and four wide, one is permitted to sow vegetables on its top, provided that the vine shoots do not touch it, in order that vegetables should not appear to rise up in the middle of the vineyard.

When does this apply? When the mound is square. If it is round, it must have a depression four handbreadths deep so that it would be distinct from the ground, and it must have a layer of soil on top of it three handbreadths thick.

23. In the case of a structure in a vineyard, the rule is as follows: If it is more than three handbreadths by three, one is permitted to sow vegetables inside it. If it is less than three handbreadths square, it is considered closed, and it is forbidden to sow in it.

24. If a single vine is planted in a cleft or a trench, one must leave a space of six handbreadths in each direction, and may

then sow in the entire trench as he would on level ground. If the trench is ten handbreadths deep and its lip above it four handbreadths wide, one is forbidden to sow within it, even if he leaves six handbreadths' space.

25. Similarly, if a single vine is surrounded by a fence ten handbreadths high and four wide, one may not sow within the enclosure, even if he leaves six handbreadths' space. If he does sow, leaving six handbreadths' space, he does not render it forfeit. How much space must he leave at the outset to be allowed to sow? Four cubits in every direction, whereupon he may sow in the rest of the trench or the rest of the space surrounded by the fence.

CHAPTER VIII

1. Vines that grow naturally, so that their branches and bunches of grapes drag on the ground—that is what is called a vineyard. If, however, one makes a frame or a network high above the ground, so that the bunches of grapes and the branches trail over them, thus raising the foliage of the vines from the ground onto this frame and making them overhang—that is called an arbor. The reeds and similar materials with which one constructs the frame or the network, causing the foliage of the vine to overhang them, are called a trellis. There are some other rules which govern the arbor.

2. If one plants one row of five vines or more and trains them over a wall ten handbreadths high, or does something similar—for example, if he plants them by the side of a trench ten handbreadths deep and four wide—this too is considered an arbor, and he must leave a space of four cubits from the arbor, the same as he would leave from a vineyard, and then sow.

3. Whence should one measure? From the base of the fence over which the vines are trained. How so? If the row of vines is one cubit away from the wall, so that the arbor extends from the vines to the wall, he must measure four cubits from the wall,

and then sow, thus making the distance between the seeds and the roots of the vines five cubits. If he comes to sow on the side of the vines, he must leave a space of four cubits from the roots of the vines, thus making five cubits from the wall. In this manner the rule applies to all arbors.

4. Whether one builds the fence first and then plants the vines or plants the vines first and then builds the fence, once he has trained the vines, it is accounted an arbor.

If the fence is demolished or the trench filled, there is no arbor, and each row is considered as single vines.

5. If an arbor has been destroyed in the middle, but five vines remain on one side of the fence and five on the other side opposite to them, it is called fragments of an arbor, and the rule is as follows: If there are eight cubits and one-sixtieth of a cubit between them, one may leave six handbreadths' space from each row, and then sow, provided that he does not sow under the trellis, as we have explained.

6. If there are exactly eight cubits between the rows, one may not sow there. If he does so nevertheless, so long as he has left six handbreadths' space from each row, he has not rendered it forfeit.

If there is no fence there, he may leave a space of six handbreadths from each row, and then sow, because here there is neither an arbor nor fragments of an arbor.

If he goes back and rebuilds the fence, the arbor is restored to its place and the fragments of the arbor are likewise restored to their places.

7. In the case of a small garden surrounded by a fence, with the vines trained around it outside over the entire fence, the rule is as follows: If on each side of the garden there is enough room for the vintager and his basket, one may sow vegetables in that space, inasmuch as it is surrounded by a fence. If there is not that much room in the garden, one may not sow within it, because otherwise it would look like an arbor with vegetables planted within it.

8. If vines are planted on an elevated terrace, so that their arbor extends to overhang the field, the rule is as follows: If one can stand on the ground and gather all the grapes, all the space underneath the arbor is regarded as if it were the place of the roots of the vines, so that it renders four cubits of the field, in every direction from the edge of the arbor, forbidden. If, however, he cannot gather the grapes unless he ascends the terrace or uses a ladder, it is forbidden to sow only underneath the arbor itself.

9. If there are two walls next to each other, and in the corner between them vines are planted, so that the arbor extends along the walls out of this corner until it ends, one must leave the required space from the roots of the vines, and may then sow in the place beyond the arbor's end where there is no arbor overhead. Even though the seed is intentionally located between the two walls which enclose the arbor, one may sow in it nevertheless between the walls, once he has left the required space.

10. In the case of a vine whose stock has grown a little above the ground, then curved to extend along the ground, and then turned upward again like a knee joint, when one measures the six handbreadths or the four cubits between the vine and the seed, he must measure from the end of the knee joint and not from the actual root of the vine.

11. We have already explained that even when one has left the required space between the seed and the vine, he must take care that the vine should not overhang the vegetable, or vice versa.

If one has sown vegetables or grain, and they sprouted, and if thereafter he has trained the vine over them, the straw is permitted but the grain must be burned.

If the roots of the vine extend into the four cubits between the vineyard and the grain, the grain must be uprooted. If the roots of the grain extend into the four cubits, it is permitted.

12. In all distances and measurements mentioned with reference to diverse kinds, the cubit comprises six handbreadths measured with the fingers spread wide. One may not use precise

measurements for diverse kinds, because such precision is used only for strictness.

13. All such measurements of spacing between vines and grain or vegetables apply only to the Land of Israel or to Syria. Outside the Land of Israel, however, it is permitted at the outset to sow alongside the vines within the vineyard. In lands outside the Land of Israel it is forbidden only to sow two species of vegetables or grain together with a grape seed with the same throw of the hand.

If one tells a heathen child to sow for him in a land outside the Land of Israel, it is permitted. He may not, however, tell this to an adult heathen, lest the latter should be mistaken for an Israelite.

14. Even though it is permitted to sow vegetables alongside a vineyard outside the Land of Israel, the vegetables sown there are nevertheless forbidden for consumption, even outside the Land of Israel, provided one is seen gathering and selling it. If there is doubt about the latter circumstance, it is permitted, as we have explained in the Laws Concerning Forbidden Foods.

CHAPTER IX

1. If one causes a male to mount a female not of its species, whether of animals, of beasts, of fowl, or even of species of beasts of the sea, he is liable to a flogging on the authority of the Torah in all places, whether in the Land of Israel or outside of it, as it is said, *Thou shalt not let thy cattle gender with a diverse kind* (Lev. 19:19).

It does not matter whether the animal, beast, or fowl is his own or his companion's. He is not, however, liable to a flogging until he actually inserts the male member with his own hand as one inserts a painting stick into a tube. If he merely causes the one to mount upon the other, or urges them on with his voice, he is liable to a flogging for disobedience.

2. It is permitted to introduce two species into the same enclosure, and if one then sees them mating with one another, he is not bound to separate them. An Israelite is forbidden, however, to lend his animal to a heathen in order that the latter should mate it for him.

3. If one transgresses and mates his animal with a forbidden species, their offspring is permitted for benefit, and if both sire and dam are of a clean species, the offspring is permitted also for consumption, as we have explained in the Laws Concerning Forbidden Foods.

4. Two species of animal or beast that look alike are accounted diverse kinds, and it is forbidden to mate them, even if they can impregnate each other and are similar in appearance, inasmuch as they are in fact two species. How so? Wolf with dog, wild dog with jackal, gazelles with goats, ibexes with ewes, horse with mule, mule with ass, or ass with wild ass—even though these resemble each other, they are regarded as diverse kinds relative to one another.

5. In the case of species which include both wild and domestic varieties, such as wild ox and domestic ox, or mule foaled by a horse dam and horse, it is permitted to mate them, because they are actually the same species. Goose and wild goose, however, are diverse kinds relative to each other, since the domestic goose has its testicles discernible from within while the wild goose has them discernible from without, hence they are two species. The *koy* is accounted diverse kinds relative to both animal and beast, but one is not liable to a flogging on its account because its status is in doubt.

6. The offspring of diverse kinds are subject to the following rules: If their dams are of the same species, they may be mated; if the dams are of a different species, they are forbidden to be mated, but if one does mate them nevertheless, he is not liable to a flogging. Even if one mates this offspring with the dam's species, he is [not] liable to a flogging. How so? A male mule whose dam is an ass may be mated with a female mule whose

dam is likewise an ass, but not with a female ass, and a male mule whose dam is a horse may not be mated with a female mule whose dam is an ass. The same rule applies to all similar cases. Therefore he who wishes to mate a male mule with a female mule, or to use them together as draft animals, must first examine their tokens, such as ears, tail, and voice. If these are similar, it is certain that their dams were of the same species, and they are therefore permitted to mate.

7. If one works two species of animal or beast together, one species clean and the other unclean, he is liable to a flogging in every place, as it is said, *Thou shalt not plow with an ox and an ass together* (Deut. 22:10). It makes no difference whether one plows with them together, or draws a wagon, or drags a boulder, or just leads them, even if only by urging them on with his voice—he is liable to a flogging, as it is said *together,* implying in any manner. He who merely pairs them, however, is exempt until he makes them draw or leads them.

8. Whether one mates an ox with an ass, or any two species, one unclean and the other clean, be it animal with animal, such such as swine with sheep, or beast with beast, such as fallow deer with elephant, or beast with animal, such as dog with goat, or gazelle with swine, and the like—for all these one is liable to a flogging on the authority of the Torah, since beasts are included in the class of animals, as we have explained in the Laws Concerning Forbidden Foods. According to Scribal law, however, two species that are diverse kinds in regard to mating are also forbidden to be used together in plowing or as draft animals, nor may they be led together, and if one does work them together, or uses them as draft animals, or leads them, he is liable to a flogging for disobedience. One is likewise forbidden to lead a terrestrial animal together with a marine beast, as for example, goat with carp, but if he does so nevertheless, he is exempt.

9. In the case of a wagon drawn by animals of diverse kinds, he who sits in it is liable to a flogging, even if he does not urge the animals on, because his mere sitting in the wagon causes the animals to pull it. Likewise, if one person sits in the wagon and

another leads the animals, both are liable to a flogging. Even if a hundred individuals lead such diverse kinds together, all of them are liable to a flogging.

10. It is permitted to work a man together with an animal or a beast, as for example, when a man plows together with an ox or pulls a wagon together with an ass, and the like, for it is said, *with an ox and an ass together,* not "with a man and an ass," or "with a man and an ox."

11. An animal that is disqualified consecrated property, though one body, is regarded by Scripture as if it were two bodies, because it was once sacred and has now become a mixture of sacred and profane, so that this animal is, as it were, a mixture of unclean and clean animals. Now Scripture says, *And if it be any unclean animal, of which they may not bring an offering unto the Lord* (Lev. 27:11), and the Sages have learned by tradition that this refers to animals that are disqualified consecrated property. Therefore, if one plows with an ox that is disqualified consecrated property, or mates it, he is liable to a flogging on account of diverse kinds. This prohibition is based on tradition.

CHAPTER X

1. In clothing only wool and linen mixed together are forbidden under the law of diverse kinds, as it is said, *Thou shalt not wear mingled stuff, wool and linen together* (Deut. 22:11). In the seashore towns there is a woollike substance that grows on the stones of the Salt Sea, has the appearance of gold, is very soft, and is called *ḵalaḵ*. It, too, is forbidden to be mixed with linen, for appearance's sake, since it resembles sheep's wool. Similarly silk is forbidden to be mixed with kalaḵ, for appearance's sake.

2. The wool of a ewe sired by a goat does not render one liable to a flogging on account of diverse kinds, but is forbidden by Scribal Law, for appearance's sake.

Once wool is joined with linen in any kind of juncture whatsoever, it constitutes diverse kinds on the authority of the Torah. How so? If one hackles wool and linen one with the other, cards them, and makes felt cloaks out of them, they are diverse kinds. If he hackles them, spins them together, and then weaves this yarn into a garment, it is likewise diverse kinds.

3. If one sews a woolen garment with twisted linen twine or even with twisted silk twine, or sews a woolen garment with single linen threads or a linen garment with single woolen threads, or ties together woolen threads with linen threads, or braids them together, or even puts wool and linen into a sack or a basket and wraps them together—all these are accounted diverse kinds. Even if he ties a braid of wool with a braid of linen around a leather strap between them, and similarly if he folds linen and woolen clothes and ties them together—all these, too, come under the law of diverse kinds, as it is said, *wool and linen together,* implying that once they are joined in any manner, they are forbidden.

4. Whence do we learn that all these prohibitions are on the authority of the Torah? From the fact that Scripture had to grant a special dispensation in the case of the fringes. For the Sages have learned by tradition that the Scriptural section dealing with diverse kinds was placed next to the one dealing with the fringes solely for the purpose of allowing diverse kinds in the case of fringes, and the fringes are merely threads tied together. From this it follows that where no religious duty is involved such a mixture is forbidden on the authority of the Torah, because the Torah cannot be used to restrict an interdict based on Scribal law.

5. Diverse kinds have no prescribed minimum—even the smallest woolen thread in a large linen garment, or vice versa, is forbidden.

6. If sheep's wool and camel's hair, or the like, are hackled together and spun into yarn, the rule is as follows: If half of it is sheep's wool, it is accounted as if it were all sheep's wool, and it forms diverse kinds if combined with linen. If, however, the

greater part of it is camel's hair, it may be mixed with linen, because the whole yarn looks like camel's hair, and no apprehension need be felt on account of the wool fibers mingled with it, inasmuch as they are not threads of wool.

7. Therefore garments made from sheepskins are permitted even though they are sewn with linen thread, and no apprehension need be felt on account of the wool fibers which become twisted with the linen thread used in sewing, because these are annulled on account of their minute quantity.

8. Similarly, if hemp and linen are hackled together, the rule is as follows: If the greater part is hemp, one may weave the yarn together with woolen threads; if they are half and half, it is forbidden.

9. If one makes a whole garment out of camel's hair, rabbit's fur, or hemp, and then weaves into it a woolen thread on one side and a linen thread on the other, it is forbidden on account of diverse kinds.

10. If a woolen garment is torn, it is permitted to fasten it together by tying it with linen threads. One may not, however, use them for sewing.

11. A person may put on a woolen garment and a linen garment together and tie each one with its own cord on the outside, provided that he does not twist the two cords together and tie them between his shoulders.

12. Garments of diverse kinds may be made for sale—it is only forbidden to wear them or to cover one's self with them, as it is said, *Thou shalt not wear a mingled stuff* (Deut. 22:11), and it is said further, *Neither shall there come upon thee a garment of two kinds of stuff* (Lev. 19:19), "coming upon" implying here the putting on of a garment which is forbidden. Where "coming upon" does not constitute wearing, it is not forbidden, so that, for example, it is permitted to sit under a tent made of diverse kinds. It is likewise permitted on the authority of the Torah to sit on mattresses made of diverse kinds, as it is said, *Neither*

shall there come upon thee, implying that one may spread it under him.

On the authority of Scribal law, however, even if there are ten mattresses one on top of the other, and only the bottom one is of diverse kinds, it is forbidden to sit on the top mattress, lest a fiber from the bottom mattress should become twisted under one's body.

13. To what does this apply? To soft materials like drapes or garments. In the case of hard materials, however, which do not get twisted, such as bolsters and cushions, it is permitted to sit or recline on them, provided that one's flesh does not touch them directly.

14. Thus also in the case of a curtain made of diverse kinds, if it is soft, it is forbidden, lest the attendant should lean against it and it should come into contact with his flesh. If it is hard and does not twist, it is permitted.

15. Footwear made of diverse kinds with only a heel may be worn, because the skin of the foot is tough and does not feel pleasure as does the skin of the rest of the body.

16. Tailors may sew garments made from diverse kinds in their usual fashion, provided that they do not deliberately use them while sewing them in hot weather as protection from the sun or in rainy weather as protection from the rain. The more scrupulous, however, deposit the garments upon the ground while they sew them.

Similarly clothing dealers may sell garments made from diverse kinds in their usual fashion, provided that they do not deliberately carry them upon their shoulders in hot weather to protect themselves from the sun or in rainy weather to keep themselves warm therewith. The more scrupulous, however, carry them hung on a staff behind their backs.

17. One may not take hold of an egg while it is hot with a garment made of diverse kinds, because he thus benefits from

diverse kinds as protection against heat or cold. The same applies to all similar cases.

18. One may not wear a garment of diverse kinds even incidentally, nor even over ten other garments, whereby he derives no pleasure from it, nor indeed in order to escape customs dues. If he does wear it in any such circumstance, he is liable to a flogging.

19. Diverse kinds are forbidden only in garments used for warmth, as for example, shirt, turban, drawers, girdle, mantle, and garments with which one covers his thighs, arms, and the like.

Small bands, however, that people make in their cuffs to hold money or spices, and rags upon which one spreads plaster, emollient, poultice, or the like, are permitted even though bare flesh touches them, since this is not the customary way of warming one's self.

20. A band of leather, silk, or the like, with threads of wool and linen attached to it to dangle over a person's face in order to drive away flies, does not come under the law of diverse kinds, since this, too, is not the customary way of warming one's self.

21. If one leads animals by holding the lead ropes in his hand, some made of linen and some of wool, it is permitted, even if he wraps the ropes together around his hand.

If, however, he ties all of them together, they become diverse kinds, and he is forbidden to wrap them around his hand.

22. Towels used to wipe one's hand, towels used to dry vessels and wipe floors, and cloths used to wrap Scrolls of Torah are forbidden on account of diverse kinds, inasmuch as the hands touch them and they always wrap themselves over the hand and warm it.

23. The marks that launderers and weavers make in garments, so that each individual would recognize his own, are subject to the following rule: If a woolen mark is on a linen garment, or

vice versa, it is forbidden, even if the mark itself is of no value to the owner of the garment.

24. A woolen garment attached to a linen garment by only one stitch is not considered firmly joined and the two garments do not come under the law of diverse kinds.

If, however, one gathers the two ends of the thread and ties them together, or makes two stitches, it is diverse kinds.

25. Shrouds for the dead may be made of diverse kinds, inasmuch as religious duty is not binding upon the dead, and one may likewise make a packsaddle for an ass and sit on it, provided that his flesh does not touch it. He may not, however, put this packsaddle on his shoulder, even if only in order to carry out manure on it.

26. One may carry on his shoulder a corpse or an animal wrapped in cloth of diverse kinds.

27. If a thread of linen woven into a woolen garment, or vice versa, is invisible, one may not sell the garment to a heathen, lest he in turn should resell it to an Israelite. Nor should one line with it a packsaddle for an ass, lest someone else should find it, strip the cloth from the saddle, and use it as a garment, seeing that diverse kinds are not recognizable in it. What then is the remedy for this garment? One should dye it, because wool and linen do not come out the same color, so that the admixture would be immediately recognizable, and he would remove the thread. If the admixture is still not recognizable, the garment is permitted, because the thread might have dropped out and gone, seeing that he has searched for it and did not find it. We have already explained in the Laws Concerning Forbidden Intercourse that all prohibitions of doubtful cases are based on the authority of Scribal law, which is why the Sages were lenient in such cases.

28. If one buys woolen garments from heathens, he must examine them meticulously, lest they should be sewn with linen thread.

29. If one sees his companion wearing a garment of diverse kinds forbidden by the Torah, even when the latter is walking in the street, he should immediately accost him and rip it off him. He should do so even if it is his master who had taught him wisdom, because respect for people's dignity does not set aside a negative commandment explicitly stated in the Torah. Why then does it so set aside in the case of the return of a lost article? Because in the latter case the negative commandment involves money. But why does it so set aside in the case of defilement by the dead? Because Scripture specifies *or for his sister* (Num. 6:7), and the Sages have learned by tradition that this means, for his sister he may not be defiled, but for an unclaimed body he may defile himself.

Anything forbidden by Scribal law, however, may be set aside everywhere when the dignity of people is involved. Even though Scripture says, *Thou shalt not turn aside from the sentence* (Deut. 17:11), this negative commandment is set aside when the dignity of people is involved. Therefore, if one's companion is wearing a garment of diverse kinds forbidden only by Scribal law, he may not rip it and take it off him in the street, but should wait until he reaches his home. If, however, it is diverse kinds forbidden by the Torah, he must remove the garment immediately.

30. He who puts on a garment of diverse kinds, or covers himself with it, is liable to a flogging. If he has worn the garment of diverse kinds all day, he is liable to only one flogging. If, however, he pulls his head out of the garment and then pulls it back, time and time again, he is liable for each act, even though he has at no time removed the garment completely.

When does the rule apply that he is liable to only one flogging for the entire day? When he had been warned only once. If, however, he had been repeatedly warned, "Take it off, take it off!", and he continued wearing the garment after each warning for a sufficient time to take it off and put it on again, he is liable for each delay after a warning, even though he did not take off the garment.

31. If A clothes B with a garment of diverse kinds, the rule is as follows: If B has submitted to it willfully, he is liable to a flogging, while A has transgressed *Thou shalt not . . . put a stumbling block before the blind* (Lev. 19:14). If B did not know that the garment was of diverse kinds, and A has dressed him willfully, A is liable to a flogging, while B is exempt.

32. Priests who wear their priestly vestments when not officiating at the Temple service, are liable to a flogging even if they do so within the Temple, because their girdle is of diverse kinds, and they are permitted to wear it only during the service, which is a positive commandment, as in the case of the fringes.

TREATISE II

LAWS CONCERNING GIFTS TO THE POOR

Involving Thirteen Commandments,
Seven Positive and Six Negative
To Wit

1. To leave a corner crop;
2. Not to reap the corner wholly;
3. To leave gleanings;
4. Not to gather the gleanings;
5. To leave the defective clusters of the vineyard;
6. Not to gather the defective clusters of the vineyard;
7. To leave the grape gleanings of the vineyard;
8. Not to gather the grape gleanings of the vineyard;
9. To leave the forgotten sheaf;
10. Not to return in order to recover the forgotten sheaf;
11. To set aside the tithe for the poor;
12. To give alms according to one's means;
13. Not to harden one's heart against the poor.

An exposition of these commandments is contained in the following chapters.

NOTE

In the list of 613 commandments prefixed to the Code, those dealt with in the present treatise appear in the following order:

Positive commandments:

[1] 120. To leave a corner crop for the poor, as it is said: *And when ye reap the harvest of your land, thou shalt not wholly reap the corner of thy field, neither shalt thou gather the gleanings of the harvest. And thou shalt not glean thy vineyard, neither shalt thou gather the fallen fruit of thy vineyard; thou shalt leave them for the poor and for the stranger* (Lev. 19:9–10);

[3] 121. To leave the gleanings for the poor;

[9] 122. To leave the forgotten sheaf for the poor, as it is said: *When thou reapest thy harvest in thy field, and hast forgot a sheaf in the field, thou shalt not go back to fetch it; it shall be for the stranger, for the fatherless, and for the widow* (Deut. 24:19);

[5] 123. To leave the defective clusters of the vineyard for the poor, as it is said: *Thou shalt not glean thy vineyard . . . thou shalt leave them for the poor* (Lev. 19:10);

[7] 124. To leave the grape gleanings of the vineyard for the poor, as it is said: *Neither shalt thou gather the fallen fruit of thy vineyard; thou shalt leave them for the poor and for the stranger* (Lev. 19:10);

[11] 130. To set aside in the third and sixth years the tithe for the poor, as it is said: *At the end of every three*

years, even in the same year, thou shalt bring forth all the tithe of thine increase (Deut. 14:28);

[12] 195. To give alms to the poor according to one's means, as it is said: *Thou shalt surely open thy hand unto thy poor and needy brother* (Deut. 15:11).

Negative commandments:

[2] 210. Not to reap the field wholly, as it is said: *And when ye reap the harvest of your land, thou shalt not wholly reap the corner of thy field* (Lev. 19:9);

[4] 211. Not to gather the gleanings, as it is said: *Neither shalt thou gather the gleaning of thy harvest* (Lev. 19:9);

[6] 212. Not to gather the defective clusters, as it is said: *And thou shalt not glean thy vineyard* (Lev. 19:10);

[8] 213. Not to gather the grape gleanings, as it is said: *Neither shalt thou gather the fallen fruit of thy vineyard* (Lev. 19:10);

[10] 214. Not to return in order to recover the forgotten sheaf, as it is said: *Thou shalt not go back to fetch it* (Deut. 24:19);

[13] 232. Not to harden one's heart against the poor, as it is said: *If there be among you a needy man . . . thou shalt not harden thy heart, nor shut thy hand from thy needy brother* (Deut. 15:7).

CHAPTER I

1. When one harvests his field, he may not harvest all of it, but should rather leave for the poor some of the standing corn at the end of the field, as it is said, *Thou shalt not wholly reap the corner of thy field when thou reapest* (Lev. 23:22).

It does not make any difference whether one reaps or merely pulls out the produce.

What is left is called corner crop.

2. Just as one is obliged to leave corner crop in the field, so is he obligated to leave it in trees: when he gathers their fruit, he must leave some for the poor.

If he transgresses and harvests the entire field or gathers all the fruit of the tree, he must take some of what he had harvested or gathered and give it to the poor, inasmuch as the giving of it is a positive commandment, as it is said, *Thou shalt leave them for the poor and for the stranger* (Lev. 23:22). Even if he has ground it into flour, kneaded it, and baked it into bread, he must give corner crop out of it to the poor.

3. If all of the harvest which one had reaped has been lost or burned before the corner crop was given to the poor, he is liable to a flogging, inasmuch as he had transgressed a negative commandment, and has thereby rendered himself unable to perform the positive commandment into which it has now been transformed.

4. Similarly as to gleanings, when one reaps and binds into sheaves, he may not pick up the ears that have dropped during the reaping, but must rather leave them for the poor, as it is said, *Neither shalt thou gather the gleaning of thy harvest* (Lev. 23:22).

If he transgresses and picks them up, he must nevertheless give them to the poor, even if he has ground and baked them, as it is said, *Thou shalt leave them for the poor and for the stranger* (Lev. 23:22).

If they are lost or burned after they have been gathered but before they were given to the poor, he is liable to a flogging.

5. The same applies to grape gleanings at the time of vintage and to defective clusters of grapes, as it is said, *And thou shalt not glean thy vineyard, neither shalt thou gather the fallen fruit of thy vineyard; thou shalt leave them for the poor and for the stranger* (Lev. 19:10).

Similarly the binder of sheaves who has overlooked one sheaf in the field may not recover it, as it is said, *When thou . . . hast forgot a sheaf in the field, thou shalt not go back to fetch it* (Deut. 24:19).

If he transgresses and does recover it, he must still give it to the poor, even if he has ground and baked it, as it is said, *It shall be for the stranger, for the fatherless, and for the widow* (*ibid.*). This is a positive commandment.

You thus learn that all these are negative commandments which transform themselves into positive ones, and that he who has not performed the positive commandments comprehended in them is liable to a flogging.

6. The rule covering overlooked sheaves applies equally to overlooked standing corn; meaning that if one has overlooked some of the standing corn and did not reap it, it belongs to the poor.

And just as this rule applies to grain and the like, so does it apply to all trees, as it is said, *When thou beatest thine olive tree, thou shalt not go over the boughs again* (Deut. 24:20). The same rule applies to all other trees.

7. You thus learn that there are four gifts to the poor in the vineyard: grape gleanings, defective clusters of grapes, gleanings, and forgotten sheaf; three in grain: gleanings, forgotten sheaf, and corner crop; and two in trees: forgotten sheaf and corner crop.

8. In all these gifts for the poor the owner has no option as to the recipient; rather the poor may come and take them regardless

of the owner's wishes. Even if he is the poorest person in Israel, these gifts must be exacted from him.

9. Wherever the term "stranger" is mentioned in connection with gifts to the poor, it refers to a righteous proselyte. This is inferred from the fact that in reference to the poor man's tithe Scripture says, *And the Levite . . . and the stranger . . . shall come* (Deut. 14:29), implying that just as the Levite is a son of the covenant, so is the stranger a son of the covenant. Nevertheless the poor of the heathens may not be excluded from these gifts; rather they may come together with the poor of Israel and take of them, for the sake of promoting ways of peace.

10. With reference to gifts for the poor Scripture says, *Thou shalt leave them for the poor and for the stranger* (Lev. 19:10), implying, so long as the poor look for them. If the poor cease to search and go about after them, whatever is left is permitted to anyone, for the corpus of these gifts is not holy like that of heave offerings, nor is one liable for their monetary value, since in reference to these Scripture says not "and he shall give to the poor" but *thou shalt leave them*. Moreover one is commanded to leave them not for beasts or fowls but for the poor, and in this case there are no poor on hand.

11. At what time is everyone permitted to take of the gleanings? After the second pickers, who had followed the first pickers, have come and gone. At what time is everyone permitted to collect grape gleanings and defective clusters? After the poor have gone into the vineyard and have come back, whatever remains thereafter is permitted to everyone else.

At what time is everyone permitted to take of the forgotten fruit of olive trees in the Land of Israel? If the owner has left it at the top of the tree, it is permitted from the beginning of the month of Kislev, which is the time of the second rainfall in a year when the rain is late. If, however, piles of olives have been left under the tree, they are permitted to everyone once the poor have ceased to look for them.

12. So long as the poor man is entitled to take of the forgotten olives that lie on the ground under the tree, he may continue to

do so, even after the forgotten olives at the top of the tree have become permitted to everyone.

And so long as everyone is entitled to take of the forgotten olives at the top of the tree, he many continue to do so, even if he has yet no right to the forgotten olives under the tree.

13. Gifts for the poor that are in the field and with which the poor choose not to concern themselves belong to the owner of the field, even if the poor have not yet ceased to look for the gifts due to them.

14. All these gifts for the poor are in force on the authority of the Torah only in the Land of Israel, just like heave offerings and tithes, since Scripture says, *And when ye reap the harvest of your land* (Lev. 19:9), and *When thou reapest thy harvest in thy field* (Deut. 24:19). It is explained, however, in the Gemara that corner crop is in force also outside the Land of Israel on the authority of Scribal law. It would therefore seem to me that this applies also to the other gifts for the poor, in that they all are likewise in force outside the Land of Israel on the authority of Scribal law.

15. What is the prescribed minimum for corner crop? According to the Torah, there is no fixed minimum, and therefore, if one has left even a single ear of corn, he has fulfilled his obligation. According to Scribal law, however, it must be not less than one-sixtieth of the total crop, whether in the Land of Israel or outside of it. Moreover, one must add to the one-sixtieth according to the size of the field, the number of the poor, and the blessing of the yield.

How so? If the field is so small that were the owner to leave one-sixtieth of the crop, it would be of no use to the poor, he should add to the minimum. Likewise, if the poor are numerous, he should add to the minimum. If he had sowed little and harvested much, having thus been blessed, he must add according to the blessing.

Whosoever adds to the corner crop, his reward is increased, for there is no maximum to this addition to the gifts for the poor.

CHAPTER II

1. Any crop that is edible, grows out of the soil, is watched over, is harvested all together, and is brought in for storage is subject to corner crop, as it is said, *And when ye reap the harvest of your land* (Lev. 19:9).

2. Anything that resembles such harvest in these five ways is also subject to corner crop. For example, grain, pulse, carobs, walnuts, almonds, pomegranates, grapes, olives, dates, whether dried or soft, and anything else that is similar to these. Woad and madder, however, and their like, are exempt, because they are not edible. Similarly truffles and mushrooms are exempt, inasmuch as they do not grow out of the soil like other produce of the soil.

Ownerless produce is likewise exempt, because it has no one to watch over it, since it is freely available to everyone.

Figs, too, are exempt, because they are not harvested all together, since a fig tree may bear some figs that are fully ripe today, while others will not ripen until many days hence.

Vegetables are similarly exempt, since they are not gathered in for storage.

Garlic and onions, on the other hand, are subject to corner crop, since when dried they are brought in for storage. So also seed onions left in the ground for seeding are subject to corner crop. The same rule applies to all similar cases.

3. Crop land of any size whatsoever is subject to corner crop, even if it is owned in partnership, as it is said, *the harvest of your land* (Lev. 19:9), *your* being plural, thus including even land that belongs jointly to several persons.

4. A field harvested by heathens for themselves, or by robbers, or gnawed by ants, or trampled down by wind or cattle, is exempt from corner crop, because the obligation of corner crop applies only to a standing crop.

5. If the owner harvests one half of the field and robbers harvest the other half, his half is exempt, since the obligation of corner crop applies only to the half harvested by the robbers.

If, however, the robbers harvest one half of the field, and the owner then proceeds to harvest the rest, he must render corner crop according to the amount of what he has harvested.

If he harvests one half of the field and then sells the yield of the other half, the buyer must render corner crop for the whole field.

If the owner harvests one-half of the field and then consecrates the yield of the other half to the Temple in Jerusalem, the person who redeems the yield of the other half from the treasurer must render corner crop for the whole field.

If the owner harvests one half of the field and consecrates its yield to the Temple, he must render corner crop from the rest of the harvest to the amount due from the whole field.

6. If one cuts off some grapes from a vineyard in order to sell them in the market place, his intention being to leave the rest of the grapes to be pressed in the wine press, the rule is as follows: If he cuts the grapes for the market on either side of the vineyard, he must render corner crop only for the grapes that he cuts for the wine press, according to what has been left. If, however, he cuts the grapes for the market on one side only, it is not the same as casual cutting, a bit here and a bit there, which would have rendered him exempt.

Similarly, if one plucks ripe ears of corn a little at a time and brings them into his house, even if he plucks his entire field in this way, he is exempt from gleanings, forgotten sheaf, and corner crop.

7. If one harvests his entire field before the crop has ripened and while it has not yet reached one-third of its full growth, it is exempt. If it has reached one-third of its growth, it is subject to corner crop. The same applies to fruits of trees—if they are one-third ripe, they are subject to corner crop.

8. If one consecrates his field of standing corn to the Temple, and then redeems it while the corn is still standing, it is subject

to corner crop. If the treasurer harvests it, and then the owner redeems it, it is exempt, because at the time when it became liable to corner crop it was sacred property, which is not subject to corner crop.

9. If a heathen harvests his field and thereafter becomes a proselyte, the yield is exempt from corner crop, gleanings, and forgotten sheaf, even though forgotten sheaf applies only at the time of the binding of the sheaves.

10. Heathen laborers may not be hired to do the harvesting, because they are not versed in the laws of gleanings and corner crop. If one has hired them, and they harvest the entire field, the yield is subject to corner crop.

11. The owner of a field who has harvested all of it without leaving corner crop must render corner crop to the poor out of the harvested ears of corn, and need not tithe it.

If he has given the greater part of the harvest to the poor as corner crop, he is exempt from tithe. Similarly, if he has done the threshing but not the winnowing, he must give them corner crop before tithing.

If, however, he has done both the threshing and the winnowing with shovel and fan, and has thus completed his work, he must tithe first and then give corner crop to the poor out of the tithed produce, to the amount due from that field. The same applies to trees.

12. Corner crop must be left only at the end of the field, so that the poor would know the place to which they are to come, and so that the place would be recognizable to passers-by, and the owner would thus not be subject to suspicion. This is also a precaution against deceitful people who intend to harvest the entire field but say to those who see them harvesting the end of the field, "I have left corner crop at the beginning of the field." Also against an owner who waits for a time when there is no one else present, and thus leaves corner crop and gives it to a poor man who is a relative of his.

If the owner transgresses and leaves corner crop at the beginning of the field or at its middle, it is valid corner crop. But he must thereafter leave enough corner crop at the end of the field to equal the amount due from the rest of the field, after he had set aside the first corner crop.

13. If after the owner has given corner crop to the poor they say to him "Give us from that side," and he does so from that other side, both are valid corner crop.

Similarly, if the owner sets aside corner crop and then says, "Let this be corner crop, and also this," or "Let this be corner crop, and this," both are valid corner crop.

14. Laborers are forbidden to harvest the entire field, and must leave the proper amount of corner crop at the end of it. The poor people, however, have no title to the corner crop until the owner, of his own will, explicitly sets it apart as such. Therefore, a poor man who sees corner crop at the end of a field is forbidden to touch it on pain of being charged with robbery, until it becomes known to him that it is such by the will of the owner.

15. Corner crop out of grain, pulse, and similar plants that are harvested, as well as out of vineyard and trees, must be given when they are still attached to the ground, and the poor people must seize them individually with their hands—they may neither cut them with sickles nor uproot them with mattocks, lest they should strike each other therewith. If the poor people wish to divide them among themselves, they may do so, but even if ninety-nine of them say "divide" while one says "seize," they must follow the one, for he has spoken according to the rule.

16. In the case of corner crop of a trained vine or a date palm, which the poor people cannot reach to seize except at the risk of great danger, the owner may take it down and distribute it among the poor. But if they wish to seize it themselves, they may do so. Nevertheless, if ninety-nine say "seize" and one says "distribute," they must follow the one, for he has spoken according to the rule, and the owner is obligated to take down the corner crop and distribute it among them.

17. Corner crop may be distributed to the poor in the field, or left for them to seize, at three times of the day, at dawn, at mid-

day, and at the time of the afternoon prayer. If a poor man comes at a time other than these, he may not be allowed to take anything, in order that there should be a fixed time for the poor to come together and take their due. Why did not the Sages fix some other time during the day? Because there are among the poor nursing mothers who must eat at the beginning of the day, and minors who are not yet awake in the morning and are therefore unable to reach the field until midday; and there are also old people who cannot reach the field until the time of the afternoon prayer.

18. If a poor man takes part of the corner crop and throws it over the rest, or deliberately falls upon it, or spreads his cloak over it, he must be punished by being barred from it. Even what he has already taken must be repossessed from him and given to another poor man. The same applies to gleanings and forgotten sheaf.

19. If A, having taken of the corner crop, says, "This is for B, a poor man," the rule is as follows: If A is himself a poor man, he acquires possession of it in behalf of B, inasmuch as he is entitled to take possession of it in his own behalf. If A is a rich man, he cannot acquire it for B, but must give it to the first poor man that he encounters.

20. If the owner of a field leaves corner crop for the poor people who stand before him, and another poor man comes up behind him and takes it, the latter acquires title to it, for a person does not acquire title to gleanings, forgotten sheaf, corner crop, or a found sela' coin until it actually reaches his hand.

CHAPTER III

1. Corner crop may not be left in one field for another. How so? If one has two fields, he may not harvest completely one field and leave in the second field the corner crop proper for both, as it is said, *Thou shalt not wholly reap the corner of thy*

field (Lev. 23:22), which implies that he must leave in each field the corner crop due from it. If he nevertheless does leave it in one field for another, it is not valid corner crop.

2. If one's field is sown entirely with one crop, and is bisected by a stream, even a stagnant one, or by a water channel so wide that he cannot harvest both sides together—provided that the channel is flowing and fixed—it is considered the same as two fields, and he must render corner crop from each side separately.

3. The same rule applies if the divider is a private road four cubits wide, or a public road sixteen cubits wide.

A private path, however—that is, a path less than four cubits wide—or a public path that is less than sixteen cubits wide, which is in regular use both in the summer and in the rainy season, constitutes a divider. If it is not used during the rainy season, it does not constitute a divider, and the entire field is regarded as one field.

4. If the divider is fallow land, neither planted nor plowed, or freshly turned land, plowed but not yet planted, or land planted with a different species of seed, as for example, if there is wheat on one side, wheat on the other side, and barley in the middle; or if one harvests the middle of the field even before the crop is one-third grown and then plows the harvested part—in all these cases the whole becomes divided thereby into two fields, provided that each one of such divisions is at least the width of three initial furrows, that is, somewhat less than the ground required to plant one-quarter ḳaḇ of seed.

Whereto does this apply? To a small field fifty cubits by two, or less. If it is larger, the fallow land or the broken ground does not divide it into two fields unless it is the width of a quarter-ḳaḇ plot. If, however, it is sown with a different species of seed, slight though the amount thereof might be, it renders the field divided.

5. If the middle of the field has been devoured by locusts or gnawed by ants, and one then plows over the blighted part, it forms a valid divider.

6. If one plants hilly ground, that is, ground not uniformly even but consisting of high hills alternating with depressions, it is accounted one field, even though he cannot plow and plant the whole of it all at once, but rather must plow the elevated places and the low places separately. He may therefore leave one corner crop at the end of the hilly ground for the whole of it.

7. In the case of terraces that are ten handbreadths high, one must render corner crop for each terrace. If the heads of the rows are intermixed, one may render corner crop out of one terrace for all of them.

If the terraces are less than ten handbreadths high, one may render corner crop out of one terrace for all of them, even if the heads of the rows are not intermixed.

If a rock extends across the entire field, the rule is as follows: If one must lift up the plow on one side of the rock and put it down on the other side, the rock constitutes a valid divider. If one does not have to do so, it does not constitute a valid divider.

8. If one plants a field that has trees in it, he may render one corner crop for the entire field, even though the seeds are planted in separate beds among the trees and are not intermixed, inasmuch as it is evident that this is one field and that the seed beds have been separated from each other because of the location of the trees.

9. When does this apply? When each group of ten trees occupies an area no larger than a one-sĕ'ah plot of ground. If the area is larger than this, one must render corner crop from each bed individually, seeing that the trees are far apart, and it was therefore not because of the trees that one had sown the seed in separate beds.

10. So also in the case of onion beds situated among vegetables, one may render one corner crop for all the onions, even though the vegetables divide them into separate beds.

11. In the case of a field planted with one species, if in some places thereof the plants begin to dry, and if one thereupon up-

roots or pulls out the dried plants on either side, with the result that the remaining green plants are separated into beds one apart from the other, the rule is as follows: If it is the custom of the people in that locality to sow such species, as for example, dill or mustard, in separate beds, one must render corner crop out of each bed, for he that observes them will be bound to say, "These were sown in individual beds originally." If it is a species with which whole fields are customarily sown, as for example, grain and pulse, one may render one corner crop for the whole field.

12. When does this apply? When the dry plants are on either side, with the green ones in the middle. If, however, the green plants are on either side, with the dry ones in the middle, one must leave corner crop out of the dry and out of the green separately.

13. If one sows his field with onions, beans, chick-peas, or the like, with the intention of selling part of them while green in the market place, and leaving part of the field to dry and be threshed for storage, he must leave corner crop separately for what he sells while green and for what he harvests after it has dried.

14. If one sows his field with a single species, he need leave but one corner crop, even if he threshes it in two installments. If he sows two species therein, he must render two corner crops, one for each species, even if he threshes both at the same time.

15. If one sows two kinds of seeds of the same species, as for example, if he sows his field with two species of wheat or two species of barley, the rule is as follows: If he threshes them all at once, he must render one corner crop; if he threshes them in two installments, he must render two corner crops. This is a rule given to Moses on Mount Sinai.

16. Two brothers who have divided their inheritance must render two corner crops. If they turn about and become partners again, they need render only one corner crop.

If two partners harvest half of their field and then divide the field, the one who takes the harvested produce need set aside nothing, while the one who takes corn that is still standing need set aside corner crop only for the half that he has taken.

If they turn about and become partners again, and then jointly harvest the remaining half of the field, each one of them must set aside corner crop out of his share of the standing corn for his partner's share, but not for the half of the field previously harvested.

17. In the case of a field subject to the following circumstances: one-half of it is one-third ripe, while the other half is not; the owner then commences harvesting the ripe half; by the time he has harvested a moiety thereof, the rest of the field becomes one-third ripe; he thereupon completes harvesting that same one-half which has ripened first—in this case he must set aside corner crop out of the initial installment of the harvest for the middle installment, and out of the middle installment for the initial and final installments.

18. If one sells his field piecemeal to several persons, the rule is as follows: If he has sold the entire field, each purchaser must give corner crop out of the share that he has bought. If, however, the owner of the field commences harvesting it, and then sells one part of it while retaining the other part for himself, he must give the corner crop due for the entire field, for once he has commenced harvesting he becomes obligated for the whole of it.

On the other hand, if he sells a part of it first, the buyer must set aside corner crop for what he has purchased, and the owner of the field must do the same for the rest of the field.

19. In the case of a field planted with trees, the only thing that constitutes a valid divider is a high fence separating the trees.

If, however, the fence separates only the lower parts of the trees, while the twigs and the upper foliage above are intertwined and touch the top of the fence, the entire field is accounted one field, and the owner must give one corner crop for the whole of it.

20. If two persons purchase one tree, they must give only one corner crop out of it. If one of the two purchases its northern

side and the other its southern side, each one of them must give corner crop for his own side only.

21. In the case of carob trees, the rule is as follows: If a man standing by a carob tree on one side of the field, and his companion standing by a carob tree on the other side of the field, can see each other, the field is accounted the same as one field, and only one corner crop is due out of all the trees.

If two men standing by trees at the opposite sides of the field cannot see each other, but can see the intervening trees, the owner must set aside corner crop out of the trees on either side for the intervening trees, and out of the intervening trees for the side trees, and may not set aside corner crop out of the trees on one side for the trees on the other side.

22. As for olive trees, all such trees located on one side of the town—for example, all the olive trees located on the western or eastern side of the town—are accounted as one field, and only one corner crop is due for all of them.

23. If one harvests part of his vineyard on each side, in order to lighten the vines so that the remaining clusters may have more room to fill out, he is said to have thinned out the vines. We have already explained that if one harvests grapes on one side only, he is not thinning them out; he must therefore give corner crop out of the rest to the amount due for all of the yield, even if the harvested grapes were meant for the market.

If, however, he has thinned them out for sale in the market place, he need not give corner crop for the grapes that he has thinned out. But if he has thinned them out in order to bring these grapes to his house, he must give corner crop out of the rest that he has left for pressing, to the amount due for all the grapes.

CHAPTER IV

1. What is "gleanings?" That which drops off the sickle during the reaping, or that which drops out of one's hand as he gathers

the ears of corn in order to cut them off; provided that only one or two ears of corn are thus dropped. If three of them drop together, all three belong to the owner of the field.

Whatsoever drops off the back of the sickle or off the back of the hand, even if it is only one ear of corn, is not gleanings.

2. If one does the reaping with his hand and not with a sickle, that which drops out of his hand is not gleanings. If, however, one plucks things which are customarily plucked, that which drops out of his hand is gleanings.

If he is reaping, or is plucking that which is customarily plucked, and if after having reaped an armful or plucked a handful, he is pricked by a thorn, causing what he holds to drop to the ground, it belongs to the owner.

3. If while one is reaping, a single ear of corn remains uncut, although all ears around it have been cut, the rule is as follows: If the tip of that ear is even with the standing corn by its side, so that it is possible to cut it off together with that standing corn, it belongs to the owner of the field; if not, it belongs to the poor.

4. In the case of two ears of corn, one next to the other, of which the inner one may be cut off together with the corn still standing, and the outer one may be cut off together with the inner one but not together with the corn still standing, the inner one saves the outer one and is in turn saved by it.

For it is the same as if it had dropped off the sickle, even though it had not yet been cut off.

The ears of corn that remain among the stubble belong to the owner.

5. If the wind scatters the sheaves, causing the owner's harvest to become intermixed with the gleanings, an estimate should be made of the gleanings that such a field would ordinarily produce, and this amount should be given to the poor, because this is a case of sheer accident. And what is this amount? Four ḳab̠ of produce for each kor plot.

6. If gleanings that had dropped to the ground have not yet been gathered by the poor, and if the owner of the field then

comes along and stacks his harvest on the ground, how should he proceed? He should remove all of his stacked produce to one place, while the remaining ears of corn that touch the ground go all to the poor. For we do not know which of all these ears are gleanings, and the principle is that doubtful gifts to the poor belong to the poor, as it is said, *Thou shalt leave* (Lev. 23:28), implying, set down for them out of your own produce.

7. But why, then, do we not in this case also make an estimate and give to the poor the proper share of gleanings? Because the owner has transgressed by stacking his produce over the gleanings, and is therefore penalized. Even if he did it unwittingly; even if the gleanings are barley and the produce stacked over them is wheat; even if he had called the poor, and they failed to come; even if others had stacked the produce without his knowledge—the rule remains that all the ears touching the ground belong to the poor.

8. If one needs to sprinkle his field before the poor have gathered the gleanings that are in it, the rule is as follows: If the damage caused to him by failure to sprinkle would exceed the damage caused by the sprinkling to the gleanings, he may sprinkle; otherwise he may not. If therefore he gathers together all the gleanings and piles them up on the fence, until such time as the poor may come and take them, this is commendable, being the way of the pious.

9. In the case of seed found in ant-holes, the rule is as follows: If the ant-holes are among the standing corn, the seeds belong to the owner, because the poor have no gift due to them out of standing corn. If the ant-holes are in the harvested part of the field, the seeds belong to the poor, since the ants may have dragged them in from the gleanings. And even if the seeds are found to be blackened, it cannot be said, "These have been left over from yesteryear," because doubtful gleanings are gleanings.

10. If an ear of gleanings becomes mixed with stacked grain, one should set aside two ears of corn and declare over one of them, "If this ear is gleanings, let it be for the poor, and if it is not gleanings, let the second ear include the tithes for which

this ear is liable." He should then turn around, make the same declaration over the second ear, and thereupon give one of these ears to a poor man, and the other will serve for tithe.

11. One may not hire a laborer with the stipulation that the latter's son is to pick up gleanings after him. In the case of sharecroppers, tenants, and purchasers of standing corn for harvesting, the other party's child may pick up gleanings after him.

A laborer may bring his wife and children to pick up after him, even if his wages are one-half of the harvest, or one-third, or one-quarter.

12. He who does not allow the poor to pick up gleanings, or admits one person and bars another, or helps one person against another, is guilty of robbing the poor.

13. One is forbidden to induce a lion or a similar wild beast to lie down in his field, so that the poor would see it and flee. If there are such poor people there as are not entitled to take gleanings, the rule is as follows: If the owner can exclude them, he may do so; if not, he must let them be, for the sake of keeping the peace.

14. If one declares the gleanings ownerless after the greater part of them has already dropped off his hand, they do not become ownerless, inasmuch as once the greater part of them has dropped off, he no longer has any authority over them.

15. What is "grape gleanings?" It is one or two grapes that become separated from the cluster during the vintage. If three such grapes drop off at the same time, it is not grape gleanings.

16. If in gathering the vintage one cuts off a cluster, but it becomes entangled in its leaves, falls to the ground, and is separated into single grapes, it is not grape gleanings. If, on the other hand, one cuts off the clusters and throws them to the ground, the rule is that if, when he comes to remove the clusters, he finds even half a cluster of separate grapes, it is grape gleanings. Indeed, if a whole cluster becomes thus separated into single grapes, it is grape gleanings.

He who places a basket under the vine the while he gathers the vintage is guilty of robbing the poor.

17. What is "defective clusters?" It is a small cluster that is not as thick with grapes as an ear of corn is thick with grains, that has no shoulder stalks, and whose grapes do not droop one over the other, but are scattered. If it has shoulder stalks but no drooping grapes, or vice versa, it belongs to the owner of the vineyard. If the case is in doubt, it belongs to the poor.

18. What are "shoulder stalks?" Small stalks attached to the spine of the main cluster, one on top of the other.

And what are "drooping grapes?" Grapes attached to the spine of the cluster and descending from it, provided that the grapes in the defective cluster are so separated that they can all touch the palm of one's hand.

And why is such a defective cluster called *'olel?* Because it is to the other normal clusters as an infant (*'olel*) is to a grown man.

19. The owner of the vineyard is not obligated to cut off the defective clusters and give them to the poor; rather the poor must cut them off for themselves. A single grape is accounted the same as a defective cluster.

20. If a vine branch has a normal cluster and on its knee another defective cluster, the rule is as follows: If the defective cluster can be cut off together with the normal cluster, it belongs to the owner of the vineyard; if not, it belongs to the poor.

21. If a vineyard consists entirely of defective clusters, it belongs to the poor, as it is said, *And from thy vineyard thou shalt not take the defective clusters* (Lev. 19:10), implying, even if it is all defective clusters.

The rules of grape gleanings and defective clusters apply only to vineyards.

22. The poor are not entitled to take grape gleanings and defective clusters until after the owner of the vineyard has commenced his vintage, as it is said, *When thou gatherest the grapes*

of thy vineyard, thou shalt not take the defective clusters (Deut. 24:21).

How much must he have gathered before the poor are entitled? Three clusters which produce a quarter.

23. If one dedicates his vineyard to the Temple before the defective clusters have become recognizable, they do not belong to the poor. If he dedicates the vineyard after the defective clusters have become recognizable, they belong to the poor, but the value of their growth must be paid to the Temple.

24. If one prunes his vines after the defective clusters have become recognizable, he may continue to do so in his customary manner, and just as he may cut off normal clusters, so may he cut off defective ones.

25. If a heathen sells his vineyard to an Israelite for the vintage, that vineyard is subject to defective clusters. If an Israelite and a heathen are partners in a vineyard, the Israelite's share is subject to defective clusters, while the heathen's share is exempt.

26. If a Levite is given tithe from which the priest's share has not yet been separated, and finds defective clusters in it, he must give them to the poor. If, however, the defective clusters can be cut off together with the normal clusters, he may set aside the heave offering of the tithe, but must designate it as heave offering for another crop.

27. If one has five vines, and having harvested their grapes, brings them into his home, the rule is as follows: If he brings them in in order to eat them as grapes, he is exempt from grape gleanings, forgotten sheaves, and fourth year's fruit, but is obligated for defective clusters.

If the vintage is intended to be made into wine, he is obligated for all these gifts, unless he leaves over some of the grapes.

CHAPTER V

1. A sheaf forgotten by the laborers but not by the owner of the field, or vice versa, or forgotten by both but seen by other

persons passing by, who have observed the laborers and the owner as they forgot it, is not accounted forgotten sheaf, until it is forgotten by everyone. Even a hidden sheaf, if forgotten, is accounted forgotten sheaf.

2. If the owner of the field, while in the city, says, "I know that the laborers have forgotten the sheaf which is in such-and-such a place," and they did indeed forget it, it is accounted forgotten sheaf.

If, however, he is in the field as he makes this declaration, and the laborers did indeed forget it, it is not accounted forgotten sheaf. For a forgotten sheaf is one forgotten initially by the owner in the field; even if the sheaf is remembered by the owner in the city and only subsequently forgotten by him, it is still accounted forgotten sheaf, as it is said, *And hast forgot a sheaf in the field* (Deut. 24:19), not in the city.

3. If the poor stand in front of the owner, or cover the sheaf with stubble, and he remembers the stubble; or if he takes hold of the sheaf in order to convey it to the city, but leaves it in the field and forgets it, it is not accounted forgotten sheaf. If, however, he moves it from place to place, even if he leaves it next to the fence gate, or next to the stacked corn, or next to cattle or implements, and forgets it, is is accounted forgotten sheaf.

4. If one takes hold of a sheaf in order to convey it to the city, places it on top of another sheaf, and then forgets both of them, the rule is as follows: If he remembers the upper sheaf before he comes accidentally upon it, the lower one is not accounted forgotten sheaf; if not, it is accounted forgotten sheaf.

5. If A's sheaves are carried away by a strong wind into B's field, and A forgets one of them, it is not accounted forgotten sheaf, as it is said, *thy harvest in thy field* (Deut. 24:19). If, however, the wind merely scatters the sheaves within A's own field, and he forgets one of them, it is accounted forgotten sheaf.

6. If one takes hold of the first, second, and third sheaves, and forgets the fourth, the rule is as follows: If there is a sixth sheaf, the fourth does not become forgotten until he takes hold of the fifth one. If there are only five sheaves, the fourth become for-

gotten once he has tarried long enough to take hold of the fifth sheaf.

7. If two sheaves become intermixed, and one of them is forgotten, it is not accounted forgotten sheaf until all the surrounding sheaves have been removed.

8. Arum, garlic, onions, and the like are subject to forgotten sheaf, even though they are buried in the earth.

If one does the reaping at night and forgets some standing corn, or ties sheaves at night and forgets a sheaf, and likewise if a blind man forgets, the law of forgotten sheaves remains applicable.

If, however, the blind man or the night reaper is of a mind to take hold of the larger plants only, there is no forgotten sheaf for him.

Whosoever says, "I am reaping with the stipulation that I am to take up later whatever I may forget," the law of forgotten sheaf remains nevertheless applicable to him, because whosoever makes a stipulation contrary to what is written in the Torah, his stipulation is null and void.

9. If one harvests his produce before it is ripe in order to feed it to cattle; and likewise if he reaps it in small bundles, without binding it into sheaves; and similarly in the case of garlic and onions that one plucks in small bundles for sale in the market place, without tying them into sheaves for storage—in all these cases the law of forgotten sheaf does not apply.

10. If a reaper commences harvesting at the beginning of a row, and forgets a sheaf in front of him and a sheaf behind him, the sheaf behind him is forgotten sheaf, while the one in front of him is not, as it is said, *Thou shalt not go back to fetch it* (Deut. 24:19), implying that it does not become forgotten sheaf until he bypasses it and leaves it behind him.

This is the general rule: Where *thou shalt not go back* applies, there forgotten sheaf applies; where *thou shalt not go back* does not apply, there forgotten sheaf does not apply.

11. If two persons commence reaping in the middle of a row, one facing north and the other south, and if each one forgets

sheaves both in front of him and in back of him, the ones in front of them are forgotten sheaves, because in each case what is in front of each one of them is behind the other. The overlooked sheaf behind each one of them at the starting point is not forgotten sheaf, because it is part also of the rows running east to west, proving that it was not really forgotten.

Similarly, in the case of rows of sheaves that have been removed to storage, two persons commencing in the middle of a row and forgetting a sheaf in the center between their backs, it is not accounted forgotten sheaf, because it is also in the middle of a row running west to east, on which they had not yet started, and this proves that it was not really forgotten.

12. If in harvesting one binds the produce into sheaves—the terms *'alummah* and *'omer* ("sheaf") are synonymous—and removes these sheaves from one place to another, and from the other to the third, and from the third to the granary, and in the process of moving forgets a sheaf, the rule is as follows: If he moves the sheaves to the place where the work of harvesting is completed, and then forgets a sheaf, the law of forgotten sheaf applies. If he moves them from that place to the granary, the law of forgotten sheaf does not apply. If he moves them to a place where the work of harvesting is not yet completed, and forgets a sheaf there, the law of forgotten sheaf does not apply. If he moves them from that place to the granary, the law of forgotten sheaf does apply.

13. Which is the place where the work of harvesting is considered completed? It is the place where one is of a mind to gather all the sheaves and thresh them, or to convey them for stacking and storage.

And the place where the work of harvesting is not yet completed is one where the sheaves are gathered in order to be made into larger sheaves for removal to another place.

14. Two bundles of produce that are separated from each other are accounted forgotten sheaf, but three are not. Two sheaves that are separated from each other are accounted forgotten sheaf, but three are not.

15. Two heaps of olives or carobs that are separated from each other are accounted forgotten sheaf, but three are not. Two stalks of flax are accounted forgotten sheaf, but three are not.

16. Two grape vines, and similarly two other trees, that are separated from each other are accounted forgotten sheaf, but three are not, for it is said, *Thou shalt leave them for the poor and for the stranger* (Lev. 19:10), implying, even if there are two, one shall be for the poor and one for the stranger.

17. If all the sheaves hold one ḳab̲ of produce each, except one which holds four ḳab̲, and if it is the latter which is forgotten, it is accounted forgotten sheaf. If it holds more than four ḳab̲, it is not forgotten sheaf.

Similarly, if all sheaves hold two ḳab̲ each, except one which holds more than eight ḳab̲, it is not accounted forgotten sheaf.

18. An overlooked sheaf holding two sĕ'ah of produce is not forgotten sheaf, as it is said, *And hast forgot a sheaf in the field* (Deut. 24:19), implying a sheaf, not a stack, even if all the sheaves hold two sĕ'ah.

If, on the other hand, one forgets two sheaves, they are forgotten sheaf, even if between them they hold two sĕ'ah. It would seem to me that they are forgotten sheaf even if they hold more than two sĕ'ah between the two of them.

19. Standing corn amounting to two sĕ'ah and forgotten by the owner is not accounted forgotten sheaf. If it does not amount to two sĕ'ah, the withered ears should be regarded as if they were healthy and long, and the blasted ones as if they were full. And if this standing corn is then fit to yield two sĕ'ah according to this estimate, and is forgotten, it is not forgotten sheaf.

20. If one forgets one sĕ'ah of plucked corn and one sĕ'ah of unplucked corn, they do not combine together, and therefore each one is considered forgotten sheaf. The same applies to garlic, onions, and tree fruits, that is to say, if one forgets them partly attached to the ground and partly already plucked, and between them they amount to two sĕ'ah, they do not combine together, and each one separately is forgotten sheaf.

21. If one forgets a sheaf by the side of standing corn that has not been forgotten, it is not forgotten sheaf, as it is said, *When thou reapest thy harvest . . . and hast forgot a sheaf* (Deut. 24:19), implying that a sheaf surrounded by harvested produce is forgotten sheaf, while a sheaf surrounded by standing corn is not forgotten sheaf.

Similarly, if one forgets some standing corn adjacent to unforgotten standing corn, be it even a single stalk of the latter, it saves the forgotten corn, and the owner may claim it.

If, however, one forgets a sheaf or some standing corn next to an unforgotten sheaf, even if the latter amounts to two sĕ'ah, it does not save the former, and the forgotten produce belongs to the poor.

The standing corn of B cannot save A's own sheaf. Nor can standing barley save a sheaf of wheat—the standing corn and the sheaf must be of the same species.

22. If one forgets one tree among several, even if it is laden with fruit amounting to several sĕ'ah, or if he forgets two trees, they are forgotten sheaf. Three trees, however, are not forgotten sheaf.

23. To what does this apply? To a tree that is not particularly known by its location, as for example, a tree standing near the wine press or near a breach; or by its productivity, as for example, one that yields an abundance of olives; or by a certain characteristic, as for example, an olive tree planted among other olive trees but known for yielding an abundance of oil, or yielding particularly rich oil, or particularly lean oil.

If, however, the tree has one of these three characteristics, it is not forgotten sheaf, as it is said, *And has forgot a sheaf in the field* (Deut. 19:24), implying a sheaf that you have forgotten permanently and would not recall unless you return and see it, which excludes the one that you would remember for a long time thereafter, even if you do not come accidentally upon it, because it is known and famous.

24. If the tree is marked in one's own mind, it is the same as if it were famous and generally known. If it stands beside a date

palm, the date palm renders it marked. If both trees are olive trees yielding abundant oil, each one marks the other.

If one's entire field is planted with olive trees yielding abundant oil, and if he forgets one or two of them, they are forgotten sheaf.

When does this apply? When he does not begin the harvest with this famous tree. If, however, he does commence the harvest with that tree and forgets part of its yield, that part is forgotten sheaf, even though it is from the famous tree, provided that the forgotten residue amounts to less than two sĕ'ah. If it amounts to two sĕ'ah, it is not forgotten sheaf, unless he forgets the entire yield of the tree, as we have explained.

25. In the case of a lone olive tree standing in the midst of rows of olive trees, three such rows surrounding it on three sides —even if each row consists of no more than two olive trees—if one forgets the middle tree, it is not forgotten sheaf, because the rows have hidden it. And why did the Sages make all these rules applicable solely to an olive tree? Because it was regarded as important in the Land of Israel at that time.

26. What is accounted forgotten sheaf in an arbor? Produce which cannot be reached and plucked by the reaper's hand.

And in a vineyard? Whatever is forgotten once one has gone past one or more vines.

And in trained plants and date palms? The produce left after one has descended from them.

And in all other trees? Whatever produce is left once one has turned to go away. When does this apply? When one has not commenced the harvesting with this particular tree. If, however, one has commenced with it and has forgotten it, it is not forgotten sheaf until he has harvested all the other trees around it.

27. If one had declared his vineyard ownerless, and has then arisen betimes, and having retaken possession of it in his own behalf, has harvested it, he is obligated for grape gleanings, defective clusters, forgotten sheaf, and corner crop, seeing that I am bound to apply to this vineyard the terms *thy field* (Lev.

19:9) and *thy vineyard* (Lev. 19:10), since it had been his and is now again his.

If, however, he had taken possession of ownerless property in someone else's field, it is exempt from all these laws. In either case he is exempt from tithes, as will be explained further on.

CHAPTER VI

1. There is another, sixth, gift that is due to the poor out of the yield of the land, namely the tithe given to the poor. This is the so-called poor man's tithe.

2. The order of heave offerings and tithes is as follows: After one has harvested the yield of the land or gathered the fruits of the tree, his work having been completed, he must set aside one-fiftieth of it—this is called great heave offering—and give it to the priest. It is to this that the Torah refers in the verse, *The first fruits of thy corn, of thy wine, and of thine oil* (Deut. 18:4).

He must then set aside one-tenth of the remainder—called first tithe—and give it to the Levite. It is to this that the Torah refers in the verse, *For the tithe of the children of Israel . . . I have given to the Levite* (Num. 18:24), and again, *And unto the children of Levi, behold, I have given all the tithe in Israel* (Num. 18:21).

3. Thereafter, out of what remains, he must again set aside one-tenth—this is called second tithe. It belongs to its owner who must consume it in Jerusalem, and is referred to in the verse, *And if a man will redeem aught of his tithe* (Lev. 27:31), and again, *Thou shalt surely tithe,* etc., *and thou shalt eat before the Lord thy God, in the place which he shall choose* (Deut. 14:22–23).

4. According to this order, these gifts are to be set aside in the first year of each septennate, as well as in the second, the fourth, and the fifth. In the third and sixth years of the septennate, however, after having set aside the first tithe, one must set aside

out of the remainder another tithe and give it to the poor, and this is called poor man's tithe. In these two years there is thus no second tithe but only poor man's tithe. The latter is referred to in the verse, *And at the end of every three years, even in the same year, thou shalt bring forth all the tithe of thine increase, and shalt lay it up within thy gates. And the Levite . . . and the stranger . . . shall come* (Deut. 14:28–29), and again, *When thou hast made an end of tithing,* etc. (Deut. 26:12).

5. The yield of the Sabbatical year is all ownerless, and out of it there is neither heave offering nor tithes, either first, or second, or poor man's. In the lands outside the Land of Israel, where the law of letting the soil rest does not apply, the rule is that in Egypt, in Ammon, and in Moab one must set aside the first tithe and the poor man's tithe, because these lands are adjacent to the Land of Israel, in order that the poor of Israel might rely on them during the Sabbatical year. It is a rule given to Moses on Mount Sinai that in the lands of Ammon and Moab the poor man's tithe must be set aside during the Sabbatical year. In the land of Shinar, however, the second tithe must be set aside during the Sabbatical year, the same as in most other years.

6. The Levite who takes the first tithe must set aside one-tenth of it and give it to the priest. This is the so-called heave offering of the tithe, referred to in the verse, *And unto the Levites thou shalt . . . say . . . When ye take . . . the tithe . . . ye shall set apart of it a gift for the Lord* (Num. 18:26).

7. The owner of a field who has a number of the poor pass by him while the poor man's tithe is still there, must give to each one enough of the tithe to satisfy his hunger, as it is said, *That they may eat within thy gates, and be satisfied* (Deut. 26:12).

8. How much is enough to satisfy his hunger? If he gives him of the wheat, he may give not less than half a ḳab̲; if of the barley, the emmer, or the dried figs, not less than one ḳab̲; if of the pressed figs, not less than the weight of twenty-five sela'; if of the wine, not less than half a log̅; if of the oil, not less than

one-quarter of a ḳab̲; if of the rice, one-fourth of a ḳab̲. If he gives him of the vegetables, he must give him one liṭra, which equals the weight of thirty-five denar; if of the carobs, three ḳab̲; if of the walnuts, ten; of the peaches, five; of the pomegranates, two; of the citrons, one. If he gives him of other fruits, he must give not less than what if sold would buy two meals.

9. If one has only a small amount of produce, and the poor are many, so that there is not enough to give the statutory amount to each person, one should set the produce before them, and they should divide it among themselves.

10. The owner has no optional right in the apportionment of the poor man's tithe that is distributed at the threshing floor; rather the poor may come and take their share, even against the owner's will. Even if the owner is himself the poorest person in Israel, the poor man's tithe may be taken out of his hand.

In the case of the tithe that is distributed in the house, however, the owner has the optional right to give it to any poor man he may choose.

11. If one has the tithe at the threshing floor, and wishes to give of it to a poor man who is his relative or his acquaintance, he may separate half of it and give it to him, but must then distribute the other half to any poor man that passes by, in the amount aforesaid.

12. When does the rule apply that the poor man must be given enough to satisfy his hunger? When the tithe is distributed in the field. If it is distributed in the house, however, one may divide it among all the poor, even if it comes to no more than an olive's bulk per person. For he is commanded to give each person enough to satisfy his hunger only in the field, since there the poor man cannot find more to take, as it is said, *That they may eat within thy gate, and be satisfied* (Deut. 26:12).

13. If a poor man and a poor woman come to the house together, one should give to the woman first, dismiss her, and then give to the man.

If the relationship between the owner and the poor man is that of father and son, or a man and his relative, or two brothers, or two partners, only one of whom is poor, the other may give him his poor man's tithe.

14. If two poor men accept a field in joint tenancy, each one may separate the poor man's tithe from his own share and give it to the other.

15. If a person undertakes to harvest the owner's field, he is forbidden to take of the gleanings, the forgotten sheaf, the corner crop, or the poor man's tithe. When does this rule apply? When he contracts to take a share of the whole field's produce, as for example, when the owner gives him one-third or one-quarter of the harvest as his wages.

If, however, the owner of the field says to him, "One-third of only what you harvest is to be yours," or "one-quarter of only what you harvest," he actually has nothing until after he has performed the harvesting, and remains therefore a poor man the while he is harvesting. Consequently he is entitled to the gleanings, the forgotten sheaf, and the corner crop. He is, however, forbidden to take of the poor man's tithe, because it becomes due to be separated only after he has done the harvesting, at which time he has already gained possession of his share in what he has harvested.

16. If a man sells his field, both land and yield, and is subsequently reduced to poverty, he is allowed to take of the gleanings thereof, the forgotten sheaf, the corner crop, and the poor man's tithe. The buyer, on the other hand, is not allowed to take of them even if he has not yet paid the purchase money. Even if he has borrowed the money for the purchase, he is forbidden to take any of the gifts to the poor.

17. One may not use the poor man's tithe to repay a debt or favor. One may use it, however, to repay an act of charity, but the recipient must be informed that it is poor man's tithe. One may not use it for ransoming captives, nor as a wedding gift, nor may one give any of it as alms. It may be given as a matter of optional

right to the town scholar. It may not be taken from the Land of Israel to a land outside of it, as it is said, *And (thou) shalt lay it up within thy gates* (Deut. 14:28), and again, *That they may eat within thy gates and be satisfied* (Deut. 26:12).

CHAPTER VII

1. It is a positive commandment to give alms to the poor of Israel, according to what is fitting for them, if the giver can afford it, as it is said, *Thou shalt surely open thy hand unto him* (Deut. 15:8), and again, *Then thou shalt uphold him; as a stranger and a settler shall he live with thee . . . that thy brother may live with thee* (Lev. 25:35–36).

2. He who seeing a poor man begging turns his eyes away from him and fails to give him alms, transgresses a negative commandment, as it is said, *Thou shalt not harden thy heart, nor shut thy hand from thy needy brother* (Deut. 15:7).

3. You are commanded to give the poor man according to what he lacks. If he has no clothing, he should be clothed. If he has no house furnishings, they should be bought for him. If he has no wife, he should be helped to marry. If it is a woman, she should be given in marriage. Even if it had been his wont to ride a horse, with a manservant running in front of him, and he has now become poor and has lost his possessions, one must buy him a horse to ride and a manservant to run before him, as it is said, *Sufficient for his need in that which he wanteth* (Deut. 15:8). You are thus obligated to fill his want; you are not, however, obligated to restore his wealth.

4. If an orphan is about to be wed, one must first rent a house for him, spread a bed for him, and provide all his furnishings, and only then have him marry a wife.

5. If the poor man comes forth and asks for enough to satisfy his want, and if the giver is unable to afford it, the latter may

give him as much as he can afford. How much is that? In choice performance of this religious duty, up to one-fifth of his possessions; in middling performance, up to one-tenth of his possessions; less than this brands him as a person of evil eye. At all times one should not permit himself to give less than one-third of a shekel per year. He who gives less than this has not fulfilled this commandment at all. Even a poor man who lives entirely on alms must himself give alms to another poor man.

6. If a poor man unknown to anyone comes forth and says, "I am hungry; give me something to eat," he should not be examined as to whether he might be an impostor—he should be fed immediately. If, however, he is naked and says, "Clothe me," he should be examined as to possible fraud. If he is known, he should be clothed immediately according to his dignity, without any further inquiry.

7. One must feed and clothe the heathen poor together with the Israelite poor, for the sake of the ways of peace. In the case of a poor man who goes from door to door, one is not obligated to give him a large gift, but only a small one. It is forbidden, however, to let a poor man who asks for alms go empty-handed, just so you give him at least one dry fig, as it is said, *O let not the oppressed turn back in confusion* (Ps. 74:21).

8. A poor man traveling from one place to another must be given not less than one loaf of bread that sells for a pondion when the price of wheat is one sela' per four sĕ'ah. We have already explained the value of all measures. If he lodges for the night, he must be given a mattress to sleep on and a pillow to put under his head, as well as oil and pulse for his repast. If he stays over the Sabbath, he must be provided with food for three meals as well as oil, pulse, fish, and vegetables. If he is known, he must be supplied according to his dignity.

9. If a poor man refuses to accept alms, one should get around him by making him accept them as a present or a loan. If, on the other hand, a wealthy man starves himself because he is so niggardly with his money that he would not spend of it on food and drink, no attention need be paid to him.

10. He who refuses to give alms, or gives less than is proper for him, must be compelled by the court to comply, and must be flogged for disobedience until he gives as much as the court estimates he should give. The court may even seize his property in his presence and take from him what is proper for him to give. One may indeed pawn things in order to give alms, even on the eve of the Sabbath.

11. A munificent person who gives alms beyond what he can afford, or denies himself in order to give to the collector of alms so that he would not be put to shame, should not be asked for contributions to alms. Any alms collector who humiliates him by demanding alms from him will surely be called to account for it, as it is said, *I will punish all that oppress them* (Jer. 30:20).

12. Orphans may not be assessed for charity, not even for the ransom of captives, not even if they have much money. But if the judge assesses them in order to have them acquire a good name, he may do so.

Collectors of alms may accept contributions from women, bondsmen, or children, provided that it is a small amount and not a large one, because the presumption is that a large amount was stolen or extorted from other persons. How much is a small amount in their case? It all depends on the wealth or poverty of their masters.

13. A poor man who is one's relative has priority over all others, the poor of one's own household have priority over the other poor of his city, and the poor of his city have priority over the poor of another city, as it is said, *Unto thy poor and needy brother, in thy land* (Deut. 15:11).

14. One who has gone on business to another city, and is assessed for alms by the inhabitants thereof, must contribute to the poor of that city.

If a large group of such visitors is assessed by the city for alms, they must contribute, but when they return home they must bring the assessment back with them and contribute it to the poor of their own city. If that city has a scholar in charge of alms, they should give it to him, to be distributed as he sees fit.

15. If a person says, "Give two hundred denar to the synagogue," or "Give a Scroll of the Torah to the synagogue," these should be given to the synagogue which he regularly attends. If he attends two synagogues, they should be given to both.

If one says, "Give two hundred denar for the poor," they should be given to the poor of that city.

CHAPTER VIII

1. Almsgiving may be included in the generality of vows. Therefore one who says, "I obligate myself to give a sela' in alms," or "This sela' is to be for alms," is obligated to give it to the poor immediately. If he tarries, he transgresses the commandment *Thou shalt not be slack to pay it* (Deut. 23:22), inasmuch as it is within his power to dispense the sela' immediately, and poor people are readily available. If there are no poor people in that place, he may set aside the sela' and leave it until poor people are available.

If he stipulates that he is to give alms only when he finds a poor man, he need not set aside the amount of the vow.

Similarly, if he stipulates at the time when he makes the vow or offers the free-will offering that the alms collectors are to be free to change the sela' or to combine it with others for conversion into gold coin, they are permitted to do so.

2. If one pledges alms by way of linking up, he becomes obligated for them the same as for any other vows. How so? If he says, "This sela' is to be like this one," the former becomes due for alms.

If one sets aside a sela' and says, "This one is to be for alms," and thereupon takes a second sela' and says, "and also that one," the second sela' is likewise pledged for alms, even though he did not say so explicitly.

3. If one has made a vow of alms but does not know how much he had vowed, he should continue giving until he can say, "I did not intend to give this much."

4. Whether one says, "This sela' is to be for alms," or "I obligate myself to give a sela' in alms," and sets it aside, if he wishes to substitute another sela' for it, he may do so. Once that sela' has reached the hand of the alms collector, however, no substitute is allowed.

If the alms collectors wish to convert small coins into denar, they are not permitted to do so. But if there are no poor people in that place to be given the alms, the collectors may convert the small coins into large ones, but only for other collectors, not for themselves.

5. If the poor would benefit by the money remaining in the hand of the collector, in order to encourage others to contribute, the collector may borrow this money belonging to the poor and pay his own debts with it, because alms money is not like money dedicated to the Temple, from which one is forbidden to derive any benefit.

6. If one donates a candelabrum or a lamp to the synagogue, no substitution is allowed. If the substitution is made for the sake of some other religious duty, it is permitted, even if the owner's name has not been dissociated from it, and people still say, "This candelabrum or lamp was donated by So-and-so." If the owner's name has been dissociated from it, substitution is permitted, even for some secular reason.

7. When does this apply? When the donor is an Israelite. If the donor is a heathen, however, substitution is not permitted even for the sake of some other religious duty, until the donor's name has been dissociated from it, lest the heathen should say, "I had consecrated an object to the synagogue of the Jews, and they have sold it for their own benefit."

8. If a heathen offers a free-will offering for the repair of the Temple, it may not be accepted from him at the outset; but if it has already been accepted from him, it may not be returned. If it is something easily identified, such as a beam or a stone, it must be returned to him, so that heathens might not have anything easily identified in the Temple, as it is written, *Ye have*

nothing to do with us to build a House unto our God (Ezra 4:3).

If his free-will offering is offered to the synagogue, it may be accepted at the outset, provided he declares, "I am setting it aside with the same intent as would an Israelite." If he does not so declare, it must be stored out of sight, since in his heart he may have intended it for the Lord. One may not accept from them free-will offerings for the upkeep of the wall of Jerusalem, nor of the water channel that is in it, as it is said, *But ye have no portion, nor right, nor memorial, in Jerusalem* (Neh. 2:20).

9. It is forbidden for an Israelite to accept alms publicly from heathens. But if he cannot sustain himself with the alms provided by Israelites, and is unable to receive alms from heathens privately, he may accept them publicly.

If a king or a prince of the heathens sends money to Israelites for alms, it should not be returned, for the sake of the peace of the kingdom. Rather it should be accepted from him and secretly distributed to the heathen poor, so that the king might not hear about it.

10. The ransoming of captives has precedence over the feeding and clothing of the poor. Indeed there is no religious duty more meritorious than the ransoming of captives, for not only is the captive included in the generality of the hungry, the thirsty, and the naked, but his very life is in jeopardy. He who turns his eyes away from ransoming him, transgresses the commandments *Thou shalt not harden thy heart, nor shut thy hand* (Deut. 15:7), *Neither shalt thou stand idly by the blood of thy neighbor* (Lev. 19:16), and *He shall not rule with rigor over him in thy sight* (Lev. 25:53). Moreover, he nullifies the commandments *Thou shalt surely open thy hand unto him* (Deut. 15:8), *That thy brother may live with thee* (Lev. 25:36), *Thou shalt love thy neighbor as thyself* (Lev. 19:18), *Deliver them that are drawn unto death* (Prov. 24:11), and many other admonitions like these. To sum up, there is no religious duty greater than the ransoming of captives.

11. If the people of the city, having collected money for the building of a synagogue, find themselves confronted with a matter of religious duty, they must divert the money to the latter. If they had already bought stones and beams, they may not sell them in order to fulfill the religious duty, unless it be the ransoming of captives. Even if they have already brought in the stones and set them up, and the beams and planed them, and thus made everything ready for construction, they must nevertheless sell everything, but only if for the ransoming of captives. If, however, they have already completed the erection of the building, they may not sell the synagogue, but should rather make a new collection from the community for the redemption of those captives.

12. Captives may not be ransomed for more than their fair value, for the sake of good world order, lest the enemies should seek them out in order to capture them. Nor may they be assisted to escape, for the same reason, lest the enemy should make their yoke heavier and guard them more vigilantly.

13. If a person sells himself and his children to heathens, or accepts a loan from them and is seized or imprisoned for his debt to them, the first time and the second time this happens it is a religious duty to ransom them; the third time, they may not be ransomed. The children, however, must be ransomed if their father dies in captivity. If the captors threaten to kill him, he must be ransomed even after several such captivities.

14. If a bondsman has been captured, he must be ransomed as if he were an Israelite, provided that he had been ritually immersed for the purpose of bondage and had accepted the commandments. If a captive has fallen into apostasy by violating even no more than one commandment, as for example, by eating nĕḇelah out of spite, or the like, it is forbidden to ransom him.

15. A woman takes precedence over a man as far as feeding, clothing, and redemption from captivity are concerned, because it is customary for a man to go begging from door to door, but

not for a woman, as her sense of shame is greater. If both of them are in captivity, and both are exposed to forcible sin, the man takes precedence in being ransomed, since it is not customary for him to submit to such sin.

16. If two orphans, male and female, are about to be given in marriage, the female should be wed before the male, because a woman's sense of shame is greater. She should be given not less than the weight of six and a quarter denar of pure silver. If the alms treasury has enough funds available, she should be given according to her dignity.

17. If there are before us many poor people or many captives, and there is not enough in the alms treasury to feed, or clothe, or ransom all of them, the procedure is as follows: a priest takes precedence over a Levite, a Levite over an Israelite, an Israelite over a profaned priest, a profaned priest over a person of unknown parentage, a person of unknown parentage over a foundling, a foundling over a bastard, a bastard over a Nathin, a Nathin over a proselyte, inasmuch as the Nathin has grown up with us in a state of holiness, and a proselyte over an emancipated bondsman, inasmuch as the latter was once included among the accursed.

18. When does this apply? When both are equal in wisdom. If, however, a High Priest is unlearned and a bastard is a disciple of the wise, the latter takes precedence. In the case of two scholars, the one greater in wisdom preceeds the other. If one of the poor or captives is a person's teacher or father, and if there is another poor man or captive greater in wisdom than one's teacher or father, so long as the latter is a disciple of the wise, he takes precedence over the one who excels him in wisdom.

CHAPTER IX

1. In every city inhabited by Israelites, it is their duty to appoint from among themselves well-known and trustworthy persons to

act as alms collectors, to go around collecting from the people every Friday. They should demand from each person what is proper for him to give and what he has been assessed for, and should distribute the money every Friday, giving each poor man sustenance sufficient for seven days. This is what is called "alms fund."

2. They must similarly appoint other collectors to gather every day, from each courtyard, bread and other eatables, fruits, or money from anyone who is willing to make a free-will offering at that time. They should distribute these toward that same evening among the poor, giving therefrom to each poor man his sustenance for the day. This is what is called "alms tray."

3. We have never seen nor heard of an Israelite community that does not have an alms fund. As for an alms tray, there are some localities where it is customary to have it, and some where it is not. The custom widespread today is for the collectors of the alms fund to go around every day, and to distribute the proceeds every Friday.

4. On fast days food should be distributed to the poor. If on any fast day the people eat all through the night without distributing alms to the poor, they are accounted the same as if they had shed blood, and it is they who are referred to in the verse in the Prophets, *Righteousness lodged in her, but now murderers* (Isa. 1:21).

When does this apply? When the poor are not given bread and such fruit as is eaten with bread, for example, dates and grapes. If, however, the alms collectors merely delay the distribution of money or wheat, they are not accounted shedders of blood.

5. Contributions to the alms fund must be collected jointly by two persons, because a demand for money may not be addressed to the community by less than two collectors. The money collected may, however, be entrusted for safekeeping to one person. It must be distributed by three persons, because it is analogous to money involved in a civil action, inasmuch as they must

give to each poor man enough for his needs over the week. The alms tray, on the other hand, must be collected by three collectors, since the contribution to it is not set, and must be distributed likewise by three distributors.

6. Contributions to the alms tray are to be collected every day, those for the alms fund each Friday. The alms tray is to provide for the poor of the whole world, while the alms fund is to provide for the poor of the town alone.

7. The residents of the town may use alms fund moneys for the alms tray, or vice versa, or divert them to any other public purpose that they may choose, even if they had not so stipulated when they collected them. If there is in that city a great Sage, whose judgment determines all collections and who distributes them to the poor as he sees fit, he too is permitted to divert these funds to any other public purpose that he may think preferable.

8. Alms collectors are not permitted to separate one from the other in the market place, except as one turns to enter a gate while the other turns to enter a shop, in order to collect contributions.

9. If an alms collector finds money in the market place, he may not put it in his own pocket, but should rather drop it in the alms pouch. After he has reached his own home, he may take it out of the pouch.

10. If an alms collector demands the repayment of a mina owned him by his companion, and the latter pays him right there in the market place, he may not put the money in his own pocket, but should rather drop it in the alms pouch, and after he arrives home, he may take it out of the pouch. He should count the alms fund money not two coins at a time, but one by one, as a precaution against suspicion, as it is said, *Then ye shall be clear before the Lord and before Israel* (Num. 32:22).

11. If the alms collectors have no poor people to distribute them to, they may convert the small coins into denar, but only for others, not for themselves. If the alms tray collectors have no

poor people to distribute the food to, they may sell it to others, not to themselves. Alms collectors may not be required to render an account of alms moneys, nor may treasurers of consecrated property be required to render an account of such property, as it is said, *Howbeit there was no reckoning made with them of the money that was delivered into their hand; for they dealt faithfully* (2 Kings 22:7).

12. One who has resided in the city for thirty days may be compelled to contribute to the alms fund, together with the other residents of the city. If he has resided there for three months, he may be compelled to contribute to the alms tray. If he has resided there for six months, he may be compelled to contribute to the clothing given to the poor of the city. If he has resided there for nine months, he may be compelled to contribute to the burial of the poor and to the other expenses connected therewith.

13. He who has food sufficient for two meals is forbidden to partake of the alms tray. If he has food sufficient for fourteen meals, he may not partake of the alms fund. He who has two hundred zuz, even if he does not use them to engage in trade, or he who has fifty zuz and uses them in trade, may not take of the gleanings, the forgotten sheaf, the corner crop, or the poor man's tithe. If he has two hundred denar less one, he may partake of all of these, even if a thousand persons give them to him at the same time. If he has money in his hand, but owes it as a debt or has it mortaged against his wife's kĕṯubbah, he is still permitted to take of these gifts to the poor.

14. In the case of a needy poor man who has his own courtyard and home furnishings, even if these include utensils of silver and gold, he may not be compelled to sell his house and his furnishings; rather he is permitted to accept alms, and it is a religious duty to give him alms. To what furnishings does this apply? To eating and drinking vessels, clothing, mattresses, and the like. If, however, he has other silver and gold utensils, such as a strigil, a pestle, and the like, he should first sell them and buy less expensive ones. When does this apply? Before he comes to ask for public assistance. If he has already asked for it, he must be com-

pelled to sell his vessels and buy less expensive ones, and only then may he be given alms.

15. If a householder traveling from town to town runs out of funds while still on the road and finds himself with nothing to eat, he is allowed to partake of the gleanings, the forgotten sheaf, the corner crop, and the poor man's tithe, as well as to benefit from alms, and when he reaches his home he is not obligated to repay, since at that particular time he was in fact a poor man. To what can this be compared? To a poor man who has become wealthy, and who is not obligated to repay past assistance.

16. A person who owns houses, fields, and vineyards, which, if sold during the rainy season would fetch a low price, but if held back until the summer would fetch a fair price, may not be compelled to sell them, and should be maintained out of the poor man's tithe up to half the worth of these properties. He should not feel pressed to sell at the wrong time.

17. If at the time when other people are buying such properties at a high price, he cannot find anyone to buy his property except at a low price, seeing that he is hard pressed to sell, he may not be compelled to sell, but is rather allowed to continue eating out of the poor man's tithe until he can sell at a fair price, with everyone aware that he is not pressed to sell.

18. If the amount collected for a poor man to provide adequately for his wants exceeds his needs, the surplus belongs to him. The rule is as follows: The surplus of what was collected for the poor must be used for other poor; the surplus of what was collected for the ransoming of captives must be used to ransom other captives; the surplus of what was collected for the ransoming of a particular captive belongs to that captive; the surplus of what was collected to pay for the burial of the dead must be used to bury other dead; the surplus of what was collected to pay for the burial of a particular person belongs to his heirs.

19. If a poor man contributes a pĕruṭah to the alms tray or to the alms fund, it should be accepted. If he does not, he may not

be constrained to do so. If when given new garments he returns to the distributors his worn-out ones, they too should be accepted, and if he does not return them, he may not be constrained to do so.

CHAPTER X

1. It is our duty to be more careful in the performance of the commandment of almsgiving than in that of any other positive commandment, for almsgiving is the mark of the righteous man who is of the seed of our father Abraham, as it is said, *For I have known him, to the end that he may command his children,* etc., *to do righteousness* (Gen. 18:19). The throne of Israel cannot be established, nor true faith made to stand up, except through charity, as it is said, *In righteousness shalt thou be established* (Isa. 54:14); nor will Israel be redeemed, except through the practice of charity, as it is said, *Zion shall be redeemed with justice, and they that return of her with righteousness* (Isa. 1:27).

2. No man is ever impoverished by almsgiving, nor does evil or harm befall anyone by reason of it, as it is said, *And the work of righteousness shall be peace* (Isa. 32:17).

He who has compassion upon others, others will have compassion upon him, as it is said, *That the Lord may . . . show thee mercy, and have compassion upon thee* (Deut. 13:18).

Whosoever is cruel and merciless lays himself open to suspicion as to his descent, for cruelty is found only among the heathens, as it is said, *They are cruel, and have no compassion* (Jer. 50:42). All Israelites and those that have attached themselves to them are to each other like brothers, as it is said, *Ye are the children of the Lord your God* (Deut. 14:1). If brother will show no compassion to brother, who will? And unto whom shall the poor of Israel raise their eyes? Unto the heathens, who hate them and persecute them? Their eyes are therefore hanging solely upon their brethren.

3. He who turns his eyes away from charity is called a base fellow, just as is he who worships idols. Concerning the worship

of idols Scripture says, *Certain base fellows are gone out* (Deut. 13:14), and concerning him who turns his eyes away from charity it says, *Beware that there be not a base thought in thy heart* (Deut. 15:9). Such a man is also called wicked, as it is said, *The tender mercies of the wicked are cruel* (Prov. 12:10). He is also called a sinner, as it is said, *And he cry unto the Lord against thee, and it be sin in thee* (Deut. 15:10). The Holy One, blessed be He, stands nigh unto the cry of the poor, as it is said, *Thou hearest the cry of the poor.* One should therefore be careful about their cry, for a covenant has been made with them, as it is said, *And it shall come to pass, when he crieth unto Me, that I will hear, for I am gracious* (Exod. 22:26).

4. He who gives alms to a poor man with a hostile countenance and with his face averted to the ground, loses his merit and forfeits it, even if he gives as much as a thousand gold coins. He should rather give with a friendly countenance and joyfully. He should commiserate with the recipient in his distress, as it is said, *If I have not wept for him that was in trouble, and if my soul grieved not for the needy?* (Job 30:25). He should also speak to him prayerful and comforting words, as it is said, *And I caused the widow's heart to sing for joy* (Job 29:13).

5. If a poor man asks you for alms and you have nothing to give him, comfort him with words. It is forbidden to rebuke a poor man or to raise one's voice in a shout at him, seeing that his heart is broken and crushed, and Scripture says, *A broken and contrite heart, O God, Thou wilt not despise* (Ps. 51:19), and again, *To revive the spirit of the humble, and to revive the heart of the contrite ones* (Isa. 57:15). Woe unto him who shames the poor! Woe unto him! One should rather be unto the poor as a father, with both compassion and words, as it is said, *I was a father to the needy* (Job 29:16).

6. He who presses others to give alms and moves them to act thus, his reward is greater than the reward of him who gives alms himself, as it is said, *And the work of righteousness shall be peace* (Isa. 32:17). Concerning alms collectors and their like

Scripture says, *And they that turn the many to righteousness* (*shall shine*) *as the stars* (Dan. 12:3).

7. There are eight degrees of almsgiving, each one superior to the other. The highest degree, than which there is none higher, is one who upholds the hand of an Israelite reduced to poverty by handing him a gift or a loan, or entering into a partnership with him, or finding work for him, in order to strengthen his hand, so that he would have no need to beg from other people. Concerning such a one Scripture says, *Thou shalt uphold him; as a stranger and a settler shall he live with thee* (Lev. 25:35), meaning uphold him, so that he would not lapse into want.

8. Below this is he who gives alms to the poor in such a way that he does not know to whom he has given, nor does the poor man know from whom he has received. This constitutes the fulfilling of a religious duty for its own sake, and for such there was a Chamber of Secrets in the Temple, whereunto the righteous would contribute secretly, and wherefrom the poor of good families would draw their sustenance in equal secrecy. Close to such a person is he who contributes directly to the alms fund.

One should not, however, contribute directly to the alms fund unless he knows that the person in charge of it is trustworthy, is a Sage, and knows how to manage it properly, as was the case of Rabbi Hananiah ben Teradion.

9. Below this is he who knows to whom he is giving, while the poor man does not know from whom he is receiving. He is thus like the great among the Sages who were wont to set out secretly and throw the money down at the doors of the poor. This is a proper way of doing it, and a preferable one if those in charge of alms are not conducting themselves as they should.

10. Below this is the case where the poor man knows from whom he is receiving, but himself remains unknown to the giver. He is thus like the great among the Sages who used to place the money in the fold of a linen sheet which they would throw over their shoulder, whereupon the poor would come behind them and take the money without being exposed to humiliation.

11. Below this is he who hands the alms to the poor man before being asked for them.

12. Below this is he who hands the alms to the poor man after the latter has asked for them.

13. Below this is he who gives the poor man less than what is proper, but with a friendly countenance.

14. Below this is he who gives alms with a frowning countenance.

15. The great among the Sages used to hand a pĕruṭah to a poor man before praying, and then proceeded to pray, as it is said, *As for me, I shall behold Thy face in righteousness* (Ps. 17:15).

16. He who provides maintenance for his grown sons and daughters—whom he is not obligated to maintain—in order that the sons might study Torah, and that the daughters might learn to follow the right path and not expose themselves to contempt, and likewise he who provides maintenance for his father and mother, is accounted as performing an act of charity. Indeed it is an outstanding act of charity, since one's relative has precedence over other people. Whosoever serves food and drink to poor men and orphans at his table, will, when he calls to God, receive an answer and find delight in it, as it is said, *Then shalt thou call, and the Lord will answer* (Isa. 58:9).

17. The Sages have commanded that one should have poor men and orphans as members of his household rather than bondsmen, for it is better for him to employ the former, so that children of Abraham, Isaac, and Jacob might benefit from his possessions rather than children of Ham, seeing that he who multiplies bondsmen multiplies sin and iniquity every day in the world, whereas if poor people are members of his household, he adds to merits and fulfillment of commandments every hour.

18. One should always restrain himself and submit to privation rather than be dependent upon other people or cast himself upon

public charity, for thus have the Sages commanded us, saying, "Make the Sabbath a weekday rather than be dependent upon other people." Even if one is a Sage held in honor, once he becomes impoverished, he should engage in a trade, be it even a loathsome trade, rather than be dependent upon other people. It is better to strip hides off animal carcasses than to say to other people, "I am a great Sage, I am a priest, provide me therefore with maintenance." So did the Sages command us. Among the great Sages there were hewers of wood, carriers of beams, drawers of water to irrigate gardens, and workers in iron and charcoal. They did not ask for public assistance, nor did they accept it when offered to them.

19. Whosoever is in no need of alms but deceives the public and does accept them, will not die of old age until he indeed becomes dependent upon other people. He is included among those of whom Scripture says, *Cursed is the man that trusteth in man* (Jer. 17:5). On the other hand, whosoever is in need of alms and cannot survive unless he accepts them, such as a person who is of advanced age, or ill, or afflicted with sore trials, but is too proud and refuses to accept them, is the same as a shedder of blood and is held to account for his own soul, and by his suffering he gains nothing but sin and guilt.

Whosoever is in need of alms but denies himself, postpones the hour, and lives a life of want in order not to be a burden upon the public, will not die of old age until he shall have provided maintenance for others out of his own wealth. Of him and of those like him it is said, *Blessed is the man that trusteth in the Lord* (Jer. 17:7).

TREATISE III

LAWS CONCERNING HEAVE OFFERINGS

Involving Eight Commandments,
Two Positive and Six Negative
To Wit

1. To set aside the great heave offering;
2. To set aside the heave offering of the tithe;
3. Not to set aside heave offerings and tithes out of order, but to do so in the prescribed order;
4. That a commoner shall not eat of the heave offering;
5. That even one who resides with the priest, or even his hired laborer, shall not eat of the heave offering;
6. That the uncircumcised shall not eat of the heave offering;
7. That the priest who is unclean shall not eat of the heave offering;
8. That a profaned woman shall not eat of the heave offering, nor of anything that has been taken from the holy things.

An exposition of these commandments is contained in the following chapters.

NOTE

In the list of 613 commandments prefixed to the Code, those dealt with in the present treatise appear in the following order:

Positive commandments:

[1] 126. To set aside the heave offering, as it is said: *The first fruits of thy corn, of thy wine, and of thine oil . . . shalt thou give him* (Deut. 18:4);

[2] 129. That the Levites shall set aside a tithe, out of the tithe which they had received from the Israelites, and give it to the priests, as it is said: *Moreover thou shalt speak unto the Levites . . . then ye shall set apart of it a gift for the Lord, even a tithe of the tithe* (Num. 18:26).

Negative commandments:

[4] 133. That a common man shall not eat of the heave offering, as it is said: *There shall no common man eat of the holy thing* (Lev. 22:10);

[5] 134. That even one who resides with the priest, or even his hired laborer, shall not eat of the heave offering, as it is said: *A tenant of a priest, or a hired servant, shall not eat of the holy thing* (Lev. 22:10);

[6] 135. That the uncircumcised shall not eat of the heave offering;

[7] 136. That the priest who is unclean shall not eat of the heave offering, as it is said: *Whosoever be of all your*

seed . . . shall not eat of the holy things until he be clean (Lev. 22:3–4);

[8] 137. That a profaned woman shall not eat of holy things, neither of the heave offering, nor of the breast, nor of the shoulder, as it is said: *And if a priest's daughter be married unto a common man, she shall not eat of that which is set apart from the holy things* (Lev. 22:12);

[3] 154. Not to separate from the produce the heave offering before the first fruits, nor the first tithe before the heave offering, nor the second tithe before the first tithe; but the separation should proceed in due order: first fruits at the beginning, then the great heave offering, then the first tithe, and last the second tithe, as it is said: *Thou shalt not delay to offer of the fullness of thy harvest and of the outflow of thy presses* (Exod. 22:28); that is to say, thou shalt not defer the offering which should be brought earlier.

CHAPTER I

1. According to the Torah, the laws of heave offering and tithes apply only in the Land of Israel, both when the Temple is standing and when it is not. The prophets, however, have ordained that these laws should apply even in the land of Shinar, because it is adjacent to the Land of Israel, and most Israelites travel there and back. The early Sages, in their turn, have ordained that these laws should apply also in the land of Egypt and in the lands of Ammon and Moab, because they border on the Land of Israel.

2. Whenever the term "Land of Israel" is mentioned, it refers to the lands conquered by an Israelite king or by a prophet with the consent of the majority of Israel. This is the so-called national conquest.

If, however, an individual Israelite family, or Tribe, has gone forth and has conquered a place for itself, even if it is part of the land given to Abraham, it is not called Land of Israel insofar as all the commandments would apply thereto.

It is for this reason that Joshua and his court have divided all of the Land of Israel among the Tribes, even though it had not yet been conquered, in order that there should be no conquest by single individuals when each Tribe proceeds to conquer its own share.

3. As for lands conquered by David outside the land of Canaan, such as Aram-Naharaim, Aram Zobah, Ahlab, and the like, although David was king of Israel and was acting with the consent of the High Court, they are in all respects neither like the Land of Israel nor like the lands outside of the Land of Israel, such as Babylonia and Egypt. Rather they have cast off the status of lands outside of the Land of Israel without assuming the status of the Land of Israel. And why have they remained below the rank of the Land of Israel? Because David had conquered them prior to his conquest of the entire Land of Israel, inasmuch

as some of the Seven Nations were still left therein. Had he captured all of the land of Canaan up to its limits, and then conquered other lands, his conquest would have assumed in all respects the status of the Land of Israel.

The lands conquered by David are consequently designated as Syria.

4. There are some things in which Syria is like the Land of Israel, and some in which it is like the lands outside of the Land of Israel. He who purchases land in Syria is accounted the same as if he had purchased land in the Land of Israel, in respect to heave offerings, tithes, and the Sabbatical year. All the rules that apply to Syria are of Scribal authority.

5. All territories seized by the Israelites who had come up from Egypt, and consecrated with the first consecration, have subsequently lost their sanctity when the people were exiled therefrom, inasmuch as the first consecration, due solely to conquest, was effective only for its duration and not for all future time. When the exiles returned and reoccupied part of the Land, they consecrated it a second time with a consecration that is to endure forever, both for its duration and for all future time. The Sages, however, left the places seized by those who had come up from Egypt, but not reoccupied by the returnees from Babylonia, in their former status, and did not exempt them from heave offerings and tithes, in order that poor people might derive sustenance from them during the Sabbatical year.

Our Saintly Rabbi, however, granted a dispensation to Beth-Shan, out of the places not seized by those who had come up from Babylonia, so that it is counted with Ashkelon and is exempt from tithes.

6. We thus find that in relation to the commandments which depend upon the land, the world is divided into three categories: the Land of Israel, Syria, and the lands outside of the Land of Israel. The Land of Israel in its turn is divided into two categories: all territories seized also by the returnees from Babylonia form one category, and all territories seized only by those who

had come up from Egypt form the other. The territories outside of the Land of Israel are likewise divided into two categories: the land of Egypt, Shinar, Ammon, and Moab, wherein these commandments are effective on the authority of the Scribes and the prophets, and the rest of the lands, wherein the laws of heave offerings and tithes do not apply.

7. What is the extent of the land seized by those who had come up from Egypt? From Rekem, which is on the eastern border of the Land of Israel, to the Great Sea, and from Ashkelon, which is on the southern border, to Acco, which is on the northern border. If one proceeds from Acco to Chezib, all the land to his right, that is, east of the road, is presumed to be outside the Land of Israel, and is considered unclean, inasmuch as it is the land of heathen nations, and is exempt from tithe and from the Sabbatical year, unless it becomes known unto you that a certain particular place is indeed part of the Land of Israel. And all the land to his left, that is, west of the road, is presumed to be part of the Land of Israel, is considered clean, not being the land of heathen nations, and is subject to tithe and to the Sabbatical year, unless it becomes known unto you that a certain particular place is indeed outside the Land of Israel.

Whatsoever slopes down from the Mountains of Ummanom inland is considered part of the Land of Israel; whatsoever slopes down from the Mountains of Ummanom outward is considered outside the Land of Israel.

As for the islands of the sea, one is to imagine a cord stretched across them from the Mountains of Ummanom to the Stream of Egypt—whatsoever is inside the cord is considered part of the Land of Israel; whatsoever is outside the cord is considered outside the Land of Israel. This is its shape.

8. From what point on did those who had come up from Babylonia seize territory? From Chezib inland toward the east. From Chezib outward, however, to Amanah, which is identical with Ummanom, up to the Stream, which is the Stream of Egypt, they did not seize. Nor did they seize Chezib itself.

9. What is the extent of Syria? From the Land of Israel outward, in the direction of Aram-Naharaim and Aram-Zobah. The entire basin of the Euphrates up to Babylonia, including such places as Damascus, Ahlab, Haran, Migbab, and the like, up to Shinar and Ṣahar, is considered the same as Syria; Acco, however, is considered outside of the Land of Israel, just like Ashkelon, both places forming the boundaries of the Land of Israel.

10. A heathen who purchases land in the Land of Israel does not thereby render it exempt from the commandments; rather it retains its status of holiness. Therefore, if an Israelite thereupon repurchases it from him, it is not considered the same as an individual conquest, rather the Israelite must set aside heave offerings and tithes and must offer first fruits, all these offerings prescribed in the Torah being due as if the field had never been sold to a heathen. In Syria, however, the heathen's purchase of land is effective in exempting it from tithes and from the Sabbatical year, as will be explained.

11. In the case of produce grown on land bought by a heathen in the Land of Israel, if its processing has been completed while it was in his possession, and if he has finally smoothed it into an even-shaped heap, it is exempt from all statutory dues, as it is said, *thy corn* (Deut. 18:4), not "the heathen's corn."

If, however, an Israelite has bought the produce after it had been plucked, but before its processing has been completed, and if the Israelite then completes the processing, the produce is subject to all the dues required by the Torah. The Israelite must therefore separate the great heave offering and give it to the priest, and the heave offering of the tithe and sell it to the priest; but the first tithe belongs to himself, inasmuch as he can say to the Levite regarding the tithe and to the priest regarding the heave offering of the tithe, "I am acting in behalf of a person from whom you are in no position to collect anything whatsoever."

Why did the Sages say that he need not give the heave offering of the tithe to the priest the same way as he gives him the great heave offering? Because concerning the heave offering of the tithe it is said, *When ye take of the children of Israel the tithe* (Num. 18:26), implying that the untithed produce acquired from an Israelite must have the heave offering of the tithe separated from it and given to the priest; in the case of untithed produce acquired from a heathen, however, you need not give the priest the heave offering of the tithe separated therefrom, rather you may sell it to the priest and retain its price.

12. If the heathen has sold his produce to an Israelite while it was still attached to the ground, the rule is as follows: If he has sold it before it has reached the stage when it became subject to tithes, and the processing of it was then completed while it was in the Israelite's possession, it is subject to all the statutory dues, and the Israelite must give heave offerings and tithes to those entitled to them. If, however, the heathen has sold the produce after it had reached the stage when it became subject to tithes, the Israelite must separate the heave offering of the tithe and the tithe and must give them proportionately to those entitled to them. How so? If the Israelite has bought from the heathen sown corn one-third grown, and it was processed while in the Israelite's possession, he must separate the heave offerings and the tithes, as we have explained, and must give two-thirds of the tithe to the Levite and two-thirds of the heave offering of the tithe to the priest, and must sell the other third to the latter.

13. If an Israelite has sold his produce to a heathen before it had reached the stage when it became subject to tithes, and the heathen then completes the processing of it, it is exempt from heave offering and tithes. If the produce was sold after it had reached the stage when it became subject to tithes, the Israelite is by Scribal law subject to all the statutory dues, even though the heathen has completed the processing of the produce.

Similarly, if the Israelite's produce has completed its processing while in the heathen's possession, it is not subject to heave offering and tithes except by Scribal law, inasmuch as its processing was completed in the heathen's possession.

14. If a heathen has sold to an Israelite produce attached to the ground after it had reached the stage when it became subject to tithes, and if the heathen then smooths it into an even-shaped heap while it is in the Israelite's possession, the produce is not subject to heave offering and tithes, inasmuch as it had reached the stage when it became subject to tithes while in the heathen's possession, and was furthermore smoothed by the heathen into an even-shaped heap, even though the latter process was accomplished under Israelite ownership.

15. He who purchases a field in Syria is subject to heave offerings and tithes by Scribal law, just as he who purchases a field in Jerusalem is subject to the same on the authority of the Torah, as we have explained.

If, however, one buys only produce from a heathen in Syria, whether already plucked or still attached to the ground, even if it has not yet reached the stage when it became subject to tithes, and even if the Israelite himself smooths it into an even-shaped heap, the Israelite is exempt from statutory dues, inasmuch as the produce is not from his own land.

16. If one purchases land from a heathen in Syria, with the produce thereof still attached to the ground, the rule is as follows: If the produce has reached the stage when it became subject to tithes while still in the heathen's possession, it is exempt. If it has not yet reached that stage, the Israelite is subject to tithing, inasmuch as he has bought the produce together with the land.

17. If an Israelite is a tenant for a heathen in Syria, his produce is exempt from tithes, because he has no share in the corpus of the land, and because the heathen's ownership right in Syria exempts his land from tithes, as we have explained.

Similarly, if one leases a field in Syria from a heathen at a rental payable in kind, in shares, or in money, he is exempt from heave offerings and tithes.

18. If an Israelite first purchases a field in Syria from a heathen with produce not yet one-third grown, and then, after

the produce is one-third grown, sells it to another heathen, the rule is that if the same Israelite subsequently repurchases the field, he is subject to heave offerings and tithes, seeing that the produce became liable to these dues while in the possession of the Israelite.

19. If an Israelite who owns land in Syria settles a tenant thereon, and if the tenant subsequently sends him some produce, that produce is exempt, for I can say that the tenant may have bought it in the market place; provided that this kind of produce is in fact available in that market place.

20. Land held in partnership with a heathen is subject to heave offerings and tithes. How so? In case an Israelite and a heathen have purchased a field in partnership, even if they then divide the field while the corn is still standing, and needless to say, if they divide the corn already stacked, it follows that untithed and profane produce are intermixed in every stalk of the heathen's share, even if it is the heathen who finally smooths the produce into an even-shaped heap. The liability of such intermixed produce is required only by Scribal law, as we have explained.

21. Where does this apply? In the Land of Israel, where tithes are obligatory by authority of the Torah, and in such matters the principle of retroactive differentiation does not apply. If however, the two partners have purchased a field in Syria, even if they divide the produce when it is stacked, the heathen's share is exempt from all dues, inasmuch as fields in Syria are subject to tithes only by Scribal law.

22. Produce of the Land of Israel exported to a land outside of the Land of Israel is exempt from dough offering, heave offerings, and tithes, as it is said, *Into the land whither I bring you* (Num. 15:18)—there, in that Land, you are obligated, but in lands outside of the Land of Israel you are exempt.

If the produce is exported to Syria, it is subject to these dues on Scribal authority.

Similarly, foreign produce imported into the Land of Israel is subject to dough offering, as it is said, *whither*—there, in that

Land, you are obligated, whether the produce is domestic or imported.

If the produce had been designated to be tithed while in the possession of an Israelite, once it has been imported into the Land of Israel, it is subject to all tithes by Scribal law.

23. If foreign earth is imported by ship into the Land of Israel, the rule is that so long as the ship remains in contact with the Land of Israel, whatsoever sprouts out of that earth is subject to heave offering, tithes, and the Sabbatical year, the same as what grows in the Land of Israel itself.

24. In the case of a tree which stands outside of the Land of Israel with its branches extended into the Land of Israel, or stands in the Land of Israel with its branches extending outside of it, it all depends on the position of the roots.

If some of the roots are inside the Land of Israel and some outside, the rule is that even if a bare rock separates them, the tree's produce is regarded as a mixture of untithed and profane.

25. In the case of a perforated pot containing seed not yet rooted in the ground underneath the pot, if the plant's roots in the pot are inside the Land of Israel and its shoots are outside, the rule follows the shoots.

26. In our time the heave offering is obligatory not by the authority of the Torah but by Scribal law, even in places seized by those who had come up from Babylonia, or even by those who had come later in the days of Ezra. For there is no heave offering authorized by the Torah except in the Land of Israel alone, and at the time when all the children of Israel are there, as it is said, *When ye are come* (Num. 15:2), implying, when all of you shall have come, as they did at the first settlement, and as they are to do again at the third settlement, not as they did at the second settlement in the days of Ezra, when only some of them returned, wherefore the Torah did not so obligate them.

It would consequently appear to me that the same rule should apply to tithes, namely that in our time they should be due by Scribal law only, just as is heave offering.

CHAPTER II

1. All human food that is watched over and that grows out of the soil is subject to heave offering, and it is a positive commandment to separate from it the first fruits for the priest, as it is said, *The first fruits of thy corn, of thy wine, and of thine oil . . . shalt thou give him* (Deut. 18:4). Just as corn, wine, and oil are human food, grow out of the soil, and have owners, as it is said, *thy corn,* so also all things similar to them are subject to heave offerings and likewise to tithes.

2. Bitter vetch, although normally not used for human food, is nevertheless subject to heave offering and tithes, seeing that people do eat it in years of famine. Savory, hyssop, and thyme, if planted at the outset for human consumption, are subject to tithe. The same applies to all similar plants. Once one has planted them for animal consumption, they are exempt, even if he later changes his mind and designates them for human consumption while they are still attached to the ground, because such a mental resolution made while the plants are still attached to the ground is of no effect.

If these plants grow spontaneously in a courtyard, the rule is as follows: If the courtyard is such as protects its produce, the latter is liable to the statutory dues, because the assumption is that it has been confined in the courtyard for human consumption. If the courtyard does not serve to protect its produce, the latter is exempt.

3. Inedible garden seeds, as for example, turnip seeds, radish seeds, onion seeds, and the like, are exempt from heave offering and tithes, because they are not human food. Fennel flower, however, is subject to heave offering and tithes.

4. Buds of fenugreek, black mustard, hyacinth bean, caper, and caper flower are exempt, because they are not produce.

When does this apply? When these have been planted for seed. If they have been planted for their greens, they are liable. Similarly caper berries are liable because they are produce.

5. Greens of coriander planted for seed are exempt from heave offering and tithes. If the coriander has been planted for its greens, one must separate heave offerings and tithes from both the greens and the seed. Similarly, in the case of dill planted for seed, its greens are exempt; but if planted for its greens, both seed and greens must be tithed, but not the pods. If it has been planted for its pods, one must separate heave offering and tithes from everything, seeds, greens, and pods.

Similarly, in the case of garden cress and garden rocket planted for seed, one must tithe both seed and greens.

How does one tithe both seed and greens? Having gathered the greens in order to eat them, he must first separate heave offering and tithes, and only then may he eat them. After the plant has dried up and he has gathered its seeds, he must separate the same from them also.

6. Vegetables, even though they are human food, are subject to tithes only by Scribal law, for Scripture says concerning the tithe, *The produce of thy seed* (Deut. 14:22), implying produce and the like, whereas vegetables are not included in the generality of produce.

It would seem to me therefore that the same rule should apply to heave offering, seeing that concerning it Scripture says, *of thy corn, of thy wine, and of thine oil* (Deut. 18:4), implying only whatsoever is similar to these; heave offering from vegetables, however, should therefore rest only on Scribal law, as in the case of tithe.

7. In lands outside of the Land of Israel vegetables are not subject to heave offerings and tithes, even in places where we have said that tithes must be separated from corn.

Similarly, vegetables imported into the Land of Israel from lands outside of it, even if they have some earth clinging to their roots, are exempt, and no precautionary dues were made obligatory from them by the Sages.

In the case of corn and pulse planted for their greens, the intent of the person who had planted them with this purpose in mind is annulled by the consensus of all people, and their greens are therefore exempt, while their seed is subject to heave offerings and tithes.

8. Fenugreek is subject to heave offering and tithe, even though it is not fit for human food once it has hardened, inasmuch as most people do eat it in the beginning.

9. The following are exempt from heave offering and tithes: gleanings, forgotten sheaf, corner crop, grape gleanings, and defective clusters. This applies even if the poor man has heaped them into a pile; if, however, he has stacked them in the field as if for threshing, they are regarded as intended for tithing, and he must separate heave offering and tithes from them.

If the recipient stacks them in the city, they are exempt, because the fact that they are gleanings, or forgotten sheaf, or corner crop is a matter of common knowledge.

10. Gleanings, forgotten sheaf, and corner crop in a field that belongs to a heathen are liable to heave offering and tithes, unless he first declares them ownerless property.

Similarly corn and olives that are not yet one-third grown are exempt from heave offering and tithes.

And how is one to determine this? Whatsoever sprouts after having been planted is ipso facto known to be one-third grown.

If, therefore, one transgresses and separates heave offering from corn and olives not yet one-third grown, it is not valid heave offering.

11. Similarly ownerless property is exempt from heave offering and tithes, even if the heathen owner has declared it so for the benefit of the recipient.

If, however, one plants an ownerless field, he is subject to heave offering and tithes.

12. If after declaring his standing corn ownerless property one retakes possession of it, and thereupon transgresses and separates heave offering from it, the offering is valid. If, however, he declares ears of corn ownerless property, repossesses them, and then transgresses and separates heave offering from them, his

offering is invalid. Similarly, anyone who separates heave offering from produce that is not subject to it, his offering is invalid.

So also in the case of tithes, anything which most people do not ordinarily plant in gardens and fields and which is presumed to be ownerless, is exempt from heave offering and tithes, as for example, pungent garlic, sharp onion, Cilician grits, Egyptian lentils, and the like.

13. If produce subject to heave offering is intermixed with produce not subject to it, as for example, if gleaners' olives are intermixed with harvested olives, or gleaners' grapes with harvested grapes, the rule is as follows: If one has also other produce, he must separate from that produce an estimated amount in proportion to what he owes for the intermixed produce. If he has nothing but the intermixed produce, he must separate heave offering and heave offering of the tithe for the entire amount thereof, as if all of it were subject to heave offering.

For first and second tithes, however, he must separate only the estimated amount in proportion to what he owes.

14. Heave offering must be given to the priest whether it is in a state of cleanness or in a state of uncleanness. Even if all of the corn or wine becomes unclean before the owner has separated the priest's due, he must nevertheless separate the heave offering from it while it is unclean and give it to the priest, as it is said, *Behold, I have given thee the charge of My heave offerings* (Num. 18:8), which implies, whether it is clean or unclean. As for the clean, the priests may consume it; as for the unclean, they may derive benefit from it by burning it. If it is oil, they may use it for illumination, and if it is corn and the like, they may use it as fuel for the oven.

15. The same applies to heave offering of the tithe: if it becomes unclean, or if the tithe itself becomes unclean, it must be separated while unclean and must be given to the priest to derive benefit therefrom by burning it.

16. Whosoever sets aside the great heave offering or the heave offering of the tithe must recite a benediction before separating

it, just as one must recite a benediction before performing any commandment, as we have explained in the Laws Concerning Benedictions.

17. One may not export heave offering from the Land of Israel to a land outside of it, even if it is unclean. Nor may one import heave offering from a land outside of the Land of Israel to the Land of Israel; if he does so nevertheless, it may not be consumed, because it has been defiled by its sojourn in the land of the heathens. It also may not be burned, lest people should say, "We saw heave offering burned notwithstanding that it had not been defiled." Neither may it be returned to the foreign land, lest people should say, "Evidently heave offerings may be exported to lands outside of the Land of Israel." Rather it should be left alone until it becomes definitely unclean with a defilement known to everyone—or until the onset of Passover eve, if it is leaven—and then it should be burned.

CHAPTER III

1. The great heave offering has, according to the Torah, no standard amount set for it, as it is said, *The first fruits of thy corn* (Deut. 18:4), implying even a minimal amount—even a single grain of wheat should release the whole heap. At the outset, however, one should set aside only the amount set by the Sages. But in our time, when heave offering stands to be burned because of uncleanness, one should at the outset set aside the minimum amount.

2. And what is the amount set for it by the Sages? For a generous person, $1/40$; for a middling person, $1/50$; and for a niggardly person, $1/60$. None should give less than $1/60$.

3. Any heave offering about which the priests are not particular, as for example, heave offerings of carobs and acorns, should amount to $1/60$ at the outset. The following heave offerings should also amount to $1/60$: from the new growth of planted heave offering; from mixed produce, part already taxed, part not

yet taxed; from produce defiled by accident or error; from consecrated produce; from produce grown outside of the Land of Israel; from fennel flower, acorns, carobs, sycamore figs, lupine, red barley, and the like; and from produce grown in an unperforated pot. Similarly guardians of orphans, when separating heave offering from produce belonging to orphans, need offer no more than 1/60.

4. This heave offering may not be set aside by measure, weight, or number, seeing that no standard amount has been stated for it in Scripture. One must rather make an estimate and set aside what in his own judgment is approximately 1/60. He must, however, set aside the exact amount of heave offering out of what has already been measured, weighed, or counted.

One may not set aside heave offering with an ordinary or oversize basket of known capacity, but may use such a basket to set aside an amount equal to one-half or one-third of its capacity. He may not, however, use a sĕ'ah basket to measure out half a sĕ'ah, because half a sĕ'ah is a measure in its own right.

5. If one sets aside an excessive amount of heave offering, it is valid, provided that he has left some profane produce. If, however, he says, "All this produce is to be heave offering," he has said nothing.

If having intended to set aside 1/10 one comes up with 1/60, his heave offering is valid. On the other hand, if his intention was to set aside 1/60, and he comes up with 1/50, his heave offering is invalid.

6. If in setting aside heave offering one comes up with even as much as 1/20 of the produce, his heave offering is valid. If having set aside the heave offering and come up with 1/60, one goes back and adds some more to it as supplementary heave offering, that supplement is subject to tithes, and the priest must first separate the tithes from it, and only then may he consume it.

If in setting aside heave offering one comes up with one part out of sixty-one, it is valid heave offering, but he must go back and set aside once more an amount sufficient to complete the total amount that he had in his mind, and this additional amount he may set aside by measure, weight, or number. He may set it aside, however, only out of produce adjacent to the original produce.

7. If one sets aside only part of the heave offering due, that part is considered not heave offering but an integral part of the heap of produce. Nevertheless, he must set aside the heave offering due from that part out of it and not out of the other produce.

8. If one sets aside part of the heave offering out of one heap and part out of another heap of the same species, he may not assign one part's offering to the other.

If one says, "The heave offering out of this heap is assigned to a component of it," the rule is as follows: If he specifies "to its northern or southern component," he has made a valid designation, and he must set aside his heave offering out of that component. If he does not designate the location of the component, he has said nothing.

9. If one says, "This heave offering is to be out of this heap, and yonder heap is to be like this heap here," the rule is that the heave offering out of the second heap ends at the same point where the heave offering of the first heap ends.

10. The heave offering of the tithe may not be set aside by conjectural estimate—its measurement must be exact, even in this time, seeing that it is stated explicitly in the Torah.

11. For this purpose things customarily measured must be measured, things customarily weighed must be weighed, and things which can be counted must be counted. If a thing can be equally well counted, weighed, and measured, the rule is that counting is satisfactory, measuring is preferred, and weighing is preferable to both.

12. The commandment of heave offering of the tithe applies to the Levite, who must set it aside out of his tithe, as it is said, *When ye take of the children of Israel the tithe* (Num. 18:26). An Israelite, however, may set it aside and give it to the priest,

and then give the balance of the tithe to the Levite, after he has set aside the latter's heave offering, which is the tithe from the tithe.

13. If an Israelite sets aside the first tithe while the corn is still in its ears, before he has threshed it and has set aside the great heave offering, and then gives it to the Levite, the Levite is not obligated to set aside therefrom the great heave offering after he himself has threshed the corn, but must set aside only the heave offering of the tithe.

If, however, the Israelite has threshed the corn and has set aside the tithe out of it before setting aside the great heave offering, and has then given the tithe to the Levite, the Levite is obligated to set aside out of it the great heave offering as well as the heave offering of the tithe, because once the corn has been turned into threshing grain, it becomes liable to the great heave offering, as it is said, *The first fruits of thy corn* (Deut. 18:4).

14. If a Levite accepts his tithe in ears of corn, he may not give his heave offering to the priest likewise in ears of corn. Rather he must be penalized for it by being made to thresh and winnow the corn, and then give to the priest a tithe of the tithe out of the grain. He is not obligated, however, to give him also a tithe of the straw or of the stalks. If the Levite has already set aside the heave offering of the tithe in ears of corn, just as it had been given to him, he must pound it and give to the priest both the seeds and the straw. And why did the Sages penalize him by compelling him to do the pounding? Because he has accepted the tithe in ears of corn and has thus exempted that batch of produce from the great heave offering.

15. If an Israelite says to a Levite, "Thus has my father spoken to me: You have some tithe due to you that is in my hand," no apprehension need be felt about the heave offering of the tithe due from it, because the father had presumably already set aside that offering out of it, and it was for this reason that he told his son that this tithe belonged to So-and-so, the Levite.

If, however, the Israelite tells the Levite, "Thus has my father spoken to me: A kor of your tithe is in my hands," apprehension should be felt about the heave offering of the tithe that is due from it.

16. Heave offering of the tithe that amounts to at least one-eighth of one-eighth must be conveyed to the priest. If it amounts to less than this, one need not bother to convey it to the priest, but may rather throw it into the fire and burn it. In the case of wine and oil, even the least amount thereof must be conveyed to the priest, provided that it is assured heave offering and undefiled. If, however, it has been defiled, or if it is doubtfully tithed, the rule is that if it is less than the minimum amount, one need not bother with it and may burn it.

17. Great heave offering may not be set aside except out of produce adjacent to the original produce. How so? If one has 50 sĕ'ah of produce in one house and 50 sĕ'ah in another house, he may not set aside 2 sĕ'ah out of one house for the total of 100 sĕ'ah, because he would be separating from one place for another. But if one does so nevertheless, his heave offering is valid, provided that what has been set aside is secure. If, however, one is carrying jugs of wine or oil, and perceiving that they are about to break says, "Let these serve as heave offering for the produce in my house," he has said nothing.

18. If produce is scattered within a house, or is kept in two storerooms in the same house, one may set aside heave offering out of one heap or one storeroom for all.

In the case of sacks of produce, round cakes of pressed figs, or jars of dried figs, the rule is that if they are kept in one compact group, one may set aside the heave offering out of one of them for all.

In the case of jars of wine the rule is as follows: Before their mouths are sealed, one may set aside the heave offering out of one of them for all; after their mouths have been sealed, one must set it aside out of each one of them individually.

19. If one gathers bundles of the same species of vegetables and leaves them in the garden, he may set aside the heave offering out of one of them for all; but if he includes another species in

them, he must set aside the heave offering out of each one individually.

If he places several species in a basket, cabbage on top, cabbage at the bottom, and another species in the middle, he may not set aside the heave offering out of the top layer for the bottom layer.

If one shifts five heaps of produce close to each other within the threshing floor, he may set aside the heave offering out of one heap for all, as long as the ground of the threshing floor is intact; if it is not intact, he must set aside the heave offering out of each heap individually.

20. Heave offering of the tithe may be set aside out of produce that is not adjacent to the original produce, as it is said, *of all your tithes* (Num. 18:28) shall you separate the heave offering, implying that even if one tithe is in one region and another in another region, one may set aside the heave offering out of one for all. Disciples of the wise, however, do not set aside the heave offering except out of adjacent produce, even in the case of the heave offering of the tithe.

21. If a Levite, having on hand first tithe produce out of which its heave offering has not yet been set aside, leaves it as it is, in order to set aside heave offering out of it later on for other produce, so that for the present it remains untaxed, what he has done is done, as it is said, *For the tithe of the children of Israel, which they set apart as a gift unto the Lord* (Num. 18:24), which implies that he may designate all of it as heave offering for other produce.

22. If he first sets aside out of it the heave offering of the tithe and then leaves it as it is, in order to set aside heave offering of the tithe out of it later on for other produce, until all of it should become heave offering of the tithe, whereupon he would give it all to the priest, he has done nothing, as it is said, *Even the hallowed part out of it* (Num. 18:29), implying that while its sacred portions are still within it, he may designate it as heave offering of the tithe for other produce; when its sacred portions are no longer within it, he may not so designate it.

Similarly, if one puts aside produce in order to set aside therefrom the great heave offering, that produce must itself be in an untaxed state as far as the great heave offering is concerned; and similarly, if he puts it aside in order to set aside therefrom the tithe, it must be in an untaxed state as far as the tithe is concerned.

23. Heave offering and tithes should be set aside in proper sequence. How so? One should first set aside the first fruits, then the great heave offering, then the first tithe, then the second tithe, or the poor man's tithe. If one advances the second tithe before the first, or the tithe before the heave offering, or the heave offering before the first fruits, even though he has transgressed a negative commandment, what he has done is done. And whence do we learn that this is enjoined by a negative commandment? From the Scriptural verse, *Thou shalt not delay to offer of the fullness of thy harvest and of the outflow of thy presses* (Exod. 22:28), implying that you may not delay that which should be advanced. Violation of this negative commandment, however, is not punishable by flogging.

24. He who wishes to set aside the great heave offering and the heave offering of the tithe at the same time should proceed as follows: He should set aside one part of produce out of 33⅓, and say, "The one-hundredth part that is included in this, let it be adjacent to what I have set aside, and let it be profane produce. What is left of what I have set aside, let it be the heave offering for the whole amount of produce. The tithe that is due from the 100 parts of this profane produce, let it be adjacent to what I have separated. What is left, over and above the heave offering of what I have set aside, let it be the heave offering of the tithe for the whole amount of produce."

CHAPTER IV

1. One may appoint an agent to set aside heave offerings and tithes in his behalf, as it is said, *Thus ye also shall set apart* (Num. 18:28), which implies that your agents also are included.

One may not, however, appoint a heathen as an agent, as it is said, *ye also,* implying that just as you are sons of the covenant, so also your agents must be sons of the covenant.

2. Five persons may not set aside heave offering, and if they do so, their offering is invalid: a deaf-mute; an imbecile; a minor; a Gentile who has set aside an Israelite's offering, even if with the latter's consent; and a person who has set it aside out of produce that is not his own, and without the owner's permission.

If, however, A separates the offering out of his own produce in behalf of B's produce, it is valid heave offering and it regularizes B's produce; A, for his part, has the optional benefit of giving the offering to any priest he may choose.

3. If A sets aside heave offering without the owner's, B's, permission, or if A goes into B's field and gathers produce without B's permission for the purpose of making it his own, and then sets aside heave offering therefrom, the rule is as follows: If B comes along and says to A, "You might as well go all the way and pick produce of superior quality," and if there is indeed in that field produce superior to that from which A had separated the offering, A's offering is valid, seeing that B was not particular about it. If there is no such superior produce in that field, A's offering is invalid, because B's remark was obviously meant to convey his protest. If, however, B comes along, gathers some produce himself, and adds it to the produce gathered by A, A's offering is valid, whether the field does or does not contain superior produce.

4. Five persons may not set aside heave offering, but if they do so nevertheless, their offering is valid: the deaf who can speak but cannot hear, because he cannot hear the appropriate benediction; the mute who can hear but cannot speak; the naked, because in his state he may not recite the benediction; the intoxicated and the blind, because they cannot direct their attention well enough to set aside the choice produce.

5. In the case of a minor who has reached the age when his vows are valid, even if he has not yet grown the statutory two hairs and has thus not yet become an adult, and who has set aside

heave offering, his offering is valid, even when it is heave offering authorized by the Torah, seeing that his vows and his dedication to the Sanctuary are valid according to the Torah, as we have explained in the Laws Concerning Vows.

6. If A says to his agent B, "Go forth and set aside heave offering in my behalf," and if B sets out to do so without A knowing whether he has indeed done so or not; and if then A comes along and finds a heap of produce from which heave offering had been set aside, no presumption exists that the heave offering due from this heap has been set aside by B, because in regard to prohibitory commandments the presumption that an agent has executed his commission may be applied not in the direction of leniency but only in the direction of greater rigor. There is therefore ground for apprehension that someone other than B has set aside the heave offering without A's consent.

7. If A says to his agent B, "Go forth and set aside heave offering," B should do so in accordance with A's disposition. If he knows that A is miserly, he should set aside $1/60$; if he knows that A is generous, he should set aside $1/40$; and if he does not know A's disposition, he should set aside the amount due from a person of middling disposition, namely $1/50$.

If while intending to set aside the amount due from a person of middling disposition, B comes up with $1/40$ or $1/60$, his offering is valid. If his intention was to set aside more than the amount due from a person of middling disposition, his offering is invalid, even if he comes up with $1/99$.

8. Produce owned in partnership is liable to heave offering and tithes, as it is said, *your tithes* (Num. 18:28), implying even if owned by two persons. The partners are not required to obtain permission from each other; rather whichever of them sets aside the offering, it is valid.

If A sets aside heave offering, and B then comes along and sets aside a second heave offering, not knowing that A had already done so, the rule is as follows: If they depend on each other in this matter, B's offering is invalid. If they do not depend on each other, and if A has set aside the required amount, B's heave offering is invalid; if A has set aside less than the required amount, the offerings of both A and B are valid.

9. If A tells his partner, or a member of his household, or his bondsman or bondswoman to set aside heave offering, and they set out to do so, and if A thereupon revokes his charge before they have executed it, the rule is as follows: If the agent has not deviated from his commission, his offering is valid. If he has deviated, as for example, if A had told him, "Set aside the offering from the northern part of the heap," and he set it aside from the southern part, his offering is invalid, seeing that he has invalidated his mission at the outset.

10. If while a tenant farmer is setting aside heave offering, the owner comes along and prevents him from doing so, the rule is as follows: If the owner intervenes before the tenant has completed the setting aside of the offering, it is invalid; if the owner intervenes after the setting aside of the offering has been completed, the offering is valid.

Guardians of orphans are empowered to set aside heave offering from the orphans' estate.

11. Heave offering set aside by a thief, a robber, or one who has seized someone else's property by some other act of violence, is valid. But if the rightful owner is in hot pursuit of them, their offering is invalid.

12. A son, a hired man, a bondsman, or a wife may set aside heave offering for what they actually consume, but not for the total produce, inasmuch as one may not set aside heave offering from what is not his.

On the other hand, when a son eats regularly with his father, and when a woman is kneading her dough, they may set aside heave offering for the total amount, because they are presumed to do so with the owner's permission.

13. Hired laborers are not permitted to set aside heave offering without the householder's permission, with the exception of those

who tread the wine press, inasmuch as had they wished, they could have rendered the wine defiled immediately. Since the householder has nevertheless charged them with this matter and has put his trust in them, they are the same as agents, and therefore, if they set aside heave offering, it is valid.

14. If the householder tells the laborer, "Gather in my harvest and set aside heave offering therefrom," and if the laborer sets aside heave offering first and then gathers the harvest, his offering is valid.

15. If a Gentile sets aside heave offering from his own produce, the law according to the Torah is that his offering is invalid, because Gentiles are not liable to it. The Sages, however, have ordained that his offering should be valid, as a precaution against holders of purses, since this commodity might really belong to an Israelite who had attached it to a heathen in order to make it exempt. The heathen who has set aside the heave offering should, however, be examined. If he says, "I have set it aside with an Israelite's consent," it should be given to the priest; if not, it should be hidden away, since his intention may have been toward heaven.

Where does this apply? In the Land of Israel. If, however, the heathen sets aside heave offering in a land outside of the Land of Israel, no precautionary measures have been ordained with regard to him. He need only be informed that he is not liable to the offering, and it is regarded as no heave offering at all.

16. If a person intending to say "heave offering'" says "tithe," or vice versa, he has said nothing, until his mouth and his heart are in accord with each other. If his thought was to set aside heave offering but he has not expressed it with his lips, it is a valid heave offering, as it is said, *And the gift which ye set apart should be reckoned unto you as though it were the corn of the threshing floor* (Num. 18:27), implying that the thought alone renders it valid heave offering.

17. If one sets aside heave offering with a condition attached, the rule is as follows: If the condition has been fulfilled, the heave offering is valid; if not, it is not heave offering.

Similarly, if one sets aside heave offering and tithes, and then regrets it, he may consult a Sage, who may absolve him, as he would absolve him from any other vow; the produce involved thereupon reverts to its former state of profaneness, until he sets aside the offerings a second time, either from the same produce as before or from other produce.

18. If one sets aside a bucketful of heave offering from a tank of wine by saying, "Let this bucketful be heave offering, provided that it comes up whole," the bucket must come up free from breakage and spillage, but not from uncleanness. If it breaks, its contents do not render forbidden other produce with which they may become mixed. If he leaves the bucket in a place where, if broken or rolled, its contents could not reach the tank, they do render other produce forbidden, since the aforementioned condition has already been fulfilled.

19. To what does this apply? To great heave offering. In the case of heave offering of the tithe, which may be set aside even out of produce that is not adjacent, the condition is regarded as fulfilled once the bucket has come up, and its contents are valid heave offering of the tithe, even if it broke, or spilled, or needless to say, was rendered unclean.

20. If one says, "What is on top is to be heave offering, and what is on the bottom is to remain profane produce," his words are effective, because the matter depends on the mind of him who does the separating.

21. He who sets aside heave offering from the threshing floor should have in mind that the heave offering is to cover the heap itself, as well as the unthreshed ears, the ears that have fallen by the side of the heap, and the grain that remains among the husks.

He who sets aside heave offering from the wine press should have in mind that it is to cover also the juice which remains in the grape seeds and in the grape skins.

He who sets aside heave offering from the wine tank, should have in mind that it is to cover also the juice that has remained in the pressing vat. If he does not have all this in mind, and sets aside the offering without further thought, the whole of the produce becomes thereby exempt, for it is a condition imposed by the court that heave offering covers the total produce.

If one sets aside a basket of figs as heave offering, and some figs are found by the side of the basket, they are exempt, because he is presumed to have had in mind to set aside heave offering for the total amount of figs.

CHAPTER V

1. One must separate heave offering only out of the very best produce, as it is said, *When ye set apart the best thereof from it* (Num. 18:30). If there is no priest resident in that locality, one should set aside the offering out of what keeps longest, even if produce that is of superior quality but does not keep as long is available there. How so? One should set aside heave offering out of fresh figs for dried figs, but where there is no priest he should set aside dried figs for fresh figs. If, however, it is his regular custom to convert fresh figs into dried figs, he may set aside fresh figs for dried figs even where there is no priest. Where there is a resident priest, one may not set aside dried figs for fresh figs, even if in that place fresh figs are customarily converted into dried figs.

2. In all places, one should set aside a whole onion, even if it is small, rather than half an onion, even if it is large.

One may not set aside one kind for another kind, as it is said, *As the increase of the threshing floor, and as the increase of the wine press* (Num. 18:30). If he does so nevertheless, the heave offering is not valid.

3. Chate melon and cucumber melon are regarded as one species; so are all species of wheat, and so also all kinds of figs, including dried figs and pressed figs. One may therefore set aside heave offering out of one of these species for the other. On the other hand, any species which, if planted together with another species, would constitute diverse kinds, may not serve as heave

offering for the other species, even if it means setting aside superior produce for inferior produce.

How so? If one has fifty sĕ'ah of wheat and fifty sĕ'ah of barley in the same house, and sets aside two sĕ'ah of wheat as heave offering for both, it is invalid.

In the case of species of produce which do not constitute diverse kinds if sown together, one must set aside heave offering out of the superior for the inferior and not vice versa. But if he does so nevertheless, it is valid heave offering, with the exception of darnel set aside for wheat, since darnel is not human food.

4. One may not set aside heave offering out of produce the preparation of which has been completed for produce the preparation of which has not yet been completed, nor out of uncompleted produce for uncompleted produce, nor out of uncompleted produce for completed produce, as it is said, *As though it were the corn of the threshing floor, and as the fullness of the wine press* (Num. 18:27), implying that it must be out of completed produce for completed produce. If one does so nevertheless, it is valid heave offering.

5. From what time on may one set aside heave offering from the threshing floor? From the time when the grain has been sifted. If only part of the grain has been sifted, he may set aside heave offering out of the sifted grain for the unsifted grain.

If one brings ears of corn into his house with the intention of parching them, he may set aside heave offering out of the unparched ears.

6. From what time on may one set aside heave offering out of the wine press? From the time when the treaders have walked the length and breadth of the press.

From what time on may one set aside heave offering out of olives? From the time when the pressing beam has been laid upon them.

7. One may not set aside heave offering out of clean produce for unclean produce, but if one does so nevertheless, it is valid heave offering. It is, however, a rule given to Moses on Mount

Sinai that in the case of a cake of pressed figs part of which has become unclean, one may at the outset set aside heave offering out of the clean part of the cake for the unclean part.

This applies not only to a cake which is in one piece, but also to a bundle of vegetables or even a stack of wheat part of which has become unclean—one may set aside heave offering out of the clean part for the unclean part.

If, however, there are side by side two cakes, or two bundles, or two stacks, one of which has become unclean while the other remains clean, one may not at the outset set aside heave offering out of the clean one for the unclean one.

One may at the outset set aside heave offering of the tithe out of the clean for the unclean, as it is said, *Even the hallowed part thereof out of it* (Num. 18:29), implying, "take out of the hallowed part thereof."

8. One may not set aside heave offering out of the unclean for the clean, but if he does so unwittingly, it is valid heave offering. If he does so deliberately, the remainder is not regularized thereby; what he has set aside retains its status of heave offering, but he must go back and set aside heave offering once more.

When does this apply? When one had not known that the unclean part was unclean. If he did know, and erroneously thought that it was permitted to set aside heave offering out of the unclean for the clean, he is considered the same as if he had acted willfully. The same applies to the heave offering of the tithe.

9. One may not set aside heave offering out of produce attached to the ground for produce detached from the ground, or vice versa.

How so? If having on hand detached produce one says, "This detached produce is to be heave offering for that attached produce"—even if he adds, "after the latter shall have been detached"—or if having two beds he says, "The detached produce of this bed is to be heave offering for the attached produce of that bed," or vice versa, it is as if he had said nothing.

If, however, he says, "The detached produce of this bed is to be heave offering for the produce of that bed after the latter shall have been detached," and the latter is thereupon detached, then inasmuch as it is in his power to detach the latter produce, no act can be deemed lacking, and once the produce of both beds has been detached, his words have been fulfilled, provided that the produce of both beds was at least one-third grown at the time he spoke.

10. One may not set aside heave offering out of fresh produce for dried produce, or vice versa, but if one does so nevertheless, his heave offering is valid. How so? If one gathers some vegetables today and some more on the morrow, he may not set aside heave offering out of the one for the other, unless it is natural for it to keep fresh for two days.

The same applies to vegetables that keep fresh for three days, as for example, cucumber melon—everything gathered within three days may be combined, and heave offering may be set aside out of the one for the other. In the case of vegetables that keep fresh for one day only, if one gathers some in the morning and some more in the evening of that same day, heave offering may be set aside out of the one for the other.

11. One may not set aside heave offering out of this year's produce for yesteryear's produce, or vice versa, and if he does so nevertheless, the heave offering is invalid, as it is said, *year by year* (Deut. 14:22).

If one gathers vegetables on the eve of New Year's Day before sunset, and then goes back and gathers some more after sunset, he may not set aside heave offering out of the one for the other, because the one is deemed new and the other is deemed old.

Similarly, if one plucks a citron on the eve of the fifteenth day of Shebat before sunset, and then goes back and plucks another citron after sunset, he may not set aside heave offering out of the one for the other, because the first day of Tishri is New Year's Day for tithes of corn, pulse, and vegetables, while the fifteenth day of Shebat is New Year's Day for tithes of trees.

12. One may not set aside heave offering out of produce of the Land of Israel for produce outside of the Land of Israel, or vice versa, nor out of produce of the Land of Israel for produce of

Syria, or vice versa. Neither may one set aside heave offering out of produce that is not liable to heave offering, as for example, gleanings, forgotten sheaf, and corner crop, nor from produce from which heave offering has already been set aside for other produce that is still liable to it, nor out of produce that is liable to heave offering for produce that is exempt from it. If one does set aside heave offering nevertheless, it is invalid.

13. If one sets aside heave offering out of first tithe from which no heave offering has yet been taken, or out of second tithe or consecrated produce that has not yet been redeemed, for other produce, the offering is invalid.

14. One may not set aside heave offering out of produce liable to it on the authority of the Torah for produce liable to it only by Scribal law, or vice versa. If he does so nevertheless, his heave offering is valid, but he must go back and set aside heave offering once more.

15. A perforated pot is deemed the same as the soil itself. How large must the perforation be? Large enough for a small root, that is, less than the thickness of an olive.

If one plants produce in an unperforated pot, and after it has reached one-third of its growth perforates the pot, so that the produce completes its growth in a perforated pot, it is the same as if it had grown in an unperforated pot, unless one perforates the pot before the produce has reached one-third of its growth.

16. If one sets aside heave offering out of produce grown in the soil for produce grown in a perforated pot, or vice versa, it is valid heave offering.

If he sets aside heave offering out of produce grown in an unperforated pot for produce grown in a perforated pot, it is valid heave offering, but he must go back and set aside heave offering once more.

If he sets aside heave offering out of produce grown in a perforated pot for produce grown in an unperforated pot, it is valid heave offering, but the priest may not consume it until heave offering and tithes have been set aside for this produce out of some other produce.

17. If one sets aside heave offering out of doubtfully tithed produce for other doubtfully tithed produce, or out of doubtfully tithed produce for produce assuredly untithed, it is valid heave offering, but he must go back and set aside heave offering once more from each one separately.

If one sets aside heave offering out of assuredly untithed produce for doubtfully tithed produce, it is valid heave offering, but it may not be consumed by the priest until one has set aside heave offering and tithes for the doubtfully tithed produce.

18. One may not set aside heave offering out of wheat ears for wheat, out of olives for olive oil, or out of grapes for wine, and if he does so nevertheless, it is not valid heave offering, as a precaution, lest he should burden the priest with the treading of the wine or the pressing of the olives.

One may, however, set aside heave offering out of olive oil for olives that are in the process of being pickled, or out of wine for grapes that are in the process of being dried into raisins. To what may this be compared? To one who sets aside heave offering out of two species which, if planted together, would not constitute diverse kinds, and substitutes the superior species for the inferior species.

Similarly, one may set aside heave offering out of oil-rich olives for pickling olives, but not vice versa; out of uncooked wine for cooked wine, but not vice versa; out of clarified wine for unclarified wine, but not vice versa; out of fresh figs for dried figs by number, and out of dried figs for fresh figs by measure, but not out of fresh figs for dried figs by measure nor out of dried figs for fresh figs by number. All this to the end that one should always be liberal in setting aside heave offering.

One may set aside heave offering out of wheat for bread but not out of bread for wheat, unless he calculates the proportion with precision.

In all these cases, if one does set aside the heave offering, it is valid.

19. One may not set aside heave offering out of olive oil for pressed olives, or from wine for trodden grapes, for this is similar to setting aside heave offering out of produce the preparation of

which has been completed for produce the preparation of which has not yet been completed. If one does so nevertheless, the heave offering is valid, but he must go back and set aside heave offering once more out of the olives and the grapes separately. The first heave offering, in itself, if it falls into other produce, renders it heave offering, and he who consumes it is as liable as he who consumes any other regular heave offering; but this is not so with the second heave offering.

20. If one sets aside heave offering out of oil for olives intended for eating, or out of one pile of olives for another pile of olives, both piles intended for eating; or out of wine for grapes intended for eating, or out of one pile of grapes for another pile of grapes, both piles intended for eating, and then changes his mind and decides to press them, and does press the olives or the grapes for which heave offering has already been set aside, he need not go back and set aside the heave offering once more.

21. One may not set aside heave offering out of vinegar for wine, but he may do so out of wine for vinegar, because wine and vinegar belong to the same species. If his intention was to set aside the offering out of one lot of wine for another lot of wine, and it turns out that he has vinegar in his hand, it is invalid heave offering. If his intention was to set aside heave offering out of one lot of vinegar for another lot of vinegar, and the vinegar set aside turns out to be wine, the heave offering is valid.

22. If one sets aside a jar of wine as heave offering for another lot of wine, and it is found to be vinegar, the heave offering is invalid, providing that the wine in the jar was known to have turned into vinegar before it was set aside as heave offering. If it turned into vinegar after it had been set aside as heave offering, the heave offering is valid. If the matter is in doubt, the heave offering is valid, but one must go back and set aside heave offering once more.

Similarly, if one sets aside a chate melon as heave offering, and it is found to be bitter, or a watermelon, and it is found to be rotten, or if they are found to have been punctured; or if one sets aside as heave offering a jar of wine, and it is found to be uncovered, so that the wine is forbidden to drink, the heave offering is valid, but he must go back and set aside a second heave offering. Neither of these two offerings, in itself, if mixed with other produce, renders it heave offering, and if one consumes one of them, he is not liable to the payment of one-fifth's fine.

23. How so? If the first of the two offerings falls into profane produce, it does not render it heave offering. If the second offering falls into another lot of produce, it likewise does not render it heave offering. If both offerings fall into the same lot of produce, it becomes heave offering in proportion to the bulk of the smaller of the two offerings.

Similarly, if a commoner consumes both offerings, he must pay for the smaller of the two, plus one-fifth thereof. And what exactly should one do with the two of them? He should give them both to the same priest, and receive from him the monetary value of the larger one.

24. If having examined the wine jar, one leaves it alone in order to use it as heave offering for other wine, when all the wine in the jar shall have become heave offering, and then give it to the priest, and if after a while he examines it again and finds that it has turned to vinegar, the rule is as follows: From the time when he had first examined it and for three days thereafter, it is deemed assured wine, and all the wine for which during these three days he intended this wine jar to serve as heave offering is considered regularized. After that period of time the matter is in doubt, and a second heave offering must be set aside out of it.

25. At three times should wine set aside for heave offering be examined, lest it should have turned into vinegar: during the east wind season following the Festival of Tabernacles, during the budding season of grape berries, and during the season when unripe grapes begin to fill with juice. Wine fresh from the press may be set aside as heave offering, under the presumption that it will remain wine up to forty days thereafter.

26. If one retains some produce in order to set aside heave offering therefrom for some other produce, until the retained produce shall have become all heave offering, the rule is that although one should not at the outset set aside heave offering except out of adjacent produce, nevertheless, if he has done so, the assumption is that the retained produce is still in existence. If, however, he finds that the retained produce has wasted away, there are grounds for apprehension about all the other produce that he had regularized—perchance when he designated the retained produce as heave offering, it had already wasted away. He must therefore set aside heave offering once more.

CHAPTER VI

1. Heave offering and heave offering of the tithe may be consumed by priests, adult and minor, male and female, by themselves, by their Canaanite bondsmen, and by their cattle, as it is said, *But if a priest buy any soul, the purchase of his money . . . they may eat of his bread* (Lev. 22:11).

A priest's fugitive bondsman or rebellious wife may eat of heave offering.

2. Heave offering authorized by the Torah may not be consumed except by a priest of proven priestly descent. Priests by presumption may consume only heave offering authorized by Scribal law. Clean heave offering, whether great heave offering or heave offering of the tithe, whether authorized by the Torah or by Scribal law, may be given only to a priest who is a disciple of the wise, because it is forbidden to eat unclean offering, and all unlearned persons are presumed to be unclean. One may therefore give unclean heave offering to any other priest he may choose.

3. A female Israelite married to a priest may eat of heave offering and of the breast and thigh, even if she is a minor only three years and one day old.

According to the law of the Torah, she may eat of it from the moment she is espoused, inasmuch as she comes at that moment into her husband's possession. The Sages, however, have forbidden her to eat of it until she enters the bridal canopy, as a precaution, lest she should give some of the heave offering to her father and brother while she is an espoused girl still residing in her father's house.

4. A female deaf-mute or imbecile married to a priest may not eat of heave offering, even if her father had given her in marriage, as a precaution, lest a deaf-mute priest should marry a deaf-mute commoner and feed her offerings. The Sages have therefore ruled that a deaf-mute female Israelite may not eat of heave offering at all.

5. A commoner is forbidden to eat of heave offerings, as it is said, *There shall no common man eat of the holy thing* (Lev. 22:10), even if he is the priest's settler-tenant or hired servant, as it is said, *A settler-tenant of a priest, or a hired servant, shall not eat of the holy thing* (*ibid.*). A settler-tenant is one hired in perpetuity, while a hired servant is one hired for a limited number of years. A Hebrew bondsman is the same as a settler-tenant or a hired servant, while a woman of priestly descent married to an Israelite is the same as a commoner, as it is said, *no common man,* implying both himself and his wife.

6. A commoner who eats of heave offering willfully, whether while he is unclean or clean, whether he eats of unclean or clean heave offering, is liable to death at the hand of heaven, as it is said, *And die therein if they profane it* (Lev. 22:9). He must be punished by flogging for eating of it, but need not pay the value of what he had eaten, because one may not be both flogged and obligated to pay.

If he has eaten of it unwittingly, he must add one-fifth of its value, as it is said, *And if a man eat of the holy thing through error, then he shall add the fifth part thereof unto it* (Lev. 22:14).

7. *And if a priest's daughter be married unto a common man, she shall not eat of that which is set apart from the holy things* (Lev. 22:12). Two matters are included in this negative com-

mandment: The first is that if she has sexual intercourse with a man forbidden to her and thus becomes a harlot or a profaned woman, as we have explained in the Laws Concerning Forbidden Intercourse, she is forever forbidden to eat of any heave offering, just like any profaned priest, since a profaned priest is the same as a commoner in every respect.

The second is that if she marries an Israelite, she is forever forbidden to eat of the parts that are set aside out of holy things, namely the breast and the thigh, even if her Israelite husband subsequently divorces her or dies.

8. She may, however, eat of heave offering after her Israelite husband has divorced her or has died leaving no son by her, as it is said, *But if a priest's daughter be a widow or divorced, and have no child, and is returned unto her father's house, as in her youth, she may eat of her father's bread* (Lev. 22:13).

9. The Sages have learned by tradition that *of . . . bread* does not imply all of the bread, meaning that she regains her right to heave offering but not to the breast and the thigh.

10. Not only a woman of priestly descent, but also a woman of Levitic descent or an Israelite woman, who has had intercourse with a man forbidden to her, is forbidden forever to eat of heave offering, even if she has a child sired by a priest, inasmuch as she has become a harlot.

11. A captive woman therefore may not eat of heave offering, even if she says, "I am clean." On the other hand, any captive woman who is to be believed when she says, "I am clean," or who has a witness thereto, so that she is permitted to her husband, may eat of heave offering. A woman who has lain with a beast is not thereby disqualified from the priesthood and may therefore eat of heave offering.

12. An Israelite woman who has a child sired by a priest may eat of heave offering on account of her child, whether male or female, even if the child is a hermaphrodite or a ṭumṭum; indeed even on account of her child's child and downward, to the end of the world, as it is said, *and have no child* (Lev. 22:13).

13. Just as the child of an Israelite man married to a woman of priestly descent disqualifies her, so the child of a priest married to an Israelite woman renders her permitted to eat of heave offering, even if the child is blemished.

How so? If A, an Israelite woman married to a priest or a woman of priestly descent married to an Israelite, gives birth to a daughter, B, by him, and if subsequently a man who is B's forbidden relative has intercourse with B, or if B marries a bastard, and then dies leaving a child, who is thus a bastard, that surviving bastard child entitles his late mother's mother, A, to eat of heave offering.

14. If his maternal grandmother is an Israelite and was married to a priest, she may, on her own account, eat of heave offering. If she is the daughter of a priest and was married to an Israelite, she may not, on her own account, eat of heave offering. Thus you learn that she may eat of it on account of her progeny, even if that progeny is blemished or even is not Israelite at all. Needless to say, in case she has a daughter sired by a priest, even if this daughter marries an Israelite, or even becomes profaned, her mother may eat of heave offering on account of her blemished daughter.

15. Similarly, a woman of priestly descent may not eat of heave offering on account of her progeny sired by an Israelite, even if that progeny is a priest. How so? If the daughter of a priest marries an Israelite and bears a daughter by him, and if this daughter goes forth and marries a priest and bears a son by him, this son is eligible to become High Priest; he therefore renders his mother permitted to eat of heave offering but disqualifies his mother's mother from it. This is so even if his mother is dead, and her mother might thus well say, "Let there not be the like of my daughter's son, the High Priest, who disqualifies me from eating of heave offering!"

16. A bondsman descendant neither disqualifies from, nor entitles to, eating of heave offering. How so? If a woman of priestly descent marries an Israelite, or a female Israelite marries a priest, and bears a son by him, and if this son subsequently gets

involved with a bondswoman who bears a son, likewise a bondsman, by him, and if the father then dies, the surviving grandson, the bondsman, remains. Nevertheless, if his paternal grandmother is an Israelite who was married to a priest, she may not eat of heave offering; if she is the daughter of a priest who was married to an Israelite, she may eat of it. For this bondsman grandson is not regarded as progeny, since bondsmen have no genealogy.

17. If an Israelite woman marries a priest, who subsequently dies leaving her with a son by him, and if she is then remarried to an Israelite, she may not eat of heave offering on account of her son by the priest. If her Israelite husband also dies and leaves her with a son by him, she may not eat of heave offering on account of her son by the Israelite. If thereupon her son by the Israelite also dies, she may eat of heave offering on account of her first son.

18. If a priest's daughter marries an Israelite and has a son by him, and is then married a second time to a priest, she may eat of heave offering. If the latter, too, dies leaving her with a son by him, she may still eat of heave offering. If her son by the priest dies, she may not eat of heave offering on account of her son by the Israelite. If her son by the Israelite dies, she may return to her father's house, as in the days of her youth, and may eat of heave offering but not of the breast and the thigh.

19. If an Israelite woman marries an Israelite first and is left with a son by him, and then is married again to a priest, she may eat of heave offering. If her priestly husband dies leaving her with a son by him, she may eat of heave offering on account of her last son, because he entitles her to eat of it the same as did his father.

CHAPTER VII

1. An unclean priest may not eat of heave offering, whether it be clean or unclean, as it is said, *What man soever of the seed*

of Aaron is a leper, or hath an issue, he shall not eat of the holy things (Lev. 22:4). What are the *holy things* whereof all the seed of Aaron, both male and female, may eat? One cannot but say, it is heave offering. Hence every unclean person who eats of heave offering is liable to death at the hand of heaven, and is therefore liable to a flogging, as it is said, *They shall therefore keep My charge, lest they bear sin for it* (Lev. 22:9). An unclean person who eats of unclean heave offering, however, is not liable to a flogging, even though this is forbidden by a negative commandment, inasmuch as unclean offering is not holy.

2. Unclean persons may not eat of heave offering until that day's sun has set, and three stars have appeared after the sun has set, as it is said, *And when the sun is down, and it is clear* (Lev. 22:7), implying that after the sky has been cleared of light the unclean person may eat of the holy things.

3. A clean priest who eats of unclean offering is not liable to a flogging, because this is forbidden by a positive commandment, as it is said, *And afterward he may eat of the holy thing* (Lev. 22:7), implying that of anything that is in a state of holiness he may eat after he has become clean, but of an unclean thing he may not eat even after he has become clean. And a negative commandment derived from a positive commandment has the force of a positive commandment.

4. A priest who, while eating of heave offering, feels his limbs trembling as a sign of an imminent discharge of semen, must take firm hold of his penis while he quickly swallows the mouthful of heave offering.

5. A deaf-mute or imbecile priest should be immersed and fed heave offering after that day's sun has set. They should be watched so that they do not fall asleep after the immersion, for should they fall asleep, they would become unclean, unless indeed they are provided with a copper sheath, in case they subsequently have a discharge of semen.

6. Camel riders are forbidden to eat of heave offering until after they have immersed themselves and that day's sun has set,

because they are presumed to be unclean, owing to the heating effect of riding upon the bare hide of the camel, which may result in the discharge of a drop of semen.

7. In the case of a woman who has just had sexual intercourse, the rule is as follows: If she did not make violent motions during intercourse, she may immerse herself and may then eat of heave offering in the evening. If she did make violent motions during intercourse, she may not eat of heave offering throughout the ensuing three days, because it is impossible for her not to discharge some semen and thus avoid becoming unclean, as shall be explained in its proper place.

8. Heave offering from outside of the Land of Israel, the root of whose sanction is Scribal authority, is forbidden only to a priest whose uncleanness is issuing out of his own body, such as men who have suffered nocturnal pollution, men or women afflicted with flux, menstruating women, and recently delivered women. All of these, once they have immersed themselves, may eat of heave offering, even if that day's sun has not yet set.

Those, however, who are unclean by reason of contact with unclean things, whether it be a dead body—an uncleanness from which it is impossible for us to be cleansed today—or a creeping thing, need not immerse themselves for heave offering originating outside of the Land of Israel.

9. Therefore a minor priestly male, who cannot yet have had a nocturnal emission, and a minor female, who cannot yet have seen menstrual blood, may also be fed heave offering without prior examination, because they are presumed not to have had any uncleanness issue out of them as yet.

A priestly leper is accounted the same as a person who has uncleanness issuing out of his body, provided that a priest of proven genealogy has declared him unclean; prior to this declaration he is considered clean.

10. An uncircumcised priest is forbidden to eat of heave offering on the authority of the Torah, as it is said, *a settler-tenant . . . or a hired servant* (Lev. 22:10) with reference to

heave offering, and *a settler-tenant or a hired servant* (Exod. 12:45) with reference to the Paschal lamb—just as the settler-tenant or the hired servant mentioned with reference to the Paschal lamb, if uncircumcised, is forbidden to eat of it, so also the settler-tenant or the hired servant mentioned with reference to heave offering, if uncircumcised, is likewise forbidden to eat of it. If he does eat, he is liable to a flogging on the authority of the Torah.

A priest who has had his prepuce drawn forward may eat of heave offering, even though he looks uncircumcised. Scribal law, however, obligates him to be circumcised a second time, so that he would look circumcised.

11. A priest born circumcised may eat of heave offering, while a ṭumṭum may not, because he is doubtfully uncircumcised. A hermaphrodite must be circumcised and may then eat of the offering.

12. Even though the uncircumcised priest and any other unclean priest are themselves forbidden to eat of heave offering, their wives and their bondservants may eat of it.

13. As for those who are crushed or maimed in their privy parts, they and their bondservants may eat of heave offering, but their wives may not. If such a priest has had no sexual intercourse with his wife since the time when he received his injury, his wife may eat of the offering.

Similarly, if such a priest marries a proselyte's daughter, she too may eat of it.

14. If a priest crushed in his privy parts weds a priest's daughter, she may not eat of heave offering. In the case of a priest who is a congenital eunuch, he, his wife, and his bondservants may eat of it. In the case of a ṭumṭum and a hermaphrodite, their bondservants may eat thereof, but not their wives.

15. If a deaf-mute, imbecile, or minor of priestly descent purchases bondservants on his own account, these bondservants may not eat of heave offering. If, however, it is the court or the

guardian who effects the purchase for these incompetents, or if the bondservants fall to them by inheritance, the bondservants may eat of heave offering.

16. A hermaphrodite who has had sexual intercourse, whether by way of his male organ or by way of his female organ, with a person disqualified from eating of heave offering, becomes himself so disqualified, the same as do women. His bondservants likewise may not eat of it.

Similarly, if he has had sexual intercourse with another hermaphrodite, who, had he had intercourse with a woman would have deprived her of the right to eat of heave offering, he becomes himself disqualified and may neither himself eat of heave offering nor feed it to his bondservants. This is so, provided he has had intercourse by way of his female organ. If he has had it by way of his male organ, the rule is that no male may disqualify another male from the priesthood.

17. A bondsman owned by several priestly partners, one of whom is disqualified from entitling his bondservants to eat of heave offering, that bondsman is forbidden to eat of heave offering.

Whosoever entitles another person to eat of heave offering entitles him also to eat of the breast and the thigh.

18. If an Israelite woman married to a priest brings in with her some bondservants who are either her mĕloḡ property or her iron sheep property, they may eat of heave offering.

Similarly, if a priest's bondservants buy bondservants, or if his wife's bondservants buy bondservants, the latter may eat of heave offering, as it is said, *But if a priest buy any soul, the purchase of his money, he may eat of it* (Lev. 22:11), implying even the purchase by a purchase; thus a purchase who is eligible to eat may render another person eligible to eat, and contrariwise, a purchase who himself may not eat may not render another person eligible to eat.

19. If a priest's daughter married to an Israelite brings him bondservants who are either her mĕloḡ property or her iron sheep property, they may not eat of heave offering.

20. In the case of a widow married to a High Priest, or of a divorcée or a woman who had formerly performed ḥăliṣah married to an ordinary priest—whether she is of priestly or Israelite descent—and similarly of other women forbidden to a priest by a negative commandment, who brings him mĕloḡ property bondservants and iron sheep property bondservants, the rule is that the mĕloḡ property bondservants may not eat of heave offering, even though the husband is responsible for their maintenance, while the iron sheep property bondservants may eat of it, because they belong to the huband. If he marries a woman of secondary degree of consanguinity, she herself may eat of it, but her mĕloḡ property bondservants may not.

21. If a priestly widow is espoused by a High Priest, or a divorcée by an ordinary priest, she may not eat of heave offering, inasmuch as she is awaiting intercourse that is blemished, according to the Torah.

Similarly, if these women enter the bridal canopy without previous espousal, they may not eat of heave offering, because the bridal chamber disqualifies them from eating of it.

If they are widowed or divorced after espousal, they regain their legitimate status, and may eat of heave offering; if after nuptials, they may not eat of heave offering, since they have already become profaned women.

22. A priestly woman whose priestly husband is dead, and who thus becomes eligible for levirate marriage to her brothers-in-law, one of whom is a profaned priest, may not eat of heave offering, because of her levirate tie to the profaned priest. This applies even if one of the unblemished brothers-in-law has already made a verbal declaration to her, because a verbal declaration does not effect a complete acquisition of the widowed sister-in-law.

23. If a priest issues a geṭ to his widowed priestly sister-in-law, thus rendering her forbidden to him while she is still tied to him by levirate tie, she may eat of heave offering, because she is awaiting intercourse that is forbidden only by Scribal law, seeing that a geṭ disqualifies a widowed sister-in-law only by Scribal law.

Similarly a priestly woman who has previously performed ḥăliṣah, or a woman of secondary degree of consanguinity, who is espoused by a priest, may eat of heave offering.

If an ordinary priest marries a barren woman, she may eat of heave offering.

CHAPTER VIII

1. An unborn child, a levir, an espousal, a deaf-mute, or a boy nine years and one day old can disqualify a woman from eating of heave offering, but cannot entitle her to eat of it.

2. How an unborn child? If an Israelite woman becomes pregnant by a priest, she may not eat of heave offering on account of the unborn child. So also, if a priest's daughter becomes pregnant by an Israelite, she is forbidden to eat of heave offering on account of the unborn child, as it is said, *And is returned unto her father's house, as in her youth* (Lev. 22:13)—*in her youth* excludes a pregnant woman.

3. If a priest's daughter has had illicit sexual intercourse with an Israelite, no apprehension need be felt that she may have become pregnant; rather she may immerse herself, and may then eat of heave offering in the evening. If she had been married to an Israelite who has since died, she may immerse herself and then eat of heave offering in the evening, for up to forty days. If in the meantime the fetus becomes recognizable, she is impaired retroactively, until the end of the forty days, because for the first forty days the fetus is regarded not as a fetus but as ordinary water.

4. In the case of an Israelite woman married to a priest who died leaving her pregnant, her bondservants may not eat of heave offering on account of the unborn child, for it is the born child that confers this right, not the unborn child.

Therefore, if the unborn child is a profaned priest, he does not disqualify the bondservants; rather they may eat of heave offering on account of his legitimate brothers, until the time when the profaned child is born, whereupon the bondservants are forbidden to eat of it.

5. How a levir? An Israelite woman who is under levirate tie to a priestly levir may not eat of heave offering, as it is said, *But if a priest buy any soul* (Lev. 22:11), and this levir has not yet acquired her.

A priest's daughter who is under levirate tie to an Israelite is forbidden to eat of heave offering on account of her levir, as it is said, *And is returned unto her father's house as in her youth* (Lev. 22:13)—*in her youth* excludes a woman awaiting levirate marriage.

6. If a priestly levir has had sexual intercourse with his widowed sister-in-law under duress, or unwittingly, or having initiated the intercourse had not completed it, the rule is that although he has acquired her, as we have explained in the Laws Concerning Levirate Marriage, he does not entitle her to eat of heave offering until he has perfect intercourse with her, with her consent. When does this apply? When she had been widowed after espousal. If she had been widowed after nuptials, she is entitled to eat of heave offering by this same act of intercourse, seeing that she had been eating of it heretofore.

7. How an espousal? A priest's daughter espoused to an Israelite is forbidden to eat of heave offering, because he has acquired her by the espousal. An Israelite woman espoused to a priest may not eat of heave offering until she has entered the bridal canopy, lest she should feed it to members of her father's household, as we have explained.

8. If an Israelite says to a priestly woman, "Behold, thou art consecrated unto me thirty days hence," she may eat of heave offering all through the thirty days, inasmuch as she is not yet espoused.

If, however, he says to her, "Behold, thou art consecrated unto me as of now, and after thirty days," she is immediately forbidden to eat of heave offering.

The same applies to all similar stipulations.

9. How a deaf-mute? A priestly woman married to a deaf-mute Israelite may not eat of heave offering, inasmuch as he has acquired her on the authority of the enactment of the Sages which makes his marriage valid.

An Israelite woman married to a deaf-mute priest may not eat of heave offering, because he cannot acquire her according to the Torah, inasmuch as he is regarded as having no mind of his own.

10. If an Israelite woman had been espoused by a normal priest who became deaf-mute just prior to wedding her, she may not eat of heave offering. If he dies, and she becomes eligible to a deaf-mute levir who contracts levirate marriage with her, she still may not eat of heave offering. If, however, he too marries her while normal but thereafter becomes deaf-mute, she may eat of it. If he too dies, and she falls to another deaf-mute levir who marries her, she may continue to eat of it, seeing that she had been eating of it previously.

If a deaf-mute priest's wife has a child by him, she may eat of heave offering on account of her child.

11. How a boy nine years and one day old? A priestly woman who has had intercourse with a male who is forbidden to her and is nine years and one day old, so that his intercourse is valid, becomes disqualified from the priesthood and is forbidden to eat of heave offering, seeing that she has become a harlot or a profaned woman, as we have explained. This is so even if the boy is impotent.

If an Israelite woman is married to a priest nine years and one day old, she may not eat of heave offering on his account, even though his intercourse is valid, because he cannot acquire her until he becomes an adult.

If there is doubt as to whether he is or is not nine years and one day old, or if she is married to a priestly male thirteen years and one day old, but there is doubt as to whether he has or has not grown the statutory two hairs, she is forbidden to eat of heave offering.

12. All of these aforementioned persons, just as they cannot confer the right to eat of heave offering that is authorized by the

Torah, so can they not confer the right to eat of heave offering that is authorized by Scribal enactment; this is a precautionary measure, lest they should feed their kin heave offering authorized by the Torah.

13. The following persons can neither deprive one of this right nor confer it upon him: a violator, a seducer, or an imbecile who has wed a wife, unless indeed the woman is forbidden to them, because they disqualify her by the act of intercourse, in which case she becomes a harlot or a profaned woman, as we have explained.

14. A woman married to an imbecile priest, or violated or seduced by a priest, who subsequently gave birth to a child, may eat of heave offering on account of her child. And even though the matter is in doubt—since there has been no betrothal, and she may possibly have become pregnant by another man—nevertheless the presumption is that the child belongs to this priest who has had intercourse with her. Provided, however, that there is no rumor involving her with another man and that everyone murmurs about her affair with that priest.

Similarly a priestly woman who has had intercourse with an Israelite imbecile, or has been violated or seduced by an Israelite, and became pregnant, may not eat of heave offering on account of the fetus. If the fetus disintegrates, she may eat of heave offering.

15. A married woman attested by witnesses to the effect that although warned by her husband she had subsequently secluded herself with another man, is forbidden to eat of heave offering until she has drunk of the water of bitterness, seeing that she is a doubtful harlot.

If her husband dies before he can cause her to drink of this water, or if she is one of those women who may neither drink nor receive their kĕṯubbah, she is forever forbidden to eat of heave offering.

Any priest's wife who says, "I have been defiled," is forbidden to eat of heave offering.

16. A minor Israelite female married to a priest without her father's consent, whether in the latter's presence or absence, even if her father had himself betrothed her, may not eat of heave offering. For should the father later protest the marriage, she would become a commoner retroactively, and the reason why he is silent now as he witnesses her marriage may be that he is indignant at her marrying without his consent.

CHAPTER IX

1. A woman may eat of heave offering until the moment when a geṭ reaches her own hand or the hand of the person whom she had appointed as her receiving agent.

Any woman who is doubtfully divorced may not eat of heave offering.

A woman who has appointed an agent to receive her geṭ in her behalf is immediately thereafter forbidden to eat of heave offering. If she says to him, "Accept my geṭ in my behalf in such-and-such a place," she is not forbidden to eat of heave offering until the agent reaches that particular place.

If she dispatches an agent to fetch her geṭ for her, she may eat of heave offering until the geṭ reaches her own hand.

If a priest says to his wife, "Here is your geṭ, to become effective one hour before my death," she is immediately forbidden to eat of heave offering.

2. A person sojourning in a besieged town, or traveling in a stormbound ship at sea, or about to be tried on a capital charge, is presumed to be still among the living. Needless to say, this applies also to a person setting out upon a sea voyage or joining a caravan.

On the other hand, a person known to have been in a besieged town captured by storm or in a ship lost at sea, or to have been led to execution out of a heathen court, or dragged away by a wild beast, or buried under a collapsed building, or washed away by a river, is subject to the more stringent rules governing both the living and the dead. Therefore, if he is an Israelite and his wife is a priest's daughter, or if he is a priest and his wife is an Israelite woman, she may not eat of heave offering.

However, a person condemned by an Israelite court and remanded to the place of stoning for execution, is presumed to be as good as dead, and his wife may not eat of heave offering.

3. If a wife—whether a priestly woman married to an Israelite or an Israelite woman married to a priest—has left her husband behind in another country in a state of imminent death, she may not eat of heave offering, inasmuch as most persons who are on the verge of death usually do die.

If one witness testifies that the husband is dead, while another testifies that he is not, she may not eat of heave offering.

4. If the wife's witness is her co-wife, or one of the five women who are not trusted to testify to her husband's death, the rule is that inasmuch as she may not remarry on the strength of their testimony, she may eat of heave offering, on the presumption that her husband is still living, until such time as a person trustworthy enough to authorize her to remarry shall have testified to her husband's death.

5. If a priest emancipates his bondservant, he disqualifies him from eating of heave offering the moment he conveys to him the writ of manumission.

Any bondservant who has been emancipated but is still awaiting his writ of emancipation—as will be explained in the Laws Concerning Bondservants—is nevertheless forbidden to eat of heave offering.

6. If A transfers title to his estate, including bondservants, in writing to B, and conveys it to him through a third party, C, the rule is that if B remains silent at first but subsequently signifies his protest, a doubt exists as to whether B's protest in the end proves his true intent at the beginning, so that the bondservants have not yet left the authority of A, or whether B's protest sub-

sequent to his silence signifies a change of mind. The bondservants therefore may not eat of heave offering, regardless of whether B is an Israelite and A is a priest, or vice versa.

7. If an Israelite hires a beast from a priest, he may feed it heave offering. On the other hand, a priest who hires an Israelite's beast may not feed it heave offering, even though he is obligated to feed it, since it is not his money purchase.

8. If an Israelite accepts a cow from a priest at an estimated value, with a view to fattening her up, on the understanding that they would share in her increased value, he may not feed her heave offering, even though the priest has a share in her increased value. On the other hand, if the priest accepts the cow from the Israelite, he may feed her heave offering, even though the Israelite has a share in her increased value, seeing that the corpus of the cow belongs to the priest, inasmuch as he had set a value upon her for himself.

9. If an Israelite's cow gives birth to a firstling, he may feed it heave offering, since the firstling belongs to the priest.

A person may store bitter vetch offering in his dovecote, and need feel no apprehension that his own doves might come and eat of it.

10. It would appear to me that if a priest sells his cow to an Israelite and accepts her sale price, he may not feed her heave offering, even if the buyer has not yet performed the act of drawing, because according to the law of the Torah, payment of money confers the right of ownership, as will be explained in the Laws Concerning Buying and Selling.

If, however, an Israelite sells a beast to a priest, it may not eat of heave offering even if money has already been paid for it, until the buyer performs the act of drawing.

CHAPTER X

1. If a commoner eats of heave offering unwittingly, he must pay the principal thereof plus one-fifth. Even if he does know

that it is heave offering and that he is subject to the warning against it, but does not know whether transgression thereof does or does not involve the death penalty, his transgression is accounted unwitting, and he must only pay the principal plus one-fifth.

2. It does not matter whether one eats something that is customarily eaten, or drinks something that is customarily drunk, or anoints himself with something that is customarily used for anointing, as it is said, *And they shall not profane the holy thing of the children of Israel* (Lev. 22:15), which includes the person who anoints himself. Whether one unwittingly eats clean or unclean heave offering, he must pay the principal thereof plus one-fifth.

One is not liable to the one-fifth until he eats an olive's bulk, as it is said, *And if a man eat of the holy thing through error* (Lev. 22:14)—"eating" cannot apply to an amount less than an olive's bulk. And just as eating of heave offering means no less than an olive's bulk, so does drinking thereof mean no less than an olive's bulk.

3. If one eats and then eats again, or drinks and then drinks again, the rule is that if the interval between the beginning of the first eating and the end of the second eating is no longer than the time it takes to eat half a loaf, or if the interval between the beginning of the first drinking and the end of the second drinking is no longer than the time it takes to drink one-quarter of a log̃, the two combine in estimating the olive's bulk.

4. Heave offering, heave offering of the tithe—whether out of doubtfully tithed or out of assuredly tithed produce—dough offering, and first fruits may all be combined in estimating the olive's bulk and in establishing liability to the death penalty and to the payment of the additional one-fifth, because the term "heave offering" is used with reference to all of them. According to the law, heave offering of the tithe out of doubtfully tithed produce should be exempt from the payment of the additional one-fifth, since one is not liable to the second tithe thereof, as

will be explained. But the Sages have said that if one were not liable to that one-fifth for it, he would make light of it.

5. If one eats of heave offering deliberately, the rule is as follows: If he was forewarned, he is liable to a flogging, and need not pay for it; if he was not forewarned, and the heave offering was clean, he must pay the principal thereof, but not the additional one-fifth; if the heave offering was unclean, he must pay its value as firewood, because it is fit only for fuel. Therefore, if he has eaten heave offering of mulberries, pomegranates, or the like that has been defiled, he is exempt from all payment, since these are not fit even for fuel.

6. If one eats leavened heave offering during Passover, whether deliberately or unwittingly, whether it is unclean or clean, he is exempt from payment. Even if it was unleavened when he had set it aside, and it became leavened thereafter, he is exempt. He need not pay even its value as firewood, since it is unfit even for fuel; for inasmuch as it is forbidden to derive any benefit from it, it has no tangible value.

7. On the other hand, one who unwittingly eats of heave offering during the Day of Atonement, or eats of punctured heave offering, or drinks heave offering wine that had been left uncovered, or anoints himself with wine and oil together, or drinks oil and vinegar together, or chews wheat grains, or swallows vinegar, must pay the principal thereof plus one-fifth.

8. If a person who has eaten to satiety, so that he would be loath to eat any more food, nevertheless stuffs himself by eating of heave offering, he need not pay the additional one-fifth, as it is said, *if a man eat* (Lev. 22:14), implying, without injuring his health. So also he who chews barley grains need not pay the additional one-fifth, since he thereby causes injury to his health.

9. If a commoner swallows jujubes of heave offering and regurgitates them, and if then another commoner comes along and eats them unwittingly, the former must pay the principal thereof plus one-fifth, while the latter must pay to the former their value as firewood.

10. If one feeds heave offering to laborers or guests, they must pay the principal thereof plus one-fifth, because they are accounted the same as those who have acted in error, while he must pay them the value of their meal, because the value of profane food is greater than the value of the heave offering that they had eaten, inasmuch as one recoils in awe from a forbidden thing.

11. An Israelite who feeds heave offering to his minor children or to his bondservants, whether adult or minor; or who eats offering imported from outside of the Land of Israel; or who eats or drinks less than an olive's bulk of heave offering; or a Nazirite who unwittingly drinks heave offering wine; or one who drinks heave offering oil or anoints himself with heave offering wine—all these must pay the principal thereof but not the additional one-fifth.

12. A priest's daughter married to an Israelite or disqualified from the priesthood, who has eaten of heave offering, must pay the principal thereof but not the additional one-fifth.

If a woman who used to eat of heave offering is told, "Your husband is dead," or "has divorced you"; and likewise, if a bondservant who used to eat of heave offering is told, "Your master is dead and has left an heir who cannot confer the right to eat of heave offering," or "Your master has sold you to an Israelite," or "has given you to him as a gift," or "has set you free"; and similarly, if a priest who used to eat of heave offering discovers that he is the son of a divorcée or of a woman who had performed ḥăliṣah—all these must pay only the principal thereof.

If the heave offering is leavened and it is Passover eve, they are exempt from all payment, since time being short, they hastened to eat of it without examining it. And all those who have anything of heave offering in their mouths, the moment they become aware that they are forbidden to eat it, must spit it out.

13. If while one is eating of heave offering he is told, "You are become unclean," or "The heave offering is become unclean," or "You were unclean," or "The heave offering was unclean"; or if it becomes known that the produce is still untithed, or that

it is first tithe from which heave offering has not yet been separated, or second tithe, or consecrated produce not yet redeemed; or if he senses the taste of a bug in his mouth—all these one must spit out.

14. If there are two baskets, one of heave offering and the other of profane produce, and if some heave offering falls into one of them without it being known which one, I say it fell into the basket of heave offering.

If it is not known which basket contains heave offering, and if a commoner eats out of one of them, he is exempt from payment, in accordance with the principle that if A has a claim against B, A bears the burden of proof. The second basket must consequently be treated as containing heave offering. If another person eats out of the second basket, he too is exempt. If a third person eats of both baskets, he must pay according to the value of the smaller of the two baskets: if he has eaten deliberately, he must pay the principal thereof; if unwittingly, the principal plus one-fifth.

15. If one unwittingly eats the additional one-fifth, he must add to it one-fifth thereof, because the additional one-fifth is subject to the same rules as the principal in every respect. The same applies to each subsequent one-fifth, ad infinitum.

Whosoever pays the principal plus one-fifth, his payment is the same as heave offering in every respect, except that if it is planted, the produce thereof is profane. Even if the priest is willing to waive his right to it, he may not do so.

Whosoever pays the principal alone, his payment is profane, and if the priest is willing to waive his right to it, he may do so.

16. In the case of an Israelite woman who had eaten of heave offering and thereafter married a priest, the rule is as follows: If she had eaten of heave offering not yet acquired by a priest, she must pay the principal plus one-fifth to herself; if it was heave offering already acquired by a priest, she must pay the principal to the owner thereof and one-fifth to herself, because whosoever must pay the principal plus one-fifth must pay the

principal to the priestly owner and may pay the one-fifth to any priest he may choose.

17. If as she is about to pay she is divorced by her priestly husband, she need not make payment to herself in either case, seeing that she is the same as if she had never been married to a priest.

18. Whosoever eats of heave offering, whether unwittingly or deliberately, must pay only out of profane produce that has been regularized by separating from it heave offerings and tithes. He may pay out of gleanings, forgotten sheaf, corner crop, ownerless property, or first tithe from which heave offering has already been separated, even if the great heave offering contained therein has not yet been set aside, as in the case of tithe set aside first before heave offering.

He may also pay out of second tithe or redeemed consecrated things, even if the redemption was made not according to the rule. He may pay with new produce for old, but not with one species for another, as it is said, *And shall give unto the priest the holy thing* (Lev. 22:14), implying that he must give the same thing as the holy thing that he had eaten.

19. He who eats late chate melons of the year preceding the Sabbatical year must wait for the early chate melons of the year following the Sabbatical year, and then pay out of them, since he may not pay his debt out of the produce of the Sabbatical year itself, as will be explained in its proper place.

20. If one eats of unclean heave offering, he must pay with profane produce, whether clean or unclean.

If he eats of clean heave offering, he must pay with clean profane produce. If he pays with unclean profane produce, whether deliberately or unwittingly, his payment is valid, but he must pay once more out of clean produce.

21. If one eats of heave offering belonging to a learned priest, he must pay him; if he eats of heave offering belonging to an unlearned priest, he must first pay to a learned priest, then take

the money from him and hand it to the unlearned priest, since it is the latter's heave offering that he has eaten. For the rule is that clean things may not be delivered to an unlearned person.

22. If one steals heave offering from his maternal grandfather, who is a priest, and eats of it, and if thereafter his maternal grandfather dies, he must pay not to himself but to another heir out of the rest of the Tribe.

Similarly, if one inherits heave offering from his maternal grandfather and eats of it, or if a creditor collects heave offering in settlement of a debt due him, or a woman collects it in payment of her kĕṯubbah, and if any one of them eats of it, they must pay the principal plus one-fifth to a learned priest, who in turn must pay them the price that the heave offering eaten by them would have fetched if sold.

23. If one steals heave offering but does not eat of it, he must pay double to the owner, but may do so with the monetary equivalent of the heave offering.

If he steals it and eats of it, he must pay double the principal thereof plus one-fifth: one principal plus one-fifth out of profane produce, and a second principal, which latter he may pay with the monetary equivalent of the heave offering.

24. If the heave offering is produce consecrated to the repair of the Temple, and one steals it and eats of it, he need not pay double, because double payment does not apply to consecrated things, as will be explained in its proper place. He must, however, pay the principal plus two-fifths, one fifth for having eaten of heave offering, and another fifth for having benefited from sacred things.

To whom should he pay? If the offering amounted to an olive's bulk but was worth less than a pĕruṭah, he must pay to the priests; if it was worth a pĕruṭah, whether or not it amounted to an olive's bulk, he must pay to the Temple Treasury.

25. Why does the prohibition of eating of consecrated things take hold upon the already existing prohibition of eating of heave

offering? Because heave offering is forbidden to a commoner but is permitted to a priest, and if the latter consecrates it, it becomes forbidden to him as well. From this follows an additional prohibition of the consecrated offering to an Israelite also, in the same manner as explained in the Laws Concerning Forbidden Intercourse and Forbidden Foods.

26. If one steals heave offering and eats of it, he must pay the principal plus one-fifth, because the one-fifth to which he is liable for eating of heave offering discharges also his liability for stealing, as it is said, *And shall give unto the priest the holy thing* (Lev. 22:14), implying that he is liable only to the one-fifth for the holy thing.

If he steals it and feeds it to someone else, the person who eats of it must pay the principal plus one-fifth.

Wherever we have said, "he must pay the principal plus one-fifth," the meaning is that if he has eaten produce worth four coins, he must pay produce of the same species as he has eaten worth five coins.

Wherever we have said, "he must pay the principal plus two-fifths," the meaning is that if he has eaten produce worth four coins, he must pay produce worth six coins. And wherever we have said, "he must pay double the principal plus one-fifth," the meaning is that if he has eaten produce worth four coins, he must pay produce worth nine coins. He must always pay according to no more than what the produce was worth at the time of his eating of it, regardless of whether it has fallen or risen in price at the time of his payment.

CHAPTER XI

1. Heave offering may be used for eating, drinking, and anointing, for anointing is like drinking, as it is said, *And it is come into his inward parts like water, and like oil into his bones* (Ps. 109:18), and drinking is included in eating; all of which signifies

eating things customarily eaten, drinking things customarily drunk, and anointing with substances customarily used for anointing. One should not use wine or vinegar for anointing, but only clean oil. Unclean oil may be used for illumination, and this oil is called everywhere "oil of burning."

2. One is permitted to press heave offering dates and form them into a cake like a fig cake, but not to make beer out of them. Similarly one may not turn dates into honey, or apples into wine, or winter fruit into vinegar. The same applies to any other fruit of heave offering—one may not alter its natural state; the only exceptions are olives and grapes.

If one transgresses and changes such food into drink, he may drink it. A commoner who eats date honey, drinks apple wine, or does something similar unwittingly is not liable to payment. If he does so deliberately, he must be flogged for disobedience.

3. One may not put fig cake or dried figs into brine, because this spoils them; but one may put wine into brine.

One may not add perfume to oil, because this removes it from the category of food and converts it into anointing oil.

One may, however, blend wine, honey, pepper, and the like in order to drink them.

4. One may not boil heave offering wine, because this reduces it. One may not soak heave offering onions in heave offering vinegar, because this spoils the vinegar.

One may not mix grain with pulse. On the other hand, substances which become separated from each other when sifted may be mixed together. Since the devastation of Judea the Sages began to allow mixing one kind of grain with another, and one kind of pulse with another, but not the mixing of grain with pulse.

5. The practice of sifting profane flour may be followed by the priest, who may sift heave offering flour, keep the edible part thereof, and throw away the coarse bran.

If he wishes to make fine flour, he may sift it through several sieves until he gets no more than one ḳab̲ or two out of one sĕ'ah.

He may not, however, throw away the rest, inasmuch as it is edible, but should deposit it in a hidden place.

6. One may not use heave offering oil to glaze an oven or a stove, nor to oil a shoe or a sandal, nor to oil one's foot while it is inside the shoe or sandal. One may, however, first oil his foot, and then put on the shoe or sandal. He may also anoint his whole body and then roll upon a new leather spread. Even though these objects are oiled thereby, one need feel no apprehension on this account. One may not, however, spread oil on a marble slab in order to roll upon it, inasmuch as he would waste the oil by doing so.

7. Whosoever eats of heave offering, even if just fruit, must wash his hands, even if his hands are clean, as will be explained in its proper place. It may not be eaten with a commoner sitting at the same table, as a precaution lest he too should eat of it. Heave offering imported from outside of the Land of Israel, however, may be eaten with a commoner sitting at the same table, and one need not wash his hands for it, as he would when about to eat profane food.

One may not anoint himself with heave offering oil while his hands are unclean, but if the oil happens to drip upon his flesh, he may rub it in with unclean hands. One may, however, anoint an infant with heave offering oil within seven days of his birth, because a newborn child is not accounted uncircumcised during the first seven days.

8. A priest may anoint himself with heave offering oil, and may also take his daughter's Israelite son and roll him against himself. If a priest anoints himself with oil and then enters the bathhouse, a commoner may massage him in the bathhouse, even though the masseur also becomes anointed thereby.

9. If a priestly woman washes her head with a solution of heave offering fenugreek, an Israelite woman is not permitted to wash her own hair with it after her. She may, however, roll her hair against the hair of the priestly woman.

Why did the Sages permit priestly persons to wash their hair in heave offering fenugreek? Because fenugreek is not fit for human consumption.

10. Stems of ripe figs, dried figs, inferior figs, carobs, and various kinds of watermelon, as well as rind of watermelon, citron, and cucumber melon—even though they have no food value—and vegetable refuse discarded by householders are forbidden to commoners. Vegetable trimmings discarded by gardeners, however, are permitted to commoners.

Bean and sesame peel, if it has food value, is forbidden; if it has not, it is permitted.

11. Citron seeds are permitted; olive, date, and carob pits, even if not gathered by a priest, are forbidden to commoners. Other kernels, if gathered by the priest and containing juice that can be sucked out, are forbidden to commoners. But if the priest has thrown them away, they are permitted.

12. Wheat bran, if new, is forbidden, because it is fit for human consumption: if it is old, it is permitted. How long is it called new? So long as people customarily continue to do their threshing on the threshing floors.

13. Rejected or decaying heave offering is forbidden, but if it raises dust, it is permitted. In the case of heave offering lees diluted with water, the first and second waters are forbidden to commoners, but the third is permitted. If no water is added to the lees, but rather the remaining wine is merely filtered out of them, even the third filtering is forbidden to commoners.

14. If a bin has been cleared of heave offering wheat, one is not obligated to sit down and pick up what remains of it grain by grain, before he puts profane produce in it. Rather, he may sweep it up as usual and then put in the profane produce.

Similarly, if a jar of heave offering oil has been overturned, one is not obligated to sit down and tap it with his hand. Rather he may proceed in the same way as he would with profane oil.

15. If one is emptying a jar of heave offering oil, he must continue pouring until the column is interrupted and the oil

begins to drip down drop by drop. Once three drops have dripped down one after another, this is sufficient, and one may pour profane oil into the same jar. If he does not pour in profane oil, but rather turns the jar on its side until the remaining oil is completely drained down, the oil residue is still heave offering.

16. A priest may fill a lamp with heave offering fuel oil and hand it to an Israelite, so that the latter might go up to the upper story or enter a dark room in order to take care of the priest's needs, but not the Israelite's needs. If, however, the Israelite is in partnership with the priest, he may do so for his own sake.

17. If an Israelite is a priest's guest, and the priest lights a lamp of heave offering fuel oil for him and then goes off, the Israelite is not obligated to extinguish it, and may let it burn until it goes out by itself. The Israelite may even dip a wick in the priest's lamp, and set it alight to light his way.

18. When a priest's cattle stand next to an Israelite's cattle, or when a priest's garment is being woven next to an Israelite's garment, the Israelite may kindle for them a light of heave offering fuel oil even without the priest's permission.

Similarly, one may light a lamp of such oil in synagogues, houses of study, and dark alleys without the permission of the priest.

If one has no profane oil to light a Hanukkah lamp, he may use heave offering fuel oil without the priest's permission.

One may light such oil also for the sick, but only with the priest's permission.

19. If an Israelite woman married to a priest is in the habit of visiting her father regularly, he may, with her permission, light a lamp of heave offering fuel oil.

A priest may light such fuel oil even in a house of mourning or in a banquet house, in the presence of a mixed group of people, and there is no cause for apprehension that the people in attendance might partake of it, inasmuch as they are not likely to touch it in the banquet house, because it would soil their dining utensils, nor in the house of mourning, because they would have no time for it on account of the mourning.

20. He who plants heave offering unwittingly must plow it under; if deliberately, he must allow it to remain. If it is already one-third grown, whether he had planted it unwittingly or intentionally, he must allow it to remain. If it is flax, even if it is already one-third grown, and even if he had planted it intentionally, he must plow it under. This is in the nature of a fine imposed by the Sages, in order that he should not plant the offering with the intention of deriving benefit from its use as firewood.

21. The outgrowth of planted heave offering is the same as profane produce in every respect, except that it is forbidden to commoners. The Sages have ordained that it should be forbidden to commoners like clean heave offering, as a precaution against unclean heave offering held by a priest, lest he should detain it in his possession in order to plant it, so that it would become common produce, and thus render himself guilty of malfeasance. It is therefore permitted to eat such outgrowth with unclean hands, and a person who has immersed himself after a day's uncleanness may eat it the same as profane produce.

22. The outgrowth of the outgrowth of heave offering is profane produce in every respect. Even in the case of plants whose seed does not dissolve, if the outgrowth of the outgrowth is greater than the basic root, the second outgrowth annuls the root, even though the seed does not dissolve, so that the entire plant is permitted to commoners.

Similarly, in the case of heave offering imported from outside of the Land of Israel, profane produce slightly intermixed with heave offering, supplementary heave offering, and inedible garden seeds—as for example, turnip and radish seeds, even though these turnips and radishes themselves are heave offering—whatever grows out of them is permitted, and is the same as profane produce in every respect.

Similarly, if one plants heave offering flax, whatever grows out of it is permitted to commoners.

23. If one plants unclean heave offering, the outgrowth thereof, even though clean, is forbidden for consumption; inasmuch as

the planted heave offering itself was to begin with forbidden for consumption, the outgrowth too is removed from consumption.

24. If one cuts off the grown leaves, and after a second growth of leaves has come out, cuts them off also, the edible part that grows thereafter is permitted for consumption.

25. Profane young plants which had become unclean and were subsequently planted and made into heave offering, are permitted, because they have been cleansed by the planting, and did not become unclean after being made into heave offering, in which case they would have been forbidden.

26. If a loose ear of corn is lying in a heap of produce, and if one smooths down the entire heap, that ear of corn becomes untithed produce, seeing that it has been smoothed down within that heap. If one then plants it and then designates it by name as heave offering, its status as such is in doubt—once it has been planted, perchance its status as untithed produce has flown away from it, and it has become like produce the preparation of which is not yet completed. If, however, one designates it as heave offering before planting it, it is assured heave offering.

Therefore, if one plucks some of it and eats it intentionally, he is liable to punishment by death, and if unwittingly, he must pay the additional one-fifth. If he bends over and eats it with his mouth right off the ground, his intention is annulled in the judgment of all people, since it is not the normal way for men to eat in this manner. He is therefore exempt from death and from the additional one-fifth if he had acted unwittingly.

27. A field planted with the outgrowth of heave offering is subject to gleanings, forgotten sheaf, corner crop, heave offering, tithes, and poor man's tithe. The poor who are Israelites and the poor who are priests are entitled to the following gifts out of it: the poor priests may eat what is due them, while the poor Israelites must sell their due to the priests at the price of heave offering, and the proceeds accrue to them. A Levite must likewise sell his tithe to the priests.

28. He who threshes such outgrowth is praiseworthy. If he lets his cattle do the threshing, how should he proceed? He

should hang baskets around the necks of his cattle and fill them with some of the same species that he is threshing; the result then is that he neither muzzles the beasts nor feeds them heave offering.

CHAPTER XII

1. One is forbidden to defile heave offering of the Land of Israel, the same as any other holy thing, nor may one cause it to become defiled, nor waste it. Rather he should consume the clean of it and use the unclean for fuel. One is contrariwise permitted to defile heave offering from outside of the Land of Israel with defilements authorized by the Torah, even though it is defiled only in foreign lands on Scribal authority, seeing that the root of its obligation is Scribal. Therefore a menstruating woman may separate her dough offering outside of the Land of Israel, because she is forbidden only to eat of it and not to touch it, as we have explained.

2. Doubtfully defiled heave offering may be neither consumed nor burned; rather one should let it lie until it becomes defiled with an assured uncleanness, whereupon it may be burned. There are, however, some doubtful cases in which heave offering may be burned, as will be explained in the discussion of uncleanness and cleanness.

3. If a doubt as to cleanness arises concerning a jar of heave offering, one should not add any new element to it: one should neither move it from its place nor uncover it, but should leave it until it becomes defiled with an assured uncleanness, whereupon it may be burned. And there is no cause for apprehension that it might be eaten in the meantime.

4. If a jar of clean heave offering breaks in the upper part of the wine press, and the lower part thereof is unclean, the rule is as follows: If one is able to save one-quarter of a log̃ of the contents in cleanness, he may do so; if not, he may save it with his hands,

without washing them, even though he thereby renders it unclean, as will be explained in the discussion of cleanness.

5. To what does this apply? To a jar of wine, provided that the lower part of the wine press contains less than 100 times the volume of the jar, in which case the whole will become subject to heave offering and be disqualified. If, however, the lower part of the wine press contains 100 times the volume of the jar, so that the proportion is 101 to 1, or if the jar contains oil, one should let it run down and become unclean, without defiling it with his hands, seeing that all of it may be used for fuel and there is thus no serious loss.

Similarly, if a jar of oil is overturned, the rule is as follows: If one can save one-quarter of a log̅ of it in cleanness, he may do so; if not, he may save it in uncleanness, for once the jar is broken, he is not cautioned against saving it in uncleanness, seeing that he must act in haste.

6. If while one is passing from place to place carrying in his hands loaves of heave offering, a heathen says to him, "Give me one of them, or else I will touch all of them and defile them," he should place one loaf on a rock before him rather than put it in the heathen's hand, so that the latter would not defile that loaf of heave offering directly with his own hands.

7. Fenugreek and bitter vetch of heave offering, being unfit for human consumption, may be handled in a state of uncleanness. Caution is required only when they are soaked in water, for if one soaks them in a state of uncleanness, he defiles them with his hands. After the soaking, however, no caution is obligatory, either when one crushes the vetch or when he feeds it to cattle. One may therefore give heave offering of fenugreek or bitter vetch to an unlearned priest.

8. One may not deposit heave offering with an unlearned priest because he is apt to consume it. One may, however, deposit it with an unlearned Israelite, in an earthenware vessel covered with a tightly fitting lid, provided that the contents are not pro-

duce open to uncleanness, as a precaution lest his wife should shake it while she is menstruating.

9. One may not handle heave offering olives of an unlearned person for him in a state of cleanness, but one may handle his profane olives in a state of cleanness, to the amount needed by the oil press operator to live on. How should one proceed? He should take an amount equal to the heave offering and put it in a vessel that is not subject to uncleanness, as for example, a stone vessel, and when the unlearned owner comes to take the profane produce and the heave offering, one should tell him, "Take care, lest you should touch the heave offering and thereby return it to its untithed state."

10. One may not pluck produce with an Israelite who handles his produce in a state of uncleanness, and needless to say, one may not tread it with him, because heave offering due out of it would likewise be thereby handled in uncleanness. One may, however, carry jars with him to and from the press.

11. If olives or grapes have become unclean, one may press them and make heave offering out of the juice. If they have become unclean after having been set aside as heave offering, one should press them in amounts less than an egg's bulk, and the liquid pressed out of them may be drunk by priests. In fact it would have been fit for libations, seeing that the juice is, in a manner of speaking, deposited within the edible produce. The requirement that they must be pressed in amounts less than an egg's bulk is merely a precaution, lest one should press them in amounts larger than an egg's bulk, in which case the juice would be rendered unclean by the egg's bulk. If the produce involved is affected by third degree uncleanness, one may treat it in the wine press or the oil press, and the juice is clean heave offering, because in the case of heave offering third degree uncleanness does not give rise to fourth degree uncleanness, as will be explained in the discussion of cleanness.

12. If heave offering bread becomes defiled, one must throw it in with the firewood, so that it might be burned. Similarly, if oil

becomes defiled, one must pour it into a foul vessel, so that it might be lighted, and would not mislead other persons into consuming it.

If wheat becomes defiled, one must boil it into pulp and pour it into a foul vessel, so that it would look clearly unfit for consumption, and then use it to heat an oven or a stove.

If liquids unfit for burning, such as wine, become defiled, they should be buried.

13. Heave offering wine left uncovered must be thrown out, inasmuch as it is forbidden to drink it.

Similarly, if figs, grapes, chate melons, pumpkins, watermelons, or cucumber melons of heave offering are found to have been punctured, they are forbidden for consumption, because they are a danger to life. What should one do with them? He should throw them into the sea or bury them.

14. Dough kneaded with water from an uncovered vessel should be burned, even though it is heave offering dough.

15. A priest is forbidden to receive heave offering or other gifts due to him until the owner has set them aside, as it is said, *The first part of them which they give to the Lord, to thee have I given them* (Num. 18:12), and it is said further, *The holy things which the children of Israel offer to the Lord, have I given thee* (Num. 18:19), implying that not until the owners have set them aside does the priest acquire title to them. Nor should the priest take them after they have been set aside except with the owner's consent, for they are the owner's property, to be given by him to any priest he may wish, as it is said, *And every man's hallowed things shall be his* (Num. 5:10). If the priest nevertheless takes them without the owner's consent, he acquires title to them, for the only right the owner has with regard to them is the option to choose the recipient, and this option is not accounted the same as money.

16. If a priest who has been given heave offering finds some other things therein, they are forbidden to him, the same as

stolen property—perchance someone else had deposited them there to be picked up later.

17. Israelites are not obligated to attend to the heave offering to the extent of transporting it from the threshing floor to the town or from the wilderness to inhabited land; rather the priests must go out to the threshing floors and receive their share there from the Israelites. If the priests fail to go out, the Israelite may set it aside and leave it on the threshing floor. If, however, a wild or domestic beast is likely to devour it, and there is no provision there to guard it against them, the Sages have ordained that the Israelite must attend to it further by bringing it into town, but may collect a cartage fee from the priest, seeing that had he set aside the heave offering and left it unprotected against domestic or wild beast, he would have been guilty of desecration of the Holy Name.

18. Priests and Levites are forbidden to serve as helpers at the threshing floor, in order to receive immediately the gifts due them. Any one of them who does so help is guilty of profaning the Holy Name, and concerning him it is said, *Ye have corrupted the covenant of Levi* (Mal. 2:8). The Israelite too is forbidden to allow them to help him; rather he should give them their share with due deference.

19. If one gives heave offering to a priest with the stipulation that he is to return it, he has fulfilled his obligation to give it. It is, however, forbidden to do so, since this makes the priest the same as a helper at the threshing floor.

Similarly priests are forbidden to grab at heave offerings and tithes or even to ask orally for their share. They should rather receive them in a dignified manner, for they are eating and drinking off the Omnipresent One's table—these gifts are the Lord's, and it is He who has bestowed upon the priests the right to them, as it is said, *And behold, I have given thee charge of My heave offerings* (Num. 18:18).

20. One may not give heave offering to his wine press watchman, nor a firstling to his sheepherder, nor gifts to his cattle

herder. If he does so nevertheless, he is guilty of degrading them, unless he first pays them their wages for their services.

One Israelite is permitted to say to another Israelite, "Here is a sela'—give heave offering," or "a firstling," or "the other gifts to So-and-so, the priest, son of my daughter," or "son of my sister," or some similar priestly relative of the payer.

21. When does this apply? When the Israelite owner wishes to give the heave offering to A, one of two priests, or one of two Levites, without any remuneration for it, and his Israelite companion tells him, "Here is a sela'—give the heave offering to B instead." If, however, the Israelite owner says to a priest or a Levite, "Here is this priestly share for you, provided that you pay me for my right of option," this is forbidden.

Similarly, it is forbidden to use heave offering as merchandise, even if one buys it from one priest and sells it to another priest.

22. Ten persons may not be allotted heave offering at the threshing floor, even though they are permitted, and may entitle others, to eat of it. They are the following: a deaf-mute, an imbecile, and a minor who does not know how to spread his hands—these three because they lack intelligence; a ṭumṭum and a hermaphrodite, because they are a separate kind of creature; a bondservant, lest passers-by in the field should see him and testify that he is a priest; an uncircumcised and an unclean person, because they are in a repulsive state; a woman, lest she should be divorced, as well as a precaution against unlawful seclusion; and a male who marries a woman unsuited to him, whom the Sages have thus penalized by depriving him of his share at the threshing floor until he has divorced her.

In all the aforementioned cases one may deliver their share privately to their houses and allot it to them there, as may be done with the other holy things of the country; the only exceptions are the male who has married a woman unsuited to him, and the unclean and uncircumcised persons, to whom nothing may be delivered.

CHAPTER XIII

1. Heave offering is neutralized in the proportion of 1 part to 101. How so? If 1 sĕ'ah of heave offering falls into 100 sĕ'ah of profane produce and all of it is intermixed, one may separate 1 sĕ'ah out of the whole and give it to the priest, and the rest may be consumed by commoners.

In the case of any heave offering produce about which priests are not particular, such as heave offering of inferior figs, carobs, or Edomite barley, one need not take up any of it for heave offering, since having fallen into 100 parts, it has become neutralized because of its minute quantity, and the whole is permitted to commoners.

2. If 1 sĕ'ah of heave offering falls into less than 100 sĕ'ah of profane produce, the whole becomes forbidden as heave offering and must be sold to the priests at the price of heave offering less the price of that 1 sĕ'ah.

When does this apply? When the ingredients of the mixture are of the same species. If they are not, it depends on whether the admixture affects the taste. If the whole tastes of heave offering, it becomes forbidden as heave offering and must be sold to the priests at the price of heave offering less the price of the heave offering ingredient. If the whole tastes of profane produce, all of it may be consumed by commoners.

3. If 1 sĕ'ah of heave offering falls into 100 sĕ'ah of profane produce, and one takes one sĕ'ah out of the whole, and if that same sĕ'ah of the mixture then falls into some other place, it can render the second mixture forbidden as heave offering only in the proper proportion. The same applies to 1 sĕ'ah of heave offering that had fallen into less than 100 sĕ'ah of profane produce, so that the whole has become forbidden as heave offering, whereupon part of the mixture has fallen into some other place—it can render the second mixture forbidden only in the proper proportion. How so? If 10 sĕ'ah of heave offering fall into 90 sĕ'ah of

profane produce and render the mixture forbidden as heave offering, and if then 10 sĕ'ah of this forbidden mixture fall into less than 100 sĕ'ah of profane produce, this second mixture too becomes heave offering, because in the 10 sĕ'ah of the first mixture there is 1 sĕ'ah of heave offering. If less than 10 sĕah of the first mixture fall into the second mixture, they do not render it forbidden as heave offering.

4. To what does this apply? To substances that do not normally blend, as for example, wheat with wheat, or flour with flour. In the case of substances that do blend, as for example, heave offering oil with profane oil, or heave offering wine with profane wine, it depends on the major ingredient: if the greater part is heave offering, it renders the mixture heave offering, as if it were all heave offering; if the greater part is profane, the whole amount is profane and cannot in turn render any further mixture forbidden as heave offering, even though the first mixture itself is forbidden to commoners.

5. If 1 sĕ'ah of heave offering falls into 100 sĕ'ah of profane produce, and one takes up therefrom 1 sĕ'ah for the priest, and if then another sĕ'ah of heave offering falls into the mixture, and one again takes up 1 sĕ'ah for the priest, the remaining profane produce is permitted, until such time as the total of heave offering shall exceed the total of profane produce, that is to say, if the 100 sĕ'ah of profane produce become mixed with 100 sĕ'ah plus a little over of heave offering, 1 sĕ'ah at a time, the final mixture becomes forbidden as heave offering.

6. If 1 sĕ'ah of heave offering falls into 100 sĕ'ah of profane produce, and if before one manages to take up the 1 sĕ'ah, another sĕ'ah of heave offering falls in, the rule is as follows: If one was aware of the first sĕ'ah before the second sĕ'ah fell in, the mixture does not become forbidden as heave offering, and he may separate 2 sĕ'ah for the priest, rendering the rest permitted—inasmuch as the first sĕ'ah could have been neutralized, it is regarded the same as if it had been neutralized. If one became aware of the first sĕ'ah only after the second sĕ'ah had fallen in, the mix-

ture is forbidden as heave offering, the same as if both sĕ'ah had fallen in simultaneously.

7. Refuse of heave offering does not combine with it to render profane produce forbidden, but refuse of profane produce does combine with it to neutralize heave offering. How so? If 1 sĕ'ah of fine quality heave offering wheat falls into 100 sĕ'ah of inferior profane wheat, and one grinds the whole mixture, the rule is that even though the coarse bran of the profane wheat amounts to much more than the bran of the heave offering wheat, once the sifted heave offering flour amounts to less than 1/100 of the unsifted profane flour, the former is neutralized, the profane flour being measured together with the coarse bran, amounting altogether to 101 sĕ'ah. If, however, 1 sĕ'ah of inferior heave offering wheat falls into less than 100 sĕ'ah of fine profane wheat, and one grinds the whole mixture, once the sifted heave offering flour is 1/100 of the profane flour, the former is neutralized in 101 parts, inasmuch as the profane flour has increased in volume while the heave offering flour has decreased.

8. If 1 log̃ of clarified heave offering wine falls into 100 log̃ of turbid wine, or vice versa, one need not first remove the lees thereof, but may rather take out the 1 log̃ for the priest as it is.

9. If one log̃ of water falls into ninety-nine log̃ of wine, and if then one log̃ of heave offering wine falls into this mixture, it renders the whole mixture forbidden, because the water does not neutralize the wine.

10. If 1 sĕ'ah of heave offering falls into less than 100 sĕ'ah of profane produce, and if then other profane produce falls into the mixture, raising the profane component to more than 100 sĕ'ah, the rule is as follows: If this was done unwittingly, the 1 sĕ'ah of heave offering is neutralized by the 100 sĕ'ah of profane produce; if deliberately, the mixture becomes forbidden as heave offering, because one may not at the outset neutralize something that is forbidden by the Torah.

11. Heave offering imported from outside of the Land of Israel may be annulled by mixing it with a larger amount of pro-

fane produce, and one may eat of it while one is ritually unclean. And not only this, but if one has such imported heave offering wine, he may take one loḡ of it and add to it two loḡ of profane wine, making three loḡ; then take another loḡ of heave offering wine and pour it into the three, and thereupon draw one loḡ out of the four and drink it. Thereafter he may again pour in another loḡ of the imported wine, again draw another loḡ and drink it, repeating this procedure over and over again, until all the heave offering wine is gone. He thus annuls several loḡ of the imported heave offering wine by means of the two loḡ of profane wine.

12. If a person plants heave offering produce next to profane produce, and it is not known which of the two is heave offering, the rule is that even if there are 100 beds of heave offering produce and only 1 of profane produce, all of them are permitted.

To what does this apply? To produce whose seed dissolves in the ground, such as wheat, barley, and the like. In the case of produce whose seed does not dissolve, such as garlic and onions, even if there are 100 beds of profane produce and only 1 bed of heave offering produce intermixed with them, the whole is forbidden as heave offering.

If all of it has already been plucked, the heave offering becomes neutralized by 101; but one should not at the outset pluck it.

13. If there are two baskets, one of heave offering and the other of profane produce, and it is not known which is which, or if one sĕ'ah of heave offering falls into one of them, and it is not known which one, and if thereafter one of these two baskets falls into profane produce, it does not render the profane produce forbidden as heave offering.

Similarly, if one plants the contents of one of these baskets, the outgrowth thereof is accounted profane produce in every respect, while the other basket must be treated as heave offering. If the second basket then falls into other produce, it does not render it forbidden as heave offering.

Similarly, if someone else plants the contents of the second basket, its outgrowth is accounted profane produce.

If both baskets fall into the same place, they render the mixture forbidden as heave offering according to the volume of the smaller basket.

If another person plants both baskets, the rule is as follows: If it is produce whose seed dissolves, its outgrowth is considered profane; if it is produce whose seed does not dissolve, its outgrowth is forbidden as heave offering.

When does this apply? When one plants the second basket before he has harvested the first. If, however, one harvests the first basket and then plants the second, the outgrowth is considered profane produce, even if its seed does not dissolve, because produce already plucked and produce still attached to the ground cannot confer liability upon each other.

14. If there are two baskets, one of profane produce and one of heave offering, and if before them there are two sĕ'ah, one of profane produce and one of heave offering, and the former two fall into the latter two, they are all permitted, for I say that the profane basket fell into the profane produce, and the heave offering basket into the heave offering produce, even if the profane produce does not exceed the heave offering by volume.

To what does this apply? To present day heave offering obligatory by Scribal law only. In the case of heave offering authorized by the Torah, it is permitted only if the profane exceeds the heave offering by volume.

15. If one sĕ'ah of heave offering falls into a heap of produce, and if the owner then says, "The heave offering for this heap is to be within it," the spot where the one sĕ'ah has fallen in marks also the location of the heave offering for the entire heap, and all of the heap is rendered forbidden as heave offering, by both the heave offering that has fallen in and by the heave offering that is due from the heap itself.

If the owner says, "The heave offering for this heap is to be at its northern part," the heap should be divided into two halves, north and south, and the northern half should again be divided into two quarters; the northernmost quarter is the one that is forbidden as heave offering.

16. If having two heaps in front of him, the owner says, "The heave offering for both heaps is to be in one of them," both are

subject to heave offering. If having before him two sĕ'ah of produce and one heap, he says, "One of these two sĕ'ah is to be heave offering for this heap," it follows that one of these two sĕ'ah is heave offering, but it is not known which one.

If having before him two heaps of produce and one sĕ'ah, he says, "This sĕ'ah is to be heave offering for one of these heaps," it follows that the sĕ'ah is heave offering and one of the heaps has been regularized and is no longer untithed, but it is not known which one.

CHAPTER XIV

1. If there are fifty black figs and fifty white figs of profane produce, and if a single heave offering fig falls in among them, the rule is as follows: If this single fig is white, the black figs are permitted while the white figs become heave offering; if it is black, the black figs are forbidden as heave offering while the white figs are permitted. If it is not known whether the single fig was white or black, it is neutralized by 101 of the total. If one had known the color of the single fig after it had fallen in, but then forgot it, all are forbidden as heave offering.

2. Similarly, in the case of round and rectangular fig cakes, if a single round or rectangular cake of heave offering falls in among them, and it is not known whether it was rectangular or round, the round cakes combine with the rectangular ones, and the heave offering cake is neutralized by 101 of the total.

In the case of large and small round cakes, among which there has fallen in a single round cake of unknown size, they neutralize it either by number or by weight.

How so? If a single fig cake falls in among 100 cakes, some large and some small, I say that the single cake was small, and one may take up one of the small cakes for the priest.

If a single cake falls in among forty cakes of which twenty weigh 4 liṭra each and the other twenty 1 liṭra each, I say that so long as their total weight comes to 101 liṭra, the single cake was a small one and one may take up one of the small cakes for the priest.

3. Ordinary flour and fine flour cannot combine to neutralize heave offering.

4. If 1 sĕ'ah of heave offering wheat falls into the mouth of a bin of profane wheat, it is not regarded as subject to neutralization by 101, because such neutralization is effective only when heave offering and profane produce are blended together or when the precise spot where the heave offering has fallen in is unknown. What should one do? The heave offering should be regarded the same as if it were wheat fallen in on top of barley, and one should scoop up the heave offering itself that has fallen into the mouth of the bin, together with a little of the profane produce upon which it has fallen, the same as one would scoop up wheat lying on top of barley.

5. If heave offering falls into one of two baskets or two bins of profane produce, becoming blended with it, and it is not known into which of the two it has fallen, the rule is as follows: If the two bins are in the same house, they combine together to neutralize the heave offering by 101 parts of the total, as if the two bins were one. In the case of two baskets, they combine together even if one of them is in one house and the other is in another house, because it is far more likely that they would be gathered into the same house. If, however, the two baskets are in two towns, they cannot combine with one another.

6. How does one neutralize the sĕ'ah of heave offering that has fallen in? If one wishes to neutralize it with one of the two baskets, he may do so. If one wishes to neutralize it with half of one basket and half of the other, he may likewise do so.

If one has pressed one liṭra of heave offering figs into the mouth of one of several jugs filled with profane figs, and it is not known into which jug, the rule is as follows: If there are 101 jugs, the heave offering becomes neutralized, and one may take one of the jugs and sell it to the priest at its price less the value of that one liṭra, and the rest is permitted to commoners. If there are less than 100 jugs, the contents of all the mouths of the jugs are forbidden as heave offering, while the contents of the lower parts of the jugs are permitted.

7. Similarly, if one presses the heave offering figs over the mouth of one of several beehive-shaped vessels or figcakes, and does not know which one, or if one presses over the mouth of a figcake, and does not know whether it was at its northern part or at its southern part, nor does he know which figcake, the contents of the vessels or figcakes are regarded as if they were separated, and the heave offering is neutralized according to the total weight, that is to say, if all the cakes together amount to 100 liṭra, the heave offering is neutralized, provided that each cake weighs more than 2 liṭra, so that the heave offering would be annulled by the major ingredient, because doubtful heave offering is annulled by a greater amount of profane produce.

8. Assured heave offering renders profane produce forbidden up to 100 times its volume. Doubtful heave offering renders profane produce forbidden up to 50 times its volume, and the mixture does not become permitted unless it contains a major amount of profane produce; if it falls into more than 50 times its volume, no major amount of profane ingredient is necessary.

9. How so? If 1 heave offering fig falls in with 99 profane figs, making the statutory amount of 100, the whole 100 is forbidden to commoners, as we have explained.

If, however, it falls in with fifty profane figs, and if one fig out of the fifty-one is then lost, a doubt arises—the lost fig may have been one of the profane figs, or it may have been the one that had fallen in; all of them are therefore forbidden, until one adds to the remaining total of fifty figs fifty-one figs from another place.

On the other hand, if one heave offering fig falls in with fifty-one profane figs, and one of the fifty-two is then lost, the rest of them is permited to commoners.

10. If 1 sĕ'ah of heave offering falls into less than 100 sĕ'ah of first tithe from which heave offering has not yet been taken, or into second tithe or consecrated produce that has not yet been redeemed, and the whole thus becomes forbidden as heave offering, the rule is as follows: If the heave offering falls into first tithe produce, the heave offering of the tithe should be designated by

name, and the whole should be sold to the priests at its price, less the value of the heave offering that has fallen in and less the value of the heave offering of the tithe that is contained in it. If the heave offering falls into second tithe or into consecrated produce, these should be redeemed and then sold to the priest at their price, less the value of the heave offering.

11. If 1 sĕ'ah of unclean heave offering falls into less than 100 sĕ'ah of profane produce, or of first tithe, or of second tithe, or of consecrated produce, whether unclean or clean, the rule is that inasmuch as the whole thus becomes forbidden as heave offering, all of it is regarded as unclean heave offering which is forbidden for consumption by anyone. All of it is consequently forbidden forever, and must be left to rot. To what does this apply? To things which are not customarily eaten raw. As for things that are customarily eaten raw, one should not leave them, lest he should stumble over them and eat of them. One should therefore set all of them on fire, as one would do with unclean heave offering.

12. If 1 sĕ'ah of clean heave offering falls into less than 100 sĕ'ah of unclean profane produce, the whole should be sold to the priests at its price, less the value of the heave offering. The priest should then eat this mixture either parched or kneaded with fruit juices which do not render produce susceptible to defilement, in order that the heave offering contained therein should not be defiled by the unclean profane produce; or else he should knead this sequestered mixture in amounts less than an egg's bulk, because unclean food cannot defile clean food unless the former amounts to an egg's bulk; or alternatively the mixture should be divided into portions less than an egg's bulk each, and added seriatim to other dough, so that the heave offering contained in it would not become defiled.

13. If 1 sĕ'ah of defiled heave offering falls into 100 sĕ'ah of clean profane produce, and similarly if 1 sĕ'ah of clean heave offering falls into 100 sĕ'ah of unclean profane produce, it is neutralized by 101, but should nevertheless be eaten parched, or

kneaded with fruit juices, or in amounts less than an egg's bulk, because the sĕ'ah that has fallen in is not the very same sĕ'ah that has been lifted out.

14. If 1 sĕ'ah of unclean heave offering falls into 100 sĕ'ah of clean heave offering, it becomes annulled because of its minute quantity, and the whole of it may be eaten by the priest in cleanness. If it falls into less than 100 sĕ'ah, he must leave all of it to rot.

15. If one sĕ'ah of heave offering falls into one of two baskets of profane produce, and it is known which basket; and if then another sĕ'ah of heave offering falls into one of these baskets, and it is not known which basket, one may say that the second sĕ'ah fell into the same basket as did the first sĕ'ah, because the presumption is that the second contamination took place in the same basket as the first.

If the first sĕ'ah falls into one of the two baskets, and it is not known which basket, and if then the second sĕ'ah falls into a known basket, one may not say that the first sĕ'ah has fallen into the same basket as did the second sĕ'ah; rather both must be considered contaminated.

16. If there are two baskets, one unclean and the other clean, and if one sĕ'ah of heave offering falls into one of them, and it is not known which basket, one may say that it must have fallen into the unclean basket.

17. If there are two baskets, one containing clean heave offering and the other containing unclean profane produce, and if one sĕ'ah of clean heave offering falls into one of them, one may say that it fell into the basket containing heave offering, but the profane produce must be eaten in cleanness, the same as heave offering.

18. If one sĕ'ah of unclean heave offering falls into one of the two baskets, one may again say that it fell into the basket containing heave offering, and the profane produce must be eaten parched, in cleanness, the same as the great heave offering.

19. If one sĕ'ah of unclean heave offering falls into one of the two baskets, one may again say that it fell into the basket containing heave offering, and the profane produce must be eaten parched or kneaded with fruit juices.

20. If there are two baskets, one containing unclean heave offering and the other containing clean profane produce, and if one sĕ'ah of clean heave offering falls into one of them, one may say that it fell into the basket containing heave offering, and the profane produce must be eaten parched.

21. If one sĕ'ah of unclean heave offering falls into one of the baskets, both become forbidden, because doubtfully unclean heave offering is forbidden, while produce doubtfully forbidden as heave offering is permitted. For the prohibition of unclean heave offering is authorized by the Torah, while the prohibition of produce doubtfully forbidden as heave offering is based upon Scribal authority, according to the principles which we have explained in the Laws Concerning Forbidden Foods.

CHAPTER XV

1. A sealed jar of heave offering renders other produce subject to the law of heave offering by the least amount. How so? If one sealed jar of heave offering is mixed with several thousand sealed jars, all become forbidden as heave offering. If the jars spring open, the jar of heave offering becomes neutralized by 101.

2. If the sealed jar is mixed with 100 other jars, and if one of them then falls into the Great Sea, the remaining jars are all permitted, and one may say that the jar that fell in was the one containing the heave offering. This is not so, however, in the case of 1 heave offering fig that falls in with 100 other figs, and one of them then falls into the Great Sea: one must set aside 1 fig for the priest, because when a jar falls in, it is noticeable, but when a fig or its like falls in, it is not noticeable.

3. If the sealed jar is mixed with 100 other jars, and one of them springs open, one may take up 1/100 of its contents for the priest, and may then drink the rest of it. The other jars, however, remain forbidden until they too spring open. From each one of them, as it springs open, one may take up 1/100 of the contents that has become forbidden as heave offering, and may then drink the remainder of the 100 parts.

If the sealed jar is mixed with 150 other jars, and 100 of them spring open, one may take up out of the 100 the amount that has become heave offering, that is, 1 jar, and drink the remaining 99 of them, but the other 50 unopened jars are forbidden. There can be no presumption that the jar of heave offering has been nullified by the major number of jars; even if there are several thousand jars, they are all considered subject to the law of heave offering, and any jar that springs open, one must take up 1/100 of its contents for the priest, and may then drink the rest of it, while the other jars remain subject to the law of heave offering.

4. We have already explained in the Laws Concerning Forbidden Foods that the smallest amount of a souring or spicing agent renders food forbidden. Therefore, if one chops up a heave offering apple and puts it into dough, so that the dough becomes leavened, all of the dough becomes forbidden as heave offering and is forbidden to commoners.

5. In the case of an egg spiced with heave offering spices, even its yolk is forbidden because it is absorbent.

6. If heave offering leaven falls into dough and one lifts it out, and if thereafter the dough becomes leavened, it is permitted.

7. We have already explained that if heave offering is mixed with profane produce of a different species, it renders it forbidden if it imparts a flavor of its own to it. Therefore, in the case of a sliced onion cooked in a broth, the rule is as follows: If the onion is heave offering and the broth is profane and tastes of onion, the broth is forbidden to commoners. If the onion is profane and the broth is heave offering, and the onion tastes of broth, the onion is forbidden.

8. If after lentils are cooked a dried onion is thrown in with them, the rule is as follows: If the onion is whole, the lentils are permitted; but if the onion is sliced, it depends on whether it has imparted a flavor of its own to the lentils. If the onion and the lentils are cooked together, whether whole or sliced, the decision depends on whether a flavor was or was not imparted by the onion. As for other broths, whether the onion is thrown in after the broth was cooked, or is cooked together with the broth, whether the onion is whole or sliced, the decision depends on whether a flavor was or was not imparted to the broth.

Why is the decision not based on flavor in the case of a whole onion thrown in with lentils that are already cooked? Because the onion, being whole, can neither absorb flavor from the lentils nor infuse its own flavor into the lentils, since they are already cooked.

If the onions are young, they are accounted the same as if they were sliced.

Similarly, if the onion's blossom end or outer peel has been removed, or if it is fresh, it is accounted the same as if it were sliced. Porret, whether fresh or dried, whether whole or sliced, renders food forbidden if it imparts a flavor of its own to it.

9. If profane vegetables are pickled with heave offering vegetables, they are permitted to commoners, with the exception of the several species of onion, leek, and garlic—if profane vegetables are pickled with heave offering onions, or profane onions with heave offering onions, they are forbidden. If heave offering vegetables are pickled with profane onions, the onions are permitted to commoners.

10. If one pickles profane olives with heave offering olives, the rule is as follows: If both are broken, or if the profane olives are broken and the heave offering olives are whole, of if one pickles them in heave offering juice, they are forbidden. If, however, both are whole, or if the heave offering olives are broken and the profane olives are whole, they are permitted, because the broken olives draw flavor from the whole olives.

11. The pickling or stewing liquid of heave offering olives is forbidden to commoners.

12. Dill is subject to heave offering so long as it does not impart a flavor of its own to the contents of the cooking pot; once it does so, it is no longer subject to heave offering.

13. If one shovels hot bread out of the oven and places it over the mouth of a jar of heave offering wine, the rule is as follows: If it is wheaten bread, it is permitted; if it is barley bread, it is forbidden, because barley is absorbent.

14. If an oven is heated with heave offering cumin and bread is baked therein, the bread is permitted, since it is not the flavor of the cumin but only its aroma that affects the bread, and aroma is not forbidden.

15. If heave offering barley falls into a cistern of water, the barley is permitted, even if the water is polluted, on the principle that any flavor which causes spoilage cannot render a substance forbidden.

16. If heave offering fenugreek, both seed and stalk, falls into a wine vat, the rule is that if the seed alone is sufficient to impart a flavor to the wine, the wine is forbidden.

17. If there are two cups of wine, one heave offering and the other profane, and if one dilutes each cup with water and then mixes them together, the profane wine is regarded as if it did not exist, so that it is as if the heave offering wine alone were mixed with water, which is a liquid of a different species. The rule therefore is as follows: If there is sufficient water to nullify the flavor of the heave offering wine, the entire mixture is permitted to commoners; if not, it is forbidden, for we have already explained that water cannot neutralize wine.

18. If heave offering wine is spilled over profane fruit, the latter may be rinsed and is then permitted.

Similarly, if heave offering oil is spilled over fruit, the latter may be rinsed and is then permitted. If the oil is spilled over wine, one should let the oil rise to the surface, whereupon the wine is permitted to commoners.

If the oil is spilled over brine, one should let it rise to the surface and then remove it together with the skin that forms on top of the brine, in order to take out all the brine that tastes of oil.

19. A pot used to cook heave offering may not be used to cook profane produce. If one does so nevertheless, it depends on whether the pot imparts a flavor of its own to its contents. But if the pot is rinsed with water or wine, it may be used to cook profane produce.

If only a part of the pot has been used in the cooking, one need not rinse the whole pot but only the part used in the cooking.

20. Great heave offering, heave offering of the tithe, dough offering, and first fruits, are all termed heave offering, for in reference to heave offering of the tithe it is said, *Then ye shall set apart of it a heave offering for the Lord* (Num. 18:26), and again, *A heave offering as the heave offering of the threshing floor* (Num. 18:20); and in reference to the dough offering it is said, *Ye shall set apart a cake for a heave offering* (*ibid.*), and it is said further, *Thou mayest not eat within thy gates the tithe of thy corn, or of thy wine, or of thine oil . . . nor the heave offering of thy hand* (Deut. 12:17). There is nothing of what must be brought to the assigned place that is not enumerated in this latter verse, except for the first fruits, and it is concerning them that it is said, *nor the heave offering of thy hand*—hence you learn that they too are called heave offering.

21. Consequently the rule governing the four of them in the matter of their consumption and their liability to heave offering when mixed with profane food is the same. They are all neutralized by 101, and they all combine with one another. If they become unclean, they must be burned. The rule governing doubtful heave offering of the tithe is the same in all these respects as the one governing assured heave offering of the tithe, except that one who eats of the former is not liable to a flogging.

22. Whosoever eats of heave offering must first recite the benediction prescribed for that particular kind of food, and must then recite the benediction "Blessed art Thou, O Lord . . . who has consecrated us with the consecration of Aaron, and has commanded us to eat of heave offering." Thus have we received it by tradition, and thus have we seen priests pronouncing the benedictions, even over dough offering originating outside of the Land of Israel. For eating of holy things in the country is the same as sacrificial service in Jerusalem, as it is said, *I give you the priesthood as a service of gift* (Num. 18:7).

TREATISE IV

LAWS CONCERNING TITHE

Involving One Positive Commandment,
To Wit

To set aside the first tithe each year of the years of sowing, and to give it to the Levites.

An exposition of this commandment is contained in the following chapters.

NOTE

In the list of 613 commandments prefixed to the Code, the positive one dealt with in the present Treatise appears as follows:

127. To set apart the tithe of the corn for the Levite, as it is said: *And all the tithe of the land, whether of the seed of the land, or of the fruit of the tree, is the Lord's* (Lev. 27:30).

CHAPTER I

1. After one has set aside the great heave offering, he must set aside one-tenth of what is left. This is what is called first tithe, and of it Scripture says, *For the tithe of the children of Israel, which they set apart as a gift unto the Lord,* etc. (Num. 18:24).

2. An Israelite is permitted to eat of the first tithe, and may eat of it while in the state of uncleanness, for it has no sanctity whatsoever. Wheresoever in reference to tithes Scripture speaks of holiness or redemption, what is meant is second tithe only.

Whence do we know that first tithe is profane produce? Because it is said, *And the gift which ye set apart shall be reckoned unto you as though it were the corn of the threshing floor, and as the fullness of the wine press* (Num. 18:27). Just as the yield of the threshing floor and the wine press is profane produce in every respect, so also the first tithe, after its heave offering has been taken up, is profane produce in every respect.

Therefore, if a Levite's daughter has been made captive or has been subjected to illicit intercourse, she may nevertheless be given her share of the tithe and may eat of it. One, however, who had heard that her husband was dead, or who had only one witness to testify for her to that effect, and who has then remarried, and thereupon her husband returned alive, has been penalized by the Sages by being prohibited to eat of the tithe.

3. Levites and priests too must set aside the first tithe in order to detach from it the heave offering of the tithe. Similarly, priests must set aside the other heave offerings and tithes for themselves. Since the priests are entitled to these gifts from all produce, one might conclude that they may eat their own produce untithed; therefore Scripture says, *Thus ye also shall set apart* (Num. 18:28), and the Sages have learned by tradition that *ye* refers to the Levites, and *ye also* includes the priests.

4. One may not collect the first tithe from the priests, as it is said, *When ye take of the children of Israel* (Num. 18:26). The

same applies to all gifts due to the priesthood—one may not collect them from one priest for another.

Ezra has penalized the Levites of his time by ruling that they should not be given the first tithe, rather that it should be given to the priests, because these Levites did not go up with him to Jerusalem.

5. If one eats of his untithed produce, and similarly if a Levite eats of the first tithe before he has set apart the heave offering of the tithe, the rule is that although they are liable to death at the hand of heaven, they need not pay the rightful recipients for these gifts, as it is said, *Which they set apart as a gift unto the Lord* (Num. 18:24), implying that not until these gifts have been actually set aside, do the rightful recipients have any claim to them.

Outside of the Land of Israel one is permitted to go on eating of the produce first, and then set aside the heave offering and the tithes therefrom.

6. One may set aside tithe from produce in one place for produce in another place, and one need not tithe only from adjacent produce. One may not, however, tithe from one species for another, nor from produce liable to tithing for produce which is exempt, or vice versa. If one does so nevertheless, there is no tithe.

7. Wheresoever we have said concerning heave offering, "one may not set aside from one for another," the same applies also to tithe, that is to say, one may not tithe from one for another. And wheresoever we have said concerning heave offering, "if he has so set it aside, his heave offering is valid," the same applies also to tithe, that is to say, if he has so set it aside, his tithe is valid. Thus also whatsoever is exempt from heave offering is also exempt from tithe, and whosoever is liable to heave offering is also liable to tithe. And wheresoever we have said, "one need not set aside the heave offering, but if he does so nevertheless, his heave offering is valid," so also if one does tithe unnecessarily, his tithes are valid; and conversely, wheresoever we have said that one's heave offering is invalid, so also are his tithes invalid.

8. If A says to B, "I am going to tithe in your behalf," B need not stay with A to see whether he will actually do so or not. If, however, B says to A, "Do tithe in my behalf," B must stay with A.

9. Carobs are liable to tithing by Scribal enactment only, because they are not used for food by [most] people. Bitter almonds, whether fully grown or still small, are exempt, because they are not fit for food.

10. A tree planted inside a house is exempt from tithes, as it is said, *Thou shalt surely tithe all the increase of thy seed, that which is brought forth in the field* (Deut. 14:22). It would appear to me, however, that it is liable to tithes by Scribal authority, since the yield of a fig tree standing in a courtyard is liable to tithe, if one gathers all of it at one time.

11. Onions rooted one beside the other, even if rooted in the floor of an upper story, are exempt from tithes. If, however, the roof collapses over them, so that they become exposed, they are the same as if they were planted in the field, and are liable to tithes.

12. If one guards his field for the sake of the grapevines growing therein, and someone else comes along and gathers up the figs that remain in that field; or if one guards his field for the sake of its chate melon and pumpkin patches, and someone else comes along and gathers up the remaining grapes scattered there in that field, the rule is as follows: So long as the owner of the field is particular about these remainders, they are forbidden under the law of robbery, and are therefore liable to tithe and heave offering. If the owner is not particular about them, they are permitted as exempted from the law of robbery, and are therefore exempt from the tithe.

13. Tithes must be set aside out of choice produce only, as it is said, *When ye set apart the best thereof from it, then it shall be counted unto you [the Levites] as the increase of the threshing floor and the increase of the wine press* (Num. 18:30), which implies that just as the tithe set aside by the Levites must be out

of the best thereof, so also the tithe set aside by the Israelites from the threshing floor and from the wine press must be out of the best thereof.

14. One may not tithe by estimate but only by measurement, weight, or count. Whosoever is scrupulous about the exact amount is praiseworthy; but he who deliberately exceeds the exact amount of tithes, his tithes are blemished, since untithed produce is intermixed with them. His own produce, however, is duly regularized nevertheless.

15. If one sets aside only a part of his tithe, it is no tithe, and it is the same as if he had merely divided the heap of produce. He must, however, set aside from this same part the tithe due from it. How so? If he has 100 sĕ'ah, and he sets aside from them 5 sĕ'ah as tithe, it is no tithe, nor may he set aside tithe for these 5 sĕ'ah from some other place; rather he must set aside from these very same 5 sĕ'ah one-half sĕ'ah, which is the tithe due from them.

16. He who sets aside this first tithe must recite a benediction, the same as he would recite before performing any other commandment. He must likewise recite a benediction over the second tithe, the poor man's tithe, and the tithe of the tithe, over each one separately. If, however, he sets aside all of these, one after another, without doing any talking in between, he may include them all in one blessing, by saying, "who has commanded us to set aside heave offerings and tithes."

CHAPTER II

1. According to the Torah, only he who has completed the preparation of his produce for his own consumption is liable to set aside tithes. He who completes its preparation in order to sell it is exempt according to the Torah but liable according to Scribal law, as it is said, *Thou shalt surely tithe . . . and thou shalt eat* (Deut. 14:22–23), which implies that one is not liable to tithing unless he completes the preparation of his produce for his own consumption.

2. Similarly one who purchases produce is exempt from tithing according to the Torah, as it is said, *the increase of* THY *seed* (Deut. 14:22), but is liable according to Scribal law.

When does this apply? When he purchases the produce after its preparation has been completed in the hand of the seller. If, however, the preparation is completed in the hand of the purchaser, he is liable to tithing even according to the Torah.

3. Produce unfit for consumption when small, such as unripe grapes and the like, is not liable to tithe until it ripens and becomes edible, as it is said, *Whether of the seed of the land, or of the fruit of the tree* (Lev. 27:30), implying that it is liable only when it is full-grown fruit. The same applies to grain and pulse, as it is said, *All the produce of thy seed* (Deut. 14:22), implying, after it has become produce. And that point of time is the season for tithing. Before grain and fruit reach that season, one may eat of them as much as one desires in any way he wishes.

4. Produce fit for consumption when small, such as chate melons and cucumber melons, which are left on the vine only in order to grow in size, and are edible at the outset, are liable to tithing when small, because they reach the season of tithing from the moment of their appearance.

5. When is the season of tithing? When the produce reaches the stage that makes the seed thereof fit to be sown and to sprout, each kind according to its particular nature. How so?

Figs, when they become soft enough to be edible twenty-four hours after they have been gathered.

Domestic grapes and wild grapes—that is, small wilderness grapes—when their seeds are visible from the outside.

Pomegranates, when their seed melts between the fingers and oozes water.

Dates, when they begin to swell like rising dough.

Peaches, when they show red veins.

Walnuts, when the edible part separates itself from the outer shell.

Sweet almonds, when their outer shell separates itself.

Bitter almonds are exempt at all times.

Other shelled nuts, such as stone pine, laudanum, and pistachio, when they develop the inner shell next to the edible part.

Olives, when they produce one-ninth of the amount of oil they would yield when fully grown, that is, when they are one-third grown.

Apples and citrons, when they assume a round shape, because they are edible even when small.

Mulberries and sumac berries, when they begin to turn red, and the same applies to other berries that customarily turn red.

Carobs, when they develop black spots, and the same applies to all other fruits that turn black.

Common pears, crustuminian pears, quinces, and hawthorn berries, when they develop white bald spots, and the same applies to all other fruits that customarily turn white.

Grains, when they are one-third grown.

Fenugreek, when its seed is fit to sprout after being sown.

In the case of vegetables, chate melons, pumpkins, watermelons, cucumber melons, and the like are liable to tithing even when small, as we have explained. Other vegetables, not edible until fully grown, are liable only when fit for consumption.

A cluster of grapes is liable to tithing even if only one grape is ripe, just as if the entire cluster had reached that stage. And not only this particular cluster is liable, but also the entire side containing the vine bearing this cluster.

Similarly, a pomegranate showing no more than one ripe seed is liable to tithing in its entirety.

6. One may not sell his produce, once it has arrived at its season of tithing, to a person who is not trustworthy in regard to tithes. If only part of the produce has ripened, one may retain that part and sell the remainder, seeing that the latter has not yet reached its tithing season.

Similarly, one may not sell his olive or grape waste to a person who is not trustworthy in regard to tithes, for the purpose of extracting the remaining juice from it, since that juice is liable to tithes. Nor may one sell his straw to such a person for the purpose of picking out the remaining grains of produce, because the produce thus picked out of the straw is liable to tithes.

7. In the case of wine lees, if one pours water over them and then filters off the liquid, the rule is as follows: If having poured in three measures of water one decants four measures of liquid, he must give tithe for the additional measure, even out of another place. He is not required, however, to set aside heave offering for this excess, because when at the outset he has set aside heave offering, his intent was that it was to cover all of his produce, as we have explained in the discussion of heave offering. If he decants less than four measures of liquid, he is exempt, even if he decants more than the amount he had poured in, and even if the decanted liquid tastes of wine.

8. If a person dedicates his produce to the Sanctuary while it is still attached to the ground, before it has reached the season of tithing, and then redeems it, and if after he has redeemed it the produce reaches the season of tithing, it is liable to tithing. If it reaches the season of tithing while it is in the hand of the Temple treasurer, and is thereupon redeemed, it is exempt.

CHAPTER III

1. Produce that has reached the season of tithing and has been detached from the ground, but has not yet been completely processed, as for example, produce that has been harvested and threshed but not yet winnowed nor smoothed down, may be eaten in a casual manner until it is completely processed. Once its preparation has been completed, it is forbidden to eat of it even casually.

2. When does this apply? When one completes the preparation of his produce in order to sell it in the market place. If, however, his intention is to carry it home, he may eat of it casually even after its preparation has been completed, until it is designated for tithing.

3. Each one of the following six things designates produce for tithing: courtyard, purchase, fire, salt, heave offering, and Sabbath. And all of them are effective only upon produce whose preparation has been completed.

4. How so? Produce intended to be carried to one's house may be eaten casually even if its preparation has been completed, until it actually enters the house. Once it has entered the house, it becomes designated for tithing and may not be eaten until tithed.

The same applies to produce sold, boiled over a fire, pickled with salt, subjected to heave offering, or overtaken by the Sabbath—one may not eat thereof until he has tithed it, even if it has not yet reached the house.

If the produce had entered the house before its preparation was completed, one may eat casually of it.

If one has commenced to complete the preparation of the produce after it had entered the house, he must tithe the whole of it. How so? If one has brought chate melons or pumpkins into the house before rubbing off the fuzz, once he has begun rubbing one of them, all become designated for tithing. The same applies to all analogous cases.

Similarly, if one has set aside heave offering for incompletely processed produce, he may eat casually thereof. The only exception is a basket of figs subjected to heave offering before their preparation has been completed, which forthwith becomes designated for tithing.

5. If fig tree branches with figs on them, or date tree twigs with dates on them, are brought into the house, the rule is as follows: If they are brought in by little children or by laborers, they are not regarded as designated for tithing. If they are brought in by the owner himself, he is liable to tithing.

If the owner brings in ears of corn to make dough therewith, they do not become designated for tithing; if to husk them for eating, they become designated for tithing.

To what does this apply? To grain. In the case of pulse, however, the designation for tithing does not take effect.

6. It is permitted to circumvent the rule governing produce by bringing it into one's house while it is still in its husks, so that his cattle might eat of it, such fodder being thus exempt from tithing. He may then winnow it a little at a time, after he had

brought it into his house, and it is forever exempt from both heave offering and tithes, for each time, as he commences winnowing, he does not intend to complete the processing of the entire amount of produce.

7. If A completes the preparation of B's produce without B's knowledge, and similarly if he designates B's produce for tithing by one of the six things which establish such a designation for tithing, without B's knowledge, the produce becomes liable to tithe.

8. What constitutes the completion of the preparation of produce? For chate melons, pumpkins, and watermelons, when one rubs them with his hand to remove the fine hairlike fuzz that makes them shine. If one does not rub them, the preparation is complete when he piles them in a heap, or in the case of watermelons, when he arranges them in their assigned place one alongside the other.

If he does rub them one by one, their preparation is accounted completed as soon as he has processed as many as he needs. One may, however, set aside heave offering from chate melons and pumpkins even if he has not yet removed the shine from them.

9. In the case of vegetables usually tied into bundles, their preparation is completed when they are so tied; if they are not usually tied, when they fill the vessel; if the vessel is not meant to be filled, one may eat casually of them until he has gathered all he needs.

10. In the case of a basket, the preparation is completed when the fruit that is in it is covered with leaves, prickly foliage, or the like; if the fruit is not so covered, when the vessel is filled; if the vessel is not meant to be filled, when one has gathered all he needs.

11. In the case of a large vessel, if one intends to fill only half of it, the preparation is completed as soon as one has half-filled it, and the fruit therein becomes liable to tithing. If one is of a mind to fill all of it, the fruit does not become liable until the

entire vessel is full. In the case of two vessels, if one is of a mind to fill both of them, the fruit does not become liable to tithing until both are filled.

12. If one ties vegetables in the field into one large bundle, they are liable to tithing, even if he is of a mind to retie them into small bundles for sale in the market place.

In the case of pomegranate seeds, raisins, and carobs, they become liable to tithing when one piles them on his roof-top to dry. In the case of onions, when one removes the leaves and the skins which are customarily discarded; if one does not remove these, the onions become liable when he piles them in a heap.

13. Grain produce becomes liable to tithing when the pile has been smoothed down. What constitutes smoothing down? Leveling the surface of the pile of produce with the winnowing shovel at the end of all the processing, the way people do when they complete all the work connected with it. If one does not smooth the pile down, it is liable to tithing when he has heaped it into a pile. In the case of pulse, it is liable once it is sifted, but one may take of it from underneath the sieve and eat; if it is not sifted, it is liable when one has smoothed down the pile. But even if one has already smoothed down the pile, he may take of the stunted ears, of the ears at the sides of the pile, or of the ears scattered among the stubble, and eat.

14. Wine becomes liable to tithing when one has poured it into jars and has skimmed the seeds and the grapeskins off the mouths of the jars. If it is still in the vat, however, and one lifts it in order to pour it into a jar, he may drink of it in a casual manner. One may also take some of it from the upper wine press, from the spout, or from any other part of the press, and drink it.

15. Oil becomes liable to tithing when it drips down into the pressing trough. But even after it has dripped down, one may take some of it from the pressing basket, from the pressing beam, or from between the pressing boards, and pour it into a small dish or a large platter of broth, even if the broth is still hot, be-

cause the oil will not boil in a vessel that is away from the fire. If, however, the broth is exceedingly hot, so that it will burn the hand, one may not pour the oil into it, because the oil will boil.

16. A round fig cake becomes liable to tithing when its surface is smoothed down; loose dried figs, when they have been stamped down; if they are gathered into a bin, when one rounds them off with his hand at the mouth of the bin.

If, as the figs are stamped down into the jar, or as the round fig cake is stamped down into the mouth of the bin, the jar breaks or the bin splits, one may not eat of these figs until they are tithed.

17. In the case of figs and grapes lying on the drying pad, one is permitted to eat of them in a casual manner while they are in their place. If, however, they are taken therefrom and conveyed to another place, one may not eat of them in a casual manner, inasmuch as their preparation has been completed, even though they are not yet sufficiently dried.

18. In the case of carobs, as long as they have not yet been taken up to the roof-top, one may fetch some of them down for his cattle, and he is exempt, because he is likely to take the remainder back for drying, so that he is thus providing only a casual feed for his beasts.

19. If one hastens to set aside his first tithe while the produce is still in its ears, he may not eat casually of it until he first sets aside its heave offering. If he does eat, he is liable to a flogging for disobedience.

What constitutes casual eating? For example, when one husks barley one ear at a time and eats it. If, however, he husks the ears and gathers the grains into his hand, he is liable to tithing. If he rubs parched ears of wheat, he may winnow them from hand to hand and eat them; but if he winnows them into his bosom, he is liable to tithing. Needless to say, if he winnows them into a vessel, that is not having a casual meal.

Similarly one may take some of the wine, pour it into a dish containing cold broth, and eat it; he may not, however, pour it into a pot, even if it is cold, because a pot is considered the same as a small vat.

One may likewise squeeze olives over his flesh, but not into his hand; and the same applies to all analogous cases.

20. Just as one may eat a casual meal out of produce the preparation of which has not yet been completed, so may one feed as much of it as he wishes to wild or domestic beasts and fowls, or declare as much of it as he wishes to be ownerless property, before he tithes it.

Once its preparation has been completed, even if it has not yet been designated for tithing, one may not declare it to be ownerless property nor feed it to domestic or wild beasts and fowls as regular fodder, until he has tithed it.

One may, however, feed untithed produce casually to cattle, even inside one's house. One may also feed bunches of fenugreek spikes prior to tying them into bundles.

21. If one finds some single plucked berries on the roadway, or even beside a field full of such single-plucked berries, they are exempt from tithe.

If one finds some dried figs, the rule is that if most people have already pressed most of their figs, he is liable to tithe them, because the presumption is that their preparation has been completed.

The same rule applies to one who finds segments of dried fig cakes, because they have obviously come from produce the preparation of which has already been completed.

22. If one finds small sheaves of produce in a private domain, he is liable to tithe; if in a public domain, they are exempt. Full-sized sheaves are liable to tithing in all places.

If one finds produce that has been smoothed down, he may use it as heave offerings and tithes for other produce, and need have no apprehension about it.

If one finds a covered basket of produce, it is liable to tithing. If he finds such a basket in a place through which most people carry produce to the market place, he is forbidden to eat a casual meal out of it, and he must treat it as doubtfully tithed produce. If he finds it in a place through which most people carry produce to their own houses, he may eat a casual meal out of it, treating it as produce assuredly not yet tithed. If the place is used equally for either purpose, the produce is considered doubtfully tithed. Once he brings it into his own house, he must treat it as produce assuredly not yet tithed.

To what does this apply? To produce whose processing has no point of completion. As for produce that has a point of completion, even if it is assuredly not yet tithed, one need not set aside heave offering from it, because the presumption is that heave offering has been set aside from it at the time when its preparation was completed.

23. In the case of ant-holes that have been in place overnight beside a heap of produce liable to tithe, whatever produce is found in them is liable to tithing, because it is evident that the ants have been carrying off all night bits out of produce the preparation of which had been completed.

24. If one finds olives under an olive tree, or carobs under a carob tree, they are liable to tithing, because the presumption is that they had dropped off from that same tree.

If one finds figs under a fig tree, there is cause for doubt, seeing that figs are likely to change their appearance by getting soiled in the dust, as to whether they had dropped off from this fig tree or from some figs that have already been tithed.

25. If one consecrates detached produce and redeems it before its preparation has been completed, he is liable to tithe. If its preparation is completed while it is in the possession of the Temple treasurer, and the owner then redeems it, it is exempt from tithes.

If one consecrates standing corn for use as meal offerings, it is exempt from tithes.

CHAPTER IV

1. According to the Torah, untithed produce becomes designated for tithing only when the owner brings it into his house, as it is said, *I have put away the hollowed things out of my house* (Deut. 26:13), provided that he brings it in by way of the gate, as it is said, *that they may eat within thy gates* (Deut. 26:12). If, however, he brings in his produce by way of roofs or outer enclosures, it is exempt from both heave offering and tithes.

2. It would seem to me that according to the Torah, one is not liable to flogging for eating of untithed produce until it is designated for tithing by virtue of its being brought into the house, as we have explained on the ground of tradition. If, however, the produce has been designated for tithing by means of one of the other six factors that we have enumerated, the transgressor is liable only to a flogging for disobedience on Scribal authority. Similarly, one who eats of produce meant to be conveyed to the market place after its preparation has been completed, is liable only to a flogging for disobedience, as we have explained, because he who completes the preparation of his produce with the intention of selling it in the market place, is liable to tithing only on Scribal authority.

3. A house that is less than four cubits square cannot effect designation for tithing. Similarly, a roof cannot effect this designation, even if the house underneath it does so. On the other hand, a roof less than four cubits square, as for example, that of a house built in a sloping shape, cannot exempt produce from tithing, such a roof being accounted as part of the house's airspace.

4. Cone-shaped roofless huts, watchmen's huts, and summer sheds consisting of four posts with a ceiling over them and without walls, as well as guard-huts erected by the owners of vineyards and orchards during summertime do not effect designation for tithing, even though the occupants reside in them all through the summer, and even though they usually contain a handmill

and are used to hold poultry. The same applies to the outer parts of potters' huts and to the festival booth used during the Festival of Tabernacles—they do not effect such a designation, because all these are not permanent dwellings.

5. Cone-shaped roofless huts and watchmen's huts render produce liable to tithing for their owners, though not for anyone else. Similarly a schoolhouse and a house of study render produce liable to tithing for him who sits there teaching, because for him it is like his own house, but not for others.

6. A house of prayer and a house of study designate the produce that is in them for tithing if a dwelling is attached to them; if not, they do not.

Storehouses and storage structures erected in the fields to hold gathered produce do not designate it for tithing. If, however, they are meant to serve also as dwellings, they do so designate it.

7. Just as a house designates produce for tithing, so does a courtyard. Once the produce enters the courtyard through the gate, it becomes designated, even if it is not subsequently brought into the house.

8. What is the definition of a courtyard which establishes the designation for tithing? Any courtyard in which watch is kept over the goods stored therein, or one wherein a person would not be loath to have his meals, or one in which a stranger, upon entering, would be asked, "What are you looking for?" Similarly a courtyard that has two tenants, or belongs to two partners, so that if one of them should open it and enter, the other would thereupon proceed to enter, or else depart and lock it; inasmuch as both are entitled to open and lock it, it designates the produce for tithing.

9. A courtyard's gateway, a porch, or a gallery are considered the same as the courtyard itself—if it designates produce for tithing, they do so also; if it does not, they also do not.

10. In the case of two courtyards, one inside the other, both designate produce brought into them for tithing. In the case of the potter's hut, composed of two huts one inside the other, the

inner hut designates produce for tithing, but the outer one does not. A merchant's shop designates produce the same as a house.

11. While a person is conveying his produce from one place to another, it does not become designated for tithing, even if along his way he brings it into and out of houses and courtyards. He may therefore eat occasionally of it until he reaches his final destination. And the same applies to his return journey.

12. Spice peddlers who go around from town to town and enter from one courtyard to another may eat occasionally of untithed produce, until they reach the place where they intend to spend the night.

13. If one carries figs from the field in order to eat them in a courtyard that exempts them from tithing, and then forgetfully brings them into his own house, he is permitted to take them out of the house and eat an occasional meal out of them.

Similarly, if one has forgetfully brought the figs up to the roof, he may eat an occasional meal out of them on the roof.

If while carrying them up in order to eat them on his rooftop, one forgetfully brings them into the courtyard of his own house, they are designated for tithing, and he may not eat of them until he has set aside the tithe.

14. A hoed courtyard is considered the same as a garden, and one may eat an occasional meal out of the produce that is in it, provided that the greater part of it has been hoed. If most of it is already seeded, he may not eat an occasional meal out of its produce. The same rule applies if most of it is already planted. If one plants the courtyard with trees for decorative purposes only, he may eat an occasional meal out of the fruit of those trees, so long as the courtyard has been hoed.

15. In the case of a fig tree standing in a courtyard, a person may eat of its figs one at a time, and he is exempt from tithe; if he plucks more than one fig at once, he is liable to tithe. When does this apply? When he stands on the ground. If, however, he climbs up to the top of the fig tree, he may fill his bosom and

eat right there, seeing that the airspace of a courtyard does not designate produce for tithing.

16. If the tree stands within the courtyard but overhangs a garden, one may eat of its fruit in the garden in his usual manner, as if the tree were planted in that garden.

If the tree is planted in the garden but overhangs the courtyard, it is the same as if it were planted in the courtyard, where one may eat of its fruit only one at a time.

17. If a grapevine is planted in a courtyard, one may not take a whole cluster of grapes and eat it, but must rather pluck one grape at a time. The same applies to a pomegranate—one may not pluck the whole pomegranate, but must rather split it while it is on the tree and eat the sections one by one as they are separated from the fruit. The same applies also to watermelon—one must slice it as it lies on the ground and eat each slice right there.

If while eating a cluster of grapes in the garden, one passes from the garden into the courtyard, he may not finish eating the cluster until he has tithed it, even if by that time he has left the courtyard.

18. In the case of coriander sown in a courtyard, one may pick off one leaf at a time and eat it; but if he picks off more than one leaf at once, he is liable to tithe. The same applies to all similar cases.

CHAPTER V

1. If one purchases produce detached from the ground for his own consumption, it is designated for tithing on Scribal authority, as we have explained. When is this designation established? When the purchaser hands over the price, even if he has not yet taken possession of the produce.

If the purchaser has not yet paid the price, but keeps picking out some of the produce and leaving some, even if he continues doing so all day, he is not liable to tithe, even if he has made up his mind to make the purchase.

If he is a heaven-fearing person, however, he should tithe the produce once he has made up his mind to purchase it, and may then return it to the seller if he decides to change his mind.

2. If one purchases produce that is still attached to the ground, or is detached but was purchased in order to be sent to one's companion, it is not designated for tithing, and the purchaser may eat an occasional meal out of it.

3. If A says to B, "Here is an 'issar for you—give me five figs for it," A may eat them one by one and is exempt from tithe; if he takes more than one at a time, he is liable to tithe.

If A says to B, "Here is an 'issar for you, for any twenty figs that I may choose for myself," A may select them one by one and eat them without tithing. If A says, "for any cluster of grapes that I may choose for myself," he may pick single grapes from it off the tree and eat them without tithing; "for any pomegranate that I may choose for myself," he may split the pomegranate while it is attached to the tree, and eat the segments one by one without tithing; "for any watermelon that I may choose for myself," he may slice it while it is attached to the ground and eat the slices without tithing.

If A cuts off the figs and eats them more than one at a time, or if he plucks the cluster of grapes or the watermelon, he is liable to tithe, since he has acquired them once they have been plucked.

If, however, A says to B, "Here is an 'issar for you for these twenty figs," or, "for these two clusters of grapes," or, "for these two pomegranates," or, "for these two watermelons," he may cut them off in his usual manner and eat an occasional meal out of them, and is exempt from tithe, since their purchase did not designate them for tithing, inasmuch as he has bought them while they were still attached to the ground.

4. If A exchanges a batch of fruit with B for each other's immediate consumption, both batches are designated for tithing, inasmuch as each one has bought the fruit detached. If A and B

have exchanged the fruit with the intention of first setting it out to dry, neither batch is designated for tithing, because sale alone cannot designate for tithing produce the preparation of which has not yet been completed, as we have explained.

If A acquires the exchanged fruit for his immediate consumption, while B acquires it for drying, the rule is that A is liable to tithe, while the fruit acquired by B is not designated for tithing.

5. If A says to B, "Go forth and pick for yourself twenty figs off my tree, while I will fill my belly out of your detached figs," both are exempt from tithe, because such an exchange is not sufficient to constitute a sale, so that if one of them had eaten more than one fig at a time he would have been liable to tithe. A gift, unlike a sale, does not designate produce for tithing.

6. If an unlearned person passes along in the market-place calling out, "Have some of my figs," the people may eat of them then and there and are exempt from tithe, because a gift does not designate produce for tithing.

If, however, people carry these figs into their houses, the rule is as follows: If the majority of the people customarily carry figs into their houses, these figs are considered assuredly untithed and liable to tithing; if most people customarily carry figs to the market place, these figs must be rectified as doubtfully tithed produce, since the unlearned person may have tithed them first and then brought them to the market place.

If the unlearned person calls out, "Have some of my figs and carry them into your houses," each recipient, having carried them into his house, must tithe them as doubtfully tithed produce.

If the recipient's share of these figs is a large one, and even if the unlearned person says, "Have some and eat," it is the same as if he had said, "Have some and carry them into your house"—the recipient may not eat of them until he has rectified them as doubtfully tithed produce.

The same rules apply to any produce that is not usually eaten raw, or to a recipient who is a distinguished person unaccustomed to eating in the marketplace—the donated produce must be rectified as doubtfully tithed.

7. In the case of two recipients, A and B, if the unlearned owner of the produce says to A, "Take and eat," and to B, "Take and carry into your house," A may eat of it and is exempt from the tithe, while B, if he eats, is liable to tithe.

8. Similarly, if several persons are seated in a gateway or in a shop, and the unlearned owner of the produce says to them, "Take and eat," they may eat of it and are exempt from tithe. The owner of the gateway or shop, however, may not eat of it until the recipients have tithed it as doubtfully tithed produce, because it is as if he himself had said to the recipients, "Take these and bring them into your houses," for the gateway and the shop are the same as their own houses, and we have already explained that a house that is not one's own does not designate produce for tithing when that produce passes through it.

9. If one hires laborers to work with him in handling produce, whether detached from the ground or attached to it, the rule is that inasmuch as according to the Torah they are entitled to eat of the produce which they are handling, they may eat of it and are exempt from tithe.

If, however, the owner stipulates that they are to eat of what the Torah has not entitled them to, as for example, if the laborer stipulates that his children are to share in eating with him, or that his son is to eat in lieu of his wages, or that he himself is to eat of the detached produce after he has finished his work, he may not eat of it until he has tithed it, seeing that he who eats on the ground of a stipulation is accounted the same as a purchaser.

10. If the owner hires the laborer to weed olives with him, and if the laborer stipulates that he is to eat of the olives, he may eat of them one by one off the tree and is exempt from tithe. If he eats several olives at a time, he is liable to tithe.

11. If the owner hires the laborer to weed onions and stipulates that the latter is to eat of the greens thereof, the laborer may pluck and eat them leaf by leaf without tithing. If he eats several at a time, he is liable to tithe.

If the laborer stipulates that he is to eat a liṭra of olives, he may eat them one by one; but if he eats more than one at a time, he is liable to tithe, because when he eats a specified amount he is the same as a purchaser, in which case, if he eats more than one at a time, he renders the produce designated for tithing.

If the laborer does not stipulate the amount and eats only what he is entitled to according to the Torah, he may gather several pieces of produce together and eat as many as he wishes, provided that he does not salt them; if he does salt them, he is permitted to eat them only one by one, but not two by two, because in the latter case the salt designates them for tithing.

12. A laborer working with poor quality figs may not eat of fine quality figs, and vice versa, unless he first tithes them. He may, however, restrain himself from eating until he reaches the place where the fine quality figs are growing.

13. If a person dispatches laborers into the field to do some work for him, the rule is that so long as he is not obligated to feed them, they may eat of the produce in that field and are exempt from tithe, provided that the preparation of the produce has not yet been completed. If, however, the owner is obligated to feed the laborers, they may not eat of the produce, even if its preparation is not yet complete, because one may not pay off a debt out of untithed produce. They may nevertheless eat figs one by one off the tree, but not out of a basket, nor out of a larger vessel, nor off the storage place.

14. Whether one boils produce, or stews it, or pickles it, he designates it for tithing. If, however, one smokes it until it becomes fit for consumption, its liability to tithe is in doubt.

15. If one deposits produce in the ground, in straw, or in compost until it becomes fit for consumption, he does not thereby render it designated for tithing.

16. If one adds wine to a hot dish, or oil to a boiling pot or stewing pan, he thereby designates the contents for tithing, and the same applies to wine decanted into hot water. Needless to say, if one boils the wine, even if it is still in the wine press, he may not drink of it until he has tithed it.

17. Garlic, garden cress, or mustard, if grated with oil while still in the field, becomes designated for tithing. Similarly, if one

squeezes a cluster of grapes into a cup, it becomes designated for tithing; but if into a tray, it does not.

18. If one salts produce in the field, he renders it designated for tithing, but if he dips olives one by one in salt and eats them, he is exempt from tithe. If one cracks olives in order to let the raw sap out of them, he is exempt. If one picks olives out of the vat, he may dip them in salt one by one and eat them, but if he salts them severally and places them before him, he is liable to tithe. The same applies to all similar cases.

19. If one sets aside heave offering out of produce in such a way that he must then set aside a second heave offering, he renders the produce designated for tithing, and he may not eat an occasional meal out of it until he has set aside both the second heave offering and the tithe.

20. Produce the preparation of which has been completed and which has been overtaken by the dusk of Sabbath eve, is rendered designated for tithing, and one may not eat of it even after the Sabbath until he has tithed it.

21. Children who had hidden figs to eat during the Sabbath and have forgotten to tithe them, may not eat them upon the expiration of the Sabbath until they have tithed them.

22. In the case of a fig tree whose fruit is intended to be eaten only on the Sabbath, if one gathers a basketful of it, he may not eat of it until he has tithed it, inasmuch as this fruit is intended for the Sabbath, and the Sabbath designates it for tithing.

If while one is eating a cluster of grapes, he is overtaken by the darkness of Sabbath eve, he may not finish eating it on the Sabbath until he has tithed it. If, however, he leaves it until after the Sabbath, he may finish eating it then.

CHAPTER VI

1. One may level the surface of a fig or raisin cake with the juice of untithed figs or grapes, respectively, and this is not considered wasteful.

Anything that commoners may not eat out of heave offering, such as seeds and the like, the same they may not eat out of untithed produce, out of tithe from which the heave offering of the tithe has not yet been set aside, out of second tithe, and out of consecrated things which have not yet been redeemed. And likewise, anything that commoners may eat out of the former, the same they may eat out of the latter.

2. One may not kindle unclean untithed produce, even on a weekday, nor, needless to say, on the Sabbath, as it is said, *The charge of My heave offerings* (Num. 18:8), implying that just as you have no right to make use of clean heave offering until it has been set aside, so have you no right to make use of unclean heave offering until it has been set aside and thereafter.

3. One may not cover untithed seed with earth, nor may one sow untithed produce; even produce the preparation of which has not yet been completed may not be sown until it has been tithed.

To what does this apply? To grain, pulse, and the like. If, however, one transplants seedlings with fruit on them from one place to another within his field, it is permitted and is not accounted as planting untithed produce, seeing that one has not harvested the fruit in the process.

Similarly, if one transplants turnips or radishes from one place to another, the rule is as follows: If his intention is to promote their further growth, it is permitted; if his purpose in transplanting them is to cause them to harden so that he might get their seeds, it is forbidden, because it is the same as if he had planted untithed wheat or barley.

4. If one liṭra of tithe from which the heave offering of the tithe has not yet been set aside is planted, and grows in yield to ten liṭra, the whole of it is liable to tithe, but for the original liṭra heave offering of the tithe must be set aside out of some other produce in proportion.

If one regularizes one liṭra of onions and plants it, he must tithe not by calculating the increase but for the entire yield.

5. In the case of seeds one-third grown, smoothed over, tithed, replanted, and increased in yield—they being seeds that are not

absorbed in the soil—it is doubtful whether they are liable to tithing by Scribal law inasmuch as they have increased, or not liable inasmuch as the seed, still extant and not absorbed, has already been tithed. Such seeds cannot be regarded as analogous to onions, because onions are not customarily grown out of seeds.

6. If one plants untithed produce, whether its seed is or is not absorbed in the soil, the rule is as follows: If the seed can still be picked up, he must be penalized by being made to pick it up. If it has already sprouted, he need not be compelled to uproot it, and the growth is regarded as profane produce.

If the seed of the produce is not absorbed in the soil, even the outgrowth of the outgrowth is forbidden, three consecutive harvests in all; the fourth is permitted. Why are the three outgrowths forbidden? Because of the heave offering of the tithe and the great heave offering which are contained in them.

The same rule applies to one who plants tithe from which heave offering of the tithe has not yet been set aside.

One may not sell untithed produce except in case of need, or to an associate scholar. It is forbidden to send untithed produce, even from one associate scholar to another, lest they should rely one upon the other, and it should be eaten untithed.

7. If A, having sold some produce to B, is reminded that it is still untithed and runs after B in order to regularize it, but cannot find him, the rule is as follows: If it is known that the produce no longer exists, having already perished or been consumed, A need not set aside tithe for it. If there is doubt as to whether the produce is or is not still in existence, A must set aside tithes for it from some other produce.

8. If after A has sold some produce to B, A claims, "I sold it with the understanding that it was untithed," while B claims, "I bought it from you with the understanding that it was nothing but tithed," A must be compelled to regularize it, being thus penalized for having sold untithed produce.

9. One may not pay a debt with untithed produce, because this is tantamount to selling it.

10. If the government seizes one's produce while it is still untithed, the rule is as follows: If it was seized in satisfaction of a debt due to the government, he must set aside tithe for it; if the government seizes it unjustly by force, he need not set aside tithe for it.

11. If one buys untithed produce from two places, he may set aside tithe out of the one batch for the other.

If A leases a field from B who is an Israelite, or a heathen, he must divide the yield and give B his share in his presence, so that B would know that he has received untithed produce. If A rents a field from an Israelite B, the rule is as follows: If A pays B with the produce of that same field, he must first set aside heave offering for all the produce, and then pay B what he had agreed to pay him, whereupon B must set aside the tithe for his received rental. If, however, A pays B with produce from another field or with produce of another species, he must set aside the tithe first, and then pay B his due.

12. If one rents a field from a heathen, he must set aside the tithe first and then pay the heathen his due. He is thus penalized in order to discourage him from renting from a heathen, so that the field might be left uncultivated and the heathen might prefer to sell it outright to an Israelite. Similarly, if one leases from a heathen a field that had formerly belonged to one's forefathers, he is penalized by being compelled to set aside the tithe first and then pay the heathen his share already tithed, in order that he should not be eager to lease the field from the heathen just because it had formerly belonged to his forefathers. Accordingly the field might be left uncultivated, with the result that the heathen might prefer to sell it outright to an Israelite.

13. What is a renter and what is a lessee? A renter is one who rents the land for a fixed amount of the produce, so-and-so many sĕ'ah, whether the land yields much or little. A lessee is one who leases the land for a share of its yield, half of it, or a third of it, whatever he and the lessor may agree upon between themselves. A hirer is one who hires the land for a hire of money.

14. If A and B jointly lease a field for a share of the produce, or inherit it, or own it in partnership, A may say to B, "You take the wheat that is in such-and-such a place, and I will take the wheat that is in such-and-such other place," or, "You take the wine that is in such-and-such a place, and I will take the wine that is in such-and-such other place." A may not, however, tell B, "You take the wheat, and I will take the barley," or "You take the wine, and I will take the oil," because this would be the same as if they sold untithed produce.

15. If a priest or a Levite has purchased produce from an Israelite after its preparation had been completed, heave offering and tithes must be collected from the purchasers and given to other priests and Levites. This penalty is imposed upon them so that they would not hasten to the threshing floors and the wine presses to buy untithed produce, in order to deprive their fellow priests of the gifts due to them.

If, however, they have bought the produce before its preparation is completed, they are not subject to compulsory payment of these gifts.

16. If a priest or a Levite had sold produce detached from the ground to an Israelite before its preparation has been completed—and needless to say, if they had sold it while still attached to the ground—the heave offering or the tithe thereof remains theirs. If they have sold the produce after its preparation had been completed, the heave offering and the tithe thereof are due from the purchaser, who must set them aside and give them to any priest or Levite that he may wish.

17. If a priest or a Levite has leased a field from an Israelite, the rule is that just as they share in the profane produce so do they share in the heave offering and the tithes; the Israelite may therefore take his own share and pay the gifts due therefrom to any priest or Levite that he may wish.

If, however, an Israelite has leased a field from a priest or a Levite, the heave offering or the tithe is due from the owner of the field; as for the other gifts, they both share in them.

18. If one has leased olives in order to produce oil from them, whether the lessee is an Israelite and the lessor a priest or a Levite, or vice versa, the rule is that just as they share in the profane produce so do they share in the heave offering and the tithes, because oil is an important item.

19. If a priest, while selling his field to an Israelite, says to him, "I am selling it on condition that the tithes thereof are to accrue to me in perpetuity," they remain his, for inasmuch as he has used the term "on condition," it is as if he had retained for himself the portion of ground where the tithe is to grow.

Should this priest die, his son is considered the same as any other priest. If, however, the priest uses the formula, "on condition that the tithes are to accrue to me and to my son," should he die, his son may take them.

If the priest says, "on condition that the tithes are to accrue to me so long as this field is yours," should the purchaser then sell it to someone else, the priest is not entitled to those tithes, even if the first purchaser subsequently repurchases that same field.

20. If an Israelite, while leasing a field from a priest or a Levite, says to him, "I lease it on condition that the tithes are to accrue to me for four or five years," he is permitted to do so. But if he says, "on condition that they are to accrue to me in perpetuity," it is forbidden, because a priest cannot make a priest.

Similarly, if a Levite is in debt to an Israelite, the latter may not claim tithes from other people and set them aside for himself until the debt is paid in full, because this Levite cannot make his creditor a Levite, thereby entitling him to collect tithe from others.

21. In the case of an Israelite who has inherited untithed produce that has been smoothed over from his priestly maternal grandfather, the latter having in turn inherited it from his own Israelite maternal grandfather, the Israelite must set aside the tithes thereof, but they accrue to him, because priestly gifts due to be set aside are accounted as already set aside, even if in fact they had not been set aside.

22. If one leases his field to a heathen or to some other person who cannot be trusted with regard to tithes, he must himself set aside tithe in their behalf, even if the crop has not yet reached the season for tithing.

If one has leased his field to an unlearned person, the rule is as follows: If the lease is executed before the crop has reached the season for tithing, the lessor need not tithe in behalf of the lessee; if the lease is executed after the crop has reached the season for tithing, the lessor must tithe in the lessee's behalf.

How should the lessor proceed? He should stand over the threshing floor and take up the tithe, and need not worry about what the lessee had already eaten, since we as lessors are not responsible for the lessees.

CHAPTER VII

1. If a person who has 100 loḡ of wine that is untithed according to the Torah says, "Two loḡ that I am going to set aside out of these are to be heave offering, ten loḡ are to be first tithe, and nine loḡ are to be second tithe," he may not commence drinking of this wine with the thought that in the end he will leave the amount needed for heave offering and tithes; rather, he must set aside these gifts first, and then drink. It cannot be said that whatever he leaves in the end may be regarded as if it had been selected in the beginning, because the obligation to set aside heave offering and tithes is a law of the Torah, and concerning such a law it cannot be said "Let us regard something as selected," until one has actually selected it.

2. If one designates that his tithes are to be located in the mouth of a jar, he may not drink out of the lower part of it, or vice versa, because the liquid therein is intermixed.

If, however, one designates that his tithes are to be in the mouth of a bin, he may eat out of its lower part, or vice versa.

3. One may designate his second tithe, redeem it, and then make it serve as first tithe for that same produce, or as heave offering of the tithe for produce from some other place.

4. If one reserves batch A of produce so that it may serve as the source of tithes for batch B of produce, he may continue to set aside tithes from batch A on the assumption that batch A is still in existence, and may eat and drink of batch B until all the produce of the reserved batch A has become tithe, whereupon he must give it to a Levite. Should he find that batch A was lost, there is ground for apprehension concerning all the produce of batch A that was set aside as tithe, and he is liable to tithe again, but only out of doubt, not out of certainty.

5. If one lends money to a priest, a Levite, or a poor man on the understanding that in satisfaction of that debt he may set aside and retain the equivalent amount of produce out of the gifts due to them, he may proceed to do so on the assumption that the debtors are still living, and need not worry that the priest or the Levite may have died or that the poor man may have grown rich.

6. How should such a creditor proceed? He should set aside the first or second tithe from his own produce and transfer title to it through a third party to that same priest, Levite, or poor man who is the debtor.

If the debtor customarily receives these gifts from the creditor, and the creditor customarily gives these gifts to no one but the debtor, the creditor need not transfer title through a third party.

After the creditor has set these gifts aside, he should calculate how much the produce set aside is worth and deduct its value from the loan, until the entire loan is repaid. Thereupon the creditor may sell the heave offering to any priest that he may choose; as for the tithes, he may eat of them himself.

7. When the creditor calculates the value of the produce that he has set aside, he should do so according to the lowest market price, and need not worry that this might be accounted the same as charging interest. The Sabbatical year cannot cancel such a debt. If the creditor wishes to retract the agreement, he may not do so, but if the debtor wishes to retract it, he may do so. If the creditor has despaired of recovering the debt, he may not set aside tithe in repayment of it, because tithe cannot be set aside for produce given up as lost.

If the debtor priest, Levite, or poor man dies, the creditor may not set aside produce out of the gifts due to the debtor until he obtains permission from his heirs, provided that the debtor has left them some land, be it even no more than a sliver the size of a needle. If, however, he has left only money, the permission of the heirs is of no avail.

If the creditor has executed the loan in court, with the explicit proviso that he may set apart produce out of the gifts due to the debtor against this money, he need not obtain permission from the heirs.

If the poor debtor has grown rich, the creditor may not set apart produce out of the gifts due to the debtor in repayment of his loan; even if the loan has been executed in court, the debtor retains title to what is in his hand.

8. If an Israelite says to a Levite, "You have due to you a kor of tithe in my keeping," the Levite may then make it heave offering of the tithe for some other produce, even though he has not yet taken possession of that kor of produce. If the Israelite has already given this kor of produce to another Levite, the first Levite can do nothing more than voice his complaint.

9. If one who has produce in a bin gives one sĕ'ah to a Levite and one sĕ'ah to a poor man, he may not thereupon reserve eight sĕ'ah from that bin for his own consumption, unless he knows that the two sĕ'ah of tithe are still in existence. If, however, the Levite and the poor man have already eaten out of them, he may reserve for himself only an amount corresponding to what still remains of the two sĕ'ah.

10. If poor men exchange their dues with the owner of the land, as for example, if they give him one sĕ'ah of gleanings, of forgotten sheaf, of corner crop, or of poor man's tithe, and take in exchange one sĕ'ah of produce off the threshing floor, the rule is as follows: What the owner receives from the poor men is exempt from tithe, even if the exchange is made with his consent; what the poor men receive from him is liable to tithe, even though it is in exchange for their dues which are exempt from tithe.

11. If one has before him two baskets of untithed produce out of which heave offering has already been set aside, and says, "The tithe due from the first basket is to be set aside from the second basket," the first basket is tithed. If he says, "The tithe due from the first basket is to be set aside from the second basket, and vice versa," only the first basket is considered tithed, because he has designated its tithe as contained in the second basket; the second basket is not considered tithed, because one may not set aside tithe from the first basket, which has already become exempt, for the second basket, which is still liable to tithe for both baskets together.

12. On the other hand, if one says, "As for the tithes from these two baskets, the tithe of each basket is designated as located in the other," his declaration is effective, the tithes of both baskets are so designated, and he must set aside their tithes accordingly, and may not set aside tithes for them from some other place.

13. How then should he tithe them? He should either take the tithe for both baskets out of one of them, or else take each basket's tithe out of it, if both baskets are equal in size. If one basket is larger than the other, he should set aside tithe out of the small basket for the large one and vice versa.

CHAPTER VIII

1. Produce from which great heave offering and heave offering of the tithe have been set aside is termed profane produce. If then the other tithes are also set aside out of it, it is everywhere termed regularized profane produce.

2. In the case of untithed produce mixed with regularized profane produce of the same species, so that it cannot be determined whether the flavor has or has not been affected by the admixture, the rule is as follows: If one has other produce, he may set aside out of it heave offering and tithes for that untithed produce, in

proportion. If he has no other produce to provide tithe for that untithed produce, the whole of it is forbidden until he sets aside the tithe. And when he does set aside the tithe, he must forfeit out of the regularized profane produce enough to cover the heave offering of the tithe contained in the untithed produce.

3. How so? If 100 sĕ'ah of untithed produce is mixed with 100 sĕ'ah of regularized profane produce, one must set aside 101 sĕ'ah out of the total, and this is considered untithed produce, leaving 99 sĕ'ah considered regularized profane produce; the owner thus forfeits 1 sĕ'ah.

Similarly, if that untithed produce has not yet had heave offering set aside out of it, the owner forfeits out of the regularized profane produce the amount equal to the great heave offering and the heave offering of the tithe that are due.

Why does he forfeit this 1 sĕ'ah? Because otherwise he could say that the 100 sĕ'ah he has set aside are the 100 sĕ'ah of profane produce, and the remaining 100 sĕ'ah are untithed produce.

Likewise, if the untithed produce amounts to less or more than the profane produce, he must lift out of the mixture the amount of the untithed produce, and out of the profane produce the amount equal to the heave offering of the tithe that is due out of that untithed produce, or the amount of the great heave offering and of the heave offering of the tithe, if they have not yet been set aside out of the untithed produce.

4. Similarly, if tithe from which heave offering of the tithe has not yet been set aside becomes mixed with regularized profane produce, the mixture is rendered forbidden, however small the amount of the tithe may be. If the owner has some other such tithe, he may take out of it the heave offering of the tithe for the original tithe, in proportion. If he has no such other tithe, he may lift out of the mixture the amount of the tithe, plus the forfeit out of the regularized profane produce to an amount equal to the heave offering of the tithe contained in the unregularized tithe.

5. How so? If 100 sĕ'ah of tithe have become mixed with 100 sĕ'ah of regularized profane produce, one must set aside out of the

total 110 sĕ'ah, all of which is tithe, and the remaining 90 sĕ'ah are profane produce as before.

Similarly, if the profane produce is more or less than the unregularized tithe, one must set aside out of the profane produce an amount equal to the tithe and to the heave offering of the tithe contained in the unregularized tithe.

6. In the case of untithed produce mixed with unregularized tithe, the rule is as follows: If the untithed produce is equal in amount to the tithe, one forfeits out of the untithed produce an amount equal to the heave offering of the tithe that is contained in it.

How so? If 100 sĕ'ah of untithed produce have become mixed with 100 sĕ'ah of unregularized tithe, one must set aside 101 sĕ'ah, which are tithe; the remaining 99 sĕ'ah are untithed produce.

If the amount of untithed produce exceeds the amount of the tithe, one need set aside heave offering of the tithe out of the tithe alone, and forfeits nothing out of the untithed produce, because if he should declare the heave offering of the tithe that is contained in the untithed produce, the tithe would consequently be mixed with the heave offering of the tithe that is lifted up with it.

If the amount of tithe exceeds the amount of untithed produce, one must declare the heave offering of the tithe which is contained in the untithed produce, and then set aside the untithed produce and an amount equal to the heave offering of the tithe due for this untithed produce, that is 1/100 of the tithe. Thus the untithed produce becomes in its entirety a forbidden mixture, and one must sell it to a priest after deducting the price of the two heave offerings that are contained in it, with the result that he would forfeit 1/100 of the tithe which equals the heave offering of the tithe that is contained in the untithed produce.

How so? If 100 sĕ'ah of untithed produce have become mixed with 200 sĕ'ah of tithe, one must set aside 103 sĕ'ah—the 3 sĕ'ah being heave offering of the tithe for the 100 sĕ'ah of untithed produce and the proportion of heave offering of the tithe for the 200 sĕ'ah of tithe. This is a precautionary measure, lest one should err when the amounts in the mixture are equal. The balance then is 197 sĕ'ah that are tithe as before.

7. If one who owns ten rows containing each ten jars of wine has designated one jar in an outer row as tithe for wine in another place, and it is not known which row is the outer row, he must take two jars at the ends of a diagonal line, mix the wine in them, and set aside tithe out of it.

8. If he has designated the tithe as part of one half of an outer row, and it is not known which half was meant, he must take four jars from the four corners of the square.

If he has designated the tithe as part of one of the rows, and it is not known which row, he must take all the jars in one diagonal row. If he has designated the tithe as part of a half-row, and it is not known which half-row was meant, he must take all the jars in two rows in diagonal lines, with the result that he would take one jar from each half-row. After each such proceeding he must mix the wine contained in these jars and set aside tithe from the mixture.

9. If he has designated one of the jars as tithe, and it is not known which jar, he must take some wine out of each one of the hundred jars, mix it together, and out of the mixture set aside one jarful of wine as tithe.

CHAPTER IX

1. In the days of Johanan the High Priest, who lived after Simeon the Just, the Great Court sent out emissaries who examined the entire territory of Israel, and found that everyone was scrupulous about setting aside great heave offering. As for the first tithe, the second tithe, and the poor man's tithe, however, the unlearned in Israel were lenient toward themselves and did not set them aside. Therefore the court decreed that only trustworthy persons should be relied upon with regard to tithes, that the produce of unlearned persons should be regarded as subject to doubt, and that their declaration, "It has been tithed," should be considered unreliable. Such produce is called "doubtfully tithed" (*dĕmay*).

2. The court therefore ordained that one must set aside out of doubtfully tithed produce only the heave offering of the tithe, because failure to do so involves the death penalty, and the second tithe, because he loses nothing thereby, since it is to be consumed by the owner of the crop. As for the first tithe and the poor man's tithe, one need not set them aside out of doubtfully tithed produce, because there is a doubt involved here, and when A makes a claim against B, the burden of proof is upon A. Therefore the owner may say to the Levite or to the poor man, "Show proof that this produce is untithed; then you will be entitled to take the tithes thereof."

3. Even though one need not set aside the poor man's tithe out of doubtfully tithed produce, he must declare it, without setting it aside, by saying, "One-tenth of what is here is poor man's tithe," in order to designate the second tithe, seeing that the poor man's tithe, set aside in the third and sixth years, takes the place of the second tithe set aside in the other years of the septennate.

4. When heave offering of the tithe and second tithe are set aside out of doubtfully tithed produce, no benediction need be pronounced over them, because of the doubt involved. Therefore one is permitted to set them aside while he is naked.

5. How should one tithe doubtfully tithed produce? He should set aside the amount due for heave offering of the tithe, which is one percent of the total, put it down by the side of the produce, and say, "This one percent is tithe, and the rest of the tithe is next to it." Thereafter he should say, "That which I have declared to be tithe, is to be heave offering of the tithe for the rest of the tithe that is next to it," and should take it up and give it to the priest, because one may not at the outset set aside heave offering of the tithe before first setting aside the tithe. Then he should set aside the second tithe.

6. In the case of doubtfully tithed produce it is permitted to set aside the second tithe before the first tithe. If one wishes to do so, he may say, "The second tithe of this produce is to be in its northern part," or "its southern part," and it may then be redeemed with money.

Similarly, one who buys a loaf from a baker must set aside out of it the heave offering of the tithe and the dough offering, and say, "One one-hundredth of what is here is to be tithe, and the rest of the tithe is next to it; what I have declared to be tithe is to be heave offering of the tithe for the rest of the tithe that is next to it, and the excess over the one one-hundredth that is in what I have set aside is to be dough offering."

The second tithe that is in the northern or southern part of the produce may then be redeemed with money, and one may then eat of it.

7. Similarly, if A invites B to eat with him, and if B does not trust A with regard to tithes, B should say on Sabbath eve, "What I am going to set apart tomorrow is to be tithe; what is next to it is to be the rest of the tithe; that which I have made tithe is to be heave offering of the tithe for the rest that is next to it, and what is to the north," or "to the south of it is to be second tithe," which latter may then be redeemed with money. For a person may stipulate these conditions in respect to doubtfully tithed produce even if it is not in his possession, but not in respect to assuredly untithed produce unless it is in his possession.

8. How so in the latter case? If a person who has in his house one hundred untithed figs finds himself on Friday in the house of study or in the field, and is apprehensive that darkness might overtake him and he would not be permitted to tithe the figs during the Sabbath, he may say, "Two figs which I am going to set aside first are to be heave offering; ten figs that I am going to set aside secondly are to be first tithe; and nine figs that I am going to set aside thirdly are to be second tithe," and on the morrow he may set all these aside and eat of the rest of the produce.

9. When verbally setting aside these gifts seriatim, one should do so in a whisper, so that he would not seem to be regularizing his produce on the Sabbath, seeing that he has stipulated the condition first. Untithed produce subject to such a condition may be moved on the Sabbath even before the tithes have been set aside; he should cast his eyes upon one side of it, and may then eat of the rest of it.

10. If one is given by an unlearned person a cup of wine to drink, he should say, "What I am going to leave at the bottom of the cup is to be tithe; the rest of the tithe is to be next to it; that which I have made tithe is to be heave offering of the tithe for the rest that is next to it; and the second tithe is to be at the mouth of the cup," and may be redeemed with money. He may then drink out of the cup, and leave enough at the bottom of it for the heave offering of the tithe.

11. Similarly, if B is invited by A to drink with him on the Sabbath, he should stipulate such a condition on Sabbath eve concerning whatsoever food he is going to partake of with him.

Similarly, if a laborer does not regard his employer as trustworthy, he should take one dried fig and say, "This fig and the nine to follow it are to be tithe for the one hundred figs that I am going to eat; this one fig is to be heave offering of the tithe for the ten figs that follow it; and the last ten figs are to be second tithe," which is to be redeemed with money.

He should then give the one dried fig that he has set aside to a priest. The laborer himself may then set aside the monetary value of the second tithe, because it is a condition imposed by the court that the heave offering of the tithe must come from the employer, while the second tithe must come from the laborer.

12. Bakers were not obligated by the Sages to set aside the second tithe out of doubtfully tithed produce, but only the heave offering of the tithe, so that they would set it aside in a state of cleanness together with the dough offering; it is the purchaser who must set aside the second tithe.

When does this apply? When the baker sells his wares inside his bakery or at its door. If, however, he sells them to a bread shop or to a shop next to a bread shop, he is obligated to set aside the second tithe as well.

13. In the case of A and B who have gathered their grapes into the same vat, A not being trustworthy with regard to tithing, and B having tithed his share of the grapes, the rule is that when B takes his share of the wine, he must additionally set aside therefrom the tithe due for the doubtfully tithed produce contained in it, out of A's doubtfully tithed grapes.

How so? If their shares are equal, and B takes 200 loḡ of wine as his half, he must set aside therefrom 1 loḡ as heave offering of the tithe and 10 loḡ as second tithe for the 100 loḡ, seeing that he has already set aside the tithe due from assuredly untithed produce for half of the total treaded in the wine press. Similarly, if his share is one-third or one-fourth, he must set aside the tithes in proportion.

CHAPTER X

1. A person who undertakes to be trustworthy with regard to tithes, so that his produce would not become doubtfully tithed, must tithe everything he eats, sells, or buys, must eschew the hospitality of an unlearned host, and must declare this undertaking publicly. Therefore, when reliable witnesses attest that he had indeed publicly undertaken these duties and that he regularly observes them always, he is to be believed when he says concerning his produce, "It has been tithed."

2. Every disciple of the wise is always to be regarded as trustworthy, and it is not necessary to investigate him. His children, the members of his household, his servants, and his wife are the same as he.

If a disciple of the wise dies leaving produce, it is presumed to have been regularized, even if he had gathered it on the same day he died.

3. An unlearned man's daughter or former wife married to an associate scholar, or his servant sold to an associate scholar, must undertake these duties before the marriage or sale, respectively. On the other hand, an associate scholar's daughter or former wife married to an unlearned man, or his servant sold to an unlearned man, retain their presumption of trustworthiness until they fall under suspicion.

An associate scholar's son or servant who customarily spends his time with an unlearned person must undertake these duties anew.

As for an unlearned person's son or servant who customarily spends his time with an associate scholar, the rule is as follows: So long as they remain with the associate scholar, they are the same as he; once they have left him, they are the same as an unlearned person.

4. If a person is not trustworthy, but one of his sons, servants, or members of his family is trustworthy, one may buy food from him on their assurance, and there is no ground for apprehension.

5. If a man is trustworthy but his wife is not, one may buy food from him but not accept his hospitality. If his wife is trustworthy but he himself is not, one may accept his hospitality but not buy food from him. May a curse befall him whose wife is trustworthy while he himself is not!

6. An associate scholar should not serve at an unlearned person's feast or banquet, unless all the provisions have been tithed and regularized under his own hand. Therefore, if one observes an associate scholar serving at an unlearned person's feast or banquet, the presumption is that all the provisions have been regularized and tithed. If, however, the associate scholar is observed dining with unlearned persons, the meal may not be presumed to have been tithed—perchance the associate scholar relies on stipulations made in his own mind.

7. Just as one may dine with an unlearned person while relying upon one's own stipulation, so must one so stipulate in behalf of one's own son, even if the son is in some other place. One need not, however, so stipulate in behalf of any person other than one's own son, even if that person is attending the same feast. Therefore, if an associate scholar's son is in attendance at an unlearned person's feast, the meal may not be presumed to have been tithed—perchance his father had stipulated in his behalf.

8. If an unlearned person hands a mĕʿah to an associate scholar and says to him, "Buy a bunch of vegetables," or "a loaf of fine bread for me," the scholar may buy it for him without explicit specification, and is exempt from tithing it. If he exchanges this

mĕʿah, however, he must tithe the purchase before handing it to the unlearned person. If the scholar, while executing the purchase, makes an explicit specification, by saying to the shopkeeper, "This bunch which I am buying from you, I am buying it for my companion, while that one, I am buying it for myself," he must tithe the bunch which he has bought for himself, but need not tithe the bunch which he has bought for his companion. If the two bunches are intermixed with one another, even if one of the scholar's bunches is mixed with 100 of his companion's bunches, the scholar must regularize all of them as doubtfully tithed produce, and may then hand 100 bunches to his companion who had sent him to purchase them for him.

9. If five persons, some of whom are associate scholars, say to a sixth one, "Go forth and bring us five bunches of vegetables," the rule is as follows: If he brings a bunch to each one of the five separately, the scholars among them need tithe their own share only; if he brings the five bunches together, the scholars among them must tithe all five bunches.

10. If an unlearned person says to an associate scholar, "Go forth and gather some figs for yourself from my fig tree," the scholar may eat an occasional meal of these figs, otherwise he must tithe them as doubtfully tithed produce.

If an associate scholar tells an unlearned person to gather figs for him, and another associate scholar hears him, the latter may eat of these figs and need not tithe them first, because an associate scholar is presumed not to let a thing which is not regularized go out from under his hand, and it is therefore presumed that he had set aside tithe for it from some other place. Even though associate scholars are not suspected of setting aside heave offering of the tithe out of produce that is not adjacent, nevertheless in this case they may do so in order to remove a stumbling block from before an unlearned person.

11. It is permitted to feed poor men and guests with doubtfully tithed produce, but one must inform them of this fact; and the poor man or guest, if he wishes to regularize the food, may do so.

12. Collectors of charity may collect from all persons without further ado, and may distribute the proceeds without further ado. The recipient who wishes to regularize his share may then do so.

13. A physician who is an associate scholar may, while serving food to a sick unlearned person out of the latter's own produce, put it into the patient's hand, but not into his mouth. If the doubtfully tithed produce belongs to the physician, he may not put it even into the patient's hand. Similarly, if the physician knows that the produce is assuredly untithed, he may not put it even into the patient's hand.

CHAPTER XI

1. It is forbidden to sell doubtfully tithed produce to an unlearned person or to send it to him as a gift, because this is the same as encouraging him to eat forbidden food. One may, however, sell or send such produce to disciples of the wise, because no disciple of the wise would eat anything until he has tithed it, or until a trustworthy person has informed him that it has been tithed.

2. Wholesale merchants, as for example, wholesale dealers in provisions and produce, may sell doubtfully tithed produce or send it as a gift, because they usually add to the exact measure. The Sages have therefore ordained that in such a case it is the buyer or the recipient who must set aside tithe for the doubtfully tithed produce.

As for retailers who deal out the exact measure, it is the seller who must set aside the tithe, inasmuch as he realizes his full profits, and he may not sell or send anything unless it has been regularized.

3. What constitutes wholesale quantity? For dry produce, one-half of a sĕ'ah; for liquids, the amount of the particular liquid that is worth one denar.

4. Baskets of olives or grapes and hampers of vegetables, even if sold in the lump, may not be sold while they are doubtfully tithed.

5. If either the seller or the buyer says to the other, "Come, let us regularize this produce," the seller must set aside the heave offering of the tithe, and the buyer the second tithe, regardless of whether the seller is a retailer or a wholesaler. This condition was imposed by the court.

6. If two brothers, one an associate scholar and the other unlearned, have inherited produce from their father who was unlearned, the former may say to the latter, "You take the wheat in this place, and I will take the wheat in that place," or "You take the wine in this place, and I will take the wine in that place." He may not, however, say, "You take the wheat, and I will take the barley," or, "You take the liquid produce, and I will take the dry produce," because this is tantamount to selling doubtfully tithed produce.

7. If a person carrying vegetables feels that the load has become too heavy for him, and wishes to discard some of them on the roadway in order to lighten his burden, he may not throw away any vegetables until he has tithed them, so that they would not entrap unlearned persons into eating of them while they are doubtfully tithed.

8. If in buying vegetables in the marketplace one performs the act of drawing, even though he did not yet weigh or measure them, nor pay money for them, and then decides to return them to the shopkeeper, he may not return them until he has first tithed them.

9. If one finds produce on the roadway, the rule is as follows: If the majority of the people customarily store their produce in their homes, he is exempt from tithing it, because it has not yet been designated for tithing. If the majority of the people are wont to sell their produce in the market place, it is accounted doubtfully tithed produce; if the two groups of people are equal, the produce is accounted doubtfully tithed.

10. If having found the produce on the roadway one picks it up in order to eat it, and then decides to store it, he may not delay tithing it, in order that it should not become a stumbling block for others. If, however, one picks it up at the outset in order to prevent it from going to waste, he may keep it until he decides to eat it, or to send it to someone, or to sell it, and must then tithe is as doubtfully tithed produce.

11. Vegetable trimmings found in the garden are not subject to the law of doubtfully tithed produce; if trimmed by the householder and found in the house, they are subject to that law. Trimmings found in the refuse heap are permitted everywhere.

12. If one hands produce to the hostess of an inn in order to have her cook it or bake it for him, he must first tithe what he hands to her, so as not to place a stumbling block before others, and must then tithe again what he receives from her, because she is suspected of exchanging one person's produce for another's. He may, however, hand produce to his mother-in-law, whether he has wed her daughter or merely espoused her, or to a woman neighbor, to bake bread or cook a dish, and need not feel any apprehension about tithe or about produce of the Sabbatical year, because these women are not suspected of exchanging produce entrusted to them.

When does this apply? When one hands to the woman leaven for the dough or spices for the pot as well. If he does not hand these to her, he should feel apprehensive about tithes and about the Sabbatical year. Therefore, if the year is a Sabbatical one, he is forbidden to hand these to the woman, lest the leaven used by her should be derived from the aftergrowth of the Sabbatical year.

13. If one brings his wheat to a miller who is an unlearned person, the flour he receives is presumed to be from his own wheat, because the miller is not suspected of exchanging one customer's wheat for another's. If he brings it to a heathen miller, the flour is accounted as from doubtfully tithed produce, because the miller might have exchanged the wheat for wheat belonging to an unlearned customer.

Similarly, if one deposits his produce with an unlearned bailee, it is presumed, when reclaimed, to be the very same produce, because the bailee is not suspected of exchanging one bailment for another.

14. An unlearned person is permitted to serve as a clerk in an associate scholar's shop, even if the shopkeeper merely goes in and out, and no apprehension need be felt that the clerk might exchange one batch of produce for another.

15. If one deposits his produce with a heathen, it is accounted the same as the heathen's own, because the latter is presumed to exchange one bailment for another.

What then is the rule governing it? If the preparation of the produce had not yet been completed, and is thereupon completed in the Israelite's hand after he has reclaimed his deposit, he must set aside his tithes, as we have explained.

If the deposited produce is untithed, and its preparation has been completed, the owner must likewise set aside tithes—perchance the heathen did not exchange it for other produce. It would appear to me therefore that in this case the tithes set aside by the owner are subject to doubt. If, however, the owner had deposited regularized profane produce, he need not set aside any tithes at all, because even if the heathen had exchanged that produce for other produce, it is exempt, as we have explained in the Laws Concerning Heave Offerings, since Scripture says, *thy corn* (Deut. 14:23), not "a heathen's corn."

CHAPTER XII

1. If having purchased produce from someone who is not trustworthy in regard to tithes, one neglects to tithe it and is overtaken by the Sabbath or by a festival, when he may not tithe, he should ask the seller whether the produce has been tithed, and if the seller says, "It has been tithed," he may eat of it on that Sabbath on the basis of the seller's statement.

Similarly, if some other untrustworthy person says to the purchaser, "This produce has been tithed," the latter may eat of it on that Sabbath on the basis of that statement, even if he has other produce of that same species that has been regularized, because the awe of the Sabbath lies upon all unlearned persons, and they are not likely to transgress against it.

2. Even though one may eat of such produce on the Sabbath on the basis of an unlearned person's statement, one may not eat of it after nightfall of the Sabbath day, until he has set aside the tithe due from doubtfully tithed produce for everything, what he had eaten during the Sabbath as well as what still remains, because the Sages' leniency in trusting the unlearned person extends only to what is needed for that Sabbath.

If the Sabbath is followed immediately by a festival day, and the unlearned person is questioned about tithing on the first of these two days, the purchaser may eat of the produce on the second day as well, because there is no opportunity to tithe the produce in the interval between the two days. The same rule applies to the two days of a festival as observed in the Dispersion.

3. If A adjures B to eat with him on the Sabbath, and B does not deem A trustworthy in regard to tithes, B may inquire of A whether the food had been tithed, and may then eat of it on the strength of A's assurance, but only on the first Sabbath. On the second Sabbath, however, even if A has vowed to enjoy no benefit from B if he will not eat with him, B may not eat with A until he has first set aside the tithe due from doubtfully tithed produce.

4. If a person who is deemed untrustworthy in the matter of tithes is observed setting aside heave offering of the tithe from his doubtfully tithed produce, and if we then see the offering drop back in our presence, either into another place or into its former place, the rule is that if he says again, "I have set aside the tithe once again," he is to be believed even on a weekday, and one may eat of the produce on the strength of his assurance, because just as the awe of the Sabbath lies upon unlearned persons, so does the fear of produce mixed with heave offering lie upon them, and they are not suspected of serving such produce to guests.

5. If an untrustworthy person seen setting aside the first tithe from some produce says that he had set aside therefrom the second tithe also, he is to be believed. If, however, having set aside the second tithe in our presence, he says that he had set aside the first tithe also, he is not to be believed, because the second tithe remains his. Thus he who is deemed trustworthy concerning the second tithe is not considered trustworthy concerning the first tithe, but he who is deemed trustworthy concerning the first tithe is considered trustworthy concerning the second tithe.

Therefore, if a person who is not trustworthy concerning tithes takes some produce out of his house and says, "This batch is first tithe," he is to be believed, and no heave offering and tithes need be set aside from it. But if he says, "This batch is second tithe," he is not to be believed regarding the first tithe and the heave offering of the tithe, and that batch has the status of doubtfully tithed produce from which heave offering and tithes are yet to be set aside. It would appear to me that he must redeem all of that produce.

6. If A says to B, who is not trustworthy in the matter of tithes, "Buy some food for me from someone who sets aside tithes," and if B accordingly goes forth, buys the food, and brings it to A, B is not to be believed. But if A says to B, "Buy it from C," B is to be believed when he says, "I bought it from that same person," because he would be afraid that A might inquire of C about it.

If B goes forth to buy the food from C for A, and upon returning says, "I did not find C, and have therefore bought it for you from someone else who is equally trustworthy," he is not to be believed.

7. If A, having entered a city where he knows no one, inquires, "Who is deemed trustworthy here? Who sets aside tithes here?", and if B replies, "I am such a one," B is not to be believed. If B says, "C is such a one," B is to be believed, and A may buy food from C, even though he does not know him, and may proceed to eat of it on B's assurance.

If A, having gone forth and bought the food from C, asks him, "Who is here that sells old wine?", and if C replies, "He who sent you to me," both B and C may be believed, even though they appear to be favoring each other.

8. When does this apply? When A does not know anyone at all in that city. If, however, he does know someone there, he may buy food only from a seller who is reputed by all to be trustworthy. If A tarries there for thirty days, he may buy food only from such a seller, even if he still does not know anyone there.

9. These procedures were declared by the Sages to be permitted only in the case of heave offerings and tithes. In cases touching Sabbatical year's produce or cleanness, one may buy food only from a seller known by all to be trustworthy.

10. In the case of ass drivers who have entered a city, if one of them says, "This my produce has not been tithed, but my companion's produce has been tithed," he is not to be believed—perchance they are acting in collusion.

11. If a person selling produce in Syria says, "It is an import from the Land of Israel," the purchaser must tithe it. If the seller says, "It has already been tithed," he is to be believed, because the mouth that forbids is the mouth that permits. If the seller says, "It is from my own field," he must tithe it, but if he then adds, "and it has already been tithed," he is to be believed, because again the mouth that forbids is the mouth that permits. If it is known that the seller has land in Syria and that most of what he sells comes from his own field, the purchaser must tithe it, because the presumption is that the produce had been brought by the seller from his own field.

12. If poor men say, "This produce is gleanings," or "forgotten sheaves," or "corner crops," they are to be believed during the entire harvest season, when gleanings, forgotten sheaves, and corner crop are available; provided that the poor man is located near enough to the threshing floor so that he can go back and forth within the same day.

If poor men say, "This produce is poor man's tithe," they are to be believed all year long, but only in regard to things which people customarily give them.

13. How so? If poor men say, "This wheat is gleanings," or "forgotten sheaves," or "corner crop," they are to be believed. If they say, "This flour is gleanings," or "forgotten sheaves," or "corner crop," they are not to be believed. Needless to say, they are not to be believed in regard to bread when they say that it is a gift to the poor; rather the bread must be tithed as doubtfully tithed produce.

14. Poor men are to be believed in regard to unhusked rice but not husked rice, whether raw or cooked. They are to be believed in regard to whole beans, but not beans, whether raw or cooked, that have been pounded. They are to be believed in regard to oil when they say, "It is poor man's tithe," but not when they say, "It was pressed from olive gleanings."

They are to be believed in regard to raw vegetables but not cooked ones, unless the quantity is very small, for it is the householder's custom to give only a few cooked vegetables out of his pot to the poor man. And inasmuch as the poor man may say, "A householder gave it to me," he may likewise say, "I have cooked it out of the gifts given to me."

15. Similarly, if a Levite says, "This produce is first tithe from which heave offering of the tithe has already been taken," he is to be believed at all times as far as the heave offering of the tithe is concerned, just as an Israelite is to believed in regard to the great heave offering. But the Levite is not to be believed in regard to the exemption of that produce from the second tithe.

16. All these rules apply only to an unlearned person who is neither suspect nor trustworthy. As for one who is suspected of selling heave offering as profane produce, it is forbidden to buy from him anything at all that involves liability to heave offering and second tithe, down to entrails of fish, because they are usually seasoned with oil.

This prohibition is limited to that which the suspected person has in front of him. If he has stored his produce, however, it is permitted to buy it from him, for he would be afraid to mingle heave offering with his stored produce, lest this should become known, and he would lose all of it.

Similarly, in the case of a person who is suspected of selling second tithe as profane produce, it is forbidden to buy from him anything that involves liability to second tithe. All these prohibitions represent a penalty imposed by the Sages.

17. If a person suspected in regard to tithes offers testimony against other people's produce, he is to be believed, on the presumption that a person—needless to say, an unlearned person—would not sin for the benefit of someone else. Therefore, an unlearned person who says, "This produce is untithed, while that produce is heave offering," or "This produce is assuredly tithed, while that produce is doubtfully tithed," he is to be believed, even if the produce involved is his own. If he says about other people's produce, "This produce has been regularized," he is to be believed, provided that there is no evidence of collusion, as we have explained.

18. In the case of A who has sold produce to B, and who after it has left his hand says to B, "The produce I sold you is untithed," or, "The meat I sold you is that of a firstling," or "The wine I sold you is wine of libation," the letter of the law is that A is not to be believed, even if he is an associate scholar, and B, if he is a scrupulous person, will be strict with himself in this matter. Nevertheless, if B prefers to believe A, he is praiseworthy, even if A is an unlearned person.

CHAPTER XIII

1. Produce presumed to be ownerless, such as wild figs, wild jujubes, hawthorn berries, white figs, sycamore figs, fennel, dates fallen off the date palm—provided they have not yet swelled up like risen dough—caper berries, coriander, and the like, is exempt from the rules governing doubtfully tithed produce. Therefore he who buys such produce from an unlearned person need set aside out of it neither heave offering of the tithe nor second tithe, because they are presumed to be ownerless. Even if the unlearned person tells the purchaser that the produce is untithed,

it is nevertheless exempt from tithes, until the purchaser learns that it is indeed part of a guarded crop.

2. Early and late produce growing in a valley is exempt from the rules governing doubtfully tithed produce; if it grows in a garden, it is liable, because it is guarded.

Early produce is produce ripened prior to the stationing of a watchman over the valley to guard its crops. Late produce is produce which remains after the drying mats in the fields have been rolled up, and which is left without a guard over it.

Similarly, vinegar made from grape refuse is exempt from the rules governing doubtfully tithed produce.

3. The decree issued concerning doubtfully tithed produce was applied only to the produce of the land occupied by those who had come up from Babylonia, which is from Chezib and inland, with Chezib itself regarded as outside the boundary. All produce found from Chezib and outward is exempt from that decree, because it is presumed to have originated in the locality in which it is found.

4. Produce known to have originated in the land occupied by the returnees from Babylonia, even if found in Syria, or needless to say, if found in land occupied only by those who had come up from Egypt, is liable to the rules governing doubtfully tithed produce, and must have heave offering of the tithe and second tithe set aside out of it.

Therefore, juicy fig cakes, the like of which is found only in the land occupied by those who had come up from Babylonia, and also large dates, straight-shaped carobs, pure white rice, and large cumin must be tithed as doubtfully tithed produce in all of the Land of Israel as well as in Syria. The same applies to all produce similar to these.

5. Ass drivers who bring produce to Tyre are subject to the rules governing doubtfully tithed produce, because the presumption is that their produce originated in the land occupied by those who had come up from Babylonia, which is adjacent to Tyre. As for rice, no apprehension need be felt about it; rather all rice found in the land outside of the Land of Israel, even if

adjacent to land occupied by those who had come up from Babylonia, is exempt from the rules governing doubtfully tithed produce, unless it is evident that the rice was imported from the Land of Israel, as we have explained.

6. He who purchases produce from storehouse keepers in Tyre is exempt from the rules governing doubtfully tithed produce, and it cannot be said, "Perchance they have stored produce from the Land of Israel."

Similarly, a single ass entering Tyre laden with produce is exempt from these rules, because the presumption is that the produce was grown in a field belonging to the city.

7. If one purchases produce from storehouse keepers in Sidon, he is liable to the rules governing doubtfully tithed produce, because Sidon is nearer to the Land of Israel than Tyre, and therefore the presumption is that they would store produce from land occupied by those who had come up from Babylonia.

If, however, one purchases produce from ass drivers in Sidon, he is exempt from these rules, because the presumption is that they import produce from land outside of the Land of Israel.

8. In the case of one who purchases produce from a heathen in the land occupied by those who had come up from Babylonia, if the heathen merchant usually buys his produce from Israelites, it is regarded as doubtfully tithed. Therefore, in early times, when most of the Land of Israel was in the hands of Israelites, the rule was that one who purchased produce from a heathen merchant had to tithe it as doubtfully tithed produce, the same as if he had purchased it from an unlearned Israelite.

9. Who is regarded as a merchant? He who brings in produce for sale twice or three times. If he brings it in only once, however, even if he brings in a triple load at one time, or if he, his son, and his laborer bring in their loads together, he is still not presumed to be a merchant.

10. When the decree concerning doubtfully tithed produce was issued, it was not extended to produce grown in lands outside of the Land of Israel and imported into the Land of Israel.

11. In the case of kinds of produce grown in the Iand of Israel that are in greater supply than those imported from lands outside of the Land of Israel, all of them come under the rules governing doubtfully tithed produce. In the case of kinds of produce imported from lands outside of the Land of Israel that are in greater supply than those grown in the Land of Israel, and similarly in the case of kinds of produce presumed to be always imported from lands outside of the Land of Israel, such as walnuts and Damascene jujubes, all of them are exempt from these rules.

12. In the case of produce imported from lands outside of the Land of Israel, the Sages have ruled not by appearance nor by flavor or odor but rather by the greater supply. When imported produce is in greater supply, it is permitted without tithing; if it is in lesser supply, it is forbidden. How so? If the imported produce is prevalent in the city but not in the countryside, or vice versa; on the hill but not in the valley, or vice versa; with shopkeepers but not with householders, or vice versa—in all these cases the rule is that where imported produce is prevalent, all such produce is permitted; where it is not prevalent, all such produce is subject to the rules governing doubtfully tithed produce.

13. If imported produce in the market place is first in greater supply and then in lesser supply, the market place reverts to its original status, and produce purchased there is subject to the rules of doubtfully tithed produce.

14. The following kinds of produce were excluded from the decree governing doubtfully tithed produce: produce purchased for seed or for cattle feed; flour for treating hides or for use as poultices and plasters; oil for lighting lamps or lubricating vessels; wine for eye salve; dough offering of the unlearned person; produce liable to heave offering that has become mixed with heave offering; produce bought with redemption money of second tithe; residue of meal offerings; and addition to first fruits—all these are exempt from the rules governing doubtfully

tithed produce. Once the purchaser is told by the unlearned seller, "This produce has been regularized," he need not tithe it.

15. If one purchases produce for his own consumption, and then changes his mind and decides to feed it to cattle, he may neither sell it to a heathen nor feed it to cattle, not even to other persons' cattle, until he has first regularized it as doubtfully tithed produce.

16. Perfumed oil is exempt from the rules governing doubtfully tithed produce, since the presumption is that it will not be used for internal consumption. The same applies to oil bought by the comber for application to wool, because it is absorbed by the wool. Oil bought by the weaver, however, for application between his fingers is liable to these rules, because it becomes absorbed in his body, and anointing is the same as drinking everywhere.

17. If a heathen asks an Israelite to apply oil to his wound, the Israelite is forbidden to use assuredly untithed oil but is permitted to use doubtfully tithed oil.

If a heathen applies oil to a board in order to roll upon it, an Israelite may sit on that board after the heathen has arisen from it.

18. If doubtfully tithed oil has spilled on one's flesh, he may wipe it off and need feel no apprehension about it.

If one purchases wine for use in fish brine or as an unguent, or pulse for grinding into meal, he is liable to the rules governing doubtfully tithed produce. If however, one buys ready-made fish brine or unguent containing wine, or ready-ground pulse, they are exempt from these rules, because these rules do not apply to such ready-made mixtures. If, however, the ingredient in the mixture that is subject to these rules is such as spices or leaven, it does not become neutralized so long as its flavor is recognizable, and therefore the entire mixture becomes liable as doubtfully tithed produce.

19. In the case of all these varieties that are exempt from the rules governing doubtfully tithed produce and have been ex-

cluded from the decree, if one has nevertheless regularized them as if they were doubtfully tithed, and has set aside heave offering of the tithe and second tithe out of them, what he has done is done. If, however, he has regularized doubtfully tithed produce as if it were assuredly untithed, and has set aside the great heave offering and the tithes out of it, or if he has regularized assuredly untithed produce as if it were doubtfully tithed, he has done nothing.

20. If all the townspeople sell assuredly untithed produce, except one person who sells doubtfully tithed produce, and if the buyer does not know from which one he had made his purchase, how should he regularize it? He should set aside heave offering and heave offering of the tithe, and give them to the priest; he should then set aside the second tithe only, which is accounted the same as the second tithe out of doubtfully tithed produce.

21. Similarly, if one had before him two baskets of produce, one untithed and the other regularized, and one of them was lost, he should set aside the great heave offering and the heave offering of the tithe out of the remaining basket, and give them to any priest that he may choose; he should then set aside the second tithe only, as he would do with doubtfully tithed produce.

CHAPTER XIV

1. If one makes a purchase from a wholesale provision merchant, and then another purchase from him once again, he may not set aside tithe out of the one purchase for the other, even if both are of the same species; not even if they have come from the same hamper, nor even if he recognizes that the jar is the same, because the wholesale merchant buys his produce from many persons and then resells it—perchance the produce sold first was bought from an unlearned person, whose produce is doubtfully tithed, while the produce sold last was bought from an associate scholar whose produce is assuredly regularized. And we have already explained that one may not tithe out of produce

that is liable to tithing for produce that is exempt, or vice versa. If, however, the wholesale merchant says that the produce had come from one and the same source, he is to be believed.

2. If the merchant is selling chate melons or vegetables as they are brought in by the growers and heaped up in front of him, the buyer must set aside tithe from each individual melon or bundle, or even from each individual date.

3. If one makes a purchase from a householder, and then another purchase from him once again, he may set aside tithe from one purchase for the other, even if they have come from separate hampers, or even from different towns, for the presumption is that the householder is selling his own produce only.

4. If a householder is selling vegetables in the market place, the buyer may set aside tithe out of one purchase for all the others, provided that the vegetables have been brought in to the seller from his own gardens. If they have been brought in to him from other people's gardens, and the buyer makes one purchase from him and then another once again, he may not tithe out of one purchase for the other.

5. If one buys loaves of bread from a baker, he may not tithe out of freshly baked bread for stale bread, for I say that yesterday's wheat may have been brought in by one grower and today's wheat by another.

6. A person who purchases bread from a bread shop may tithe out of one loaf for all the others, even if the loaves are baked in different molds, for the baker who sells his bread to the bread shop casts his dough in several different molds. He, however, who buys from the monopolist must tithe the loaves from each mold separately, for the monopolist buys his bread from at least two bakers.

7. In the case of nine monopolists who buy their bread from ten bakers, the rule is that inasmuch as at least one of the monopolists must have bought his bread from two bakers, the person who buys bread even from only one of the nine monopolists must tithe the loaves from each mold separately.

8. If a person buys bread pieces from a poor man, and likewise if a poor man is given pieces of bread, or slices of fig cake, he must tithe every piece. In the case of dates and dried figs, however, he may lump them together and then tithe them. When may he do so? When the total of gifts is large. If the total is small, he must tithe each individual gift separately.

9. If the recipient has broken up the bread into crumbs, or pressed the dried figs into fig cakes, he may tithe out of one piece for all others.

10. If laborers or guests who had been reclining at the table and eating have left some pieces, the recipient must tithe each piece separately.

TREATISE V

LAWS CONCERNING SECOND TITHE AND FOURTH YEAR'S FRUIT

Involving Nine Commandments,
Three Positive and Six Negative
To Wit

1. To set aside the second tithe;
2. Not to spend the redemption money of this tithe for any necessaries other than food, drink, and oil for anointing;
3. Not to eat of the second tithe while in a state of uncleanness;
4. Not to eat of it while in mourning;
5. Not to eat of the second tithe of grain outside Jerusalem;
6. Not to consume the second tithe of the vintage outside Jerusalem;
7. Not to consume the second tithe of oil outside Jerusalem;
8. That the fourth year's fruit shall all be holy, and by law shall be eaten by its owner in Jerusalem, in all respects like the second tithe;
9. To pronounce the confession of the tithe.

An exposition of these commandments is contained in the following chapters.

NOTE

In the list of 613 commandments prefixed to the Code, those dealt with in the present treatise appear in the following order:

Positive commandments:

[1] 128. To set aside the second tithe, to be eaten by its owner in Jerusalem, as it is said: *Thou shalt surely tithe all the increase of thy seed* (Deut. 14:22). The Sages have learned by tradition that this refers to the second tithe;

[8] 119. That the fourth year's fruit shall be holy, as it is said: *And in the fourth year all the fruit thereof shall be holy, for giving praise unto the Lord* (Lev. 19:24);

[9] 131. To pronounce the confession of the tithe, as it is said: *Then thou shalt say before the Lord thy God, I have put away the hallowed things* (Deut. 26:13).

Negative commandments:

[2] 152. Not to spend the redemption money of the second tithe on anything other than food and drink, as it is said: *Nor given thereof for the dead* (Deut. 26:14). Anything other than things necessary for the needs of the living body comes within the term "given for the dead";

[3] 150. Not to eat of the second tithe, even in Jerusalem, in a state of uncleanness, until it has been redeemed, as it is said: *Neither have I put away thereof, being unclean* (*ibid.*);

[4] 151. Not to eat of the second tithe while in mourning, as it is said: *I have not eaten thereof in my mourning* (*ibid.*);

[5] 141. Not to eat of the second tithe of grain outside Jerusalem, as it is said: *Thou mayest eat within thy gates the tithe of thy corn* (Deut. 12:17);

[6] 142. Not to consume the second tithe of the vintage outside Jerusalem, as it is said: *Or of thy wine* (*ibid.*);

[7] 143. Not to consume the second tithe of oil outside Jerusalem, as it is said: *or of thine oil* (*ibid.*).

CHAPTER I

1. After the first tithe has been set aside each year, one must set aside also the second tithe, as it is said, *Thou shalt surely tithe all the increase of thy seed* (Deut. 14:22).

In the third and sixth years of each septennate the poor man's tithe must be set aside instead of the second tithe, as we have explained.

2. The first of Tishri is New Year's Day for the tithing of grain, pulse, and vegetables, and wherever New Year's Day is mentioned, it refers to the first of Tishri. The fifteenth of Shebat is New Year's Day for the tithing of trees.

How so? Grain and pulse that have reached the season of tithing before the New Year of the third year, even if fully ripened and harvested in the third year, are liable to second tithe; if they do not reach the season of tithing until after the New Year of the third year, they are liable to poor man's tithe.

Similarly, fruit of trees that has reached the season of tithing before the fifteenth of Shebat of the third year, even if fully ripened and gathered thereafter at the end of that year, is tithed as of the past year, and is liable to second tithe. Likewise, if it has reached the season of tithing before the fifteenth of Shebat of the fourth year, even if fully ripened and gathered in that year, it is liable to poor man's tithe; if it reaches the season of tithing after the fifteenth of Shebat, it is tithed as of the following year.

3. Carob trees are tithed as of the following year, even if their fruit is fully formed before the fifteenth of Shebat, because they are liable to tithing by Scribal law only.

It would appear to me that this rule applies solely to the carobs of Zalmonah and similar varieties which are considered unfit for consumption by most people, and these are the ones which are liable to tithing by Scribal law only. The other varieties of carobs, however, it would seem to me, are like any other fruit of trees.

4. A vegetable is tithed as of the time when it is gathered. How so? If it is gathered on New Year's Day of the third year,

it is liable to poor man's tithe, even if it had reached the season of tithing and was completely ripe in the second year. If it is gathered in the fourth year, it is liable to second tithe.

5. Similarly the citron, alone of all the fruit of trees, is considered the same as a vegetable and is subject to the time of its gathering in regard to both tithe and Sabbatical year. How so? If it is gathered in the third year after the fifteenth of Shebat, it is liable to poor man's tithe, even if it was fully ripe in the second year. Likewise, if it is gathered in the fourth year before the fifteenth of Shebat, it is liable to poor man's tithe. If it is gathered in the fourth year after the fifteenth of Shebat, it is liable to second tithe.

6. Although the citron is subject to the time of its gathering, nevertheless, if it grows in the sixth year but enters the seventh, it is liable to tithes, even if it was no larger than an olive at the end of the sixth year and thereafter grew to be as large as a loaf of bread.

7. The berries of the caper bush are subject to the more rigorous rules affecting both trees and seed plants, so that if they are of the second year going into the third, and have been gathered after the fifteenth of Shebat, one must set aside first tithe out of them and thereafter second tithe. One must then redeem the latter, and after redemption give it to the poor; while he himself may use the redemption money for his own consumption as if it were the redemption money of second tithe. The result is that he has as much as set aside both second tithe and poor man's tithe.

8. Rice, durra, millet, and sesame, even if rooted before the New Year, are subject only to the time when the fruit is fully grown and are tithed as of the following year. The same applies to cowpea, even if it takes root partly before New Year's Day and partly after—one may collect the whole crop into one heap and set aside heave offering and tithes out of all of it together, since all of it is subject to the time when the fruit is fully grown.

9. Seedless onions cut off from water thirty days before New Year, and seed onions deprived of water for three periods of

watering before New Year, are tithed as of the past year. If the water has been withheld from them for a lesser time, they are tithed as of the following year, even if they had started to dry before New Year's Day.

10. In the case of cowpea that is one-third grown before New Year's Day, the rule is as follows: If it is planted for seed, it is tithed as of the past year; if it is planted for consumption as a vegetable, it is tithed as of the following year. If it is planted both for seed and for vegetable, or if one plants it for seed but thinks of using it also as a vegetable, one may tithe out of its seed for its vegetable and vice versa.

If it is not one-third grown before New Year's Day, the rule is as follows: If it is planted for seed, the seed is tithed as of the past year, and the vegetable is tithed as of the time when it is gathered, provided that some of it is gathered before New Year's Day; if it is gathered after New Year's Day, both the seed and the vegetable are tithed as of the following year.

If one plants it for seed but then thinks of using it entirely as a vegetable, the rule follows his thought. If he plants it as a vegetable but then thinks of using it entirely for seed, the thought of seed does not supersede the original intent, unless the plant is deprived of water for three periods of watering, and provided also that it is one-third grown before New Year's Day. If, however, it is not one-third grown until after New Year's Day, and even if it is not deprived of water for three periods of watering, the rule is that if it was planted for seed and has brought forth fully developed pods before New Year's Day, the seed is tithed as of the past year and the vegetable as of the time when it is gathered. If only some pods are fully developed while some are not, the rule of "one may collect the whole crop into one heap and set aside heave offering and tithes out of all of it" comes into effect, with the result that one may set aside tithes out of the seed for the vegetable and vice versa.

11. If second-year produce is intermixed with third-year produce, or third-year produce with fourth-year produce, the rule follows the majority. If the mixture is half-and-half, one must set

aside second tithe for the entire mixture, but not poor man's tithe, because second tithe is the more stringent one, since it is holy, whereas poor man's tithe is profane.

Similarly, when there is doubt as to whether some produce is of the second year or of the third year, one must set aside second tithe out of it.

12. Produce exempt from first tithe is exempt also from second tithe and from poor man's tithe; and contrariwise, produce liable to first tithe is liable also to the other two tithes. Whosoever is eligible to set aside heave offering is eligible also to set aside tithes, and whosoever is not eligible to set aside heave offering may not likewise set aside these tithes. In the case of ineligible persons whose heave offering, should they nevertheless set it aside, is valid, so are their tithes, second and poor man's, valid, should they nevertheless set them aside. If their heave offering is invalid in any case, so are their aforementioned tithes invalid, should they set them aside nevertheless.

13. If one sets aside first tithe out of produce that has not yet been designated for tithing, he may eat an occasional meal out of this produce before he sets aside second tithe, because first tithe does not designate that produce for second tithe. Once the produce has been designated for tithing, however, one may not eat an occasional meal out of it, even if first tithe has already been set aside, until he has set aside also second tithe or poor man's tithe.

14. The second tithe, inasmuch as it involves bringing it to Jerusalem, may not be imported from outside of the Land of Israel, the same as the firstling of cattle. The Sages have therefore not made it obligatory to set aside second tithe in Syria.

It would appear to me that second tithe set aside in the Land of Shinar and in Egypt should be redeemed and the redemption money should be brought to Jerusalem.

It would appear to me also that the Sages have made it obligatory to set aside second tithe in these two countries for the sole purpose of thereby causing produce to be designated for poor man's tithe, so that the poor in Israel might depend upon it.

CHAPTER II

1. The second tithe must be consumed by its owner inside the wall of Jerusalem, as it is said, *And thou shalt eat before the Lord thy God, in the place which He shall choose to cause His name to dwell there* (Deut. 14:23).

Second tithe is obligatory both when the Temple in Jerusalem is in existence and when it is not, but the tithe may be consumed in Jerusalem only while the Temple is extant, as it is said, *The tithe of thy corn, of thy wine, and of thine oil, and the firstlings of thy herd and of thy flock* (*ibid.*). The Sages have learned by tradition that just as the firstling may be eaten only while the Temple is in existence, so may the second tithe be eaten only while the Temple is in existence.

2. It is the practice of piety in these days to redeem the second tithe at its full value, just as was done while the Temple was standing. The Geonim, however, have ruled that today, if one wishes at the outset to redeem it at the rate of one pĕruṭah for one mina, he may do so, since this should not be subject to greater stringency than consecrated things. One must thereupon cast the pĕruṭah into the Great Sea.

3. Similarly, if one renders tithe worth one mina profane by substituting for it other produce worth one pĕruṭah, the substitution is valid, and he must then burn the substitute produce, so that it might not become a snare for others, the same as is the case of the redemption of the fourth year's fruit at the present time, as we have explained in the Laws Concerning Forbidden Foods.

4. Just as today one may not eat second tithe in Jerusalem, so also may one not redeem it there, nor render it profane, nor sell it. If one enters Jerusalem with second tithe, even today, he may not take it out of the city, but must leave it there until it rots. Similarly, if one transgresses and does take it out, he must leave it until it decays.

Therefore, at present, second tithe may not be set aside in Jerusalem; rather one must remove the untithed produce out of the city, and there set aside second tithe and redeem it. If one nevertheless does at present set it aside in Jerusalem, he must leave it to rot.

5. Any person who eats an olive's bulk of second tithe, or drinks one quarter of a loḡ of wine of second tithe, outside of the wall of Jerusalem is liable to a flogging, as it is said, *Thou mayest not eat within thy gates the tithe of thy corn, or of thy wine, or of thine oil,* etc. (Deut. 12:17). He is liable to a flogging for each of these three varieties separately. Therefore, if he consumes the minimum quantity of all three of them together outside of the wall, he is liable to a triple flogging, as it is said further on, *And thou shalt eat before the Lord thy God . . . the tithe of thy corn, of thy wine, and of thine oil* (Deut. 14:23). Now in the earlier verse, *Thou mayest not eat within thy gates the tithe of thy corn,* etc., why does Scripture enumerate the three varieties once more, instead of saying "Thou mayest not eat them within thy gates"? In order to make one liable for each one separately.

6. One is not liable to a flogging, according to the Torah, unless he eats second tithe after it had been brought within the wall of Jerusalem, as it is said, *Thou mayest not eat within thy gates . . . but thou shalt eat them before the Lord thy God in the place which the Lord thy God shall choose* (Deut. 12:17–18), implying that once second tithe has been brought into the place where it should be eaten, he who eats of it outside of that place is liable to a flogging. If, however, he eats of it before it is brought into Jerusalem, he is liable only to a flogging for disobedience, on Scribal authority.

7. If part of the second tithe is inside Jerusalem and part outside, the rule is that he who eats of the part that has not yet been brought into Jerusalem is liable to a flogging for disobedience, while he who eats outside of Jerusalem of the part that has entered Jerusalem is liable to a flogging on the authority of the Torah.

8. One may not redeem second tithe in Jerusalem unless it has become unclean, as it is said, *And if the way be too long for thee* (Deut. 14:24), implying that it may be redeemed if the owner has come to Jerusalem from afar, but not if he has come from a place nearby. If the owner is inside Jerusalem and his load of second tithe is outside, even if he holds it suspended from his staff, he may redeem it right there next to the wall, inasmuch as it has not yet entered the city.

9. Second tithe brought into Jerusalem, even if taken out of doubtfully tithed produce, is forbidden to be taken out of the city, inasmuch as the city wall has already encompassed it. The same rule applies to produce bought with the redemption money of second tithe, as it is said, *And thou shalt eat before the Lord thy God* (Deut. 14:23).

If one transgresses and takes the produce out of Jerusalem, or if it is taken out unwittingly, it must be returned to the city and eaten there.

The rule concerning containment within the city wall is based on Scribal authority. Even tithe one-fifth of which is worth less than one pĕruṭah, which is obligatory only on Scribal authority, is regarded as encompassed, and it is forbidden to remove it. Redemption money of second tithe, however, may be brought into, and out of, Jerusalem.

10. In the case of fully processed produce that has passed through Jerusalem and out of the city, the owner may not set aside second tithe for it out of other produce which has not yet entered Jerusalem. He must rather return the second tithe out of the same produce to Jerusalem and eat it there. Nor may it be redeemed outside Jerusalem, even if the owner makes all of this produce, after it has been taken out of Jerusalem, second tithe for some other produce which has not yet entered Jerusalem—he must return all of it to Jerusalem and eat it there. It is a stringent rule that governs the walls of Jerusalem, namely that once they have encompassed something, it remains encompassed.

11. In the case of incompletely processed produce that has passed through Jerusalem and out of the city, such as baskets of grapes

going to the wine press or baskets of figs going to the drying shed, second tithe thereof may be redeemed outside Jerusalem.

Similarly, second tithe of doubtfully tithed produce may be redeemed outside Jerusalem, even if the produce has been completely processed and had passed through Jerusalem and gone out of the city.

12. If second tithe becomes unclean in Jerusalem, and the owner redeems it, the rule is as follows: If the uncleanness was caused by a secondary source of defilement, the tithe may not be taken out of Jerusalem, and the owner must consume its redemption money within the city, because secondary defilement is based on Scribal authority only. If the uncleanness was caused by a primary source of defilement, or outside of Jerusalem even by a secondary source of defilement, the tithe may be redeemed and the redemption money consumed in any place, even if the tithe had already entered Jerusalem.

13. When does this apply? When the tithe is brought into Jerusalem with the stipulation that the city walls are not to encompass it. If this condition was not stipulated, once the tithe has entered Jerusalem, and inasmuch as according to the Torah it is clean, since according to the Torah a secondary source of defilement cannot render a third object unclean, it is consequently regarded as having been encompassed by the city walls, and may not leave the city.

14. One may eat second tithe fenugreek in its budding stage, because it is only in this stage that it is fit for human consumption. The same applies to second tithe bitter vetch. If vetch is kneaded into dough, one may bring it into, and out of, Jerusalem, because vetch is not included in the term "produce." If it becomes unclean in Jerusalem, one must redeem it and consume its redemption money outside of the city.

15. In the case of a tree standing within the wall of Jerusalem with its foliage extending outside, one may not eat of second tithe underneath its foliage. Second tithe that has come under its foliage may not be redeemed, inasmuch as it is accounted the same as if it had entered Jerusalem.

16. In the case of houses adjacent to the wall of Jerusalem, with their entrances inside the wall and their inner space outside, the part from the wall inward is regarded as within the wall in every respect. As for the part from the wall outward, one may not eat of second tithe there nor redeem it, as a matter of stringency.

If the inner space of the houses is inside the wall and their entrances are outside, the part from the wall outward is regarded as outside, and one may redeem second tithe there but not eat of it. As for the part from the wall inward, one may not eat of second tithe there nor redeem it, as a matter of stringency.

The windows and the thickness of the wall itself are regarded as inside the wall.

CHAPTER III

1. He who while unclean eats of second tithe is liable to a flogging, as it is said, *Neither have I put away thereof, being unclean* (Deut. 26:14), whether the tithe was unclean and the consumer thereof clean, or vice versa, provided that he eats of it in Jerusalem prior to its redemption, for one is not liable for eating of it while unclean unless he does so in the place where it must be eaten. If, however, he eats of it while unclean outside of Jerusalem, he is liable only to a flogging for disobedience.

2. One may not even light a lamp with second tithe oil that has become unclean until it has been redeemed, as it is said, *Neither have I put away thereof, being unclean* (Deut. 26:14).

3. We have already explained that second tithe which has become unclean, even in Jerusalem, may be redeemed and its redemption money consumed. The produce bought with this money must be consumed in cleanness, as if it were original second tithe, as will be explained. Even if all of the produce has become unclean while untithed, one must set aside second tithe out of the unclean produce and redeem it.

4. The rule applies to an uncircumcised person as well as to an unclean person, that if he eats of second tithe, he is liable to a

flogging according to the Torah, the same as if he had eaten of heave offering. For heave offering is termed holy and second tithe is likewise termed holy, as it is said concerning it, *It is holy unto the Lord* (Lev. 27:30).

An unclean person who has immersed himself may eat of second tithe, even if the sun of that day has not yet set.

5. If one eats of second tithe while in mourning enjoined by the Torah, he is liable to a flogging, as it is said, *I have not eaten thereof in my mourning* (Deut. 26:14), provided that he eats of it in Jerusalem, where it must be eaten. If, however, he eats of it while in mourning outside of Jerusalem, or if he eats of it inside Jerusalem while in mourning enjoined by the Scribes, he is liable only to a flogging for disobedience.

6. Who is a mourner? He who mourns for one of those relatives for whom he is obligated to mourn. On the day of death he is a mourner according to the Torah; in the ensuing night he is a mourner according to Scribal law, as it is said, *And if I had eaten the sin offering today, would it have been well-pleasing in the sight of the Lord?* (Lev. 10:19), implying that during the day it is forbidden, but at night it is permitted.

If the deceased is kept several days and is only then buried, during all the days after the day of death up to the day of burial one is a mourner according to Scribal law. The day of burial, however, does not include the following night.

7. Not only in the case of second tithe but also in the case of all holy things, if one eats of them while in mourning enjoined in the Torah, he is liable to a flogging; if while in mourning enjoined by Scribal law, he is liable only to a flogging for disobedience.

8. One may not give second tithe to an unlearned person, nor produce bought with redemption money of second tithe, nor even such money itself, because an unlearned person is presumed to be unclean.

One is permitted, however, to eat of second tithe of doubtfully tithed produce while in mourning, or to give it to an unlearned person, provided that one eats a corresponding amount of produce in Jerusalem.

One may not deposit second tithe, not even with an associate scholar, lest he should die, and the tithe should come into the possession of his heir who might be an unlearned person. One may, however, deposit second tithe of doubtfully tithed produce with an unlearned person.

9. It is forbidden to throw away second tithe, be it even a small amount thereof, on the roadways—one must rather carry even a minimal amount of it, or its monetary equivalent, to Jerusalem. One may, however, throw away on the roadways a small amount of second tithe of doubtfully tithed produce. What constitutes a small amount? Less than the bulk of a dried fig, whether it is a complete item of food or a piece thereof. If, however, it is the size of a dried fig, it may not be thrown away.

He who sets aside second tithe of doubtfully tithed produce, even if it is less than a dried fig's bulk, may give it to an unlearned person, but must eat a corresponding amount of produce in Jerusalem. He should not, however, at the outset set it aside in order to throw it away, for one may not set aside tithe with the intention aforethought of wasting it.

10. Second tithe has been appointed for use as food and drink, as it is said, *And thou shalt eat before the Lord thy God* (Deut. 14:23). Anointing is considered the same as drinking.

One may not, however, spend it for any other need, as for example, to buy vessels, clothing, or servants therewith, as it is said, *Nor given thereof for the dead* (Deut. 26:14), meaning "I did not spend it for things that do not sustain the body." If one nevertheless does spend of it for other things, even such as involve a religious duty, as for example, if he purchases with it a coffin and a shroud for an unclaimed corpse, he must consume a corresponding amount of it as second tithe.

11. In the case of second tithe, one may eat that which is usually eaten, drink that which is usually drunk, and anoint himself with that which is customarily used for anointing. One may

not use wine or vinegar for anointing, but only oil. One may not squeeze fruit in order to obtain its juice, except only olives and grapes.

One may not spice oil, but one may spice wine.

One is not obligated to eat tithed bread that has become moldy or use tithe oil that has become rancid; rather, once the tithe has become unfit for human consumption, its holiness has departed from it.

12. Whatever a commoner may eat of heave offering, he may eat also of second tithe, as if it were profane produce.

In the case of lees of second tithe over which water has been poured, the first pouring is forbidden as second tithe, while the second pouring is permitted as profane produce. In the case of doubtfully tithed produce, even the first pouring is permitted.

13. If honey and spices that have fallen into second tithe wine act to increase its bulk and improve its quality, the appreciation is calculated proportionally.

Similarly, if fish are cooked with second tithe porret, which acts to increase their bulk and improve their quality, the appreciation is calculated proportionately.

14. If second tithe dough, when baked, increases in value, the appreciation accrues to the second tithe. This is the general rule: Wherever the appreciation is recognizable, if there is increase in bulk, the appreciation is calculated proportionately; if there is no increase in bulk, the appreciation accrues to the second tithe only. Wherever the appreciation is not recognizable, it accrues to the second tithe only, even if there is increase in bulk.

15. How is the appreciation to be calculated proportionately? If honey and spices worth one zuz fall into second tithe wine worth three zuz, adding both to its bulk and to its flavor, so that now the total is worth five zuz, the appreciated worth is calculated as four and one-quarter zuz. The same proportion applies to all other produce.

16. Even though one may use second tithe oil for anointing, he may not pour it over a callus or a lichen eruption, or use it as an

ingredient in an amulet and the like, for it is not meant to be used as a medicament.

17. Second tithe is sacred property, as it is said, *And all the tithe . . . is the Lord's* (Lev. 27:30). Therefore it may not be acquired as a gift, unless the donor gives the produce to the donee while it is still untithed, and the donee then sets aside the tithe. One may not use it to betroth a woman, nor sell it, nor take it in pawn, nor exchange it, nor pawn it.

18. How may one not take it in pawn? The pledgee may not enter the pledgeor's house and seize his second tithe as a pledge, and if he transgresses and does so, the pledge may be repossessed from his hand.

How may one not pawn it? The pledgeor may not say to the pledgee, "Here is this second tithe; let it lie in your possession, and give me money in lieu of it."

How may one not exchange it? A may not say to B, "Here is second tithe wine; give me second tithe oil in exchange for it." A may, however, tell B, "Here is second tithe wine, seeing that I have no oil." If B then wishes to give A oil, he may do so, inasmuch as A did not request an exchange with B, but merely informed B that he had no oil, and if B then wishes to give oil to A, he may do so.

19. Second tithe produce may not be used as a counterweight, even in weighing gold denars, not even in order to redeem with that gold some other second tithe. This is a precaution, lest one should deliberately use it as a regular counterweight; for the true weight of the produce might grow short, the while he is weighing against it money to render profane therewith some other second tithe, with the result that he would be rendering second tithe profane at less than its fair value.

20. Brothers who divide second tithe among themselves may not weigh their shares one against the other.

Similarly redemption money of second tithe may not be used as counterweight, nor sold, nor exchanged, nor pledged, nor given to a money-changer to advertise himself with it, nor loaned in order to exhibit the lender's wealth. If one does lend it in order to prevent the coins from tarnishing, it is permitted.

21. One may not use second tithe to pay a debt, nor to make a wedding gift, nor to reciprocate a favor, nor to make a compulsory contribution to charity as assessed in the synagogue. He may, however, use it to repay a voluntary charitable gift, but he must inform the recipient to that effect.

22. A may not say to B, "Carry this second tithe produce to Jerusalem, and take your share therefrom," because this amounts to B being paid with second tithe for carrying it to Jerusalem. A may, however, tell B, "Carry it up to Jerusalem, so that we might eat and drink of it there together."

23. A may ask B in Jerusalem to anoint him with second tithe oil, even though in the process B's hand is also anointed, for this does not constitute payment for B's service to A.

24. We have already explained that second tithe is sacred property. We say therefore that he who steals second tithe is not obligated to repay double, and he who obtains it by robbery is not obligated to pay a penalty of one-fifth.

25. If one consecrates his second tithe for the upkeep of the Temple, whosoever wishes to redeem it may do so, with the proviso that he must give to the Temple what is due to it, and to second tithe what is due to it.

CHAPTER IV

1. He who wishes to redeem second-tithe produce must redeem it at its fair value, while saying, "These moneys are to be in lieu of this produce," or, "This produce is to be rendered profane in lieu of these moneys." If he does not declare this explicitly, but merely sets aside the money against the produce, it is sufficient; he need not declare it explicitly, and the produce involved becomes thereby profane, while the money must be carried up to

Jerusalem, where he must spend it, as it is said, *And if the way be too long for thee, so that thou art not able to carry it . . . then shalt thou turn it into money,* etc. (Deut. 14:24–25).

2. Similarly, if one wishes to render profane second tithe produce by substituting other produce for it, the latter must be carried up to Jerusalem and consumed there. He may not, however, substitute one species for another, nor better quality produce for inferior quality produce, even of the same species. But if he does make such a substitution, it is valid.

3. He who redeems second tithe must recite the benediction, "Blessed art Thou, O Lord . . . who has sanctified us with His commandments and has commanded us concerning the redemption of second tithe." If he renders it profane by substituting other produce or money, he must say, "concerning rendering second tithe profane." He who redeems or renders profane second tithe of doubtfully tithed produce need not recite a benediction.

4. When second tithe is redeemed, it should be redeemed not in the name of second tithe but in the name of profane produce, and one should say, "How much is this profane produce worth?", even though everyone knows that it is second tithe produce, in order not to cast contempt upon sacred property.

5. One may not render profane second tithe redemption money by substituting for it other money, whether both moneys are silver, or both are copper, or the first money is silver and the second copper, or vice versa. But if one transgresses and performs such a substitution, it is valid.

6. Second tithe redemption money may not be rendered profane by substituting produce for it, and if one does so nevertheless, he must bring the produce up to Jerusalem and eat it there. Nor may one substitute for it live domestic or wild animals or fowl; if he does so nevertheless, the domestic animals do not become second tithe, lest he should breed herd upon herd out of them. If, however, he substitutes slaughtered animals, they are considered the same as any other produce, and the carcasses must be

brought up to Jerusalem and consumed there, while the original redemption money becomes profane.

7. In case of emergency it is permitted to exchange silver coins for copper coins, not that one should leave them thus, but rather until his situation is eased, when he must reverse his procedure and render the copper money profane by substituting silver money for it.

8. In the case of second tithe of doubtfully tithed produce, one may render it profane at the outset by substituting silver for silver, silver for copper, copper for copper, and produce for copper. The produce must then be brought up to Jerusalem and eaten there.

9. According to the Torah, one may not redeem second-tithe produce with any money except silver, as it is said, *And bind up the silver* (Deut. 14:25).

Similarly, if one has redeemed his own tithe and added the required one-fifth, that fifth must also be silver, the same as the principal.

One may not redeem with any silver that is not coined, but only with silver which has engraved upon it a figure or a legend, as it is said, *And bind up the silver*. If he redeems with a bar of silver, or the like, which is called plain bullion, he has done nothing.

One may not redeem with silver coins worth less than one pĕruṭah, for this would be the same as redeeming with plain bullion.

10. One may not redeem second tithe produce with coins that are not current at that time and in that place, as it is said, *And thou shalt bestow the silver for whatever thy soul desireth* (Deut. 14:26), implying that it must be acceptable for expenditure. Coins of past kings, if still current in their name, may be used for redemption.

11. One may not redeem second tithe produce with money that is not in his own possession, as it is said, *And bind up the silver in thy hand* (Deut. 14:25). If his purse has fallen into a

cistern, and he is able to recover it, he may redeem with the money that is in it, since it is regarded as still in his possession.

12. If one is walking along the road holding money in his hand, and sees a person given to violence coming toward him, the rule is as follows: If he can save the money by exercising great effort, he may redeem with it the produce that is in his house; otherwise, should he say, "The produce that I have in my house is to be rendered profane in lieu of this money," he has said nothing.

13. If a person sets aside money to redeem second-tithe produce therewith, he may redeem it with this money on the presumption that it is still there. If he then finds that the money is lost, he is subject to apprehension retroactively concerning all the produce that he had redeemed with it.

14. If a person located in Tiberias has money of Babylonian coinage in Babylonia, he may not render second tithe profane in lieu of that money. On the other hand, if he has money of Tiberian coinage in Babylonia, he may use it for such redemption. The same applies to all similar cases.

15. If one says, "This second tithe produce is to be rendered profane in lieu of the sela' that my hand will produce out of this purse," or "in lieu of the sela' that I will realize from changing this gold denar," or "in lieu of the pondion that I will realize from changing this sela'," the redemption is valid, and the sela' produced by his hand or received by him as change becomes second tithe.

16. On the other hand, if one says, "This second tithe produce is to be rendered profane in lieu of the sela' that is in my son's hands," the redemption is not valid—perchance the sela' was not in his son's hand at that moment.

17. If one redeems the second tithe before he has set it aside, as for instance, if he says, "The second tithe of this as yet untithed produce is to be redeemed with these moneys," he has said nothing, nor did he designate any second tithe.

If, however, he designates the tithe by saying, "The second

tithe of this produce, which tithe is on the northern side," or "on the southern side thereof, is to be rendered profane in lieu of these moneys," the second tithe is considered redeemed.

18. When second tithe is redeemed, it should be redeemed at its fair value. It should be redeemed at the lower price paid by the shopkeeper when he buys his merchandise, and not at the higher price at which he sells it. And the money should be calculated at the rate charged by the money-changer when changing large coins for small, and not at the rate charged by him when changing small coins for large. But if one transgresses and redeems a mina's worth for a pĕruṭah, or renders profane a mina's worth of produce with a pĕruṭah's worth of other produce, the redemption is valid.

19. If the particular sela' is one-sixth or less short of its full weight, the rule is as follows: If it is spendable, even though only with difficulty, one may use it at the outset to render profane produce valued at one sela', and need not feel any apprehension on that score. If he redeems the produce with a sela', and the coin proves to be bad, he must replace it with another coin.

20. Second tithe may not be redeemed by guesswork—one must rather be exact as to its measure or weight and must pay its fair value. If its value is known, it may be redeemed at the valuation of one appraiser. If it is a thing of unknown value, as for example, wine that has begun to sour, or fruit that has begun to rot, or coins that have begun to tarnish, it must be redeemed according to the valuation of three merchants. Even if one of these three is a heathen, or the owner of that same second tithe, or even if the three consist of a man and his two wives, it may be redeemed at their valuation. The owner should be compelled to make the first bid, the law being more stringent in the case of second tithe than in the case of consecrated property.

21. One may not carry second tithe produce from one place to another in order to redeem it there. If one does carry it from a place where such produce is dear to a place where it is cheap, or vice versa, he must redeem it at the price in the place where it is

redeemed. If it is doubtfully tithed produce, it may be redeemed at the lower price, inasmuch as it might have been sold at that price.

22. If one who has second tithe produce at his threshing floor goes to some expense in order to convey it from his household to the city, where its price is higher, he must redeem it at the price prevailing in the city, so that he loses his transportation expenses.

CHAPTER V

1. He who redeems his second tithe for himself, whether it is his own, or has fallen to him by inheritance, or has been given to him as a gift while the original produce was still untithed, as we have explained, must add one-fifth to its value, as for example, if it is worth four zuz, he must pay five, as it is said, *And if a man will redeem aught of his tithe, he shall add unto it the fifth part thereof* (Lev. 27:31).

2. A woman who redeems her own second tithe for herself need not add one-fifth. The Sages have learned by tradition that the words *a man . . . of his tithe* (Lev. 27:31) indicate that the law applies only to a man, not to a woman.

Similarly, one who redeems for himself the redemption money of the second tithe must add one-fifth.

3. One who having redeemed his second tithe and added one-fifth, goes back and once more redeems for himself the redemption money, must now add one-fifth of the principal only, without adding also one-fifth of the first one-fifth.

4. If one-fifth of the second tithe amounts to less than one pĕruṭah, it need not be added.

Similarly, if the value of the second tithe is not known, it is sufficient to say, "It, plus one-fifth thereof, is to be rendered profane in lieu of this sela‘."

In the case of any second tithe that one redeems with money other than his own, he need not add one-fifth. So also in the case of doubtfully tithed produce one need not add one-fifth.

5. If one who is redeeming one batch of second tithe owns also another batch not yet redeemed, the rule is that if the value of one-fifth of the latter is less than one pĕruṭah, it is sufficient for him to say, "The latter second-tithe produce, plus one-fifth thereof, is to be rendered profane in lieu of the first money," for it is impossible for a man in such a case to apportion his money with exactitude.

6. If one redeems the second tithe at more than its fair value, the excess does not assume the status of tithe.

7. If the owner of the second tithe bids one sela' for its redemption, and another person also bids one sela', the owner has priority, because he must add one-fifth.

If the owner bids one sela', and another person bids one sela' and one pĕruṭah, the other person has priority, since he has added to the principal.

8. It is permitted to use subtlety in the redemption of one's own tithe. How so? A man may say to his adult son or daughter, or to his Hebrew bondsman, "Take this money and redeem with it this tithe for yourself," in order that the owner himself should not have to add the one-fifth. He may not, however, say to him, "Redeem for me with this money."

Similarly, if he says to him, "Redeem for me with your money," he need not add the one-fifth.

9. One may not, however, give money for redemption to his minor son or daughter, or to his Canaanite bondsman or bondswoman, because their hand is the same as his own hand.

If he gives the money to his Hebrew bondswoman, the rule is as follows: If the particular second tithe is obligatory only by Scribal law, as for example, tithe from a plant in an unperforated pot, his words are valid, because a Hebrew bondswoman can only be a minor, and a minor cannot acquire title for other persons except in matters enjoined only by Scribal law.

10. A similar subtlety is for the owner to give the second tithe as a gift while the original produce is still untithed, and then say, "This produce is to be rendered profane in lieu of the money which is in my house."

11. Two brothers, two partners, or father and son, may redeem the second tithe for each other, in order to avoid adding one-fifth.

If a woman brings in second tithe produce to her husband, he does not acquire title to it, inasmuch as it is sacred property, as we have explained. Therefore, if he redeems it, he need not add the one-fifth.

12. If a person redeeming the second tithe for himself pays the principal but not the one-fifth, the rule is that although the non-payment of the one-fifth does not prevent the tithe from becoming profane, and it thus is rendered profane, nevertheless he may not eat of it until he has paid the one-fifth, even on the Sabbath, this being a precaution lest he should transgress and not pay it at all.

13. If one wishes to exchange second tithe redemption money for gold denars, which are less burdensome to carry, he may do so. If he does it for himself, he need not add the one-fifth, because this does not constitute another redemption.

14. If one changes one sela' of second tithe redemption money, whether in Jerusalem or outside of Jerusalem, he should not change it all into copper coins, but rather into one shekel in silver coins and one shekel in copper coins.

15. It is permitted to render profane silver together with second-tithe produce in lieu of silver, providing that the produce is worth less than one denar. If, however, the produce is worth one denar, it may not be rendered profane jointly with the silver, but only separately. How so? If one has second tithe produce worth one denar, and also three denar of second tithe redemption money, he may not render it all profane in lieu of one sela'. If, however, he has one half-denar's worth of produce and one half-denar in coin, he may render both profane together in lieu of one denar.

CHAPTER VI

1. If both profane money and second tithe redemption money are spilled, and are then picked up piece by piece here and there,

the rule is as follows: Whatsover is picked up is deemed second-tithe money, until the full amount thereof is complete; the rest is considered profane money. If the coins are picked up indiscriminately by the handful, or are picked up from one side only, and the total is found to be less than it should be, they should be divided proportionately.

How so? If 200 zuz of second tithe redemption money and 100 zuz of profane money are spilled, and are then picked up indiscriminately by the handful, and are found to total only 270 zuz, 180 are second-tithe redemption money and 90 are profane money.

This is the general rule: Whatsoever is picked up coin by coin is allotted first to second tithe; and whatsoever is picked up indiscriminately is divided proportionately.

In either case one should pronounce the following stipulation: "If what is in my hand is second-tithe money, the rest is to be profane money; if what is in my hand is profane money, the tithe money, wheresoever it may be, is to be rendered profane, in lieu of this money on hand."

2. If only one sela' of second-tithe money is intermixed with one sela' of profane money, one may bring a sela''s worth of other money, even copper coins, and say, "One sela' of second tithe, wheresoever it may be, is rendered profane in lieu of these coins." He should then select the better of the two sela' and render the copper coins profane in lieu of it, so that it once again becomes second-tithe money.

3. If a man says to his son, "There is second-tithe money in this corner," and it turns out to be in another corner, it is considered profane money.

If he says to him, "There is one mina of tithe money there," and it turns out to be 200 zuz, the surplus is profane money. If he says, "There are 200 zuz there," and it turns out to be one mina, it is profane money.

If he has deposited one mina of second-tithe money, and later finds 200 zuz, or vice versa, all of it is considered profane money, even if it is contained in two bags.

4. If the father says to the son, "I have a bag of second tithe money in the house," and if the son goes forth and finds three bags, the rule is that the largest of the three is considered second tithe, and the rest are profane. Nevertheless he should not consume the money in the small bags until he has rendered it profane in lieu of the money in the large bag.

5. If a person who has lost his speech is asked, "Is your second tithe in such-and-such a place?", and nods his head in assent, he should be examined three times, the same as is done in the case of divorce, and thereafter his assent is valid.

6. If one is told in a dream, "Your father's second tithe, which you are looking for, is in such-and-such a place," the rule is that even if he actually finds what he was told, it is not considered second tithe, because things connected with dreams do not matter one way or the other.

7. If a man says to his sons, "Even if you are on the verge of death from hunger, do not touch this corner," and money is found there, it is profane money. If he hides some money from them, but tells them, "It belongs to So-and-so," or, "It is second tithe money," the rule is as follows: If this is a ruse on his part, no attention need be paid to his words; if he meant it literally, his words are valid.

8. If one finds a vessel inscribed with a single letter, the nature of its contents is presumed to accord with the following table:

letter *mem* (*ma'ăśer*)—second tithe
letter *dalet̲* (*dĕmay*)—doubtfully tithed produce
letter *ṭet̲* (*ṭeb̲el*)—assuredly untithed produce
letter *taw* (*tĕrumah*)—heave offering
letter *ḳof* (*ḳorban*)—sacrificial offering

If it is a metallic vessel, both the vessel and its contents are presumed to be a sacrificial offering. For in times of danger people used to inscribe only the initial letter of the word.

9. Money found in Jerusalem, even if it includes gold denars as well as silver and copper coins, is considered to be profane, inasmuch as the streets of Jerusalem are swept every day.

If a potsherd inscribed "tithe" is found with the coins, they are deemed second tithe.

When does this apply? On most days of the year. On festivals, however, all is deemed second tithe.

10. Money found in front of cattle dealers in Jerusalem is always deemed second tithe, because the presumption is that most people bring along second-tithe money to buy cattle therewith. Money found on the Temple Mount is always deemed profane, because the presumption is that it came from the Fund of the Chamber, and was rendered profane by the treasurers in lieu of cattle.

11. If money is found in a chest used for both profane money and second-tithe money, the rule is as follows: If most people usually put second-tithe money in it, the found money is regarded as second tithe; if most people usually put profane money in it, the found money is considered profane; if it is a case of half-and-half, the found money is regarded as profane.

12. If one finds some produce between piles of second-tithe produce and piles of heave-offering produce, it is regarded as belonging to the nearest pile. It it is equidistant, it must be consumed according to the stricter rules governing both kinds, that is to say, it is forbidden to commoners, one must wash his hands before handling it, and an unclean person may not eat of it until after sunset, all these being restrictions governing heave offering; and it is forbidden to a mourner, and must be brought to Jerusalem, these being restrictions governing second tithe. The same rule applies to moneys found between piles of profane money and piles of second-tithe money.

13. If second tithe of doubtfully tithed produce and second tithe of assuredly tithed produce are intermixed, the mixture must be consumed according to the stricter of the rules governing the two kinds.

14. In the case of second-tithe produce intermixed with profane produce, the mixture must be eaten while one is in the state of

cleanness, and in Jerusalem, or else he must redeem the second-tithe produce. Therefore, if they are intermixed in Jerusalem, the mixture, when both components are of the same species, is forbidden, no matter how small the amount of the second-tithe component, because being in Jerusalem, the mixture has the status of a substance that can eventually be made permitted, and therefore it must all be eaten in a state of cleanness.

15. If one plants second-tithe produce after it has entered Jerusalem, the outgrowth thereof is likewise second tithe. If it is planted before reaching Jerusalem, the outgrowth thereof is profane, even in the case of plants whose seeds are not absorbed in the soil. Such plants must be redeemed when planted.

16. Second-tithe produce is neutralized by a greater quantity of profane produce. To what second tithe does this rule apply? To second tithe that had already entered Jerusalem and has left it, the rule of enclosure being no longer applicable, since there are no longer any enclosures within which one might return it, nor can it be redeemed, inasmuch as it had already entered Jerusalem. It is thus not a substance which can eventually be made permitted, and therefore it is neutralized by a greater quantity of profane produce, even if it is worth less than one pĕruṭah, as we have explained in the Laws Concerning Forbidden Foods.

CHAPTER VII

1. Produce bought with second-tithe redemption money may not be redeemed at a far distance from Jerusalem, unless it has become unclean by contact with a primary defilement; otherwise it must be brought to Jerusalem and consumed there.

2. This stringency applies only to produce bought with second-tithe redemption money, and not to the original second-tithe itself. If such produce becomes unclean by contact with a secondary defilement enjoined by Scribal law only, it may be redeemed and consumed in Jerusalem.

3. One may buy with second-tithe redemption money only such comestibles as are fit for human consumption and grow out of the earth, or out of their outgrowth, as for instance, the varieties enumerated in the Torah, *And thou shalt bestow the money . . . for oxen, or for sheep, or for wine, or for strong drink* (Deut. 14:26).

4. Therefore one may not buy with second-tithe redemption money water, salt, truffles, or mushrooms, because these do not grow out of the earth; nor produce that is still attached to the ground, nor produce that cannot reach Jerusalem, because these are not similar to cattle and sheep.

5. Honey, eggs, and milk are the same as cattle and sheep, for though they do not grow out of the earth, they are the products of the outgrowth of the earth.

6. Grape skin wine that has not yet soured may not be bought with second-tithe redemption money, because it is the same as water, but once it has soured, it may be bought, the same as wine and beer. If one buys it before it has soured and it then becomes sour, the purchase becomes second tithe.

7. When does this apply? When one pours in three measures of water and obtains less than four measures of grape skin wine. If, however, he pours in three measures of water and obtains four measures of wine, it is regarded as regular mixed wine, and may be bought with second-tithe redemption money.

8. Young sprouts of the service tree and of the carob tree that cannot yet be sweetened may not be bought with second-tithe redemption money; once they can be sweetened, they may be bought. Arum, mustard, lupine, and all the other seeds that are pickled, whether they can be sweetened or not, may be bought. Terminal palm buds may be bought with second-tithe redemption money.

9. Saffron may not be bought with second-tithe redemption money, because it is used only for coloring. The same applies to similar things that are used solely to impart aroma, color, or

flavor: inasmuch as they are eaten not for their own sake but only for the flavor which they impart, they may not be bought with second-tithe money. Therefore spice tips, black pepper, costus root, asafetida, lozenges of safflower, and all similar things may not be bought with second-tithe money.

10. Dill, if used to flavor a dish, may not be bought with second-tithe redemption money; but if used in milk sauce or the like, where it itself is food, it may be bought with second-tithe money.

If one mixes water and salt and adds oil to the mixture, it is considered the same as fish brine and may be bought with second-tithe money, but the price of the water and salt must be included in the price of the oil.

11. One may not buy heave offering with second-tithe redemption money, since this would reduce the number of persons who may eat of either second tithe or heave offering. For heave offering may be eaten only by priests who have shed their uncleanness at sunset, is permitted to a mourner, and may be eaten in any place, whereas second tithe is permitted to commoners, and prior to sunset to an unclean person who has immersed himself, but is forbidden to a mourner and may be eaten only in Jerusalem. Thus purchase of heave offering with second-tithe money would reduce the number of persons who may eat of either second tithe or heave offering.

12. One may buy cattle for peace offerings with second-tithe redemption money, because peace offerings may be consumed by commoners.

At first it was the custom to buy cattle with second-tithe money for consumption as profane meat, in order to avoid sacrificing them on the altar. Later the court decreed that one may buy cattle with second-tithe money only for peace offerings. One may, however, buy wild beasts and fowl, which are not fit for peace offerings.

13. Seventh-year produce may not be bought with second-tithe redemption money, because one must clear it out, as will be explained.

14. If one buys with second-tithe redemption money water, salt, produce still attached to the ground, or produce that cannot reach Jerusalem, the purchase does not become second tithe, even though the second-tithe money paid for it has become profane.

15. If one buys produce outside of Jerusalem with second-tithe redemption money, the rule is as follows: If he does it unwittingly, the seller must be compelled to return the money to the purchaser, and it becomes second-tithe money as it was before; if he does it deliberately, the purchased produce must be carried up to Jerusalem and eaten there. If the Sanctuary is not in existence, the purchaser must leave the produce to rot.

16. Similarly one may not buy cattle outside of Jerusalem with second-tithe redemption money. If he does so nevertheless, the rule is as follows: If he does it unwittingly, the money must be restored to its former place; if deliberately, the cattle must be taken up to Jerusalem and consumed there; and if the Sanctuary is not in existence, the beast must be buried together with its hide.

17. If one buys servants, land, or an unclean animal, whether deliberately or unwittingly, the rule is that if the seller has absconded, the buyer must consume their equivalent in monetary value in Jerusalem as second tithe.

This is the general principle: If a person spends second-tithe redemption money for anything other than food, drink, or ointment, and the seller has absconded or died, he must consume its monetary equivalent in Jerusalem as second tithe. If the seller is still present, the money must be restored to its original place.

Similarly, if one brings burnt offerings, sin offerings, or guilt offerings bought with second-tithe money, he must consume their monetary equivalent in Jerusalem as second tithe.

18. If one buys with second-tithe redemption money a wild beast for a peace offering or a domestic animal for human consumption, it is the same as if he had bought an ox for plowing, and the purchased beast does not become a peace offering. If he buys a domestic animal for a peace offering, and a blemish befalls it, the sanctity of second tithe departs from it, and he must redeem it, but the redemption money does not become second

tithe. Nevertheless, if he redeems it for himself, he must add one-fifth.

19. If one assigns second-tithe redemption money to peace offerings, the money does not become peace offering, because the sanctity of peace offering cannot be superimposed upon the sanctity of second tithe, for second tithe is sacred property. Needless to say, if he assigns the second tithe itself to peace offerings, it does not become peace offering.

20. If one eats second-tithe produce as if it were profane produce, even if he does so deliberately, the rule is as follows: If he eats the produce itself, let him cry out to heaven. If he eats produce bought with second-tithe redemption money, the money must be restored to its former place, and must then be taken up to Jerusalem and consumed there; or else, if the money cannot be restored, he must consume its equivalent in Jerusalem.

CHAPTER VIII

1. If one buys with second-tithe redemption money a domestic animal for a peace offering or a wild beast for human consumption, from a person who is not a merchant and is not particular, the hide becomes profane property, even if its value exceeds the value of the flesh. If, however, he buys it from a merchant, the hide does not become profane.

2. Similarly, if one buys sealed jugs of wine with second-tithe redemption money, from a place where they are customarily sold sealed and from a person who is not a merchant, the vessels themselves become profane. The seller must therefore unseal the necks of the jugs, if he wishes that the vessels should not become profane. If the seller wishes to be strict with himself and sell the contents by exact measure, the jug becomes profane.

3. If one buys the jugs either unsealed or sealed, in a place where they are customarily sold unsealed, or if he buys from a

merchant who is particular in his selling, the jug does not become profane.

If one buys baskets of figs or grapes together with the vessel containing them, the monetary value of the latter does not become profane.

4. If one buys walnuts, almonds, or the like with second-tithe redemption money, the shells become profane. If he buys a bale of dates, the wrapping becomes profane. If he buys a hamper of dates, the rule is as follows: If the dates are pressed, the hamper becomes profane; if the dates are not pressed, the hamper does not become profane.

5. If one has second-tithe wine, and assigns his vessels by way of a loan to contain this second-tithe wine, they do not assume the status of second tithe, even if he has sealed their mouths.

If he pours the wine into these vessels without any assignment, the rule is as follows: If he designates this wine as second tithe before sealing the mouths of the vessels, the vessels do not become second tithe; if he designates the wine as second tithe after sealing the mouths of the vessels, they too become second tithe.

If he pours into the vessel one-quarter loḡ of profane wine, or some second-tithe oil, vinegar, fish brine, or honey, without assigning the vessel as a loan, it does not become second tithe, whether he then seals the vessel or not.

6. A gazelle that was bought with second-tithe redemption money and died must be buried together with its hide. If having been bought live and slaughtered it becomes unclean, it must be redeemed, the same as any other produce that has been rendered unclean.

In the case of one who sets aside one denar as second-tithe redemption money, in order to consume a corresponding amount of produce until the denar shall have been rendered profane, the rule is that if at that time the rate of exchange is twenty mĕʿah for one denar, and he consumes ten mĕʿah worth of produce, and if after a while the mĕʿah depreciates to the rate of forty mĕʿah for one denar, he must consume additionally twenty mĕʿah worth of produce, and his denar then becomes profane. If the mĕʿah appreciates to the rate of ten mĕʿah for one denar, he must consume additionally five mĕʿah worth of produce, and his denar then becomes profane.

7. If a person buying produce with one selaʿ of second-tithe redemption money takes possession of the produce, but before he can hand over the selaʿ to the seller the produce goes up in price to two selaʿ, he must pay the seller only one selaʿ of second-tithe money, as it is said, *Then he shall give the money . . . and it shall be assured to him* (Lev. 27:19), implying that the handing over of the money constitutes transfer of title, and any profit accrues to the second tithe.

8. If a person takes possession of produce worth two selaʿ, and before he can hand over the money to the seller the produce depreciates to one selaʿ, he must pay to the seller only one selaʿ of second-tithe redemption money, but must add another selaʿ out of profane money and give it to the seller. If the seller is an unlearned person, the purchaser may give him the second selaʿ out of second-tithe money with which doubtfully tithed produce had been redeemed.

If the purchaser pays the seller one selaʿ of second-tithe money, but before he can take possession of the produce it goes up in price to two selaʿ, the rule is that what the purchaser has redeemed is redeemed, and as for the sale of the produce, the two parties must settle it between themselves.

9. If the buyer pays the seller two selaʿ of second-tithe redemption money, and before he can take possession of the produce it depreciates once more to one selaʿ, the rule is that what he has redeemed is redeemed, and as for the sale of the produce, the two parties must settle it between themselves, because as far as second tithe is concerned, its redemption takes effect the moment the buyer has taken possession of the produce.

10. A person who has some profane produce in Jerusalem and some second-tithe redemption money outside of Jerusalem may say, "That money is to be rendered profane in lieu of this pro-

duce." He may then eat of the produce there in a state of cleanness, while the money becomes profane in the place where it is.

11. If one has some second-tithe redemption money in Jerusalem and some produce outside of Jerusalem, he may say, "This money is to be rendered profane in lieu of that produce." The money then becomes profane, while the produce must be brought up to Jerusalem and consumed there, for the money and the produce need not be in the same place when one is rendered profane in lieu of the other.

12. If A has some second-tithe redemption money in Jerusalem and is in need of spending it, and if B has some profane produce that he wishes to consume, A may say to B, "This money of mine is to be rendered profane in lieu of your produce." Thus the result is that the produce has been purchased with tithe money, and B may consume it in a state of cleanness. In this manner B sustains no loss whatsoever, while A's money has become profane.

13. When does this apply? When B is an associate scholar, because produce that is assuredly second tithe may be handed over only to an associate scholar. Therefore, if the money had been used to redeem doubtfully tithed produce, A may make the aforecited declaration even to an unlearned person.

It is permitted to render second tithe of doubtfully tithed produce profane in lieu of produce or money belonging to an unlearned person, and no apprehension need be felt that these too might be second tithe.

14. If a person deposits one denar of second-tithe redemption money in order to go on consuming produce or money against it, the rule is that once he has consumed enough to leave a balance of less than one pĕruṭah, that denar becomes profane.

To what does this apply? To a denar used to redeem doubtfully tithed produce. If the denar was used to redeem assuredly tithed produce, it does not become profane until the balance is less than one pĕruṭah after the addition of the statutory one-fifth, as for example, until the balance is less than four-fifths of one pĕruṭah.

15. If unclean and clean persons are eating or drinking together in Jerusalem, and if the clean ones wish to eat of their second tithe, a person must lay down one sela' of second-tithe redemption money and say, "Whatsoever the clean persons are eating and drinking, this sela' is to be rendered profane in lieu of it," and this sela' is thereby rendered profane, since the clean persons have eaten and drunk its equivalent in cleanness. Provided that the unclean persons do not touch the food, thereby defiling it.

CHAPTER IX

1. Fourth-year's fruit is holy, as it is said, *And in the fourth year all the fruit thereof shall be holy, for giving praise unto the Lord* (Lev. 19:24), and the rule is that it must be consumed by its owner in Jerusalem, the same as second tithe.

And just as there is no second tithe in Syria, so is there no fourth-year's fruit in Syria.

It is concerning fourth-year's fruit that Scripture says, *And every man's hallowed things shall be his* (Num. 5:10), for there is no hallowed thing concerning which the Torah does not specify to whom it shall belong, except fourth-year's fruit.

2. Whosoever wishes to redeem fourth-year's fruit may redeem it, the same as second tithe, and if he redeems it for himself, he must add one-fifth. It may not be redeemed until it reaches the season of tithing, as it is said, *That it may yield unto you more richly the produce thereof* (Lev. 19:25), implying that it is redeemable only when it has become produce.

It may not be redeemed while it is still attached to the ground, the same as second tithe, and like second tithe it is sacred property. Therefore it may not be acquired as a gift unless given while yet unripe.

In all other respects touching eating, drinking, and redemption, fourth-year's fruit is likewise analogous to tithe.

3. He who redeems fourth-year's fruit may redeem grapes while they are still grapes or after they have been pressed into wine, whichever he prefers, and the same applies to olives. Other fruit, however, may not be changed from its natural state.

4. A fourth-year's vineyard has neither forgotten sheaf, nor corner crop, nor grape gleanings, nor defective clusters. Nor is one obligated to set aside heave offering and tithes out of it, the same as in the case of second tithe. Rather all of its yield must be brought up to Jerusalem, or else it must be redeemed and the redemption money brought up to Jerusalem and consumed there, the same as second tithe.

5. The court has ordained that grapes from a fourth-year's vineyard must be brought up to Jerusalem if the vineyard is located up to one day's journey from it on each side, in order to decorate the streets of Jerusalem with such fruit. Since the Temple was destroyed these grapes may be redeemed even close to the city wall. Other fruits may be redeemed close to the city wall even when the Temple is in existence.

6. How should one redeem fourth-year's fruit? The owner should set down a basket of it before three appraisers, who should estimate how many such basketfuls a person would be willing to redeem for himself for one sela', on condition that the outlay for watchmen, ass drivers, and laborers shall be defrayed by his own household. After they have set the estimated price, he should lay down the money and say, "Whatsoever is gathered from this field is to be rendered profane in lieu of this money, at the rate of so many baskets per sela'." During the Sabbatical year he should redeem it at its full value, because in that year there are no expenditures for watchmen and laborers. If the field is ownerless property, the person who harvests the yield can claim only the cost of harvesting.

7. If one has a fourth-year's plantation during the Sabbatical year, when everyone has an equal claim to its fruit, he must mark it with clods of earth, so that people would recognize it and re-

frain from eating of it until he redeems it. If the new plantation is within its years of *'orlah,* it should be marked with potsherds, so that people would keep away from it, for should it be marked with clods of earth, they might crumble. This extra precaution is required because the interdict of 'orlah is more stringent, since it forbids any benefit therefrom whatsoever.

Scrupulously pious persons used to set down money during the Sabbatical year and say, "Whatsoever shall be gathered of this fourth-year's fruit is to be rendered profane in lieu of this money," since it cannot be redeemed while still attached to the ground, as we have explained.

8. The first of Tishri is New Year's Day for 'orlah and for fourth-year's fruit. From what time does the count of years commence? From the time of planting, not from New Year's Day to New Year's Day; but thirty days of one year count as a full year, provided that the planting has taken root before the thirty days. How much time do all trees need to take root? Two weeks.

9. You thus learn that if one plants forty-four days before New Year's Day, it counts as a full year. Nevertheless the fruit of this planting does not shed its status as 'orlah or as fourth-year's fruit until the fifteenth of Shebat, which is New Year's Day for trees.

10. How so? If one plants a tree yielding edible fruit on the fifteenth of Ab of the tenth year of a Jubilee period, it remains within its 'orlah years until the fifteenth of Shebat of the thirteenth year. Whatsoever the tree yields within that time is 'orlah, even if the fruit ripens many days thereafter. From the fifteenth of Shebat of the thirteenth year to the fifteenth of Shebat of the fourteenth year the tree is fourth-year's planting, and whatsoever it yields within that time is fourth-year's fruit and must be redeemed. If that year is a leap year, the extra month thereof is included in the years of 'orlah or in the fourth year.

11. If one plants the tree on the sixteenth of Ab of the tenth year, that year does not count, and its fruit is 'orlah during the eleventh, the twelfth, and all of the thirteenth year, and is fourth-

year's fruit from New Year's Day of the fourteenth year to year-end.

12. If one plants the tree between the first of Tishri and the fifteenth of Shebat, he must count three years from day to day for 'orlah and one year from day to day for fourth-year's planting.

I have seen some statements ascribed to the Geonim concerning the reckoning of 'orlah and fourth-year's fruit which do not merit detailed citation and refutation, and must certainly be due to scribal error. We have already explained the truth as it should be.

13. Leaves, sprouts, sap, and buds of the vine are exempt from the law of 'orlah and fourth-year's fruit.

On the other hand, grapes parched and damaged by the east wind, grape skins and seeds, and wine made therewith, pomegranate shells and blossoms, walnut shells, and various fruit seeds are forbidden as 'orlah but are permitted as fourth-year's fruit. Unripe fruit fallen from the tree is in all cases forbidden.

CHAPTER X

1. Every tree subject to the law of 'orlah is subject also to the law of fourth-year's fruit; and conversely, every tree exempt from the law of 'orlah is not liable to the law of fourth-year's fruit, as it is said, *Three years shall it be as forbidden unto you,* etc., *and in the fourth year all the fruit thereof shall be holy* (Lev. 19:23–24).

2. If one plants a fruit-bearing tree but is of a mind that it is to serve as fencing for the garden, or if he plants it for its timber and not for its fruit, it is exempt from the law of 'orlah.

If he plants the tree for fencing but changes his mind and decides to use it for food, or vice versa, the rule is that once he has involved in it an intention which renders it liable, it is liable. If he plants it with the intention that during the first three years it is to serve as fencing and thereafter for food, it is not subject to the law of fourth-year's fruit, because anything that has no ʿorlah has no fourth-year's fruit either.

3. If the person planting the tree is of a mind that its inward part is to serve for food and its outward part is to serve as fencing, or that its lower part is to serve for food and its upper part is to serve as a fence, the rule is that the part intended for food is subject to the law of ʿorlah, while the part intended for a fence or for firewood is exempt, because the matter depends entirely upon the planter's intention.

In the case of the caper bush only the berries are liable to ʿorlah, the flowers being exempt.

4. If an individual plants a tree inside his own field but for the benefit of the general public, it is liable to ʿorlah, as it is said, *And when ye . . . shall have planted* (Lev. 19:23), implying even for the use of the many.

Where does this apply? In the Land of Israel. Outside of the Land of Israel, however, the tree is exempt.

5. A tree planted in a public domain or in a ship, a tree grown spontaneously in a private domain, a tree planted by a heathen, whether for an Israelite or for himself, and a stolen seedling planted by a robber—all these are liable to ʿorlah and to fourth-year's fruit.

6. A tree grown spontaneously in a craggy place is exempt, and so is even a tree deliberately planted in an uninhabited place, provided that it does not yield enough to induce a person to tend the fruit until he can bring it over to a settled place. If the tree does yield enough to warrant such care, it is subject to ʿorlah.

7. If one plants a tree for a religious purpose, as for example, if he plants a citron tree in order to use its fruit jointly with the lulaḇ, or an olive tree in order to use its oil in the candelabrum of the Temple, it is subject to ʿorlah.

If he first consecrates the tree to the Temple and then plants it, it is exempt from ʿorlah; if he first plants it and then consecrates it, it is subject to ʿorlah.

8. If one plants a tree in an unperforated pot, it is subject to ʻorlah, for even though such a pot is not accounted the same as soil in regard to seeds, it is accounted the same as soil in regard to trees.

9. A tree planted inside a house is liable to ʻorlah. If it was planted by a heathen before our ancestors' arrival in the Land, it is exempt. After their arrival, even trees planted by heathens are liable, as it is said, *And when ye shall come into the Land, and shall have planted . . . trees* (Lev. 19:23), implying all trees planted after their arrival.

10. A heathen who grafts a fruit tree to a tree that bears no fruit is liable to ʻorlah. A Gentile can be subject also to the law of fourth-year's fruit, for if he adopts the practice of this commandment, his fruit is just as holy as the fruit of an Israelite's fourth-year's plant.

11. Whether one plants a seed or a shoot of a tree, or uproots the whole tree from its place and transplants it in another place, he is liable to ʻorlah and must count the time from the moment of planting. If he merely shakes the tree without uprooting it, and then fills in the earth around it, the rule is as follows: If it could have gone on living even if he had not filled in the earth around it, he is exempt; if not, it is the same as if he had uprooted and transplanted it, and he is therefore liable.

12. Similarly, if a tree is uprooted, but one of its roots, be it no thicker than the pin upon which the weaver wraps the scarlet thread, is left, and if the tree is then returned to its place and replanted, the owner is exempt from ʻorlah, because the tree can still go on living.

If the whole tree is uprooted together with the clod of earth covering its roots, and is then replanted as it was, with all the earth around the roots, the rule is as follows: If the tree could have gone on living out of that clod of earth without being replanted, it is the same as if it had not been uprooted. If it could not, the owner is liable to ʻorlah.

13. If one cuts down a tree, and the stump grows again into a new tree, he is liable to ʿorlah, and the count of time must begin from the moment the original tree was cut down.

14. Whether one plants a twig, or bends it into the ground, or grafts it, he is liable to ʿorlah. When does this apply? When he severs the twig from the tree and bends it into the ground or grafts it to another tree. If, however, he merely stretches the twig from the mother-tree and bends it into the ground, or grafts the twig to another tree while the base of the twig is still integral with the mother-tree, he is exempt.

15. If this daughter-tree bent into the ground grows up and brings forth fruit, and if one then severs its root that it attached to the mother-tree, he must count the time from the moment the root is severed. The fruit already grown is therefore permitted, because it has grown while the tree was not subject to ʿorlah. If after the root is severed one leaves the fruit until it increases by one two-hundredth, the fruit also is forbidden.

16. If the daughter-tree is subsequently intertwined by grafting with the mother-tree, and there is fruit on the daughter-tree, the rule is that even if that fruit has increased two-hundredfold since the grafting, it is forbidden, because permitted growth cannot neutralize forbidden primary growth.

17. If the twig of a tree is bent into the ground, and if the entire tree is then uprooted, so that it continues to live only in that bent twig, the tree is regarded as if it were planted now, and is subject to ʿorlah. The count of time for the tree and for what grows out of the twig must begin from the moment when the tree was uprooted.

18. If one bends the twig of a tree into the ground, and it grows, and if he then bends another twig from that second growth into the ground, and it too grows, and if thereupon he once more bends a twig from this third growth into the ground, and so forth, be there even one hundred such twigs attached to each other, the fruit of them all is permitted, so long as they

have not been severed from the original root. If the original root is severed, the time must be counted from the moment of severance.

19. A tree growing out of the trunk of another tree is exempt from ʿorlah; if it grows out of the other tree's roots, it is liable to ʿorlah.

A daughter-tree that is not higher than one handbreadth is liable to ʿorlah throughout its years of life, because it looks like a one-year-old plant.

To what does this apply? To a single plant or to two plants facing two others with a fifth plant extending like a tail. If the whole vineyard, however, is less than one handbreadth in height, such an unusual vineyard becomes a matter of common knowledge, and the count of time for it must be the same as for other normal trees.

20. One may plant a shoot, but not a nut, from an ʿorlah tree, because a nut is a fruit, and one is forbidden to derive any benefit from fruits of ʿorlah, as we have explained in the Laws Concerning Forbidden Foods.

If he transgresses and does plant such a nut of ʿorlah, what grows out of it is permitted, the same as any other trees.

21. Similarly one may not graft branches bearing early berries from an ʿorlah date palm to another date palm, because the berries are accounted the same as fruit. But if he transgresses and does so, the fruit is permitted, for the rule is that the produce of two causes, one forbidden and the other permitted, is permitted. Therefore what grows out of ʿorlah fruit is permitted, since its growth was caused by both the fruit, which is forbidden, and the earth, which is permitted.

CHAPTER XI

1. It is a positive commandment to make confession before the Lord after one has taken out all the hallowed gifts due from

the yield of the land. This is the so-called "confession of the tithe."

2. One may not make this confession until after the year in which the poor man's tithe is set aside, as it is said, *When thou hast made an end of tithing all the tithe of thine increase . . . then thou shalt say before the Lord thy God, I have put away the hallowed things out of my house* (Deut. 26:12–13).

3. When must one make this confession? In the afternoon of the last festival day of Passover in the fourth and seventh years of each septennate, as it is said, *When thou hast made an end of tithing* (Deut. 26:12), implying, on the festival during which all the tithes end, and Passover of the fourth year does not arrive until after all the produce of the third year, whether fruit of the trees or fruit of the land, has been tithed.

4. One may make confession during daytime only, but any part of daytime is valid for the confession of the tithe.

One is obligated to put away hallowed gifts and make confession, regardless of whether the Temple is or is not in existence.

5. This confession may be made in any language, as it is said, *Then thou shalt say before the Lord thy God* (Deut. 26:13), implying, in any language in which you may say it. Each person may say it individually, but if several persons wish to make confession jointly, they may do so.

6. The commandment as it stands requires the confession to be made in the Temple, as it is said, *Before the Lord* (Deut. 26:13), but if one makes it in any other place, he has fulfilled his obligation.

7. One may make confession only after none of the hallowed gifts remains with him, since the confession states expressly *I have put the hallowed things out of my house* (Deut. 26:13). The putting away takes place on the eve of the last day of the Passover festival, and the confession in the afternoon of the following day.

8. How should one proceed? If one has any heave offering or heave offering of the tithe remaining with him, he should give it to the priest; if any first tithe, he should give it to the Levite; if any poor man's tithe, he should give it to the poor.

If one has any assured second-tithe produce, or any fourth-year's fruit, or their redemption money remaining with him, he should put them away by throwing them into the sea or burning them.

If one has second tithe of doubtfully tithed produce remaining with him, he is not obligated to put it away. If he has any first fruits remaining with him, they should be put away wherever they are.

9. When does the rule of burning in order to put away apply? When one has an amount of produce remaining with him that he cannot completely consume before the onset of the festival. If, however, it is a cooked dish of second tithe or of fourth-year's fruit, one need not put it away, because the leftover of a cooked dish is considered the same as put away. Similarly, wine and spices are considered the same as put away.

10. If one has produce which at the time of putting away has not yet reached its season of tithing, he need not be prevented from making confession, and is not obligated to put the produce away.

11. If one's produce is located far away from him, and the day of putting away has arrived, he may designate the several gifts due by name, transfer title to them to whom they are due, by dint of land, or to a third party who acquires them in behalf of those to whom they are due, and then make confession on the morrow. For such transfer of title to movables by dint of land is merely a gift that has been provided with a reenforcement. He may not, however, transfer title to the tithe by means of symbolic barter, because this looks the same as a sale, whereas Scripture expressly states that tithes, heave offerings, and other gifts must be given and not sold.

12. Whence do we know that one may not make confession until he has put away all the gifts? From the verse, *I have put away the hallowed things out of my house* (Deut. 26:13): *hallowed things* refers to second tithe, and also to fourth year's fruit, which is likewise called holy; *out of my house* refers to dough offering, which is the gift out of the house due to the priests; *and also have given them to the Levite* (*ibid.*) refers to first tithe; *and also have given them* implies that this has been preceded by other gifts, which must refer to the great heave offering and to the heave offering of the tithe; *unto the stranger, to the fatherless, and to the widow* (*ibid.*) refers to poor man's tithe, gleanings, forgotten sheaf, and corner crop, although failure to put away gleanings, forgotten sheaf, and corner crop does not hinder one from making confession.

13. One must set aside the gifts in their proper sequence and then make confession, as it is said, *According to all Thy commandment which Thou hast commanded me* (Deut. 26:13). Hence, if one advances the second tithe before the first, he cannot make confession. If his untithed produce is burned, he likewise cannot make confession, inasmuch as he has not yet set aside the gifts nor given them to those to whom they are due. During the time when first tithe used to be given to the priests, one did not make confession, as it is said, *And also have given them unto the Levite* (*ibid.*)

14. One who has nothing but second tithe may make confession, because confession refers basically to the tithe.

Similarly, if one has nothing but first fruits, he may make confession, as it is said, *I have put away the hallowed things* (Deut. 26:13), implying the first hallowed thing to be set aside, which is first fruits.

On the other hand, he who has nothing but heave offering itself need not make confession, for heave offering requires confession only when it is included with the other gifts.

15. *I have not transgressed any of Thy commandments* (Deut. 26:13) implies that one has not set aside one species for another,

nor produce detached from the ground for produce still attached to the ground, or vice versa, nor new produce for old, or vice versa. *Neither have I forgotten them* (*ibid.*) implies that one has not forgotten to bless Him and to mention His name over them. *I have not eaten thereof in my mourning* (Deut. 26:14) implies that if one has eaten it while in mourning he need not make confession. *Neither have I put away thereof, being unclean* (*ibid.*) implies that if one has done so while being unclean, he likewise need not make confession. *Nor given thereof for the dead* (*ibid.*) implies that one has not used aught thereof to purchase a coffin or a shroud, nor given aught thereof to other mourners. *I have hearkened to the voice of the Lord my God* (*ibid.*) implies that one has brought it to the chosen House. *I have done according to all that Thou hast commanded me* (*ibid.*) implies that one has himself rejoiced and has made others to rejoice therein, as it is said, *And thou shalt rejoice in all the good* (Deut. 26:11).

16. *Look forth from Thy holy habitation, from heaven,* etc., to *as Thou didst swear unto our fathers, a land flowing with milk and honey* (Deut. 26:15) constitutes a plea that He might give sweet savor to the produce.

17. Israelites and bastards may make confession but not proselytes and freedmen, because the latter have no share in the Land, since it is said, *And the land which Thou hast given us* (Deut. 26:15). Priests and Levites may make confession, for though they did not take a share in the Land, they have cities with open land around them.

TREATISE VI

LAWS CONCERNING FIRST FRUITS AND OTHER GIFTS TO THE PRIESTHOOD WITHIN THE BOUNDARIES OF THE LAND OF ISRAEL

Involving Nine Commandments,
Eight Positive and One Negative
To Wit

1. To set aside first fruits and bring them to the Temple;
2. That the priest shall not eat of the first fruits outside of Jerusalem;
3. To read the prescribed declaration over them;
4. To set aside the dough offering for the priest;
5. To give the shoulder, the two cheeks, and the maw of a beast killed for food to the priest;
6. To give to the priest the first of the fleece at the shearing;
7. To redeem the first-born son, and to give the redemption money to the priest;
8. To redeem the firstling of an ass, and to give the redemption lamb to the priest;
9. To break the neck of the firstling of an ass, if the owner does not wish to redeem it.

An exposition of these commandments is contained in the following chapters.

NOTE

In the list of 613 commandments prefixed to the Code, those dealt with in the present treatise appear in the following order:

Positive Commandments:

[1] 125. To bring the first fruits to the Temple, as it is said: *The choicest first fruits of thy land,* etc. (Exod. 23:19);

[3] 132. To read the prescribed declaration over the first fruits, as it is said: *And thou shalt speak and say before the Lord thy God,* etc. (Deut. 26:5);

[4] 133. To set aside the dough offering for the priest, as it is said: *Of the first of your dough ye shall set apart a cake for a gift* (Num. 15:20);

[5] 143. To give to the priest out of the carcass of cattle the shoulder, the two cheeks, and the maw, as it is said: *That they shall give unto the priest the shoulder, and the two cheeks, and the maw* (Deut. 18:3);

[6] 144. To give the first of the fleece at the shearing to the priest, as it is said: *And the first of the fleece of thy sheep shalt thou give him* (Deut. 18:4);

[7] 80. To redeem the first-born of man, as it is said: *Howbeit the first-born of man shalt thou surely redeem* (Num. 18:15);

[8] 81. To redeem the firstling of an ass, as it is said: *And every firstling of an ass thou shalt redeem with a lamb* (Exod. 13:13);

[9] 82. To break the neck of such a firstling if it is not to be redeemed, as it is said: *And if thou wilt not redeem it, then thou shalt break its neck* (*ibid.*).

Negative commandment:

[2] 149. That a priest may not eat of the first fruits prior to their deposit in the Temple court, as it is said: *Thou mayest not eat within thy gates,* etc., *nor the offering of thy hand* (Deut. 12:17), which refers to the first fruits.

CHAPTER I

1. Twenty-four gifts were granted to the priests; all are explicitly stated in the Torah, and concerning all a covenant was made with Aaron. Any priest who does not acknowledge them has no share in the priesthood, and may not be given anything out of these gifts.

2. Whosoever eats of a gift that is endowed with sanctity must pronounce the benediction, "Blessed art Thou, O Lord . . . , who has sanctified us with the sanctity of Aaron and has commanded us to eat of such-and-such."

3. Of these gifts eight may be eaten by the priests only in the Temple, within the wall of the Temple court; five may be eaten only in Jerusalem, within the city wall; five accrue to them, according to the Torah, only in the Land of Israel; five accrue to them both in the Land of Israel and outside of the Land of Israel; and one gift accrues to them directly from the Temple.

4. What are the eight gifts that may be eaten only in the Temple? The flesh of the sin offering, whether the offering is of fowl or of cattle; the flesh of the guilt offering, whether the guilt is conjectural or certain; congregational heave offerings; the residue of the 'omer; the residue of Israelite meal offerings; the two loaves; the shewbread; and the leper's loḡ of oil—these may be eaten only in the Temple.

5. What are the five gifts that may be eaten only in Jerusalem? The breast and the thigh of the peace offering; what is set aside out of the thank offering; what is set aside out of the Nazirite's ram; the firstling of clean cattle; and the first fruits—these may be eaten only in Jerusalem.

6. What are the five gifts that apply only in the Land of Israel? The heave offering, the heave offering of the tithe, the dough offering—these three are holy; the first of the fleece and

the field of possession—these two are profane. According to the Torah, these five accrue to the priests only in the Land of Israel. Heave offerings and dough offerings of the Land of Israel may be eaten only in the Land of Israel.

7. What are the five gifts that accrue to the priests in all places? The lay gifts, the redemption money of the first-born son, the firstling of an ass, things illegitimately seized from a proselyte, devoted things—all five of them are profane in every respect.

8. The gift that accrues to the priests directly from the Temple is the hides of burnt offerings. The same rule applies to the hides of other offerings that are of the same highest degree of sanctity—all accrue to the priests.

9. The eight gifts that may be eaten only in the Temple are all of the highest degree of holiness, and may be eaten only by the male members of the priesthood, as will be explained in its proper place. Concerning them it is said, *Every male among the priests may eat thereof; it is most holy* (Lev. 6:22).

10. The five gifts that may be consumed only in Jerusalem are things of a minor degree of sanctity and may be eaten by both males and females. Concerning them it is said, *I have given them unto thee, and to thy sons and to thy daughters with thee, as a due for ever* (Num. 18:11). Nevertheless they are given only to the males of the priesthood, seeing that they belong to the priests of the division on duty in the Temple at the particular time.

As for the firstling, its fat and its blood must be offered up, and only a male priest may do so.

The same rule applies to the hides of sacrifices of the highest degree of sanctity, to the field of possession, to devoted things, and to things illegitimately seized from a proselyte—all accrue only to the male priests on duty at the particular time, as will be explained.

Similarly the redemption money of the first-born son accrues to the males of the priesthood, as it is said concerning it, *And thou shalt give the money . . . unto Aaron and to his sons* (Num. 3:48).

The firstling of an ass likewise accrues to the males of the priesthood, because the rule concerning firstlings applies equally to all of them, in that they accrue to the males of the priesthood and not to the females.

11. You thus learn that five gifts accrue to females as well as to males, to wit: heave offering, heave offering of the tithe, dough offering, lay gifts out of cattle, and the first of the fleece.

And whence do you know that the first of the fleece accrues also to priestly women? From the verse, *The first fruits of thy corn, of thy wine, and of thine oil, and the first of the fleece of thy sheep, shalt thou give him* (Deut. 18:4), implying that just as the first of the corn is given to females as well as to males, so is the first of the fleece.

12. The Sages have enumerated these gifts in a different fashion, by saying: Twenty-four priestly gifts were granted to Aaron, to wit: ten in the Temple, four in Jerusalem, and ten within the boundaries of the Land of Israel outside of Jerusalem.

13. The ten in the Temple are: sin offering of cattle, sin offering of fowl, guilt offering for guilt certain, guilt offering for guilt conjectural, congregational peace offerings, the leper's log̱ of oil, the two loaves, the shewbread, the residue of the meal offering, and the residue of the ʻomer.

14. The four in Jerusalem are: the firstling, the first fruits, that which is set aside out of the thank offering and of the Nazirite's ram, and the hides of consecrated animals.

15. The ten within the borders of the Land of Israel are: heave offering, heave offering of the tithe, dough offering, the first of the fleece, the lay gifts, the redemption money of the first-born son, the redemption money of the firstling of an ass, the devoted field, the field of possession, and things illegitimately seized from a proselyte.

According to this count, everything set aside out of things of a minor degree of sanctity is counted as one gift, and these are: the breast and the thigh set aside out of every peace offering, together with the loaves set aside with them—if the peace offering is a thank offering, including the shoulder that is set aside from the Nazirite's ram, together with the loaves that are set aside with it—that is to say, together with the breast and the thigh. Inasmuch as all three of these are peace offerings, what is set aside out of them is counted as one.

16. All the gifts that depend upon offerings will be explained, each in its proper place, in the Laws Concerning Offerings.

Similarly, things illegitimately seized from a proselyte will be explained in the Laws Concerning Robbery. As for the rules governing the field of possession and devoted property, we have already explained them in the Laws Concerning Valuations. The rules governing heave offering and heave offering of the tithe have been likewise explained in the Laws Concerning Heave Offerings.

In the following rules I will explain the regulations concerning gifts that do not depend upon offerings, to wit; first fruits, dough offering, lay gifts, the first of the fleece, the redemption money for the first-born son, and the redemption money for the firstling of an ass.

CHAPTER II

1. It is a positive commandment to bring first fruits to the Temple. This commandment is in effect only during the time that the Temple is in existence, and only in the Land of Israel, as it is said, *The choicest first fruits of thy land thou shalt bring into the House of the Lord thy God* (Exod. 23:19).

First fruits authorized by Scribal law must be brought from the cities of Sihon and Og and from Syria, for one who acquires land in Syria is the same as if he had acquired land in Jerusalem. In the case of Ammon, Moab, and Babylonia, however, even though these lands are liable to heave offering and tithes according to Scribal law, one need not bring first fruits from them. If one does bring first fruits from outside the Land of Israel, they are not regarded as first fruits.

2. First fruits must be brought only out of the seven species mentioned in praise of the Land of Israel, to wit: wheat, barley, grapes, figs, pomegranates, olives, and dates. If one brings first fruits out of species other than these seven, they do not become consecrated.

3. One may not bring first fruits out of dates grown on hills, or fruits grown in valleys, or oil olives that are not of choice quality; he must rather bring them out of dates grown in valleys and fruits grown on hills, for they are of choice quality. If one brings first fruits that are not of choice quality, as for example, dates grown on hills, or figs that are spoiled and worm-eaten, or grapes that have been dusted and fumigated, they do not become consecrated.

4. Liquids may not be brought as first fruits, except only the liquid of olives and grapes, as it is said, *The fruit of the ground* (Deut. 26:2), implying the fruit itself, not the liquid thereof. If one does bring liquids, they cannot be accepted from him.

5. Those who live near Jerusalem must bring fresh figs and grapes; those who live far away may bring them as dried figs and raisins.

6. First fruits may not be brought before the Feast of Weeks, as it is said, *The feast of harvest, the first fruits of thy labors* (Exod. 23:16). If one does bring them at that time, they may not be accepted from him; rather he must leave them there until the arrival of the Feast of Weeks, and then he may recite the declaration over them.

Similarly one may not bring the first fruits after Hanukkah, because first fruits ripened after Hanukkah are counted as of the following year, and he must therefore leave them until after the next Feast of Weeks.

7. One may not bring first fruits out of new produce for old, or vice versa.

How so? One may not bring first fruits of produce formed before the fifteenth day of Shebat for produce formed after the fifteenth day thereof.

It does not matter whether the produce was planted or grew of itself, as it is said, *The first ripe fruits of all that is in their land* (Num. 18:13).

8. Produce owned in partnership is liable to first fruits, as it is said, *The first fruits of all that is in their land* (Num. 18:13).

9. Produce grown in a pot, even a perforated one, or in a ship, is not liable to any first fruits, as it is said, *In their land* (Num. 18:13). On the other hand, produce grown on a roof or in a deserted building is liable.

10. If A plants a tree in his own field, and then sinks a shoot of it into B's field or into a public domain; or if A plants a tree in B's field or in a public domain, and then sinks a shoot of it into his own field; or if the root of the tree is in A's field, and A sinks part of the tree within that same field, but a public or private road separates it in the middle, between the root and the part sunk into the ground—in all these cases A is not obligated to bring first fruits out of the produce grown in either domain, as it is said, *First fruits of thy land* (Exod. 23:19), implying, only when all the outgrowth is from your own land.

11. If B gives A permission to sink a shoot into B's field, even if only temporarily, A must bring first fruits. If the tree is adjacent to B's boundary, or overhangs B's field, A, even though he is obligated to train the tree away from the boundary, must bring first fruits out of it and recite the declaration, inasmuch as it was on this condition that Joshua had distributed the land.

12. Tenants, lessees, strong-arm men who force owners to sell their land for less than it is worth, and robbers need not bring first fruits, even if the rightful owners have despaired of recovering the land, as it is said, *First fruits of thy land* (Exod. 23:19).

13. If A buys a tree in B's field, A need not bring first fruits, because he owns no land. If he buys three trees, he does own

land, for even though in fact he owns only the trees, it is the same as if he had acquired the land occupied by them as well. If he buys one tree together with its land, he must bring first fruits.

14. If one buys detached produce together with the land, he must bring first fruits, inasmuch as he owns both the land and its produce.

If the owner has sold the produce but not the land—even if the produce is still attached to the ground—the seller need not bring first fruits, since he owns no produce, nor need the buyer, since he owns no land.

If the seller in turn buys the produce back from the buyer, he must bring first fruits, inasmuch as he now owns both the land and its produce.

15. If one sells his field to a heathen, and then turns around and buys it back from him, he is liable to first fruits by authority of the Torah, for the field is not released from this commandment by the heathen's purchase, as we have explained.

16. One may not bring first fruits from an asherah that has ceased to be worshiped, because first fruits are as holy as the hallowed things of the Temple.

17. No prescribed amount is given in the Torah for first fruits; Scribal law, however, prescribes $1/60$ of the yield. If one wishes to designate the entire yield of his field as first fruits, he may do so.

18. If a person, having set aside his first fruits, turns around and adds to them, or adorns them with a wreath, the addition is also regarded as first fruits.

When does this apply? When he brings the fruits from the Land of Israel. If, however, he brings them from Transjordan or Syria, the addition is not regarded as first fruits.

But even though it is not so regarded, it may be eaten only in a state of cleanness.

One may wreathe first fruits in any place with the seven species of produce only.

19. How are first fruits to be set aside? When a person goes down to his field and sees a fig that has ripened, or a cluster of grapes that have ripened, or a pomegranate that has ripened, he should tie reed grass around them and say, "These are to be first fruits," and they become first fruits while they are still attached to the ground, as soon as he has so designated them, even if they have not yet completely ripened. When they are fully ripe, and he plucks them from the ground, he need not so designate them once more.

If he has not set them aside while still attached to the ground, nor designated them, but has first plucked them, he must set them aside after they have been plucked.

If all the fruits have become unclean, one may not set aside first fruits out of them, but must rather, at the outset, set aside first fruits out of other clean fruits for the unclean. If he has no other clean fruits, it would appear to me that he may not set aside first fruits that are unclean, for one may not set aside first fruits that are bound to go to waste.

Similarly, it would appear to me that one may not heat an oven with unclean first fruits, as he would with unclean heave offering, because the former are as holy as the hallowed things of the Temple.

20. If one sets aside his first fruits, and they rot, or are taken from him by force, or are lost or stolen, or become unclean, he must bring others in their stead, as it is said, *Thou shalt bring into the House of the Lord thy God* (Exod. 23:19), implying that he is responsible for them until he brings them to the Temple Mount.

21. If one sets aside his first fruits with the intention of bringing them to Jerusalem by his own hand, he may not send them through an agent instead. If, however, he had gathered them at the outset with the intention of sending them through an agent, he is allowed to send them in this manner.

CHAPTER III

1. First fruits must be given to the priests of the division on duty in the Temple at the particular time, and they are to divide them among themselves, as they do with all the hallowed things of the Temple.

We have already explained that first fruits are called heave offering. Therefore a commoner who eats of first fruits anywhere is liable to death at the hands of heaven, provided that he eats of them after they had entered inside the wall of Jerusalem.

2. If some of the first fruits are within the wall and some without, he who eats of those within the wall is liable to death, seeing that they are hallowed things in every respect; those without the wall, however, are profane in every respect.

3. A priest who eats of first fruits outside of Jerusalem, after they had entered within its walls, is liable to a flogging according to the Torah, as it is said, *Thou mayest not eat within thy gates,* etc., *the offering of thy hand* (Deut. 12:17), the latter referring to first fruits, as we have explained in the Laws Concerning Heave Offiering.

Similarly, if a priest eats of them in Jerusalem before they have been deposited in the Temple court, he is liable to a flogging according to the Torah, just as if he had eaten of them outside of Jerusalem, inasmuch as they must be first deposited in the Temple court, as it is said, *And thou shalt set it down before the altar of the Lord thy God* (Deut. 26:10).

4. Once the owner has deposited his first fruits in the Temple court, they become permitted to the priest, even if the owner has not yet made confession over them, for failure to recite the confession cannot prevent the eating of the fruits.

First fruits that have gone out of their precinct, and were then returned thereto, are allowed to be eaten.

5. The eating of first fruits is the same as the eating of heave offering in every respect, except that first fruits are subject to the

following additional rules: they are forbidden to a mourner; they must be brought to the prescribed place; and a clean priest who eats of unclean first fruits is liable to a flogging, the same as a clean Israelite who eats of unclean second tithe—all of which does not apply to heave offering.

6. Whence do we learn that first fruits are forbidden to a mourner? From what is said concerning them, *And thou shalt rejoice in all the good* (Deut. 26:11), which implies that one must eat of them in joy and not in mourning. He who eats of them while in mourning is liable to a flogging for disobedience.

7. First fruits require the use of a container, as it is said, *And thou shalt put it in a basket* (Deut. 26:2). The preferable way is to bring each species in its own basket, but if one brings them all in the same basket, he has fulfilled his obligation. He should not, however, bring them mixed together, but rather barley at the bottom, then wheat, then olives, then dates, then pomegranates, and finally figs at the top of the container. Something else should separate one species from another, such as palm foliage, young shoots, leaves, and the like. The figs should be surrounded on the outside by grape clusters.

8. If one brings first fruits in a metallic vessel, the priest should take out the first fruits and return the vessel to its owner. If one brings them in a basket made of willow, young shoots, or the like, both first fruits and baskets accrue to the priests. If the first fruits become unclean, the baskets do not accrue to the priests.

9. When they used to bring first fruits they would bring in their hands also turtledoves and young pigeons, and would moreover suspend turtledoves and young pigeons from the sides of the baskets, to form a wreath around the first fruits. The birds suspended from the baskets were offered up as burnt offerings, while those carried in the people's hands were given to the priests.

10. It is a positive commandment to make confession in the Temple over first fruits at the time that one brings them. One should begin by reciting, *I profess this day unto the Lord thy*

God that I am come unto the land (Deut. 26.3), etc., *A wandering Aramean was my father* (Deut. 26:5), and go on to the end of the Scriptural lesson, up to *Which Thou, O Lord, hast given me* (Deut. 26:10). It may be recited only in the holy tongue, as it is said, *And thou shalt speak and say* (Deut. 26:5), implying, in that same tongue.

11. Originally all those who knew how to read would recite the confession, while those unable to read would have it read to them. As a result those who could not read refrained from bringing first fruits in order to avoid embarrassment. The Sages consequently ruled that both those who could read and those who could not should have it read to them.

12. He who brings first fruits to Jerusalem may let his bondsman or his relative carry them all the way until he reaches the Temple Mount. Once he reaches the Temple Mount, he himself must take the basket upon his shoulder, even if he is a great king in Israel, enter, and proceed until he reaches the Temple court, whereupon he must recite the confession, with the basket still upon his shoulder, *I profess this day unto the Lord thy God* (Deut. 26:3). He must then take the basket off his shoulder and hold it by its rim, while the priest puts his hand underneath it and waves it, as the owner recites, *A wandering Aramean was my father, and he went down into Egypt,* etc. (Deut. 26:5), until he completes the whole lesson. The owner must thereupon set the basket down by the side of the altar, in the southwestern corner, by the southern side of the corner, bow down, and depart.

Whence do we know that first fruits require waving? From what is said, *And the priest shall take the basket out of thy hand* (Deut. 26:4), which implies the additional requirement of waving the first fruits. And just as they require waving, so do they require a peace offering and a psalm, as it is said of them, *And thou shalt rejoice in all the good* (Deut. 26:11). Inasmuch as of the festivals it is said, *And thou shalt rejoice in thy feast* (Deut. 16:14), it follows that just as the rejoicing in the festival is marked with peace offerings, so here too rejoicing in first fruits must be marked with peace offerings. Failure to bring these offerings, however, cannot prevent the bringing of first fruits.

13. When is a psalm to be recited over them? When the owner has reached the Temple court, the Levites would commence reciting, *I will extol Thee, O Lord, for Thou hast raised me up* (Ps. 30:2).

14. First fruits require an overnight stay in Jerusalem. How so? If one brings his first fruits into the Temple, recites the confession, and offers his peace offerings, he may not leave Jerusalem that same day to return home, but must remain there overnight and on the morrow return to his city, as it is said, *And thou shalt turn in the morning, and go unto thy tents* (Deut. 16:7), implying that all departures from the Temple, whenever you come there, shall take place only the next morning.

You thus conclude that first fruits involve seven requirements: bringing to the required place, container, recitation, offering, psalm, waving, and overnight stay.

CHAPTER IV

1. Whosoever brings first fruits to Jerusalem is required to accompany them with offering, psalm, waving, and overnight stay in the Holy City. The requirement to make confession, however, does not apply equally to everyone, for there are some persons who are obligated to bring first fruits but not to recite the confession over them.

2. The following are required to bring first fruits but not to recite the confession: a woman; a *ṭumṭum* and a hermaphrodite, since they are doubtfully female, and therefore cannot say, *Which Thou, O Lord, hast given me* (Deut. 26:10); also a guardian, a bondsman, and an agent, because they too cannot say, *Which Thou, O Lord, hast given me.*

3. A proselyte must bring first fruits and recite the confession, since Abraham was told, *The father of a multitude of nations have I made thee* (Gen. 17:5), implying that he is the father of everyone who enters under the wings of the Presence; and the Lord's oath was given first to Abraham that his children shall

inherit the Land. Similarly priests and Levites must bring first fruits and recite the confession, because they were given open land cities.

4. If A buys two trees in B's field, he must bring their first fruits but need not recite the confession, because there is doubt as to whether he has or has not acquired some land as well. What then should he do? He should first consecrate the first fruits to the repair of the Temple, because they are doubtfully profane, and one may not bring profane produce into the Temple court. The priest may then redeem them from the Temple treasury, and may thereupon eat of them. A must then set aside out of them heave offering and tithe, because they are doubtfully profane, and must give their tithes to the priests, since they may be first fruits, and are therefore forbidden to commoners. A should not bring them himself, but should send them along by the hand of an agent, so that failure to recite the confession would not prevent them from being eaten, because in the case of anything that is not suitable for recital of the confession out of doubt, failure to recite the confession prevents it from becoming permitted for consumption.

5. If one sets aside his first fruits and then sells his field, he must bring the first fruits to Jerusalem but need not recite the confession, for he cannot say, *Which Thou, O Lord, hast given me* (Deut. 26:10), seeing that he no longer owns the field. The buyer need not set aside any other first fruits of the same species, inasmuch as the seller has already set them aside. If the buyer does set them aside nevertheless, he need only bring them to Jerusalem, and need not recite the confession. Anything he sets aside out of another species, however, he must bring to Jerusalem and recite the confession.

6. If one sells only the yield of his field, the buyer must bring first fruits but need not recite the confession, because acquisition of the yield is not equivalent to the acquisition of the corpus of the field. A husband, however, must bring first fruits from the estate of his wife, and must recite the confession, even though he does not own the corpus of the wife's land, as it is said, *Which*

the Lord thy God hath given unto thee, and unto thy house (Deut. 26:11).

Even if his wife dies after he has set aside the first fruits, while he is on the way to Jerusalem, he must bring them there and recite the confession.

7. If A sells his field to B, or sells some trees and the land underneath them, the rule is as follows: When the Jubilee law is in force, B must bring first fruits and recite the confession only during the first Jubilee period, for the reason that A's mind is not yet set that the land is to return to him. If, however, A then sells it again to B in the second Jubilee period, B must then bring first fruits but need not recite the confession, inasmuch as A's mind is now set that B is to have only the yield, and the ownership of the yield is not tantamount to the ownership of the corpus of the land.

8. If A, having set aside his first fruits, falls critically ill, his heir apparent may bring them to Jerusalem but need not recite the confession. If A, having set aside his first fruits, sends them off by the hand of B, and B dies on the way, the rule is that even if A himself thereupon brings them to Jerusalem, he need not recite, *That thou shalt take . . . and thou shalt come* (Deut. 26:2–3), since this implies taking and bringing simultaneously.

9. If a person, having set aside his first fruits, loses them before he reaches the Temple Mount, and then sets aside other fruits in their stead, he may bring the latter but need not recite the confession, inasmuch as he cannot say, *The first of all the fruit of the ground* (Deut. 26:2), since they are not the first. These replacement fruits are not liable to the additional one-fifth, whereas the original first fruits are so liable.

10. If one brings in his first fruits, and they become unclean in the Temple court, he may scatter them there and need make no confession.

11. If one brings his first fruits out of a certain species and recites the confession, and then again brings first fruits out of another species, he need not recite the confession over them, as

it is said, *I profess this day* (Deut. 26:3), implying that he must make confession once a year and not twice.

12. If after one has set aside his first fruits the well that waters the tree dries up, or the tree is cut down, he must bring in the first fruits but need make no confession, since he is now the same as if he had no land, seeing that his land has perished.

13. If one brings first fruits after the Festival of Tabernacles, up to Hanukkah, he may do so but need not recite the confession, even if he had set them aside before Tabernacles, as it is said, *And thou shalt rejoice in all the good* (Deut. 26:11), implying that the recital may be made only during the time of rejoicing, from the Feast of Weeks until the end of Tabernacles. All others who bring first fruits, except those previously mentioned, must bring them and make confession.

14. First fruits, heave offerings, dough offering, the principal and the additional one-fifth, and the gifts out of cattle are the property of the priest. He may purchase with them bondsmen, land, and unclean animals; his creditor may seize them for his debt, and his wife for her kĕṯubbah; and he may buy a Scroll of the Torah with them.

15. Even though first fruits and heave offerings may not be consumed by commoners, and even though first fruits, like heave offering, are neutralized by 101 times their bulk of profane produce, nevertheless, if the slightest amount of them becomes mixed in Jerusalem with profane produce of the same species, they render the mixture forbidden, the same as second tithe, for inasmuch as Jerusalem is the place where they must be consumed, they are regarded by the Sages as a forbidden thing which will eventually become permitted. Therefore, even though first fruits are forbidden to commoners even in Jerusalem, they nevertheless render the mixture forbidden by the slightest amount of them.

Even if one plants the first fruits after they had entered Jerusalem, their outgrowth is forbidden, and even the slightest admixture thereof, if it occurs in Jerusalem, renders the mixture forbidden. If, however, first fruits are planted before they enter Jerusalem, their outgrowth is accounted profane produce.

16. How are first fruits to be brought to Jerusalem? All the inhabitants of the towns that constitute the particular post would assemble in the principal city of that post, in order that they should not go up separately, as it is said, *In the multitude of people is the king's glory* (Prov. 14:28). They would then come and spend the night in the open place of the city, taking care not to enter any houses, as a precaution against uncleanness conferred by being under the same roof with a dead body.

Early in the morning the appointed officer would say, *Arise ye, and let us go up to Zion, unto the Lord our God* (Jer. 31:6). An ox would lead the way, its horns overlaid with gold and an olive wreath around its head, to make known that the first fruits are brought out of the seven species. A flute would play before them until they reached the vicinity of Jerusalem, while they would walk all the way reciting, *I rejoiced when they said unto me, Let us go unto the House of the Lord* (Ps. 122:1). They would walk only during two-thirds of the day.

When they reached the vicinity of Jerusalem, they would send messengers ahead to notify the people of Jerusalem, and they would wreathe their first fruits and decorate them. If they had both fresh and dried fruit, they would exhibit the fresh on top of the dried. The chiefs of the priests and of the Levites, and the treasurers of the Temple, would go out of Jerusalem to meet them, in corresponding numbers: if many people were arriving, many would go out to meet them; if few, few. As they all entered the gates of Jerusalem, they would commence reciting, *Our feet are standing within thy gates, O Jerusalem* (Ps. 122:2).

17. All the craftsmen of Jerusalem would rise up before them and greet them, saying, "Brethren, people of such-and-such a place, welcome!"

They would then walk through Jerusalem, the flute playing before them, until they reached the Temple Mount. Once they reached the Temple Mount, each person would take his basket upon his shoulder and would recite, *Hallelujah. Praise God in His Sanctuary* (Ps. 150:1), etc., up to *Let everything that hath breath praise the Lord. Hallelujah* (Ps. 150:6). They would then walk upon the Temple Mount reciting until they reached the Temple court, whereupon the Levites would sing the psalm, *I will extol thee, O Lord, for Thou hast raised me up* (Ps. 30:2).

CHAPTER V

1. It is a positive commandment to set aside a heave offering for the priest out of dough, as it is said, *Of the first of your dough ye shall set apart a cake for a gift* (Num. 15:20). This *first* has no set amount stated for it in the Torah, so that even if one sets aside as little as the bulk of one barleycorn, he has made the rest of the dough exempt. If one turns all of his dough into dough offering, he has done nothing, unless he has left some of it over.

2. According to Scribal law, however, one should set aside 1/24 of the dough, in order to form an amount fit for a gift to the priest, as it is said, *shalt thou give him* (Deut. 18:4), implying, give him something suitable for a gift.

A baker who bakes goods for sale in the market place must set aside 1/48—his dough being ample, this amount is sufficient for a gift.

3. A person who prepares dough for his son's wedding feast must set aside 1/24 of it, even though his dough is ample, in order to avoid variations in the rule for householder's dough. Likewise a baker who prepares only a small amount of dough must set aside 1/48 of it, in order to avoid variations in the rule for baker's dough.

4. If the dough becomes unclean by error or unavoidably, one must set aside 1/48 of it, but if one renders it unclean deliberately, he must set aside 1/24, so that he would not profit by his transgression.

Unclean dough offering is permitted to a priest for use as fuel, the same as unclean heave offering.

5. One is liable to dough offering, according to the Torah, only in the Land of Israel, as it is said, *When ye eat of the*

bread of the land (Num. 15:19), and only when all Israel are there, as it is said, *When ye come into the land* (Num. 15:18), implying, when all of you come and not just some of you. Therefore at this time, and in the days of Ezra even in the Land of Israel, one is liable to dough offering only according to Scribal law, as we have explained in regard to heave offering.

6. Produce from lands outside of the Land of Israel imported into the Land of Israel is liable to dough offering, while produce of the Land of Israel exported abroad is exempt, as it is said, *Whither I bring you* (Num. 15:18), implying that there, in the Land of Israel, you are liable, whether the produce is domestic or imported.

7. Dough offering must be set aside outside of the Land of Israel on Scribal authority, in order that the practice of dough offering should not be forgotten in Israel. However, dough offerings from lands outside of the Land of Israel may not be imported into the Land of Israel, just as heave offering and first fruits may not be so imported. If one does so nevertheless, he must leave it until the eve of Passover, when it must be burned the same as heave offering.

8. There are three rules governing dough offering in three respective territories. In all the territory up to Chezib, occupied by those who went up from Babylonia, one dough offering must be set aside in the prescribed amount and is to be consumed by the priests. In the remainder of the Land of Israel, occupied by those who went up from Egypt and not by those who went up from Babylonia, which is from Chezib up to Amanah, two dough offerings must be set aside, one to be burned and one to be consumed. And why must one set aside two dough offerings? Because the first dough offering is unclean, inasmuch as that territory was not reconsecrated in the days of Ezra, while its original sanctity had lapsed once the children of Israel were exiled; but since it is an integral part of the Land of Israel, dough offering in the amount of $\frac{1}{48}$ must be set aside and burned. Thereafter one must set aside a second dough offering and give it to the

priest for consumption, so that people would not say that clean heave offering may be burned, seeing that the first offering was burned, even though it did not contract any uncleanness that was manifest to everyone. This second dough offering has no prescribed amount, rather one may set aside as much as he may choose, seeing that it is required by Scribal law only.

In all the territory from Amanah and beyond, whether now situated in Syria or in the other lands, one must set aside two dough offerings, one to be burned, so that people would not say, "We saw unclean heave offering being consumed," and one to be consumed, so that the practice of dough offering should not be forgotten in Israel.

Inasmuch as both these offerings are prescribed by Scribal law only, it is preferable to increase the amount of the one that is consumed; the one to be burned, therefore, has no prescribed minimum, and the least amount of it is sufficient. The prescribed amount of the one to be consumed is 1/48, and today, when there is no clean dough, it is permitted to men and women afflicted with flux and, needless to say, to all other unclean persons.

9. Because of the prevalence of uncleanness due to contact with dead bodies, a uniform amount of dough offering must be set aside in all of the Land of Israel, namely 1/48; it must be burned, because it is unclean and because it is authorized by the Torah.

From Chezib up to Amanah a second offering must be set aside for the priest's consumption; this one has no prescribed amount, and represents a continuation of the original offering.

10. Dough offering of lands outside of the Land of Israel, even though unclean, is not forbidden for consumption, inasmuch as the root of the liability to it is Scribal, except to priestly persons whose uncleanness stems from within their own bodies, to wit, males who have suffered nocturnal pollution, males and females afflicted with flux, females who are menstruating or have been delivered, and lepers. Other unclean persons, however, who are unclean by contact with other unclean things, even with dead bodies, are permitted to eat of it.

Therefore, if a priestly minor in that territory outside of the

Land of Israel, whether in Syria or elsewhere, wishes to set aside one dough offering, he may set aside 1/48 of the dough, which may then be consumed by a minor male who has not yet experienced any nocturnal pollution, or by a minor female who has not yet menstruated, and he need not set aside a second offering.

11. And likewise, if there is an adult priest in a land outside of the Land of Israel who has immersed himself in order to purify himself from a nocturnal pollution or from flux, he may eat of the first dough offering, even if it is not yet sunset, and even if his uncleanness is due to contact with a dead body, and he need not set aside a second dough offering.

12. Whosoever sets aside dough offering must first pronounce the benediction, "Blessed art Thou, O Lord . . . who has sanctified us by His commandments and has commanded us to set aside dough offering," whether he does it in the Land of Israel or outside of it.

Just as he must pronounce the benediction over clean dough offering, so must he pronounce it over unclean dough offering. Therefore a naked person may not set aside dough offering, because he may not pronounce the benediction. A naked woman, however, who is seated with her private parts completely pressed to the ground, may pronounce the benediction and set aside dough offering.

13. A menstruating woman, and her like, may pronounce the benediction and set aside dough offering outside of the Land of Israel, for she is cautioned not against touching it but only against eating of it.

If there is a priestly minor in a land outside of the Land of Israel, or an adult priest who has immersed himself—both of whom are permitted to eat of dough offering, as we have explained—he may eat of it together with a commoner at the same table, inasmuch as it does not render other dough intermixed with it forbidden, even if the two ingredients are equal in amount. It may be given to an unlearned priest, because it has already been rendered unclean by the air space of the foreign land. And there need be no apprehension here that this may be a case of a priest assisting in the preparation of gifts due to himself, which is forbidden.

If one wishes to eat of the dough first, and then set aside dough offering, in a land outside of the Land of Israel, he may do so, inasmuch as the root of this liability is in a Scribal enactment.

14. Dough offering is called "heave offering." Therefore it may be taken only from dough that is situated nearby, as is the rule for heave offering; and it may not at the outset be taken from clean dough for unclean dough.

15. Wherever we have said in connection with heave offering, "one may not set it aside, and if he does so nevertheless, it is not heave offering," the same rule applies also to dough offering; and the same is true wherever we have said in the same connection, "one may not set it aside from one batch for another." Whosoever may not eat of heave offering, may not eat of dough offering; and contrariwise, whosoever may eat of heave offering, may eat also of dough offering.

16. Blind and intoxicated persons may at the outset set aside dough offering, because in the case of dough there is no distinction between inferior and choice, and therefore no prior intention is required to set aside the offering out of choice dough.

CHAPTER VI

1. If one buys bread from a baker, he is liable to dough offering, and may set it aside out of freshly baked bread for stale bread, and vice versa, even if the loaves are from several molds.

2. Only five species of produce are liable to dough offering, to wit: wheat, barley, emmer, two-rowed barley, and spelt, as it is said, *When ye eat of the bread of the land* (Num. 15:19), and only bread made out of these is called bread. He, however, who makes bread out of rice, durra, or the like of the pulse species, is not at all liable to dough offering.

3. Gleanings, forgotten sheaf, corner crop, ownerless produce, and produce not yet one-third grown, even though exempt from heave offering, are liable to dough offering.

Similarly, first tithe set aside anticipatorily while still in its ears, out of which heave offering of the tithe has been set aside, even though it still contains the part which is great heave offering; second tithe and consecrated produce which have been redeemed; the residue of the ʻomer, of the two loaves, and of the shewbread, once all these residues have been redeemed—all these are liable to dough offering.

4. Dough of second tithe in Jerusalem, dough of seventh-year's produce, and dough doubtfully mixed with heave offering are liable to dough offering; dough assuredly mixed with heave offering is exempt from dough offering.

5. Loaves of thank offering and the Nazirite's wafers prepared by him for his own use are exempt from dough offering, inasmuch as they are hallowed; if prepared to be sold in the market place to Nazirites and offerers of thank offerings, they are liable to dough offering, because the maker's intention is, should they not be sold, to eat them himself.

6. Dough held in partnership and dough prepared for public consumption are liable to dough offering.

7. If one makes dough out of untithed produce, whether he then sets aside dough offering first and heave offering second, or vice versa, what he has done is done. If he sets aside dough offering first, it may not be eaten until he has set aside also the heave offering and the heave offering of the tithe that are due out of it. If he sets aside heave offering first, it may not be eaten until he has set aside the dough offering that is due out of it.

8. If one makes dough in order to feed the bread thereof to cattle or to wild beasts, it is exempt from dough offering. Dough made to feed dogs is liable to dough offering if herdsmen also eat of it.

The dough of a heathen is exempt.

9. If an Israelite and a heathen own dough in partnership, the rule is that if the Israelite's share amounts to as much as the minimum that is subject to dough offering, all of the dough is liable to dough offering.

10. If a heathen sets aside dough offering, even in the Land of Israel, it is not dough offering; he should be informed that he need not do so, and it may be consumed by a commoner.

Why should we not be apprehensive that the dough might in fact belong to an Israelite, who had assigned it to the heathen in order to exempt it? Because had the Israelite wished to do so, he could have claimed exemption by making his dough in several batches, each smaller than the minimum that is subject to dough offering.

11. If one mixes wheat flour with rice flour to make dough, the rule is as follows: If the dough tastes of wheat, it is liable to dough offering; if not, it is exempt. Even if wheaten leaven is added to rice dough, the rule is the same: If the dough tastes of wheat, it is liable to dough offering; if not, it is exempt.

12. Whether one kneads the dough with wine, oil, honey, or hot water, or spices it, or boils the water first and then pours the flour into it and kneads it; and then bakes it, whether in an oven or in a hole in the ground, whether in a shallow pan or in a deep pan, whether he first plasters the dough in the shallow pan or in the deep pan and then heats it over the fire until the bread is baked, or first heats the pan and then plasters the dough in it—in all these cases the dough is liable to dough offering.

If, however, one makes dough in order merely to dry it in the sun or boil it in a pot, it is exempt from dough offering, inasmuch as sun-dried dough is not bread, whether it is kneaded with water or with other liquids. Similarly, parched corn kneaded with water or honey and eaten without baking is exempt, for only dough intended to be baked into regular bread for human consumption is liable to dough offering.

13. If one kneads dough at first with the intention of sun-drying it, but ends by baking it as bread, or vice versa, and like-

wise if one kneads parched corn at first with the intention of baking it as bread, but fails to do so, he is liable to dough offering.

14. In the case of bread made for use in a milk sauce, its preparation indicates its status: If made like thin cakes, it is liable to dough offering; if made like shingles, it is exempt.

15. What is the minimum amount of dough that is liable to dough offering? An ʿomerful of flour, whether of one of the five species or of all of them together, for they all combine to form the minimum.

And what is the volume of an ʿomer? Two ḳab̲ less one-fifth of a ḳab̲. One ḳab̲ is four log̅; one log̅ is four quarters; one quarter is two fingers by two fingers, raised to the height of two fingers plus one-half of a finger plus one-fifth of a finger; and one finger is the width of the thumb. You thus learn that the volume of ten fingers by ten fingers raised to the height of approximately three and one-ninth fingers is an ʿomer. A volume that is seven fingers less two-ninths by seven fingers less two-ninths raised to the height of seven fingers less two-ninths is likewise an ʿomer, the two volumes being approximately equal.

How much does this volume contain? About 43⅕ middle-sized eggs, corresponding to the weight of 86⅔ selaʿ of Egyptian wheat flour, equal to 520 current Egyptian zuz. This measure, containing the aforementioned weight of such wheat flour, is the one to be used in measuring for dough offering everywhere.

16. One is forbidden to make his dough in quantities less than the minimum required to make it liable to dough offering, in order to render it exempt from it. On the other hand, he who sets aside dough offering from dough of less than the minimum amount has done nothing, and it remains profane as it was before.

If one kneads less than the minimum amount of dough, bakes it, puts the bread in a basket, and then bakes some more bread and puts it in the same basket, and so forth, the rule is that if enough bread accumulates in the basket to constitute the minimum liable to dough offering, the basket combines it for the dough offering, and one must set aside dough offering from that bread, as it is said, *Then it shall be that when ye eat of the bread of the land* (Num. 15:19), which teaches us that dough offering must be set aside out of dough that is already baked. The oven, however, cannot combine several batches of dough for the purpose of dough offering.

17. If the loaves bite into one another and all together amount to the required minimum, they are liable to dough offering, even if they are not yet in the same basket.

If one bakes the dough piece by piece, and then gathers all the pieces on a flat board, the matter is in doubt. If it is a case of dough offering required only by Scribal law, one is not obligated to set it aside until all the pieces are deposited in a concave vessel.

18. Unsifted flour kneaded with its bran is liable to dough offering, provided that the whole flour amounts to an 'omer. If, however, one first removes the coarse bran from the flour, and then goes back and replaces it in order to bring the dough up to the minimum volume, it is not liable to dough offering.

19. If a baker makes dough for distribution as leaven, it is liable to dough offering, seeing that if it is not sold, he will turn it into bread.

If, however, one makes dough in order to distribute it as kneaded dough, it is exempt.

20. In the case of several women who give flour to a baker to make leaven for them, the rule is that if each one's share is less than the required minimum, it is exempt, even if the combined shares amount to that minimum.

CHAPTER VII

1. In the case of two batches of dough amounting together, but not separately, to the minimum liable to dough offering, which touch or bite into each other, they are exempt from dough offering if they belong to two persons, even if they are of the same

kind of grain, because ordinarily two persons are particular about intermixing their dough. But if it is known that they are not so particular, the two batches combine together.

2. If the two batches of dough belong to the same person, the rule is as follows: If they are made of the same species of grain, they combine together and are liable to dough offering; if they are made of different species of grain, they do not combine together, because ordinarily one person does not mind if his batches of dough are mixed together. If he is known to be particular that his batches of dough should not touch each other or be mixed with one another, they do not combine together even if they are made of the same species of grain.

3. How do batches of dough belonging to the same person and made of the same species of grain combine together? If a batch of wheaten dough touches a batch of emmer dough, they combine together; if it touches dough made of the other species of grain, they do not combine together.

Similarly, barley dough that touches dough of emmer, two-rowed barley, or spelt, and likewise batches of dough of emmer, two-rowed barley, and spelt that touch each other, do combine together.

4. Dough made of new grain cannot combine with dough made of old grain, even if they are of the same species, lest people should say that one may set aside heave offering out of new produce for old. Nor should one set aside dough offering from the adjoining portions of the two batches—he should rather bring a third batch made of old or new grain and combine it with one of the two, respectively, to complete the required minimum.

When does this apply? When one batch of dough touches the other. If, however, a person mixes flour of the five species of grain to make one batch of dough, the five ingredients combine together to make up the required minimum liable to dough offering, as we have explained.

5. If batch A of dough, less than the prescribed minimum, is on one side, and batch B, likewise less than the prescribed minimum, is on the other, and batch C, which is not liable to dough

offering—as for example, rice dough, or heave offering dough, or dough mixed with heave offering, or heathen's dough—is in the middle, the three batches do not combine together even if they touch each other, inasmuch as batch C, which is exempt from dough offering, separates batches A and B.

6. If batch C is one out of which dough offering has already been set aside, batches A and B do combine together, inasmuch as batch C between them had been liable to dough offering.

Similarly, if batch C is consecrated dough, batches A and B combine together, because batch C is fit to be redeemed, after which it will become liable to dough offering.

The same rule applies if batch C is made of a different species of grain, or belongs to another person, or is dough of new produce—batches A and B on the two sides combine together as far as dough offering is concerned.

7. If two batches of dough, each one less than the required minimum, which have had dough offering set aside out of them, subsequently touch each other, so that together they constitute that minimum, they are now jointly liable to dough offering, because the first dough offerings were invalid.

8. In the case of two heathens who had jointly made dough up to the required minimum, divided it between themselves, and then became proselytes, whereupon, after having become proselytes, each one added to his own share to bring it up to the required minimum, each share is liable to dough offering, inasmuch as it was not liable while they were heathens, since at that time each one of them had an amount less than the minimum.

9. If, however, two Israelites do likewise, and after dividing the dough each one goes back and adds to his share enough to bring it up to the required minimum, each share is exempt, inasmuch as the dough was potentially liable at one time, although the partners were then exempt because they had made the dough with the intent aforethought to divide it.

10. If the dough is held in partnership by a heathen and an Israelite, is divided, and then the heathen becomes a proselyte,

whereupon both he and the Israelite add to their shares to bring each share up to the required minimum, the rule is that the Israelite's dough is liable, while the heathen's dough is exempt.

11. A person who has taken leaven from dough out of which no dough offering has yet been set aside, and has put it into dough out of which dough offering has already been set aside, should fetch another batch of dough, such as together with the leaven would come up to the minimum liable to dough offering, put it next to the dough out of which dough offering has already been set aside, and set aside out of this new batch the amount of dough offering due out of it together with the leaven, in order to take the offering out of adjacent dough. If he has no other dough, it all becomes unregularized dough, and he must set aside dough offering for the entire amount of it, since even the smallest amount of unregularized produce of the same species renders a mixture forbidden.

12. Dough still liable to dough offering is not like dough offering itself but rather like profane produce as far as uncleanness is concerned, since profane produce that is unclean in the second degree cannot render other produce that it touches unclean in the third degree, as will be explained in its proper place.

It is permitted to cause profane produce in the Land of Israel to become unclean. Therefore, if one has two batches of dough, one unclean and the other clean, he may take the amount of dough offering due out of both of them, out of other dough from which the offering has not yet been set aside, place it in the middle next to the clean dough, and draw an egg-sized piece of the unclean dough over the clean dough, in order to set aside dough offering out of adjacent dough.

13. A person may make a batch of clean dough without setting aside the dough offering due out of it, and may put it aside in whole or in part, in order to set aside out of it dough offering for other batches of dough, even if these latter become unclean, until all of it becomes dough offering and may be given to the priest, provided that it does not become unfit for human con-

sumption. If it does become putrid, he may not use it for such vicarious offering.

When does this rule apply? When there is doubt as to whether dough offering has or has not been set aside out of these batches of dough, because in the case of such doubtful batches the offering may be taken at the outset out of clean dough for unclean dough, and out of dough which is not adjacent.

CHAPTER VIII

1. If one sets aside his dough offering out of flour, it is not valid dough offering; if it is already in the hand of the priest, it is the same as something obtained by robbery; and the rest of the dough remains liable to dough offering. As for the flour which had been set aside as dough offering, if it amounts to an ʿomer and is made into dough, it is liable to dough offering, like all profane flour.

2. When is dough offering to be set aside? As soon as water is added and the flour is mixed with it. One must set aside dough offering out of the first dough that is kneaded, as it is said, *Of the first of your dough* (Num. 15:20), provided that there is not an ʿomer of unkneaded flour left in the trough.

If one says, "This is to be dough offering for the mixed flour, for the leaven, and for the flour that is left unmixed," in order that when all of it becomes one batch of dough, that which is in his hand should become holy as dough offering, he is permitted to do so.

3. If one leaves the mixture until all of it is kneaded and mixed together, and then sets aside dough offering, it is still valid.

If he does not set aside the offering while the bread is still in the form of dough, but bakes it all, he may set aside the offering out of the baked bread, as we have explained.

4. When does the mixture become liable to dough offering? If it is made of wheat flour, when it has been rolled out and the

flour has become thoroughly mixed with the water; if of barley flour, when it has been well blended so that it becomes one cohesive mass. One may make a casual meal out of the mixture until it has been rolled out in the case of wheat flour, or until it has become a cohesive mass in the case of barley flour. Emmer is like wheat, and two-rowed barley and spelt are like barley, in this respect.

5. If wheaten dough has been rolled out, or barley dough kneaded into a cohesive mass, he who eats thereof before dough offering has been set aside is liable to the death penalty, because it has the status of unregularized produce. Therefore, if the dough is liable to dough offering on the authority of the Torah, he who eats thereof is liable to a flogging, like all who eat of unregularized produce. If it is liable only by Scribal law, he is liable to a flogging for disobedience.

6. If dough is mixed with heave offering, the rule is as follows: If it is so mixed before being rolled out, it is exempt; if after, it is liable.

Similarly, if one consecrates his dough or declares it ownerless before it is rolled out, then redeems it or retakes possession of it, and thereupon rolls it out; or if he consecrates it or declares it ownerless after it has been rolled out, and then redeems it or retakes possession of it, it is liable to dough offering.

7. If one dedicates the dough before it is rolled out, and it is then rolled out while in the possession of the Sanctuary, and he thereafter redeems it, it is exempt, inasmuch as at the time when it would have been liable it was exempt.

8. Similarly, if a heathen hands over flour to an Israelite to make dough for him, and then, before it is rolled out, gives it to him as a gift, it is liable to dough offering. If it is given to the Israelite after being rolled out, it is exempt.

9. In the case of an owner of dough who becomes a proselyte, the rule is as follows: If the dough was rolled out before he became a proselyte, it is exempt; if after, it is liable; if the fact is in

doubt, the dough is liable to dough offering, because here transgression involves the penalty of death.

A commoner who eats of such doubtful dough offering, and its like, is not liable to the additional one-fifth.

10. If suspicion of uncleanness arises concerning a batch of dough, the rule is as follows: If the suspicion arises before the dough is rolled out, one may process it in uncleanness, inasmuch as it is permitted to render profane produce unclean in the Land of Israel, and then one must burn its dough offering. If the suspicion arises after the dough is rolled out, and the suspected uncleanness is such that were it certain, it would have rendered profane produce unclean on the authority of the Torah, one may complete the processing in cleanness, because the Sages have ordained that where assured uncleanness would render profane dough unclean, doubtful uncleanness does not render unclean profane dough out of which dough offering has not yet been set aside, inasmuch as it has already become liable to dough offering. The dough offering thus remains in suspense—it may be neither eaten nor burned.

11. A person should not at the outset make his dough in uncleanness, but should rather exercise all possible care to make his person and his vessels clean, in order to set aside dough offering that is likewise clean. If, however, there is more than four miles between him and a body of water, he may make the dough in uncleanness and separate dough offering that is unclean.

12. One may not process dough offering for an unlearned person in cleanness, but one may make profane dough in cleanness.

How so? An associate scholar may knead this dough, set aside the offering due out of it, and put it in vessels made of baked dung, stone, or earth, since such vessels do not contract uncleanness. When the unlearned person comes along, he may take both the dough and the offering, but should be told, "Be careful not to touch the offering, lest it should once more become liable to dough offering." Why did the Sages allow him to do so? In order to safeguard the professional kneader's livelihood.

13. An associate scholar's wife may sift and pick over flour jointly with an unlearned person's wife, but not once the latter has poured water into the dough, inasmuch as she is then making her dough in uncleanness.

Similarly one should not knead or shape dough jointly with a baker who does his work in uncleanness, for no help should be given to those who engage in transgression. One may, however, assist him in conveying the bread to the retail dealer.

14. If one buys bread from an unlearned baker in Syria, and is told by him, "I have already set aside dough offering," the buyer need not set aside another dough offering out of doubt, because just as all Israelites in the Land of Israel are not under suspicion in regard to great heave offering, so are unlearned persons in Syria not under suspicion in regard to dough offering.

15. If one buys bread from a baker in a land outside of the Land of Israel, he must set aside heave offering out of doubt. If, however, one buys from a householder, and needless to say, if one is his guest, he need not set aside dough offering out of doubt.

CHAPTER IX

1. It is a positive commandment for anyone who slaughters a clean animal for food to give to the priest the shoulder, the two cheeks, and the maw, as it is said, *And this shall be the priests' due . . . the shoulder, and the two cheeks, and the maw* (Deut. 18:3). These are termed everywhere "gifts."

This commandment is binding at all times, both during the time of the Temple and after the time of the Temple, and in all places, both in the Land of Israel and in the lands outside of it. It applies to profane animals, but not to consecrated ones.

2. All consecrated animals that had suffered a permanent blemish prior to their consecration and have therefore been redeemed, are liable to these gifts. If the prior blemish was only temporary, or if the animals had been unblemished when consecrated, but subsequently became permanently blemished and were redeemed, they are exempt from the gifts.

3. A doubtful firstling is liable to the gifts in either case: if it is indeed a firstling, all of it belongs to the priest; if it is not a firstling, its gifts are due to the priest.

If there are two animals, and there is doubt as to which one is the firstling, and if the priest takes one of them out of doubt, the other animal is exempt from the gifts, the same as if the priest had acquired it first, and then returned it to its owner on account of its blemished state.

On the other hand, an animal that is tithe out of doubt is exempt under any circumstances, on the principle that if A claims something from B, the burden of proof is on A.

4. If a consecrated animal disqualified because of a blemish, and thus not liable to the gifts, is intermixed with other animals, even in the proportion of 1 to 100, they are all exempt, so long as each animal belongs to a different person, for inasmuch as each animal is subject to the doubt, it comes under the principle that if A claims something from B, the burden of proof is upon A.

If the same person slaughters them all, only one of them is exempt from the gifts.

5. Only a clean animal is liable to the gifts, as it is said, *Whether it be ox or sheep* (Deut. 18:3). A hybrid that is the offspring of a sheep and a goat is liable to the gifts. As for the *koy*, even though its nature is doubtful, all the gifts should be set aside out of it. If a male gazelle is mated with a she-goat, the offspring is liable to half the gifts, as it is said, *or sheep,* implying, or even part sheep. If a he-goat is mated with a hind, the offspring is exempt from the gifts.

6. Whether one slaughters an animal for Israelite consumption, or for heathen consumption, or for dog food, or for medicinal purposes, it is liable to the gifts.

7. An animal owned in partnership is liable to the gifts, as it is said, *From them that offer a sacrifice* (Deut. 18:3).

8. If one buys an animal with produce of the Sabbatical year, he is liable to the gifts.

Priests and Levites are exempt from the gifts, as it is said, *From the people* (Deut. 18:3), and there is doubt as to whether the Levites are or are not included in the term "people"; therefore the gifts are not taken from them. But if a priest has already taken them, he need not return them.

9. When does this apply? When the priest slaughters an animal for his own use. If, however, a priest who is a professional slaughterer slaughters and sells the meat in the market place, he should be given two or three weeks' grace, and thereafter the gifts should be seized from him and given to other priests. If he sets up a regular butcher shop to sell his meat, he should be allowed no period of grace, and the gifts should be seized from him immediately. If he refuses to give them up, he should be put under a ban until he does give them up.

10. If one slaughters an animal for a heathen or for a priest, he is exempt from the gifts. If he owns the animal jointly with a priest, he must mark his own share with some sort of sign, so that the gifts might be deposited by him with the share of the priest. If he fails so to mark his own share, he is liable to the gifts, inasmuch as not everyone knows that the priest is his partner.

Therefore, if the priest stands by him in the slaughtering place and transacts the business with him, he need not mark his own share.

If one owns an animal jointly with a heathen, he likewise need not mark his own share, inasmuch as a heathen would generally spread abroad the fact that he is a partner, even if he is not present during the sale of the meat.

11. If the priest stipulates to his Israelite partner that the partnership is not to cover the gifts, they are due to the priest, inasmuch as having said, "You may sell your own share, with the exception of the gifts that are due to me," the priest has reserved these gifts to himself as his share, and therefore they are his.

If, however, the priest has said to his partner, "You may sell your own share, on condition that the gifts are to accrue to me," the gifts remain the property of the Israelite partner, and he may give them to any priest he wishes. For even though the priest had stipulated that the gifts are to accrue to him, the Israelite is not thereby exempted from setting them aside, since when one says, "on condition," he does not reserve for himself anything of the corpus of the gifts. Having reserved no partnership for himself in them, the priest has acquired no title to them by this stipulation.

12. If the priest is joint owner of the animal's head, the Israelite partner is exempt from giving the cheeks; if the priest is joint owner of the whole leg, the Israelite is exempt from giving the shoulder; if the priest is joint owner of the entrails, the Israelite is exempt from giving the maw.

If the priest says to his Israelite partner, "The whole animal is to be mine, but the head is to be yours," the Israelite is liable to the cheeks, inasmuch as the part that is liable belongs to the Israelite.

13. If a proselyte has a slaughtered animal, the rule is as follows: If it was slaughtered before he became a proselyte, he is exempt; if after, he is liable; if the time of slaughtering is in doubt, he is exempt, for the burden of proof lies upon the claimant.

14. One may eat of the flesh of an animal out of which the gifts have not yet been set aside, for it is not like untithed produce, inasmuch as the gifts due to the priesthood are distinct. The gifts themselves are forbidden for consumption by an Israelite except with the priest's permission.

If the Israelite transgresses and does eat of them, or damages them, or sells them, he is not liable to restitution, because they constitute property that has no specific claimant. He who buys them, though forbidden to do so, may consume them, because gifts due to the priesthood are amenable to being seized by robbery.

15. If one says to a butcher, "Sell me the cow's entrails," and if these include the gifts, the buyer must give them to the priest,

and the seller need not allow any discount from the purchase price on that account. If, however, they are bought by weight, the buyer must give the gifts to the priest, and the seller must allow a discount from the price on that account.

16. If A sends to B some meat which includes the gifts, B need not be apprehensive that A might have transgressed and misappropriated them. In a place where there is no priest, B must estimate the monetary value of the gifts, and only then eat of them, so as to avoid any loss to the priest, and thereupon give the money to any priest he may choose.

17. If one wishes to give all the gifts to the same priest, he may do so. If he wishes to divide them, he should not give half the maw or half the shoulder to one priest and half to another, but rather the whole shoulder to one, the whole maw to another, and one cheek to each of the two, as it is said, *Shalt thou give him* (Deut. 18:4), implying that there must be enough of it to constitute a gift.

If the gifts are those of an ox, he may slice them, provided that each slice is large enough to constitute a gift.

18. What is the definition of the shoulder? The right shoulder from the knee joint up to the shoulder socket of the foreleg, which comprises two joints, one attached to the other.

What is the definition of the cheeks? From the joint of the jaw to the knob of the windpipe, including the great ring of the windpipe, and the part of the tongue that is between them—all these accrue to the priest.

19. The shoulder and the cheeks may not be scalded to remove the hair, nor flayed; rather they must be given with their hide and hair.

The maw must be given with the fat that is both on it and inside it. The priests, however, have long ago established the custom of leaving the fat of the maw to the owner.

20. A priest's daughter, even if married to an Israelite, may eat of the gifts, for there is no holiness in them. Moreover, an Israelite

husband may eat of them on account of his priestly wife. A profaned woman, however, may not eat thereof, because profaned persons related to priestly families are not included in the priesthood.

If the priest wishes to sell the gifts, or give them away as presents, even to a heathen, or feed them to dogs, he may do so, because no holiness whatsoever is inherent in them.

21. If a priest whose Israelite companions regularly give him the gifts wishes to assign them to one of his Israelite companions, he may do so, even if these gifts have not yet come to his hand. His companions may then slaughter the animals and give the gifts directly to the Israelite assignee, provided that the latter is in dire need and does not have the wherewithal to buy meat, and that the priest assignor is the assignee's companion.

If, however, the priest assignor is the assignee's servant, or his hired man, or his gleaner, he may not assign such gifts to him until they have first come to his own hand, lest he should assign them under duress.

22. A priest should neither snatch the gifts nor himself ask for them, but should accept them only when they are given to him in a dignified manner. When there are several persons in the butcher shop, modest priests should withdraw their hands from the gifts, since only greedy priests would seize them then and there.

In the case of the modest priest whose priestly status is not known to everyone, he may take hold of the gifts in order to let it be known to all that he is a priest.

The priests may eat the gifts only roasted, with mustard, as it is said, *By reason of the anointing* (Num. 18:8), implying that they are to be eaten in a royal manner.

CHAPTER X

1. It is a positive commandment to give the first of the fleece to the priest, as it is said, *And the first of the fleece of thy sheep shalt thou give him* (Deut. 18:4). In regard to this commandment the Levites are included with the Israelites.

There is no amount prescribed in the Torah for the first of the fleece, but a Scribal enactment defines it as not less than one-sixtieth.

It is binding only in the Land of Israel, both when the Temple is in existence and when it is not, as is the rule for the first of the corn; and it applies to profane animals but not to consecrated ones.

2. How so? For example, if one consecrates animals for the repair of the Temple, and then shears them, would he be obligated to redeem the first of their fleece and give the money to the priest? Or if he consecrates a single animal except for its fleece, would he be liable to the first of its fleece? The answer is that Scripture states expressly, *thy sheep*—these, once consecrated, are no longer his sheep.

3. All consecrated animals permanently blemished prior to their consecration and redeemed, are liable to the first of the fleece. If, however, their consecration antedates their blemish, or if the prior blemish is temporary and is later superseded by a permanent blemish, and they are redeemed, they are exempt from the first of the fleece.

4. Only sheep, male as well as female, are liable to the first of the fleece, for only their wool is suitable for clothing. If their wool is stiff and unsuitable for garments, they are exempt from the first of the fleece, because this gift is given to the priest for the sole purpose of making clothing out of it.

5. Inasmuch as the Holy One, blessed be He, has bestowed upon the priest the heave offerings, which are his bread and his wine, and the gifts out of both profane animals and those consecrated to the Temple, which are his meat, He has also bestowed upon him the first of the fleece for his clothing, as well as things unlawfully seized from a proselyte, things devoted to the priests, the field of possession, and the redemption money of firstlings, for his other expenses and needs, seeing that he has no share in Israel's inheritance and in the spoils of war.

6. If the sheep's wool is naturally red, black, or dark brown, it is liable to the first of the fleece. If, however, one shears the wool

and then dyes it before giving the first of the fleece to the priest, he is exempt from this duty. If he bleaches it before giving it, he is obligated to set aside the first of the fleece after bleaching it. If one plucks the wool with his hand instead of shearing it, he is liable to the first of the fleece.

7. The first of the fleece is due from hybrid animals, from a *ḳoy,* and from a *ṭĕrefah* animal. He who shears a dead animal, however, is exempt.

8. If one sets aside the first of the fleece and it is lost, he remains responsible for it until he gives it to the priest. If one says, "All my shearings are to be the first of the fleece," his words are valid.

9. If one buys the fleece of a heathen's sheep after they have been shorn by the heathen, he is exempt from the first of the fleece. If he acquires the sheep solely for their fleece, he is liable, even though it had grown while the sheep were in the heathen's possession, and even though the sheep are to revert to the heathen after the shearing—inasmuch as the shearer is an Israelite and the shearings are his, he is liable, since the obligation takes effect only at the time of the shearing.

10. If A buys the shearings of B's sheep, the rule is as follows: If B has retained some of his sheep to be sheared, B is obligated to set aside enough of the fleece of these retained sheep to cover the whole flock, even if he was not the one who began the shearing; for the presumption is that one may not sell the gifts due to the priesthood. If B does not retain any sheep, A is obligated to set aside the first of the fleece.

11. If there are two kinds of shearings—for exemple, light and dark, or of males and females—and B sells one kind and retains the other, each one must give out of his fleece for his own account, A out of what he has bought, and B out of what he has retained.

12. If a proselyte has fleece concerning which it is not known whether it was sheared before he became a proselyte or after, he

is exempt, and the burden of proof rests upon him who claims the first of the fleece out of it.

13. How many sheep must one have in order to be liable to the first of the fleece? Not less than five, provided that their combined fleece amounts to not less than sixty sela' by weight, and the fleece of each one of the five is not less than twelve sela' by weight. If one of them yields less than twelve sela' of fleece, they are exempt from the first fleece, even if the total fleece of the five sheep does amount to sixty sela' or more.

14. Holders in partnership are liable to the first of the fleece, provided that the share of each partner amounts to the required minimum. If, however, two partners own only five sheep, they are exempt.

15. The commandment of the first of the fleece comes into force at the beginning of the shearing, but if one performs it in the middle or at the end of the shearing, he has fulfilled his obligation.

If one who has five sheep shears only one of them and sells the fleece, then shears the second sheep and sells its fleece, and then shears the third sheep and sells its fleece, they all combine together as far as the first of the fleece is concerned, and he is liable to it even after many years.

One may set aside the first of the fleece out of new fleece for old fleece, or out of one kind for another. If, however, he owns a ewe, shears it, and sets the fleece aside, then buys a second ewe, shears it, and sets its fleece aside, they do not combine together.

16. If one has several batches of the first of the fleece and wishes to distribute them among several priests, he may not give to each priest less than the weight of five sela', bleached and sufficient to make a small garment. This does not mean that he must bleach it first and then give it to the priest; rather he must give him enough unbleached fleece to weigh five sela' or more after bleaching, as it is said, *Shalt thou give him* (Deut. 18:4), which implies, give him enough to make a usable gift.

17. The first of the fleece is profane in every respect. I say therefore that one may give it to a priest's daughter, even if she is married to an Israelite, as is the rule with gifts due to the priest out of cattle. It would appear to me that both kinds should be subject to the same rule in this respect.

CHAPTER XI

1. It is a positive commandment for each Israelite to redeem his son who is first-born to his Israelite mother, as it is said, *All that openeth the womb is Mine* (Exod. 34:19), and again, *Howbeit the first-born of man shalt thou surely redeem* (Num. 18:15).

2. The mother is not obligated to redeem her first-born son, for it is he who is obligated to redeem himself that is obligated to redeem his first-born son.

If the father transgresses and does not redeem his son, the son must redeem himself when he grows up.

3. If one is liable to redeem both himself and his first-born son, he must redeem himself first and then his son. If he has enough money for only one redemption, he must redeem himself.

4. If one's first-born son needs to be redeemed just as the time has come for the pilgrimage to Jerusalem, and if one does not have enough money to cover both obligations, he must redeem his son first, and then endeavor to make the pilgrimage, as it is said first, *All the first-born of thy sons thou shalt redeem* (Exod. 34:20), and only thereafter, *And none shall appear before Me empty* (*ibid.*).

5. He who redeems his first-born son must recite the benediction, "Blessed art Thou, O Lord . . . who has sanctified us by His commandments and has commanded us concerning the redemption of the first-born son," and then the benediction, "Blessed art Thou, O Lord . . . who has kept us alive." After that he must pay the redemption money to the priest.

If one redeems himself, he must say, "and has commanded us to redeem the first-born," and then pronounce the "kept us alive" benediction.

6. This commandment is in force in all places and at all times.

With how much money must one redeem his first-born son? With five sela', as it is said, *And their redemption money—from a month old shalt thou redeem them—shall be, according to thy valuation, five shekels of silver* (Num. 18:16). The five sela' are payable in silver or in equivalent movables whose corpus has monetary value, the same as other shekel payments. Therefore one may not redeem the first-born with land, nor with bondsmen, because they are accounted the same as land, nor with writs whose corpus has no monetary value. If one does redeem with these, the redemption is invalid.

7. If the father executes a writ of indebtedness to the priest for five sela', the indebtedness is effective but the son is not thereby redeemed.

If the father gives the priest a vessel which, though not worth five sela' in the market place, is accepted by the priest as worth five sela', the son is thereby redeemed.

If the father distributes the five sela' to ten priests, whether simultaneously or seriatim, he has discharged his obligation.

8. If the priest wishes to refund the redemption money to the father, he may do so, but the father may not first give it to the priest with the understanding that it is to be refunded. If he does so nevertheless, and the priest then refunds it to him, the son is not redeemed, until the father resolves in his heart to give it to the priest in perpetuity. If thereafter the priest chooses to refund it, he may do so.

Similarly, if the father states explicitly that the money is being given to the priest on condition that it be refunded, and the priest then refunds it, the son is redeemed.

9. Priests and Levites are exempted from the redemption of their own first-born sons by an inference *a minore ad maius:* If in the desert of Sinai the Levites had exempted the first-born sons

of the Israelites, surely the Levites are entitled to exempt their own first-born sons.

10. A first-born Israelite whose mother is of priestly or Levitical descent is exempt from redemption, because the obligation depends not upon the father but upon the mother, as it is said, *That openeth the womb in Israel.*

11. If a Levite woman is impregnated by a heathen, her son is exempt from redemption. Contrariwise, if a priestly woman is impregnated by a heathen, her son is liable, inasmuch as his mother is disqualified from the priesthood by her intercourse with a heathen.

12. In the case of a priest who has fathered a profaned son, the rule is as follows: If the father dies within thirty days, the son is obligated to redeem himself, since the father had not yet gained title to his redemption money. If the father dies after thirty days, the son is not obligated to redeem himself, since the father had gained title to the redemption money.

13. In the case of a bondswoman who was emancipated, or a heathen woman who became a proselyte, while pregnant, and then gave birth, the rule is that although conception has taken place not in holiness, the male child is liable to redemption, inasmuch as he was born in holiness, as it is said, *That openeth the womb in Israel,* and the opening of the womb did indeed take place in Israel.

If it is not known whether parturition had taken place before or after the mother became a proselyte, the case comes under the principle that the burden of proof is upon the claimant.

14. In the case of a heathen woman or a bondswoman who had given birth previously and then become proselytized or emancipated, and thereafter gave birth to another male child, he is exempt from redemption, as it is said, *That openeth the womb,* and here there is no opening of the womb.

The same rule applies to a male child born subsequently to a stillbirth. In the case of any miscarriage which renders the mother unclean by reason of parturition, the child that follows thereafter is not considered as having opened the womb. In the case of any miscarriage which does not render the mother unclean by reason of parturition—as for example, when a woman drops a fetus resembling a fish or a locust, or if the miscarriage occurs on the fortieth day, and the like—the child born thereafter is accounted first-born as far as the priest is concerned, and must be redeemed.

15. If the fetus was dismembered in the mother's womb and was extracted piece by piece, what comes after it is not considered as having opened the womb.

In the case of a living fetus eight months old that had thrust out its head, then retracted it and died—and likewise a nine-months-old fetus that, even if dead, thrust out its head and then retracted it—whereupon its brother emerged and was born alive, the second-born infant is not regarded as having opened the womb, inasmuch as the womb was already opened by the head of the first infant, the rule being that once the first child's forehead emerges, it renders the child coming after it exempt.

16. A male child born out of its mother's side and the one born normally after it are both exempt, the former because he did not emerge out of the womb, and the latter because he was preceded by another child.

17. At what age does redemption become obligatory? When the child is a full thirty days old, as it is said, *And their redemption money—from a month old shalt thou redeem them* (Num. 18:16).

If the son dies within the thirty days, or even on the thirtieth day, and similarly if he is afflicted with a fatal malady, he is not liable to the five sela'. If the father hastens to pay the redemption money to the priest prematurely, the latter must refund it. If the child dies after thirty days, the father is liable to redemption, and if he has not yet paid the redemption money, he must pay it.

18. If the father, in redeeming his first-born son within the thirty days, says to the priest, "as of now," his son is not redeemed.

If he says, "after the thirty days," his son is redeemed, even if the money is no longer in existence after the lapse of the thirty days.

19. A person subject to doubt as to whether he is or is not liable to redemption, is exempt, on the principle that the burden of proof lies upon the claimant.

If the father dies within the thirty days, the presumption is that the child had not been redeemed, unless proof is furnished from his father that he had redeemed him prior to his death.

If the father dies after the expiration of the thirty days, the son is presumed to have been redeemed, unless he is told that he had not been redeemed.

20. If a man's primipara wife is delivered of both a male and a female child, and it is not known which one emerged first, the priest has no claim to redemption money. If she is delivered of two male children, the father must pay five sela' to the priest, even though it is not known which one is first-born.

If one of the boys dies within the thirty days after birth, the father is exempt, on the principle that the burden of proof is upon the claimant.

If the father dies, either within the thirty days or after, five sela' must be paid out of his estate to the priest, inasmuch as the estate has already become obligated, regardless of whether it has or has not been divided between the two infants.

21. If a man's two primiparae wives give birth to two male children, he must give ten sela' to the priest. If one of the children dies within the thirty days after birth, the rule is as follows: If the father had paid the ten sela' to the same priest, the latter must refund five sela'; if the father had paid them to two priests, he cannot reclaim the five sela', since he did not designate, "This half of the total is for this child," wherefore each one of the two priests can say, "Get it back from my fellow-priest."

22. If a man's two primiparae wives give birth to a male child and a female child, or to two males and one female, the father must pay five sela' to the priest, since it is impossible that one male child among them should not be one *that openeth the womb*.

23. If the two wives give birth to two female children and one male, or to two females and two males, and it is not known which one was born first, the priest has no claim to redemption money, since I can say, "A female was born first and a male second."

24. If a man has two wives, one multipara and the other primipara, and they give birth to two male children who are subsequently intermixed, he must pay five sela' to the priest.

If one of the infants dies within the thirty days, the father is exempt.

If the father dies, five sela' must be paid to the priest out of the estate.

25. If the two wives give birth to a male child and a female child, or to two male children and one female, the priest has no claim to redemption money, since I can say: The primipara gave birth first to the female child and then to the male child, and the multipara gave birth to the other male child.

26. In the case of two primiparae wives of two men giving birth to two male children who are subsequently intermixed, each father must pay five sela'. If they have done so, and then one of the children dies within the thirty days after birth, the rule is as follows: If the redemption money was paid to two priests, it cannot be reclaimed from them. If the money was paid to the same priest, one of the fathers should execute a writ of authorization for the other, who should then go and reclaim the five sela' from the priest on the strength of it.

27. If these wives give birth to a male child and a female child who are subsequently intermixed, both fathers are exempt, but the son is obligated to redeem himself later.

Similarly in the case of a primipara woman who had remarried without waiting three months after separation from her first husband, so that it is not known whether her male child is a nine-months child by her first husband or a seven-months child by her second husband, both husbands are exempt, but the son must redeem himself later.

28. If these wives give birth to two female children and one male, or two females and two males, the priest has no claim to redemption money.

29. In the case of two wives of two men, one a multipara and the other a primipara, who give birth to two male children, the rule is that the husband of the primipara must pay five sela' to the priest.

If they give birth to a male child and a female child, the priest has no claim to redemption money.

30. If these wives give birth to two male children and one female, the husband of the primipara must pay five sela', because he would be exempt only when two doubts are involved: If his wife has given birth to only one male child, he is liable; if she has given birth to both a male and a female child, he is also liable, and is exempt only if the female child was born first. Inasmuch as this possibility is remote, he must pay the redemption money.

CHAPTER XII

1. It is a positive commandment for each person in Israel to redeem the firstling of an ass with a lamb, or if he does not wish to redeem it, to break its neck, as it is said, *And the firstling of an ass thou shalt redeem with a lamb: and if thou wilt not redeem it, then thou shalt break its neck* (Exod. 34:20).

These two commandments are in force in all places and at all times.

The commandment to redeem the firstling has priority over the commandment to break its neck.

2. The lamb used for redemption must be given to the priest, as it is said, *Every thing that openeth the womb . . . shall be thine; howbeit the first-born of man shalt thou surely redeem, and the firstling of unclean beasts shalt thou redeem* (Num. 18:15).

3. The term *unclean beasts* in the aforecited verse refers to the ass alone.

4. One is forbidden to derive any benefit from the firstling of an ass until it is redeemed. If one sells it before its redemption, the money received is forbidden. If it dies before its redemption, or if its neck is broken, it must be buried, because it is forbidden to derive any benefit from it even after its neck is broken, so long as it was not redeemed.

Therefore, if the owner does not redeem the firstling of an ass and instead hands it over to the priest, the latter may make no use of it until he himself redeems it with a lamb and takes the lamb as his due, or else breaks the firstling's neck and buries it. Priests may not, however, be trusted in this matter, and an Israelite is therefore forbidden to give the firstling of an ass to the priest, unless the priest himself redeems it in his presence.

5. If one has set aside the redemption lamb, and it dies before it is given to the priest, the owner is not answerable for it, and may give the lamb's carcass to the priest for whatsoever benefit the latter may derive from it.

If the firstling of an ass dies after its redemption, the owner, after giving the lamb to the priest, may derive benefit from the dead firstling, since it has already been redeemed.

6. When is the owner obligated to redeem the firstling of an ass? From the time of its birth until the lapse of thirty days. After the thirty days, if he wishes to break its neck, he may do so, and if he prefers to redeem it, he may likewise do so—the only drawback is the delay in the performance of the commandment.

7. If one does not wish to redeem the firstling of an ass, he must break its neck from its back with a butcher's hatchet, as it is said, *And if thou wilt not redeem it, then thou shalt break its neck* (Exod. 34:20).

One may not kill it with a cane, a reed, a mattock, or a saw, but only with a butcher's hatchet. Nor may one bring it into an enclosure, lock the gate after it, and leave it to die, as it is said, *then thou shalt break its neck.*

8. One may not redeem the firstling of an ass with a calf, a wild beast, a slaughtered lamb, a *ṭĕrefah* animal, a hybrid animal, or a *ḳoy,* as it is said, *Thou shalt redeem with a lamb* (Exod. 34:20), and only live young of sheep or goats are called lambs.

9. One may not redeem the firstling of an ass with a lamb that resembles some other animal, but if one does so nevertheless, the redemption is valid.

One may redeem with a lamb extracted alive out of its slaughtered dam's womb, but not with consecrated animals that have become unfit for the altar, as it is said concerning them, *As the gazelle and as the hart* (Deut. 12:15; 15:22), implying that just as one may not redeem with the gazelle or with the hart, so may one not redeem with dedicated animals that have become unfit.

10. One may redeem the firstling of an ass with any lamb, male or female, unblemished or blemished, small or grown up.

11. A lamb bought with money realized from the sale of Sabbatical year produce may not be used for the redemption of an assured firstling of an ass, but may be used for a doubtful firstling.

If one has no lamb for such redemption, he may use something else of equivalent monetary value, and must give the money to the priest. The Torah prescribes a lamb not in order to be severe toward him but rather in order to be lenient with him. For if he has a firstling of an ass worth ten sela', he may redeem it with a lamb worth even as little as one denar; and this case should not be any more stringent than the case of consecrated things which may be redeemed with their equivalent in money.

12. When does this apply? When the value of the firstling of the ass is three zuz or more; if, however, its value is less than three zuz, it may be redeemed only with a lamb or with three zuz. A generous person should give not less than one sela', a miserly person not less than half a sela', and an average person not less than three zuz.

13. If A redeems the firstling of B's ass, it is considered redeemed, and the firstling accrues to B.

14. Priests and Levites are exempt from the redemption of the firstling of an ass, as it is said, *The first-born of man . . . and the firstling of unclean beasts shalt thou redeem* (Num. 18:15), which indicates that whatever applies to the first-born of man applies also to the firstling of an unclean beast; consequently whosoever is exempt from the redemption of the first-born of man is also exempt from the redemption of the firstling of an unclean beast.

15. If an Israelite buys the embryo of a heathen's ass, or sells the embryo of his own ass to a heathen—although he is not permitted to do so—he is exempt from the law of the firstling, and may not be fined for doing so.

If the heathen is a partner in the dam or in the firstling, even if his share is only $1/1000$ of it, the Israelite partner is exempt.

If the heathen's share in the firstling or in its dam is one of the members of its body, such as a foreleg, a hindleg, or even an ear, meaning anything which if cut off would make the animal blemished, the Israelite is exempt from the law of the firstling; if it will not make the animal blemished for the altar, he is liable.

Similarly, if an Israelite accepts an ass from a heathen, or vice versa, in order to look after it and to share the offspring between them, both shares of the firstling are exempt from the law of the firstling, as it is said, *Whatsoever openeth the womb among the children of Israel, both of man and of beast* (Exod. 13:2), implying that all of it must belong to an Israelite.

16. A proselyte concerning whom it is not known whether his she-ass had dropped a firstling before he became a proselyte or after, must either break the firstling's neck or redeem it. If he redeems it with a lamb, the lamb belongs to him, for the burden of proof lies upon the priest claimant.

17. If a heathen sets aside the firstling of an ass, he should be informed that he is not subject to the pertinent law, and that he is permitted to shear it and work it.

18. If a cow drops a young resembling an ass, or if an ass drops a young resembling a horse, they are exempt, as it is said twice,

And the firstling of an ass thou shalt redeem with a lamb (Exod. 13:13; 34:20), implying twice that the dam must be an ass and the firstling must be an ass.

If the firstling has some of the tokens of an ass, it is subject to law of the firstling.

19. If one's primipara she-ass drops two males, he must give one lamb to the priest. If it drops a male and a female, he must set aside one lamb in order to release the male from the sanctity of the firstling, so that it would be permitted for benefit, since it might have been born first. The lamb set aside belongs to the owner and not to the priest, on the principle that the burden of proof lies upon the priest claimant.

20. If one's two primiparae she-asses drop two males, he must give two lambs to the priest. If they drop a male and a female, or two males and a female, he must give only one lamb to the priest.

21. If the two she-asses drop two females and one male, or two males and two females, the priest receives nothing, and the owner need not set aside a lamb for himself, inasmuch as there are several doubts involved here: one she-ass may have dropped a male, and the other two females; or the one may have dropped a female, and the other first a male and after it a female, or first a female and after it a male. Similarly there are several doubts when the young are two males and two females.

If one of the she-asses is a primipara and the other a multipara, and they drop two males that are subsequently intermixed, the owner must give one lamb to the priest. If they drop a male and a female, the owner must set aside one lamb for himself but need not give it to the priest, because the matter is in doubt, and the burden of proof is upon the priest claimant.

22. Similarly, if one buys a she-ass from a heathen without knowing whether it had dropped young before or not, and it subsequently drops a male, he must redeem it with a lamb, which, however, belongs to him, seeing that the matter is in doubt.

23. If one has ten lambs, each set aside for a doubtful firstling of an ass, they are accounted profane in every respect, and must be

tithed the same as other animals—he must set aside one of them as tithe, and the rest are his as before.

24. If an Israelite has in his possession ten assured firstlings of asses acquired by him by inheritance from his priestly maternal grandfather, who in turn had inherited them from his own Israelite maternal grandfather, the grandson must set aside ten lambs in their stead, and these lambs remain his and are liable to tithe.

TREATISE VII

LAWS CONCERNING THE SABBATICAL YEAR AND THE YEAR OF THE JUBILEE

Involving Twenty-Two Commandments,
Nine Positive and Thirteen Negative
To Wit

1. That the land shall remain untilled in the Sabbatical year;
2. That one should not engage in agricultural work in that year;
3. That one should not engage in arboricultural work in that year;
4. That one should not reap the aftergrowth after the manner of the reapers;
5. That one should not gather the fruit of the undressed trees after the manner of those who gather the fruit in vineyards and orchards;
6. That one should leave what the land produces in that year free to all;
7. That one should release all loans;
8. That one should not press, or make demands upon, the debtor;

9. That one should not refrain from making loans prior to the Sabbatical year for fear of losing one's money;
10. To count the years by sevens;
11. To hallow the fiftieth year;
12. To sound the ram's horn on the tenth of Tishri of that year, as a signal that bondsmen shall go out free;
13. That the soil shall not be tilled in that year;
14. Not to reap the aftergrowth of that year after the manner of the reapers;
15. Not to gather the fruit of the untended trees in that year after the manner of those who gather the fruit in vineyards and orchards;
16. To grant redemption to the land in that year, which applies to both inherited and purchased land;
17. That the land shall not be sold in perpetuity;
18. The law concerning houses in walled cities;
19. That the entire Tribe of Levi shall receive no inheritance in the Land of Israel, but only cities to dwell in shall be given to them as a gift;
20. That the Tribe of Levi shall take no share of the spoils of war;
21. To give to the Levites cities to dwell in, together with the open spaces surrounding them;
22. That the open space surrounding any of their cities shall never be sold in perpetuity, and that the Levites may redeem it at any time, before the Jubilee or after.

An exposition of these commandments is contained in the following chapters.

NOTE

In the list of 613 commandments prefixed to the Code, those dealt with in the present treatise appear in the following order:

Positive commandments:

[1] 135. To cease from tilling the land in the Sabbatical year, as it is said: *In plowing time and in harvest thou shalt rest* (Exod. 34:21);

[6] 134. To let the land lie fallow in the Sabbatical year, as it is said: *But in the seventh year thou shalt let it rest and lie fallow* (Exod. 23:11);

[7] 141. To release loans in the Sabbatical year, as it is said: *Every creditor shall release that which he hath lent to his neighbor* (Deut. 15:2);

[10] 140. To count the years of the Jubilee by years and by cycles of seven years, as it is said: *And thou shalt number seven Sabbaths of years unto thee* (Lev. 25:8);

[11] 136. To hallow the Jubilee year by resting and letting the land lie fallow, as it is said: *And ye shall hallow the fiftieth year* (Lev. 25:10);

[12] 137. To sound the ram's horn in the Jubilee year, as it is said: *Then shalt thou make proclamation with the blast of the horn* (Lev. 25:9);

[16] 138. To grant redemption to the land in the Jubilee year, as it is said: *And in all the land of your possession ye shall grant a redemption for the land* (Lev. 25:24);

[18] 139. That houses sold within a walled city may be redeemed within a year, as it is said: *And if a man sell a dwelling-house in a walled city, then he may*

redeem it within a whole year after it is sold (Lev. 25:29);

[21] 183. To give the Levites cities to dwell in, together with the open spaces surrounding them, as it is said: *That they give unto the Levites of the inheritance of their possession cities to dwell in; and open land round about the cities shall ye give unto the Levites* (Num. 35:2).

Negative commandments:

[2] 220. Not to till the ground in the Sabbatical year, as it is said: *Thou shalt neither sow thy field* (Lev. 25:4);

[3] 221. Not to do any work on trees in the Sabbatical year, as it is said: *Nor prune thy vineyard* (*ibid.*);

[4] 222. Not to reap the aftergrowth of the Sabbatical year as it is reaped in other years, as it is said: *That which groweth of itself of thy harvest thou shalt not reap* (Lev. 25:5);

[5] 223. Not to gather the fruit of the tree in the Sabbatical year as it is gathered in other years, as it is said: *And the grapes of thy undressed vine thou shalt not gather* (*ibid.*);

[8] 230. Not to demand the return of a loan after the Sabbatical year has elapsed, as it is said: *He shall not exact it of his neighbor and his brother* (Deut. 15:2);

[9] 231. Not to refrain from making a loan to a poor man because of the Sabbatical year, as it is said: *Beware that there be not a base thought in thy heart, saying: The seventh year, the year of release, is at hand* (Deut. 15:9). The general rule is that wherever the terms "be-

ware," "lest," or "do not" occur, it is in reality a negative commandment;

[13] 224. Not to cultivate the soil nor do any work on the trees in the Jubilee year, as it is said concerning that year: *Ye shall not sow* (Lev. 25:11);

[14] 225. Not to reap the aftergrowth of the Jubilee year in the same way as in other years, as it is said: *Neither reap that which groweth of itself in it* (*ibid.*);

[15] 226. Not to gather the fruit of the tree in the Jubilee year in the same way as in other years, as it is said: *Nor gather the grapes in it of the undressed vines* (*ibid.*);

[17] 227. Not to sell a field in the Land of Israel in perpetuity, as it is said: *And the land shall not be sold in perpetuity* (Lev. 25:23);

[19] 169. That none of the Tribe of Levi shall take any portion of territory in the Land of Israel, as it is said: *All the Tribe of Levi shall have no portion nor inheritance with Israel* (Deut. 18:1);

[20] 170. That none of the Tribe of Levi shall take any share of the spoil at the conquest of the Promised Land, as it is said: *The priests the Levites . . . shall have no portion* (*ibid.*);

[22] 228. Not to change the character of the open spaces surrounding the cities of the Levites or their fields, as it is said: *But the fields of the open land about their cities may not be sold* (Lev. 25:34). It has been learned by tradition that this is a prohibition of any such change.

CHAPTER I

1. It is a positive commandment to desist from working the land or tending the trees in the seventh year, as it is said, *Then shall the land keep a Sabbath unto the Lord* (Lev. 25:2), and elsewhere, *In plowing time and in harvest thou shalt rest* (Exod. 34:21). He who performs any labor in working the land or tending the trees in that year not only nullifies a positive commandment but also transgresses a negative commandment, as it is said, *Thou shalt neither sow thy field, nor prune thy vineyard* (Lev. 25:4).

2. According to the Torah, one is liable to a flogging only for sowing or pruning, and for reaping or gathering the vintage. This rule applies equally to a vineyard and to other trees.

3. Pruning is included in sowing, and gathering the vintage is included in reaping. Why then did Scripture enumerate them separately? To tell you that for the transgression of only these two secondary prohibitions is one liable to a flogging. For the transgression of the other secondary prohibitions affecting the working of the land, as well as of the other primary prohibitions not specified in this discussion, one is liable not to a Scriptural flogging but to a flogging for disobedience.

4. How so? He who digs or plows for the benefit of the soil, clears away stones, spreads manure, or performs analogous tasks relating to the working of the land, and similarly he who sinks a vine shoot into the ground, grafts, plants, or performs analogous tasks relating to the tending of trees, is liable to a flogging for disobedience on the authority of Scribal law.

5. One is forbidden to plant even a nonfruit-bearing tree in the Sabbatical year. Nor may one cut warts off trees, strip off dry leaves and branches, sprinkle a treetop with powder, or make smoke under the tree to kill worms.

One may not smear plants with an evil-smelling substance to

prevent birds from eating them while tender, or oil unripe figs, or pierce them.

One may not tic plants together, or clip them, or prop up young trees with stakes, or perform analogous tasks relating to the tending of trees.

If one performs any of these tasks during the Sabbatical year, he is liable to a flogging for disobedience.

6. One may not light a fire in a thicket of reeds, because this prepares the soil for planting. One may not teach a cow to pull the plow except on sandy ground, nor test seeds in a pot filled with earth; one may, however, do so in a pot filled with manure.

One may soak seeds during the Sabbatical year in order to plant them after its expiration. One may store aloe on the roof top, but not water it.

7. One may paint a tree with red paint or load it with rocks, or hoe the soil under grapevines. As for shaking olive trees, if one does so to improve the tree, it is forbidden; if to plug the gaps, it is permitted.

8. One may water an irrigated field, that is, a sown field that is exceedingly thirsty, during the Sabbatical year.

The same applies to an orchard, provided that the trees therein are spaced so that there are more than ten to a *beṯ sĕ'ah*.

9. One may lead the water from tree to tree, but may not water the whole field. If the trees are so close to each other that there are ten to a beṯ sĕ'ah, one may water the entire field for their sake.

Similarly the soil of a grain field may be sprinkled with water in the Sabbatical year, so that the trees in that field would not deteriorate. One may also make grooves around vines, or dig a canal at the outset and fill the connecting channels with water therefrom.

10. Why did the Sages permit all these things? Because should one fail to water the soil, it would become salty, and every tree on it would die. Moreover, inasmuch as the prohibition of these

things and their like is only Scribal, the Sages did not decree it effective here, since the Torah forbids only the two primary acts and the two derivatives thereof, as we have explained.

11. When men of violence became numerous, and the heathen kings compelled the children of Israel to prepare food stores for their armies, the Sages permitted planting in the Sabbatical year, but only of those things that are required by the king's servants.

Similarly, anyone forced by a man of violence to perform work without pay, such as the king's corvée, in the Sabbatical year, may do so.

12. He who plants anything in the Sabbatical year, whether unwittingly or deliberately, must uproot it, inasmuch as the children of Israel are suspect in regard to matters of the Sabbatical year. Should you say, "If he has planted it unwittingly, let him retain it," the deliberate planter is bound to say, "I, too, planted unwittingly."

13. If one plows his field, or lets cattle manure it, or manures it himself in the Sabbatical year, in order to prepare it for planting in the next year, he is liable to a fine, and may not seed it in the eighth year. Nor may someone else rent the field from him in order to seed it; rather it must be left fallow before him. If he dies, however, his son may seed it.

14. In the case of one who removes thorns or stones from his land in the Sabbatical year in order to prepare it for the next year, even though he is not permitted to do so, the Sages did not hold him liable to a fine, and he is allowed to seed it in the eighth year.

15. If one buries turnips, radishes, or their like during the Sabbatical year, the rule is as follows: If some of the leaves are exposed, he need not be concerned; if not, it is forbidden.

If one buries arum or its like, he must bury not less than four ḳab̲ of it, one handbreadth high, with one handbreadth of earth on top of it. He must moreover bury it in a place well trodden by passers-by, so that it would not grow.

It is permitted to stir the soil of a rice field in the seventh year, but not to trim the rice plants.

16. Originally the Sages used to say: A may pick up wood, stones, or grass in his own field, provided that he picks up only large pieces, so that his intention is not to clear the land; in B's field, however, A may pick up both large and small pieces.

When, however, transgressors who intended to clear their fields but said, "We are picking up only large pieces," became numerous, the Sages forbade A to pick up anything in his own field, but permitted him to do so in B's field, provided that he does not do it as a favor to B, and does not tell B, "Look at the favor which I have done for you, by clearing your field."

17. If one's cattle is standing in his field, he may pick up fodder in it and place it before the cattle, since his animals are proof of his intention.

Similarly, if his oven is standing there, he may pick up pieces of any size for use as fuel, because his oven is evidence of his intention.

18. If one intends to cut down one or two trees for their wood, he may uproot them. If, however, he cuts down three trees or more, one alongside the other, he may not uproot them, inasmuch as he would be thereby preparing the land for sowing; rather he should cut off the tree above the ground, and leave the roots in the ground.

Where does this apply? In one's own field. In his companion's field he is permitted to uproot trees.

19. If one trims branches off an olive tree for use as wood, he may not cover the cut with earth, because this constitutes cultivation; he may, however, cover it with stones or straw.

20. If one trims grapevines or cuts reeds in the Sabbatical year, he may do so in the usual manner, with a mattock, a sickle, a saw, or any other tool that he may wish to use.

21. One may not cut down a virgin sycamore in the Sabbatical year in the same manner as one does in other years, because such

cutting down constitutes cultivation of the tree, since as a result of this cutting down it will grow and increase its yield. If one has need of its wood, he may cut it down, but in a manner other than that of its usual cultivation.

22. How should one cut it down? Either level with the ground or more than ten handbreadths above the ground.

A split tree may be tied up in the Sabbatical year, not to heal the split but to prevent it from widening.

CHAPTER II

1. One may not remove manure from his courtyard and dump it in his field in the Sabbatical year, because it looks as if he were manuring his field to make it fit for planting. If he does remove it but makes a regular dunghill out of it, it is permitted.

He should not, however, make such a dunghill in his field until the time for using the dung in the process of cultivation of the land has passed, namely after the moist dung has hardened.

One should make the dunghill not less than 150 sĕ'ah of dung, so that it would be recognizable as a dunghill. If one wishes to add to this amount, he may do so. If he already has a little dung pile in his field, he may go on adding to it.

One may not set up more than three dunghills in each beṯ sĕ'ah.

2. If one wishes to set out his entire field three dunghills to each beṯ sĕ'ah, he may do so. If he wishes to dump all of his manure into one large storage pile, he may likewise do so.

3. If one piles up his manure on rocky ground, or digs three handbreadths deep below the surface of the field and dumps the manure there, or erects a platform three handbreadths high above the ground and deposits the manure on it, there is no prescribed minimum for the pile. Even if he makes several such dunghills, large or small, per beṯ sĕ'ah, it is permitted, inasmuch as it is evident that they are meant not for the cultivation of the land but for the storage of manure.

4. A person is permitted to remove manure from a sheepfold and deposit it in his field, in the manner of all those who collect manure.

He who makes a sheepfold in his field in the Sabbatical year may make it no larger than two bet̲ sĕ'ah, and may then drive his sheep into it. When they have manured the entire ground of the fold, he may leave one of the sides of the fold standing, and build another fold next to it. In this manner he will have manured four bet̲ sĕ'ah of his field.

5. If one's entire field measures but four bet̲ sĕ'ah, he must leave a little of it unenclosed for appearance's sake, so that all may know that the sheep had manured it while penned in the fold, and would not say, "This one has manured his entire field in the Sabbatical year."

6. One may not, at the outset, open a stone quarry in his field in the Sabbatical year, lest people should say, "His purpose was to improve his field by removing the rock from it." If, however, he has begun before the Sabbatical year, and has hewed from the rock twenty-seven stones totaling three cubits square by three cubits high, each stone being one cubit square by one cubit high or larger, he is permitted to quarry as many stones from it, in the Sabbatical year, as he may wish.

7. In the case of a stone fence ten handbreadths or more high, if one wishes to remove all its stones, he may do so, provided that it contains ten or more stones each of which forms a load for two or more men, inasmuch as anyone who sees him do it would know that he is removing the stones because he needs them.

If the fence is less than ten handbreadths high, or if there are less than ten heavy stones, or if these stones are too small to form a load for two men, he may remove them only until one handbreadth's height is left close to the ground.

8. When does this apply? When one's intention is to improve his field, or when one commences removing the stones during the Sabbatical year.

If, however, it is not his intention to improve his field, or if he had commenced before the Sabbatical year, he may remove during that year as many stones as he may wish, from any place, down to the level of the soil.

Similarly, if one is removing stones from his companion's field, he may go down to the level of the soil, even if he does it under contract.

9. In the case of stones that are loose enough to be turned up by the plow, or were covered up and are now laid bare, the rule is as follows: If there are among them two stones each of which forms a load for two men, one is permitted to remove them. If they are smaller than this, they may not be removed.

10. If one clears away stones from his field in the Sabbatical year because he needs them, he may remove the top layer, but must leave those touching the ground.

Similarly, if one has in his field a pile of pebbles or a heap of small stones, he may remove the top layer, but must leave those touching the ground. If, however, there is rock or straw underneath them, he may remove all of them.

11. One may not fill a ravine with earth, or level it partially with earth, because this is the same as improving the ground for cultivation. He may, however, make an embankment against the ravine. In doing this, any stone that he can take hold of by stretching his hand while standing at the edge of the ravine, may be taken up.

12. Shoulder-weight stones, that is, stones that may be carried two or three at once on the shoulder, may be brought from any place, whether from a companion's field or from one's own.

Similarly, a lessee may remove stones, even small ones, from any place, whether from his leased field or from any other field.

13. A breach in the fence blocked with earth may not be built up if it does not trip up the public. If it does trip up the public, or if it is not blocked with earth and opens into the public domain, it is permitted to build it up.

14. One may not build a fence between his own field and his companion's field in the Sabbatical year, but may do so between

his own field and the public domain. In the latter case one may dig down to rock level, remove the earth, and heap it up within his own field, in the manner of all those who pile up manure.

Similarly, if one digs a cistern, a ditch, or a cave in the Sabbatical year, he may heap up the earth within his own field, as do all diggers.

CHAPTER III

1. The cultivation of the land during the thirty days of the sixth year immediately preceding the Sabbatical year is forbidden by authority of a rule revealed to Moses on Mount Sinai, because one would be thereby improving it for planting in the Sabbatical year.

This is what is forbidden by tradition while the Temple is in existence. The Sages, however, have decreed further that while the Temple is in existence, a field planted with trees may be plowed up in the sixth year only up to Pentecost, and a cornfield only up to Passover. When the Temple is not in existence, it is permitted to cultivate the land until New Year's Day, as is the rule according to the Torah.

2. What is the definition of a field planted with trees? Any field having at least three trees per beṯ sĕ'ah. Whether they are nonfruit-bearing trees or fruit-bearing trees, even if they belong to three persons, they are considered the same as fig trees; and if such fig trees can yield a cake of pressed figs weighing sixty minas, the whole beṯ sĕ'ah may be plowed up to Pentecost on their account, provided that there is a clearance between each tree and the one next to it wide enough for yoked oxen to pass through.

3. If there are less than three trees per beṯ sĕ'ah; or if of these three trees one is fit to yield 60 minas or more, while the other two are unfit to yield anything; or if two of the trees are fit to yield even as much as 100 minas, while the third one yields nothing—in all these cases one may plow up to Pentecost only such an area

as is essential for them, namely a clearance wide enough for the fruit picker and his basket.

4. If there are more than three and up to nine trees per beṯ sĕ'ah, and all together they are fit to yield sixty minas, one may plow up the entire field up to Pentecost for their sake, even if some of them are unfit to yield anything at all.

5. If there are ten or more trees per beṯ sĕ'ah, whether yielding the required minimum of fruit or not, one may plow the entire beṯ sĕ'ah up to Pentecost for their sake.

If there are ten saplings spread out over the entire beṯ sĕ'ah, one may plow the entire beṯ sĕ'ah for their sake up to New Year's Day. This also is a rule revealed to Moses on Mount Sinai.

6. If the trees are set out in one row or in a circle like a crown, only such space may be plowed up to New Year's Day as is needful for each one. Pumpkins planted with the saplings may be included in the count of ten.

7. What is the definition of a sapling? A young tree, so long as it is still called a sapling.

8. In the case of a tree that had been cut down and produced new shoots, the rule is as follows: If it had been cut down one handbreadth or more from the ground, it is accounted a tree; if from one handbreadth down to the ground, it is accounted a sapling.

9. All these rules apply while the Temple is in existence, as we have said. At the present time, however, it is permitted to cultivate such land up to New Year's Day.

Even while the Temple is in existence, it is permitted to clear away rocks, to manure the soil, to hoe beds of melons and pumpkins, and irrigated fields, up to New Year's Day.

One may also manure, strip, fertilize with powder, fumigate, trim, prune, lop off dry twigs, oil plants, wrap them, clip them, build shelters for them, water them, oil unripe figs and pierce them—all these services are permitted in the sixth year up to New Year's Day of the Sabbatical year, even while the Temple is in existence.

10. Unripe figs of the sixth year which remain on the tree into the Sabbatical year, as well as those of the Sabbatical year which remain on the tree into the eighth year, may not be oiled or pierced.

While the Temple is in existence, one may not build steps into the sides of ravines in the sixth year, after the rains have ceased, since this would constitute improvement of the ravines for the Sabbatical year.

11. At the present time, too, one may not plant trees, graft them, or sink vine shoots in the sixth year, except when there is enough time for the plant to take hold and go on thereafter for thirty days before New Year's Day of the Sabbatical year. Taking hold generally takes two weeks. This prohibition is permanent, for appearance's sake, lest anyone who sees such work should say, "This planting was done in the Sabbatical year."

You must thus conclude that he who plants a tree, sinks a vine, or grafts a tree in the sixth year forty-four days before New Year's Day of the Sabbatical year, may retain the plant; if less than that period of time, he must uproot it. If he does not uproot it, the fruit is permitted. If he dies before uprooting it, the heir is required to uproot it.

CHAPTER IV

1. Whatever the land produces in the Sabbatical year, whether from seed that has dropped into it before that year, or from roots whose produce has been harvested before and which yield new produce—both are called aftergrowth, whether it be grass or vegetables that have grown spontaneously without planting—all such are permitted for consumption according to the Torah, as it is said, *And the Sabbath produce of the land shall be for food for you* (Lev. 25:6).

Even if a field has been improved in the Sabbatical year and brings forth plants, its produce is permitted to be eaten. As for the Scriptural prohibition, *That which groweth of itself of thy harvest thou shalt not reap* (Lev. 25:5), it means only that one may not harvest in the same manner as he does every year. If he does harvest in the customary manner, he is liable to a flogging—as for example, if he harvests the whole field, sets up a heap of produce, and threshes it with his cattle, or harvests in preparation for the cultivation of the land, as we have explained. Rather he must harvest a little at a time, beat it out, and eat of it.

2. It is a Scribal enactment, however, that all aftergrowths are forbidden for consumption. Why was this enactment made? Because of the transgressors, so that one should not go forth and secretly plant corn, pulse, or garden seeds in his field, and after they have sprouted eat them, saying, "They are aftergrowths." That is why the Sages have forbidden all aftergrowths that sprout in the Sabbatical year.

3. You thus learn that one may not eat any of the produce of the Sabbatical year except the fruit of trees and grasses that are not planted by most people, such as rue, wild amaranth, and the like.

As for vegetables planted in gardens by most people, as well as species of grain and pulse, whatever grows of them spontaneously is forbidden by Scribal enactment, and he who gathers them is liable to a flogging for disobedience.

4. Aftergrowths that come up in an uncultivated field or in a field prepared for plowing, for grapevines, or for sowing, may be eaten. Why did the Sages enact no precautionary measure to forbid them also? Because a person would not do any planting in such places: as for an uncultivated field, no one would pay any attention to it; as for a field prepared for plowing, one would be satisfied to improve it for plowing; as for a field prepared for grapevines, one would not wish to render his vineyard forbidden; and as for a field readied for sowing, the aftergrowth would damage it.

The same applies to straw of the Sabbatical year—it is permitted everywhere, and no precautionary measure was enacted against it.

5. Aftergrowths of the Sabbatical year which continue growing into the eighth year may not be consumed, nor be plucked by

hand; rather one should plow them under in his customary manner, and cattle may be pastured therein in the usual way.

6. Until what time are aftergrowths of the Sabbatical year forbidden in the eighth year? From New Year's Day until Hanukkah. From Hanukkah onward they are permitted.

If one plants the aftergrowths of the Sabbatical year after the expiration of that year, their yield is permitted.

7. In the case of Sabbatical year's onions which continue growing into the eighth year, the rule is as follows: If they have grown in the eighth year as much as onions grown entirely in that year, they are permitted; if not, they are forbidden.

The same applies to other Sabbatical year's produce—one may buy it in the eighth year only after it has grown in the eighth year as much as produce grown entirely in that year. Once the early crop is ripe, the late crop is permitted.

It is permitted to buy vegetables immediately after the expiration of the Sabbatical year.

8. From what time on may one buy arum after the expiration of the Sabbatical year? Once the new crop is available in abundance.

9. The first of Tishri is the beginning of the year in regard to the Sabbatical year and the Jubilee year. If the produce of the sixth year is left over into the Sabbatical year, the rule is as follows: If it is grain, pulse, or fruit of trees, and if it has reached the season of tithing before New Year's Day, it is permitted, and even if gathered in the Sabbatical year, it is accounted produce of the sixth year in every respect. If it has not reached the season of tithing until after New Year's Day, it is accounted produce of the Sabbatical year.

10. Grain and pulse are forbidden for consumption as aftergrowths. Fruit of trees may be consumed as food partaking of the sanctity of the Sabbatical year.

11. In the case of rice, durra, millet, sesame, and cowpea planted for seed, the rule depends upon their ripening: If their yield is fully ripened before New Year's Day, it may be eaten in

the Sabbatical year as produce of the sixth year; if their yield is fully ripened after New Year's Day, even if they have struck root before New Year's Day, it is forbidden as aftergrowth.

12. The rule affecting vegetables depends upon the time of their gathering. The citron, even if no larger than a bean before New Year's Day but grown as large as a loaf of bread in the Sabbatical year, is liable to tithing as produce of the sixth year. And contrariwise, even if it is the size of a loaf of bread in the sixth year, it is considered the same as produce of the Sabbatical year if it is gathered in the Sabbatical year, but must still be tithed as produce of the sixth year in order to be on the stringent side.

13. Similarly, in the case of Sabbatical year's produce which continues growing into the eighth year, namely grain, pulse, and fruit of trees, the rule depends upon the season of tithing. In the case of millet, sesame, rice, durra, and cowpea sown for seed, the rule depends upon the ripening of the yield, and in the case of vegetables, upon the time of its gathering.

14. In the case of cowpea planted for seed in the sixth year, whose yield has ripened before New Year's Day of the Sabbatical year, both its leaves and its seeds are permitted in the Sabbatical year; if planted for its leaves, and overtaken by the Sabbatical year, both its leaves and its seeds are forbidden the same as aftergrowths of the Sabbatical year; if planted for seed as well as for leaves, both are forbidden.

15. If one transgresses and plants cowpea in the Sabbatical year, and it continues growing into the eighth year, the rule is as follows: If planted for seed, both seed and leaves are forbidden after the expiration of the Sabbatical year, the same as other aftergrowths; if planted for its leaves, both seeds and leaves are permitted, provided that they are gathered in the eighth year; if planted for seed as well as for leaves, the seeds are forbidden because they are aftergrowths, while the leaves are permitted.

16. White figs, if arrived at the season of tithing before New Year's Day of the eighth year, may be eaten in the second year of

the next septennate as produce of the Sabbatical year, inasmuch as it takes them three years to complete their ripening.

17. Seedless onions and cowpeas deprived of irrigation for thirty days prior to New Year's Day, and the same planted in a rain-watered field from which three waterings have been withheld, are accounted produce of the sixth year; if less than this, they are accounted aftergrowth of the Sabbatical year.

18. Pumpkins reserved for seed may be left growing during the Sabbatical year, provided that they have hardened before New Year's Day and have become unfit for human consumption, inasmuch as they are obviously produce of the sixth year; otherwise they are forbidden as aftergrowth of the Sabbatical year.

The same applies to all vegetables that have hardened before New Year's Day—they may be left growing during the Sabbatical year; but if they are soft, it is forbidden to leave them growing, inasmuch as they are aftergrowths of the Sabbatical year.

One is not obligated to uproot arum, but may leave it in the ground as it is, and if it continues growing into the eighth year, it is permitted for consumption. Nor is one required to uproot artichoke—he must only clip its leaves, and if it continues growing into the eighth year, it is likewise permitted.

19. Arum of the sixth year, summer onions of the sixth year, and madder ripened before the beginning of the Sabbatical year may be uprooted in the latter year even with metal mattocks, and this is not accounted cultivating the land.

20. Onions wetted by rain in the Sabbatical year and sprouted are permitted, so long as their leaves are green; once the leaves have darkened, they are regarded the same as planted in the ground, and such leaves are forbidden as aftergrowths. In either case the bulbs of the onions remain permitted.

21. In the case of an onion uprooted in the Sabbatical year and replanted in the eighth year, whose outgrowth exceeds its root, the outgrowth neutralizes its root and the whole becomes permitted—just as the produce of the Sabbatical year is rendered

forbidden by the soil, so does its neutralization come through the soil.

22. Fruit yielded by a tree in the Sabbatical year may not be gathered in the same manner as in other years, as it is said, *And the grapes of thy undressed vine thou shalt not gather* (Lev. 25:5). If one does gather it as part of the regular cultivation of the tree, or in the usual manner of grape gatherers, he is liable to a flogging.

23. How then should one proceed? Figs of the Sabbatical year may not be spread for drying in the regular drying place, but should be dried in a deserted spot. Nor may one tread grapes in a vat, but only in a kneading trough. One may not process olives in a regular olive press or in a small press, but should first crush them and then place them in a miniature press. One may, however, first grind them in a regular olive press and then place them in a small olive press. The same applies to other produce—wherever one can vary the usual procedure, he should do so.

24. It is a positive commandment to leave ownerless all that the land produces in the Sabbatical year, as it is said, *But the seventh year thou shalt let it rest and lie fallow* (Exod. 23:11). Therefore whosoever locks his vineyard or fences in his field in the Sabbatical year has thereby nullified a positive commandment. The same applies to anyone who gathers all his produce into his own house. One must rather leave everything free to all comers, and everyone everywhere has an equal right to it, as it is said, *That the poor of thy people may eat* (*ibid.*).

One may, however, bring a small amount of this produce into his own house, the same as people would bring out of ownerless property, such as five jugs of oil or fifteen jugs of wine. But if he brings in more than this, it is allowed.

25. The Sabbatical year applies only to the Land of Israel, as it is said, *When ye come into the land,* etc. (Lev. 25:2). It is in force both while the Temple is in existence and when it is not.

26. All parts of the Land of Israel occupied by those who came up from Babylonia, as far as Chezib, may not be cultivated in the

Sabbatical year, and all aftergrowths that sprout spontaneously therein are forbidden for consumption.

As for the parts of the Land occupied only by those who came up from Egypt, which are from Chezib to the River of Egypt and Amanah, although it is forbidden to cultivate them in the Sabbatical year, the aftergrowths that sprout in them are permitted for consumption. The territory from the River and Amanah and beyond may be cultivated in the Sabbatical year.

27. Even though according to the Torah the Sabbatical year does not apply to Syria, the Sages have decreed that the cultivation of the land there is forbidden in that year the same as in the Land of Israel, in order that people should not leave the latter and go and settle in the former. The lands of Ammon, Moab, Egypt, and Shinar, however, although liable to tithes according to Scribal law, are not subject to the law of the Sabbatical year.

28. The Sabbatical year applies to Transjordan by Scribal enactment, but aftergrowths in Syria and Transjordan are permitted for consumption, in order that these lands should not be under a stricter rule than the parts of the Land of Israel occupied by those who came up from Egypt.

29. If a heathen buys land in the Land of Israel and plants it in the Sabbatical year, its produce is permitted, for the prohibition of aftergrowths is aimed at Israelite transgressors, and heathens, not being subject to the Sabbatical year, do not come under that prohibition.

30. Trusted guards should be appointed in towns of the Land of Israel situated near the border, to prevent heathens from raiding them and looting the produce of the Sabbatical year.

CHAPTER V

1. Sabbatical year's produce may be used for food, drink, anointing, kindling a lamp, and dyeing. The Sages have learned by tradition that the words *shall . . . be* (Lev. 25:7) include lighting a lamp and dyeing.

2. How for food and drink? To eat that which is customarily eaten, and to drink that which is usually drunk, as is the rule for heave offering and second tithe.

3. One may not divert produce from its natural use, just as one may not divert heave offering or second tithe. Thus a thing customarily eaten raw may not be eaten cooked, and vice versa. Consequently one may not boil cattle feed, nor may one treat spoiled broth or moldy bread to make it fit for human consumption, as is the rule for heave offering and second tithe.

4. One may not cook Sabbatical year's vegetables in heave offering oil, lest he should render the latter unfit. But if he cooks a small amount and eats it immediately, it is permitted, seeing that he does not let it stand and haply become unfit.

5. Produce intended exclusively for human consumption may not be fed to domestic or wild beasts, or fowls. If, however, the beast of its own volition goes to browse under a fig tree and eats of the figs, its owner is not obligated to chase it away, as it is said, *And for thy cattle, and for the beasts that are in thy land, shall all the increase thereof be for food* (Lev. 25:7).

6. How for anointing? To anoint with what is customarily used for anointing. One may not anoint with wine or vinegar, but one may do so with oil. One may not perfume the oil, nor anoint himself therewith in the bathhouse, but one may do so outside the bathhouse and then enter it.

7. One may not seal an oven or a stove with Sabbatical oil, nor may one oil a shoe or a sandal with it, nor may one do the oiling with unclean hands. If some of the oil is spilled on one's flesh, he may rub it in with unclean hands. One may not anoint his foot inside the shoe, but should anoint his bare foot first and then put on the shoe. One may anoint his body and then roll on a leather blanket.

8. How for kindling a lamp? By lighting the lamp with the Sabbatical year's oil itself. If one sells this oil and with the money buys other oil, or exchanges it for other oil, both are

forbidden to be used for kindling the lamp, since one may not light it with oil bought with money realized from Sabbatical year's produce. Furthermore, one may not pour the oil into a burning fire, but may only kindle it in a lamp.

9. How for dyeing? One may dye for human use with substances regularly used for dyeing, even if they are used also for human food, but one may not dye for animal use with Sabbatical year's produce, even if it is used also for animal feed, inasmuch as Sabbatical year's sanctity does not devolve upon dyes for animal use.

10. Laundering substances, such as soap and aloe, are subject to the sanctity of the Sabbatical year, and one may launder with them, as it is said, *And the Sabbath produce of the land shall be . . . for you* (Lev. 25:6), implying for any of your needs.

One may not, however, launder with edible Sabbatical year's produce, nor may one make a poultice out of it, as it is said, *And the Sabbath produce of the land shall be for food for you,* implying not for a poultice, a spray, or an emetic, nor for soaking or laundering.

11. The Sages have laid down the following important principle concerning Sabbatical year's produce: Anything used exclusively for human food, such as wheat, figs, grapes, and the like, may not be used as poultice or plaster, even for man, as it is said, *For food for you* (Lev. 25:6), implying that whatsoever is used exclusively by you shall be used for consumption and not for medicinal purposes.

Anything not used exclusively for human food, such as tender thorns and thistles, may be used as poultice for man but not for beast.

As for anything not used exclusively as food for either man or beast, such as savory, hyssop, and thyme, the rule depends upon one's intention: If one intends to use it for wood, it is subject to the rule governing wood; if for food, it is the same as produce; if for both human food and animal feed, it is subject to the more stringent rule affecting both human food, in that one may not use it as poultice, and animal feed, in that it may not be boiled.

12. One may sell human food or animal feed and buy other human food with the proceeds; but one may not sell one kind of animal feed in order to buy another kind, and needless to say, one may not sell human food in order to buy animal feed. If one does buy human food with the proceeds of animal feed, or exchanges the latter for the former, the food is considered as human food which may not be used as poultice for man.

13. One may not export Sabbatical year's produce from the Land of Israel to a land outside of the Land of Israel, even to Syria; and one may not feed it to a heathen or to a hired man. If, however, the man was hired by the septennate, or by the year, or by the month, or if the employer has obligated himself to feed him, he is the same as a member of the employer's household and must be fed accordingly.

Boarders may be fed with the produce of the Sabbatical year.

14. The court may not assign maintenance to a woman out of Sabbatical year's produce, because this is the same as discharging a debt with such produce. She may, however, maintain herself out of her absent husband's produce.

15. One may not gather Sabbatical year's produce while it is still unripe, as it is said, *Ye shall eat the increase thereof* (Lev. 25:12), implying that it may be eaten only when it is ripe produce. One may, however, occasionally eat in the field a small amount of it when it is young, in the same manner as one eats it in the other years of the septennate. But one may not bring the produce into his house in order to eat it until it has reached the season of tithing.

16. From what time on is one permitted to eat of the fruit of a tree in the field in the Sabbatical year? In the case of young figs, as soon as they assume a lustrous appearance, when one may eat his bread with them in the field, and the same applies to all similar produce.

In the case of unripe grapes, as soon as they become juicy, when one may eat them in the field, and the same applies to all similar produce.

In the case of olives, as soon as each sĕ'ah of olives can produce one-quarter loḡ of oil, when one may crack them and eat them in the field. If they can produce half a loḡ of oil, one may press them and use the oil for anointing in the field. If they can produce one-third of a loḡ of oil, one may bring them into his house, inasmuch as they have already reached their season of tithing.

17. One may cut down trees for wood in the Sabbatical year before any fruit appears on them, but once they begin to produce fruit, one may not cut them down, inasmuch as he would be wasting food, and it is said, *For food for you* (Lev. 25:6), not to be wasted.

Once a tree has brought forth fruit, and the fruit has reached the season of tithing, it is permitted to cut down the tree, inasmuch as it has already produced its fruit and is no longer subject to the law of the Sabbatical year.

18. From what time on is it forbidden to cut down trees in the Sabbatical year? In the case of carob trees, once the branches have begun to droop; in the case of grapevines, once they produce globules; in the case of olive trees, once they blossom; in the case of any other tree, once it yields unripe fruit.

One may not cut down trees with date berries in the Sabbatical year, because this is waste of fruit. If these berries regularly ripen not into soft dates but into stony dates, one may cut down such trees.

19. One may not burn straw and stubble of the Sabbatical year, because it is feed fit for cattle. One may, however, use olive or grape refuse of the Sabbatical year for fuel.

20. One may bathe in a bath heated with Sabbatical year's straw or stubble, even if he pays a fee for it; but if one is an important person, he is forbidden to do so, lest other things should be burned for his sake in order to perfume the bath, thus wasting produce of the Sabbatical year.

21. Shells and kernels which in the case of heave offering produce are permitted to commoners, are not subject to the holiness of the Sabbatical year and are accounted the same as wood, unless they may be used for dyeing. The edible terminal bud of a palm tree, however, is subject to the holiness of the Sabbatical year.

22. If one ties Sabbatical year's seasoning into a bunch and puts it into cooked food, the rule is as follows: If the seasoning's flavor is neutralized, it is permitted to be used for any purpose whatsoever; if it retains its flavor, it retains also its holiness as produce of the Sabbatical year.

23. One may not stuff Sabbatical year's straw or stubble into a mattress or mix it with clay. If one does so nevertheless, it is regarded the same as destroyed.

An oven heated with Sabbatical year's straw or stubble should be first allowed to cool.

Once the second rain has fallen after the expiration of the Sabbatical year, one may use that year's straw or stubble for fuel.

CHAPTER VI

1. One may not use produce of the Sabbatical year as merchandise, but if one wishes to sell a small amount of it, he may do so. The money realized therefrom has the same status as Sabbatical year's produce; it must be used only to purchase food, which must be consumed in the sanctity of the Sabbatical year, and the produce sold retains its sanctity.

2. One may not gather vegetables of the Sabbatical year in order to sell them. Nor may one, for a fee, do any dyeing with dye made of fruit rind of that year, because this constitutes trading with that year's produce.

If one gathers vegetables for his own consumption and has some left over, he may sell the leftover, but the money realized is the same as produce of the Sabbatical year.

Similarly, if one gathers vegetables for his own use, and his son or daughter's son takes some and sells it, this is permitted, but the money realized is the same as the proceeds of Sabbatical year's produce.

3. When one does sell produce of the Sabbatical year, he should not sell it by measure, by weight, or by number, in order that it should not appear as if he were trafficking with Sabbatical year's produce. He should rather sell the little that he does sell in the lump, to let it be known that it is ownerless, and that he takes the money in order to buy other food therewith.

4. One may tie into bundles such produce as is customarily tied for use in the house, in order to sell it in the market place in the lump, but he must tie it in the way it is tied for use at home, and not for sale in the market place, so that it should not appear as if he were selling it in exact amounts.

Produce customarily tied into bundles only for the market may not be tied into bundles at all.

5. Produce of a land outside of the Land of Israel imported into the Land of Israel may not be sold by measure, by weight, or by number, but rather in the lump, the same as the produce of the Land of Israel. But if it is obvious that it was imported from abroad, it is permitted to be sold in the regular manner.

6. Sabbatical year's produce is subject to a stricter rule than consecrated things: if one redeems a consecrated thing, it becomes profane, while the redemption money becomes sacred in its stead. Sabbatical year's produce, however, is not so: rather, if one sells it, the money realized therefrom assumes the sanctity of Sabbatical year's produce, but the produce sold does not become profane like the produce of other years, as it is said, *shall be* (Lev. 25:6), in the future tense, implying that it retains its status forever, and since it is termed holy, it imparts its sanctity to the money paid for it. You thus conclude that the final item assumes the sanctity of Sabbatical year's produce, while the produce itself remains as it was.

7. How so? If one buys meat with Sabbatical year's produce, or with money realized from its sale, the meat becomes the same

as the produce, and he must eat it as such, and must clear it out when he clears out all Sabbatical year's produce.

If he buys fish with that meat, or with money realized from its sale, the meat sheds its sanctity, while the fish assumes it in the meat's stead. If he then buys oil with the fish, or with the money realized from its sale, the fish sheds its sanctity, while the oil assumes it. If he finally buys honey with the oil, or with the money realized from its sale, the oil sheds its sanctity, while the honey assumes it. The original produce and the final honey must be cleared out in the same manner as all Sabbatical year's produce.

One may not use either of them to make a poultice, nor waste them, the same as other Sabbatical year's produce.

8. The produce of the Sabbatical year may be rendered profane only by way of purchase. To what does this apply? To the first produce. As for the second produce, it may be rendered profane both by way of purchase and by way of exchange.

9. When one renders profane produce bought the second time, one may not do so by exchanging it for live domestic or wild animals or fowls, lest one should keep them and raise flocks out of them. Needless to say, one may not do so with Sabbatical year's produce itself. One may, however, render such produce profane with slaughtered animals.

10. One may not discharge a debt with money realized from Sabbatical year's produce, nor use it for a wedding gift or to repay a favor, nor assign some of it to alms for the poor in fulfillment of a promise made in a synagogue. One may, however, send it out as an act of charity, provided that the recipient is informed that it is money realized from Sabbatical year's produce.

Similarly, one may not use it to buy bondsmen, land, or unclean animals. If he does so nevertheless, he must consume its equivalent, as he would in the case of second tithe.

Nor may one use it to procure the bird offerings due from men or women afflicted with flux, or from delivered women, or sin offerings, or guilt offerings. If one does so nevertheless, he must consume its equivalent.

One may not oil vessels or hides with Sabbatical year's oil. If one does so nevertheless, he must consume its equivalent.

11. Money realized from Sabbatical year's produce may not be used to pay a bathhouse attendant, a barber, a sailor, or any other artisan. One may, however, use it to pay the person who draws water from a cistern in order to give him a drink.

It is permitted to give Sabbatical year's produce, or money realized from it, to artisans as a free gift.

12. If one says to a laborer, "Here is an 'issar for you—gather some vegetables for me today," his fee is permitted and is not considered money realized from Sabbatical year's produce. The laborer may therefore spend it for whatsoever he may wish, and is not penalized by having it count the same as Sabbatical year's money.

If, however, one says to the laborer, "Gather some vegetables for me in return for this 'issar," it is the same as Sabbatical year's money, and the laborer may spend it only for food and drink to be consumed as that year's produce.

13. In the case of ass drivers who while conveying Sabbatical year's produce perform work that is forbidden in that year, as for example, by transporting more produce than is enough for one's needs, their fee is the same as money received in payment for Sabbatical year's produce, and this is a penalty imposed upon them. Why then did the Sages impose this penalty upon ass drivers and not upon the laborer? Because the laborer's pay is small, he was not penalized, in order to allow him to earn his livelihood.

14. If while buying from a baker a loaf worth one pondion, one says to him at the time of the purchase, "When I gather vegetables in my field, I will bring you some for it," this is permitted, and that loaf is accounted the same as Sabbatical year's produce.

If the loaf is bought without specifying any terms, it may not be paid for with money realized from Sabbatical year's produce, for one may not discharge a debt with it.

15. One may eat of the produce of the Sabbatical year by favor or not by favor. How by favor? When the owner himself gives him that year's produce in the manner of doing him a favor, or when he brings him into his garden to eat of its produce in the manner of doing him a favor.

One who acquires produce of the Sabbatical year as a gift or by inheritance may eat of it as if he were eating of produce that he himself had gathered in his field.

CHAPTER VII

1. Produce of the Sabbatical year may be eaten only as long as the particular species is available in the field, as it is said, *And for thy cattle, and for the beasts that are in thy land, shall all the increase thereof be for food* (Lev. 25:7), implying that so long as the wild beast can eat of that species in the field, you may eat of what there is of it in the house. Once there is no more of it in the field for the wild beast, you must clear out what there is of that species in the house. This is what is referred to as the removal of Sabbatical year's produce.

2. How so? If, for instance, one has dried figs of the Sabbatical year in his house, he may eat of them so long as there are figs on the trees in the field. Once the figs in the field have been exhausted, he may not eat of those in the house, and must clear them out.

3. If one has a large quantity of produce, he may apportion it at the rate of three meals for each person in his household.

Both poor and rich are forbidden to eat of it after its removal. If no one is available to consume the produce at the time of its removal, one must burn it in the fire, or throw it into the Dead Sea, or destroy it by other means that would cause it to perish.

4. If one has Sabbatical year's raisins at the time when grapes that are ownerless have been exhausted in the fields, gardens, and orchards, even though there are still grapes on the vines in

his courtyard, he may not eat of the raisins on account of these grapes in his courtyard, for these latter are not accessible to wild beasts.

If, however, these are exceedingly hard grapes which do not ripen until the end of the year, one may eat of the raisins on their account. The same rule applies to all similar cases.

5. In the case of a tree that yields fruit twice a year, if one has fruit of its first yield, he may eat of it so long as fruit of the second yield is still available in the field, since there is in fact fruit of the same species still in the field. One may not, however, eat of earlier fruit on account of autumn fruit, because the latter looks like next year's fruit.

6. If one preserves three species of produce in the same jar, whichever species is exhausted in the field, that same species must be cleared out of the jar. If he has begun eating out of that jar, all of its contents is liable to removal.

Just as onc must clear out human food, so must he clear animal feed out of his house, and may not feed it to his cattle once that species has been exhausted in the field.

7. Just as one must clear out produce, so must he clear out money. How so? If, for instance, one has sold some Sabbatical year's pomegranates and has purchased other produce with the money realized, and if the pomegranates on the trees in the field are exhausted while he has some money left out of the proceeds of the sold pomegranates, that money must be cleared out.

8. How should he proceed? He must either purchase food with the leftover money and apportion to each person in his household enough of it for three meals, or else cast the money into the Dead Sea if he cannot find anyone to consume it.

9. Three regions are to be distinguished in respect to the law of removal. All of Judea, including hill country, plains, and valleys, is considered one region. All of Transjordan, including the plain of Lydda, the hilly part of that plain, and the country

from Beth Horon to the sea, is likewise considered one region. All of Galilee, upper and lower, as well as the area of Tiberias, is also considered one region. One may eat of Sabbatical year's produce of each region until that produce in the fields of the last of the three areas of each region is exhausted.

10. How so? If one has produce in Judea, he may eat of it so long as there is any of that same species in all of Judea. And the same applies if one has produce in Galilee or in Transjordan.

11. In regard to carobs, olives, and dates, however, these three regions are accounted as one. One may eat dates until the last date is exhausted in Zoar. When does this happen? Up to the Feast of Purim of the eighth year. Olives may be eaten until Pentecost of that year. Grapes may be eaten until Passover of the eighth year, and dried figs until Hanukkah of that year.

12. If one conveys Sabbatical year's produce from a place where it has come to an end to a place where it has not yet come to an end, or vice versa, he is obligated to clear it out, because he is subject to the more stringent rules governing either place.

The produce of the Land of Israel exported abroad must be cleared out where it is and may not be carried from place to place.

13. The Sages have formulated an important general principle concerning Sabbatical year's produce: Whatsoever is used as human food, or as animal feed, or as dyestuff, whose roots do not remain underground, is subject to the law of the Sabbatical year, both itself and the money realized from its sale, and both are liable to removal. For example, leaves of wild arum, ceterach fern, and chicory are human food; thorns and thistles are animal feed; woad and safflower are species of dyestuff. If its roots remain underground—for example, madder and dyer's rocket, which are species of dyestuff—even though both it and the money realized from its sale are subject to the law of the Sabbatical year, neither is liable to removal, inasmuch as its roots remain underground; rather one may use it as dyestuff until New Year's Day.

14. As for whatsoever is not used exclusively for human food, animal feed, or dyestuff, both it and the money realized from its sale are subject to the law of the Sabbatical year, so long as it is not used for wood. It is not, however, liable to removal, even though its roots do not remain underground; rather one may derive benefit from both it and the money realized from its sale until New Year's Day. This applies, for example, to the roots of wild arum, of ceterach fern, and of heliotrope.

15. In the case of husks and blossoms of pomegranates, walnut shells, and fruit kernels, both they and the money realized from their sale are subject to the law of the Sabbatical year, but neither is liable to removal.

Sprouts of service and carob trees, and the money realized from their sale, are both subject to the law of the Sabbatical year and are liable to removal.

Sprouts of terebinth, pistachio, and boxthorn trees, and the money realized from their sale, are both subject to the law of the Sabbatical year, but are not liable to removal. Their leaves, however, are liable to removal.

16. What is the time of removal? For leaves, once they wither and drop off from their trees. Leaves of olive trees, reeds, and carob trees, however, are not liable to removal, inasmuch as they do not wither and waste away.

17. Until what time is one permitted to gather fresh herbs in the Sabbatical year? Until the ground moisture is dried up. One may pick dried herbs until the second rainfall of the eighth year.

18. Until what time in the eighth year are poor people permitted to enter orchards in order to gather Sabbatical year's produce? Until the second rainfall.

19. Rose, henna, and laudanum, and the money realized from their sale, are subject to the law of the Sabbatical year. Balsam, that is the sap secreted by trees, namely by their leaves or stems, is not subject to the law of that year. The sap of unripe figs and the money realized from its sale are subject to the law of that year.

20. To what does this apply? To fruit-bearing trees. In the case of trees that do not bear fruit, even that which issues from leaves and stems is accounted the same as fruit, and both it and the money realized from its sale are subject to the law of the Sabbatical year.

21. If one preserves Sabbatical year's rose in sixth-year's oil, he may extract the rose, and the oil is then permitted. If he preserves it in eighth-year's oil, the oil, too, is liable to removal, inasmuch as the rose was dry and has already become liable to removal.

22. In the case of Sabbatical year's carobs, however, preserved in sixth-year's or eighth-year's wine, the wine is liable to removal, since it is flavored with Sabbatical year's produce. This is the general rule: Sabbatical year's produce, however small its amount, mixed with other produce, renders the latter forbidden, if it is of the same species; if not of the same species, it renders the mixture forbidden if it imparts a flavor to it.

CHAPTER VIII

1. Just as it is forbidden to cultivate the land in the Sabbatical year, so is it forbidden to encourage Israelites who do cultivate it or to sell them implements for cultivation, inasmuch as it is forbidden to encourage transgressors.

2. The following are the implements that a craftsman may not sell in the Sabbatical year to anyone suspected of violating the laws of that year: a plow with all its appurtenances, a yoke, a winnowing fan, and a mattock.

This is the general principle: Any implement designed exclusively for a kind of work forbidden in the Sabbatical year, may not be sold to anyone who is under suspicion; if it is designed for a kind of work that may or may not be forbidden, it may be sold to the suspected person.

3. How so? One may sell him a sickle or a wagon with all its appurtenances, for he might reap only a small amount with the

sickle and convey that small amount in the wagon, which is permitted. If, however, he is known to reap in the manner of regular reapers or to transport the entire yield of his field, one may not sell these implements to him.

4. It is permitted to sell without any stipulation, to a person who is not suspect, even an implement that is used exclusively for work forbidden in the Sabbatical year, inasmuch as it is possible that he is buying it in that year in order to work with it after that year.

5. A potter may sell no more than five oil jugs or fifteen wine jugs, but to a heathen he may sell more than this and need not fear that the heathen might resell them to an Israelite. Outside of the Land of Israel the potter may sell to an Israelite any number of jugs, and need not fear that the latter might import them to the Land of Israel.

6. A may sell in the Sabbatical year a plowing cow to B, who is suspected of not observing the laws of that year, because B may slaughter her; or a field, since B may let it lie fallow. A may not, however, sell to B a field of trees, unless he stipulates that B is to have no share in the trees themselves.

A may lend to B a sĕ'ah measure to measure with it, even if A knows that B has a threshing floor, since it is possible that B will use it only for measuring inside his house.

A may give B small coins in exchange for large ones, even if he knows that B employs laborers.

All these transactions, however, are forbidden if B has declared beforehand that his purpose is such as would violate the laws of the Sabbatical year.

7. Similarly a woman may lend to her companion, suspected of violating the laws of the Sabbatical year, a sifter, a sieve, a handmill, or an oven; but she may not do the actual sifting or grinding with her.

8. Encouragement may be offered to heathens in the Sabbatical year only verbally. For example, if one sees a heathen plowing or sowing, he may say to him, "May you wax strong," or, "May you prosper," or use similar expressions, inasmuch as heathens

are not required to let the land rest. One may not, however, help them physically.

It is permitted to collaborate with them in taking honey out of the beehive, or to rent plowed fields from them, inasmuch as they are not subject to the obligation affecting such actions and hence cannot be penalized for plowing in the Sabbatical year.

9. In Syria it is permitted to process produce already cut, but not produce still attached to the ground. How so? One may thresh, winnow, tread, and bind into sheaves; but one may not reap crops, cut off grapes, or harvest olives. The same rule applies to all similar cases.

10. Just as it is forbidden to use Sabbatical year's produce as merchandise or to retain it, so is it forbidden to buy it from an unlearned person, for one may not hand over Sabbatical year's money to such a person, not even the smallest amount of it, lest he should fail to eat its equivalent in the state of sanctity of that year.

11. If A buys a lulaḇ from B, an unlearned person, in the Sabbatical year, B must give A a citron as a gift, or else he must include the price of the citron in the price of the lulaḇ.

12. When does this apply? When B is selling similar produce, out of guarded crops, such as figs, pomegranates, and their like. If, however, B is selling produce presumed to be ownerless, such as rue, wild amaranth, purslane, mountain coriander, and their like, it is permitted to buy only a small amount from B, no more than enough for three meals, so as not to endanger his livelihood.

13. Anything that is not liable to tithes, such as Baalbek garlic, dyer's rocket bulb, Cilician pounded beans, and Egyptian lentils, as well as garden seeds that are not edible, such as seeds of turnips, radishes, and their like, may be purchased from anyone in the Sabbatical year.

14. To whom does this apply? To an unlearned person in general. As for one who is suspected of trading with Sabbatical year's produce, or of retaining his produce past the time of re-

moval and selling it, it is forbidden to buy from him anything whatsoever that comes under any restriction at all on account of that year; nor may one buy from him flax, even if it is combed, but one may buy it from him if it is already spun and twisted.

15. A person under suspicion concerning the Sabbatical year need not be suspected also concerning tithes, and vice versa, for although both are authorized by the Torah, tithe involves conveyance of the produce to a certain place, which is not the case with Sabbatical year's produce. Moreover produce of the Sabbatical year is not redeemable, whereas tithe is redeemable.

16. A person suspected of not abiding by the laws of cleanness need not be equally suspected concerning tithes and Sabbatical year's produce, because the unclean food sold by him as clean can render other persons unclean by Scribal enactment only, and a person suspected of not abiding by Scribal enactments need not be suspected of not abiding by the laws of the Torah as well.

17. In the case of a person suspected concerning a certain thing, even though he is not to be trusted in regard to this thing that is his own property, he may be trusted in regard to the property of others, for the presumption is that a person would not sin in order to benefit others. Therefore one who is suspected concerning a certain thing may nevertheless act as judge or witness in the case of another man suspected concerning the same thing.

18. Priests are under suspicion concerning Sabbatical year's produce, for they might say, "Since heave offerings are permitted to us, although forbidden to commoners under the penalty of death, how much more so Sabbatical year's produce." Therefore, if one sĕ'ah of heave offering produce falls into one hundred sĕ'ah of Sabbatical year's produce, it becomes neutralized. If it falls into less than that, the whole mixture must be left to rot and may not be sold to a priest, as the law requires for all produce which has become forbidden by intermixture with heave offering, because priests are suspected concerning Sabbatical year produce.

19. Dyers and fatteners of animals may buy coarse bran from any place, and need not be apprehensive that it may have come from Sabbatical year's aftergrowth.

20. Collectors of food for charity in the Sabbatical year need not be particular while in the courtyards of those who eat of that year's produce. If the collectors are given bread, it is permitted, and they need not feel apprehensive that it may be the produce of aftergrowths of the Sabbatical year, inasmuch as the children of Israel may be suspected only of giving either money realized from the sale of that year's produce or eggs bought with such money.

It is permitted to borrow Sabbatical year's produce from the poor, and to repay them with produce of the eighth year.

CHAPTER IX

1. It is a positive commandment to cancel debts in the Sabbatical year, as it is said, *Every creditor shall release that which he hath lent unto his neighbor* (Deut. 15:2).

He who claims a debt passed over by the Sabbatical year transgresses a negative commandment, as it is said, *He shall not exact it of his neighbor and his brother* (*ibid.*).

2. According to the Torah, the cancellation of debts is in force only while the law of the Jubilee year is in force, which involves the repossession of land, when the land is restored to its original owner without any payment.

This is a matter of oral tradition, and the Sages have said accordingly: "Whenever you must release the land, you must also release cash debts, in any place, whether in the Land of Israel or outside of it. Whenever there is no repossession of land, you need not cancel debts in the Sabbatical year, even in the Land of Israel."

3. Scribal law, however, requires the cancellation of debts at the present time in all places, even though the Jubilee is no longer in force, in order that the law of the cancellation of debts should not be forgotten among the children of Israel.

4. The Sabbatical year effects the cancellation of debts only at its conclusion, as it is said, *At the end of every seven years thou*

shalt make a release. And this is the manner of the release, etc. (Deut. 15:1–2), and again, *At the end of every seven years, in the set time of the year of release, in the Feast of Tabernacles* (Deut. 31:10). Just as in the latter verse the conclusion of seven years is meant, so is the cancellation of debts to take effect at the expiry of seven years. Therefore, if one lends money to his companion within the Sabbatical year itself, he may claim the debt throughout that entire year; but once the sun has set on the nights of New Year's Day of the eighth year, the debt is forfeit.

5. In the case of one who slaughters a cow and divides the meat among purchasers under the impression that it is New Year's Day of the eighth year, and it develops that the preceding month of Elul has been intercalated, so that the day turns out to be the last day of the Sabbatical year, the money owed for the meat is forfeit, since the Sabbatical year has passed over the debt.

6. The Sabbatical year cancels a loan, even one secured by a bond constituting a lien on the debtor's property. If, however, the debtor has assigned a particular field as security for his loan, the debt is not cancelled.

The Sabbatical year dismisses the debtor's obligation to swear an oath, as it is said, *He shall not exact it* (Deut. 15:2), implying dismissal in any case, whether the lender demands payment or an oath.

7. To what does this apply? To an oath imposed by a court of judges, and to similar oaths, in cases where, if the debtor admits the whole debt, the Sabbatical year would cancel it. In the case of oaths imposed on bailees, partners, and their like, where, if the debtor admits the debt, he must pay it, he may swear the oath even after the expiration of that year.

8. If A, having lent money to B, demands payment, and B denies the debt, and as the Sabbatical year arrives, persists in his denial, but subsequently, after the expiration of that year, admits his indebtedness, or else witnesses come forth and testify against B, the Sabbatical year does not cancel B's debt.

9. If A lends money to B for the fixed period of ten years, the intervening Sabbatical year cannot cancel the indebtedness, for even though eventually it will come under the command, *He shall not exact it* (Deut. 15:2), A cannot in fact exact payment during that year.

If, however, B stipulates that A should not demand payment, the Sabbatical year cancels the debt.

10. If A lends money to B with the stipulation that the Sabbatical year is not to cancel the debt, it is nevertheless cancelled, for A cannot annul the law of that year. If A stipulates that B is not to claim cancellation of the debt even in the Sabbatical year, the stipulated condition is valid, for all stipulations concerning money matters are valid, and it follows that B, having obligated himself for a money debt that the Torah has not made him liable for, is therefore fully obligated for it.

11. Store credit is not cancelled by the Sabbatical year, but if converted into a loan, it is cancelled. Hired man's wages likewise are not cancelled, but if formalized into a loan, they are cancelled.

12. Fines imposed upon the violator, the seducer, or the defamer are not cancelled, but if formalized into a loan, they are cancelled. At what time are they formalized into a loan? From the time the case is brought into court.

13. If a man divorces his wife before the Sabbatical year, her kĕṯubbah is not cancelled. If she has already impaired it, or has formalized it into a loan against him, it is cancelled.

14. A loan secured by a pledge is not cancelled, provided that the amount of the loan and the value of the pledge are equal. If the loan exceeds the value of the pledge, the excess amount is cancelled.

15. If the creditor surrenders his bonds to the court and says to the judges, "You collect this debt owed to me," it is not cancelled, as it is said, *Whatsoever of* THINE *is with thy brother* (Deut. 15:3), whereas now it is the court that claims this debt.

Similarly, if the court delivers a written verdict, to wit, "To

So-and-so: You are herewith obligated to pay to this person such-and-such a sum," it is not cancelled, for the debt is now accounted as already collected and in the creditor's possession, and is no longer an uncollected loan.

16. When Hillel the Elder saw that people were refraining from lending to one another, thus transgressing what is written in the Torah, *Beware that there be not a base thought in thy heart* (Deut. 15:9), he instituted the *prosbul,* in order that debts might not be cancelled and people might continue to grant loans one to another.

The prosbul is effective only in the cancellation of cash debts at the present time, when the law of the Sabbatical year is in force solely by Scribal authority. When the law is in force by the authority of the Torah, the prosbul has no effect.

17. A prosbul may be written only by exceedingly great Sages, such as the members of the court of Rabbi Ammi and Rabbi Assi, who are empowered to enforce payment of the people's money. Any other court may not write it.

18. This is the form of the prosbul: "I hereby declare before you, So-and-so and So-and-so, judges in such-and-such a place, that touching any debt owed to me, I am to have the right to collect it at any time I please." The judges or the witnesses must sign below it.

19. A prosbul may be written only for a loan secured by real estate. If the debtor has no land, the creditor may sell him a piece, however small, of his own land, within his own field, even if it is an area no larger than that occupied by a cabbage stalk.

If the creditor lends the debtor enough ground to hold an oven or a stove, a prosbul may be written with it as security.

If the debtor has land held as pledge, a prosbul may likewise be written with it as security.

20. A prosbul may be written for a husband with his wife's property as security, and for orphans with their guardian's property as security. If the debtor has no land, but his guarantor has some, a prosbul may be written with it as security.

If A, who owns land, is indebted to B, C may write a prosbul against B with A's land as security, inasmuch as B has a lien upon A's land.

21. If one debtor borrows money from five creditors, each one needs his own prosbul. If five debtors borrow from the same creditor, one prosbul is enough for all.

22. If the creditor first writes the prosbul and then lends the money, the prosbul has no effect, and the debt is cancelled, unless the prosbul is written after the execution of the loan. You thus conclude that a loan antedating the prosbul is affected by it; and contrariwise, a prosbul antedating the loan cannot affect it.

23. Consequently a predated prosbul is valid, while a postdated one is void. How so? If one writes a prosbul in the month of Nisan and predates it to Adar, it is valid, inasmuch as its term has been reduced, since it affects debts contracted only up to Adar. If he postdates it, however, as of the month of Iyar, it is void, since its term has been extended to affect debts contracted up to Iyar, which is contrary to the law, for according to the law only debts contracted up to Nisan, when the matter is delivered to the court, should be affected.

24. If one produces a note of indebtedness after the Sabbatical year, but can show no prosbul with it, he forfeits his claim. If he says, "I had a prosbul, but lost it," he is to be believed, for since the time of peril and onward a creditor may exact payment without producing a prosbul. Not only that, but when the creditor produces his note, or demands payment of a loan executed verbally, the debtor should be told, "Pay him." If the debtor objects, saying, "Where is his prosbul?", the creditor should be asked, "Did you have a prosbul and lose it?" If he answers, "Yes," he is to be believed. If he admits that he had written no prosbul, he forfeits his claim.

Orphans need no prosbul.

25. If the creditor produces a prosbul, but the debtor objects, saying, "This loan which he demands was contracted after the writing of this prosbul," the creditor replying, "It was contracted

before the writing of the prosbul," the creditor is to be believed, inasmuch as had he said, "I had a prosbul but lost it," he would have been believed, even though we do not know the date of the lost prosbul.

26. If the debtor argues, "What I owe him is a loan," while the creditor says, "Not so—it is a store credit, which is not cancelled in the Sabbatical year, seeing that I did not formalize it into a loan," the creditor is to be believed, inasmuch as he had wished, he could have said, "It was a loan, and I had a prosbul for it, but it was lost." For since the Sages have established the prosbul, the presumption is that a person will not leave what is permitted and eat of what is forbidden.

27. In the case of disciples of the wise who have lent money one to another, if the creditor makes a declaration before the disciples, saying, "I hereby declare before you that touching any debt that is owed to me, I am to have the right to collect it at any time I please," he need write no prosbul, because the disciples know that the cancellation of cash debts at the present time is based only on a Scribal enactment, and may be set aside by a mere verbal declaration.

28. Whosoever repays a debt passed over by the Sabbatical year causes the Sages to be well pleased with him. The creditor, however, should say to him, "I am cancelling your debt, and you are acquitted with me." If the debtor then says, "Nevertheless it is my wish that you accept the money," the creditor may accept it, as it is said, *He shall not exact it* (Deut. 15:2), and here the creditor did not in fact exact it. The debtor should not say to the creditor, "I give it to you in payment of my debt," but rather, "These monies are mine, but I give them to you as a gift."

29. If the debtor returns the money to the creditor without pronouncing the formula aforesaid, the creditor should keep on conversing with him until he says, "These monies are mine, but I give them to you as a gift." If he still fails to say so, the creditor should not accept the money from him, but should rather let him retain his money and go his way.

30. He who refrains from lending money before the Sabbatical year, for fear that the payment of his loan might be delayed and the loan cancelled, transgresses a negative commandment, as it is said, *Beware lest there be a base thought in thy heart* (Deut. 15:9). This is a grievous sin, inasmuch as here the Torah has warned against it with a double caveat, *beware lest,* and wherever it is said *beware,* or *lest,* or *thou shalt not,* a negative commandment is implied. Moreover the Torah is very particular about this evil thought, since it designates it as *base* (*belial*). And furthermore, Scripture adds a warning and an injunction that one should not refrain from giving but should give, as it is said, *Thou shalt surely give him, and thy heart shall not be grieved when thou givest unto him* (Deut. 15:10). And finally, the Holy One, blessed be He, has promised a reward for the observance of this commandment in this world, as it is said, *Because that for this thing the Lord thy God will bless thee in all thy work* (*ibid.*).

CHAPTER X

1. It is a positive commandment to count seven times seven years and to sanctify the fiftieth year, as it is said, *And thou shalt number seven Sabbaths of years unto thee . . . And ye shall hallow the fiftieth year* (Lev. 25:8–10). These two commandments were entrusted to the Great Court only.

2. At what time was the count begun? Fourteen years after the Israelites had entered the Land, as it is said, *Six years thou shalt sow thy field, and six years thou shalt prune thy vineyard* (Lev. 25:3), implying, until each person could identify his own parcel of land. Now they had spent seven years in conquering the Land, and another seven years in distributing it. You thus conclude that they began counting from the year 2503 after the creation of the world, beginning with the following New Year's Day, reckoning from the birth of the first man, which took place in the second year after creation, thus making the year 2510 after creation, which is the twenty-first year after they had entered the Land, a

Sabbatical year. They then counted seven septennates and hallowed the following fiftieth year, which was the sixty-fourth since they had entered the Land.

3. The Israelites counted seventeen Jubilees from the time they entered the Land until they left it. The year in which they left it, when the First Temple was destroyed, was the year following a Sabbatical year, and the thirty-sixth year of that Jubilee period, the first Temple having stood for 410 years. When the First Temple was destroyed, this count of Jubilees ceased, and from that time on the Land remained desolate for 70 years. The Second Temple was then built and stood for 420 years. In the seventh year after its erection Ezra went up to the Land, and this is called the second entrance into the Promised Land. From that year the Israelites began a new count and declared the thirteenth year after the building of the Second Temple a Sabbatical year. They then counted seven septennates and hallowed the following fiftieth year. Even though the Jubilee was not in force during the time of the Second Temple, it was nevertheless counted in order to hallow the Sabbatical years.

4. You thus learn that the year in which the Second Temple was destroyed, which begins with the month of Tishri, about two months after the actual destruction of this Temple—the reckoning of both Sabbatical and Jubilee years beginning with Tishri—was the year following a Sabbatical year, and the fifteenth year of the ninth Jubilee period. According to this reckoning, the present year, which is the year 1107 since the destruction of the Second Temple, corresponding to the year 1487 of the Seleucid Era, and 4936 of the Era of Creation, is a Sabbatical year, and the twenty-first year of the Jubilee period.

5. All the Geonim, however, have said that they had a tradition handed down from person to person, to the effect that during the seventy years between the destruction of the First Temple and the building of the Second Temple only Sabbatical years were counted, but not Jubilee years. Similarly, since the Second Temple was destroyed, the fiftieth year was no longer counted, only

the septennates, reckoning from the beginning of the year of the destruction of that Temple. The same reckoning, which is traditional, follows also from the Gemara in the tractate 'Aḇoḏah Zarah.

6. The Sabbatical year was known and widely published among the Geonim and the people of the Land of Israel. All of them counted only the years since the destruction of the Second Temple, in groups of seven. According to this reckoning, the present year, which is the year 1107 since the destruction of the Second Temple, is the year following a Sabbatical year. This is the reckoning upon which we rely, and in accordance with it we issue instructions concerning tithes, Sabbatical year, and cancellation of cash debts, for tradition and actual practice are great pillars in legal decisions, and it is proper to rely upon them.

7. The Jubilee year itself is not included in the count of the years of the septennate, so that the forty-ninth year is a Sabbatical year, the fiftieth year is the Jubilee year, and the fifty-first year is the first of the six years of the next septennate. The same applies to all Jubilees.

8. When the Tribes of Reuben and Gad and half of the Tribe of Manasseh were exiled, Jubilees ceased, as it is said, *And proclaim liberty throughout the land unto all the inhabitants thereof* (Lev. 25:10), implying, as long as all its inhabitants are there, and furthermore, as long as the Tribes are not intermixed, and each Tribe dwells as at the time the settlement was established.

Whenever the Jubilee is in force in the Land of Israel, it is in force also outside of it, as it is said, *It shall be a Jubilee* (*ibid.*), implying, in all places and at all times, whether the Temple is or is not in existence.

9. Whenever the Jubilee is in force, the law of the Hebrew bondsman, of houses in walled cities, of the devoted field, and of the field of possession are also in force; resident aliens may be accepted; the Sabbatical year is operative in the Land of Israel; and the cancellation of cash debts is effective in all places—all this according to the Torah.

Whenever the Jubilee is not in force, none of these are in force, excepting the Sabbatical year in the Land of Israel, and the cancellation of cash debts in all places, both only according to Scribal law, as we have explained.

10. It is a positive commandment to sound the ram's horn on the tenth of Tishri of the Jubilee year. This commandment was originally entrusted to the Great Court, as it is said, *Then shalt thou make proclamation with the blast of the horn* (Lev. 25:9). But each individual is likewise obligated to sound the horn, as it is said, *Shall ye make proclamation with the horn* (*ibid.*).

Nine blasts must be sounded with the ram's horn, the same as on New Year's Day, and proclamation must be made with the horn throughout the boundaries of the Land of Israel.

11. The ram's horn used for the Jubilee and the one used for New Year's Day are subject to the same rules, and the same number of blasts is required, except that in the Jubilee year the horn may be sounded either at the court in which the new month had been sactified or at the court in which it had not been sanctified. Each individual is under obligation to sound the horn as long as the court is in session, even if he himself is not in the presence of the court.

12. If New Year's Day fell on the Sabbath, the ram's horn was sounded only at the court in which the new month had been sanctified, and an individual might sound it only in the presence of the court.

13. Three things are essential for the Jubilee year: sounding the ram's horn, emancipation of bondsmen, and restoration of fields to their original owners termed "release of the land."

14. From New Year's Day to the Day of Atonement, bondsmen were neither released to their homes nor retained as subjects of their masters, nor were fields restored to their original owners. Rather the bondsmen ate, drank, and rejoiced, with their wreaths upon their heads.

Once the Day of Atonement arrived and the court caused the

ram's horn to be sounded, the bondsmen were released to their homes and the fields were restored to their original owners.

15. The law enjoining cessation of cultivation of the land is the same for the Jubilee year as for the Sabbatical year in all respects: whatsoever work on the land is forbidden in the Sabbatical year is forbidden also in the Jubilee year, and whatsoever is permitted in the Sabbatical year is permitted also in the Jubilee year. Work involving liability to a flogging in the Sabbatical year involves the same liability in the Jubilee year. So also is the law governing the consumption, sale, and removal of produce of the Jubilee year, the same as the law of the Sabbatical year in all respects.

16. The differences between the Sabbatical year and the Jubilee year are as follows: the Sabbatical year cancels cash debts, while the Jubilee year does not; the Jubilee year emancipates bondsmen and restores land to its original owner—this rule affects the sale of fields referred to in the Torah, and is a positive commandment, as it is said, *Ye shall grant a redemption for the land* (Lev. 25:24)—while the Sabbatical year does not; the Jubilee year releases the land at its beginning, while the Sabbatical year cancels cash debts only at its end, as we have explained.

CHAPTER XI

1. No part of the Land of Israel that was divided among the Tribes may be sold in perpetuity, as it is said, *And the land shall not be sold in perpetuity* (Lev. 25:23). If it is sold in perpetuity, both buyer and seller have transgressed a negative commandment, their act avails nothing, and the field must be restored to its original owner in the Jubilee year.

2. If one sells his field for a term of sixty years, it is not restored to him in the Jubilee year, inasmuch as only those fields are restored that are sold either without any specification or in perpetuity.

3. One may not sell his house or his field of possession, even though they will be restored to him eventually, unless he has been reduced to destitution, as it is said, *If thy brother be waxen poor, and sell some of his possession* (Lev. 25:25). One is therefore not allowed to sell them in order to put the money in his pocket, or engage in trade, or buy vessels, bondsmen, or cattle therewith; he may use the money only for his own maintenance. If one transgresses and sells them nevertheless, the properties are validly sold.

4. In this respect the rule governing a house is the same as the one governing houses in walled cities, and the rule governing a field is the same as the rule governing the field of possession. The rule governing the person who sells his field of possession is that he must reckon the price according to the years remaining until the Jubilee year; and whenever he wishes to redeem it, he must make an accounting with the buyer of the number of years in which the latter has enjoyed the produce of the field, deduct the corresponding amount from the money received from the buyer, and return the balance to him.

5. How so? If, for example, with ten more years remaining until the Jubilee year, A had sold the field to B for 100 denar, B has enjoyed it for three years, and now A wishes to redeem it, A must pay B 70 denar, and B must restore the field to A. Similarly, if B has enjoyed it for six years, A must pay B 40 denar, and B must restore the field to A.

If A does not redeem the field but leaves it in B's hands until the Jubilee year, it goes back to A, the original owner, without any payment, as it is said, *And according unto the number of years of the crops he shall sell unto thee* (Lev. 25:15).

6. If A, having sold the field when it was laden with produce, redeems it after two years, he may not say to B, "Restore it to me full of produce, as it was when I sold it to you."

Therefore, if A has sold the field to B laden with produce before New Year's Day, and redeems it two years later, B has enjoyed three crops in two years, but must account with A for only two years, as it is said, *According to the years* (Lev. 27:18), implying, not according to the crops.

7. Stems, twigs, and fruit of the sycamore tree in the field belong to the buyer, as does any other produce thereof. As for a cut-down or dried-up tree, both buyer and seller are forbidden to enjoy it. How then should one proceed? It should be sold, land should be bought with the proceeds, and the buyer may then enjoy the produce of that land until the field is redeemed from him.

8. If one buys a field of possession and plants it with trees, thereby increasing its value, the rule is that when it is restored in the Jubilee year, the appreciation effected by the trees planted in it must be estimated, and the original owner of the field must pay to the buyer the amount of this appreciation, as it is said, *The house that was sold . . . shall go out* (Lev. 25:33), implying that what was sold is to be restored, not the appreciation thereof.

9. If one sells his field at the time when the Jubilee is in force, as we have explained, he is not permitted to redeem it before the lapse of at least two years, as it is said, *According unto the number of years of the crops he shall sell unto thee* (Lev. 25:15); even if the buyer is willing, it is still forbidden, as it is said, *According to the number of years after the Jubilee thou shalt buy of thy neighbor* (*ibid.*), the plural *years* implying not less than two full years from the day of the sale.

10. It is necessary for the buyer to consume two crops in two years, and only then may the field be redeemed, as it is said, *years of the crops* (Lev. 25:15). Therefore, if one of these two years is a Sabbatical year, or is stricken with blight or mildew, it is not included in the reckoning.

11. If the buyer leaves the field uncultivated for one year and consumes its crops in the other year, or consumes its crops for one year and plows it in the other year but does not plant it,

both years are taken into account. If he sells it in the Jubilee year itself, the sale is ineffective, and the money must be returned to the original owner.

12. If the owner sells the field one year before the Jubilee, the buyer is entitled to enjoy its crops for another year after the Jubilee, as it is said, *years of the crops* (Lev. 25:15).

13. If the owner sells ravines filled with water or stony ground unfit for planting, he may redeem the ground in less than two years, as it is said, *According unto the number of years of the crops* (Lev. 25:15), implying that only a field fit for growing crops may not be redeemed before two years have passed. If he does not redeem it, and even though it is unfit for planting, it must be restored to its original owner in the Jubilee year.

14. If one sells fruit trees, they may not be redeemed before the lapse of at least two years, inasmuch as they are fit to produce. If he does not redeem them, they need not be restored to their original owner in the Jubilee year, as it is said, *And he shall return unto his field of possession* (Lev. 25:27), implying not unto trees.

15. If A sells his field to B, B sells it to C, C sells it to D, and so forth, up to one hundred consecutive sales, the field must be restored to A, its first owner, in the Jubilee year, as it is said, *In the year of Jubilee the field shall return unto him of whom it was bought, even to him to whom the possession of the land belongeth* (Lev. 27:24).

16. If A sells the field to B for 100 denar, and B sells it to C for 200, and A then wishes to redeem it, he need negotiate only with B, as it is said, *Unto the man to whom he sold it* (Lev. 25:27).

If A sells the field to B for 200 denar, and B sells it to C for 100, A need negotiate only with C.

Similarly, if A sells it to B for 100 denar, and it rises in value in B's hands until it is fit to be sold for 200, A need reckon only according to the price that he had sold it for. On the other hand, if A sells the field for 200 denar, and it depreciates, so that it is fit to be sold for only 100, A need reckon only according to its present value. In general, in the case of a field of possession, the power of the seller is to be strengthened, while the power of the buyer is to be weakened.

17. If the person who has sold his field of possession has other fields, and proposes to sell some of them in order to redeem the field sold first, no attention is to be paid to him, as it is said, *And find sufficient means to redeem it* (Lev. 25:26), implying that he may redeem it only if he finds something that was not available to him at the time of the first sale. Similarly, if he proposes to borrow money in order to redeem it, no attention is to be paid to him, as it is said, *And he be waxen rich* (*ibid.*), implying, not if he has to borrow.

18. If he finds a little money and wishes to redeem one-half of the field sold, no attention is to be paid to him, as it is said, *Sufficient means to redeem it* (Lev. 25:26), implying that he must either redeem the entire field or not redeem it at all. If his kinsmen wish to redeem it, they may do so, as it is said, *Then shall his kinsman that is next unto him . . . redeem* (Lev. 25:25).

19. If one gives away his field as a gift, it must be restored to him in the Jubilee year, as it is said, *Ye shall return every man unto his possession* (Lev. 25:13), which is meant to cover gifts as well.

20. Brothers who had divided their field of inheritance are regarded vis-à-vis each other the same as buyers, and must return their shares to each other in the Jubilee year, but their original division of the estate is not rendered void thereby.

Similarly the first-born brother, or the brother who has taken his widowed sister-in-law in levirate marriage, must return in the Jubilee year the share that he had received and take its corresponding equivalent instead.

21. The husband, however, who has inherited his deceased wife's estate, need not restore it in the Jubilee year—even though

the husband's right to his wife's estate is based only on Scribal law, the Sages have given the same force to their decree as if it were in the Torah.

If he has inherited a burial plot from her, he must restore it to the family, in order not to disgrace them. The family in turn must pay him the plot's value minus the value of his wife's grave, seeing that he is obligated to bury her.

CHAPTER XII

1. He who sells a house in a city surrounded by a wall may redeem it within twelve months from the day of sale, at any time he wishes, even on the same day he sold it. Whenever he wishes to redeem it, he must return to the purchaser all the money which he received, and may not deduct anything from it.

2. The seller's kinsmen may not redeem the house—only the seller himself may do so, if he has the means. He may sell some of his other property in order to redeem it, but he may not borrow money for that purpose, nor may he redeem the house by halves.

3. If the buyer dies, the seller may redeem the house from his son. Similarly, if the seller dies, his son may redeem it within the twelve months.

4. If A sells his house to B, and B resells it to C within the year of grace, which is counted from the date of the sale to B, once the first year has passed, the house remains in the permanent possession of C, inasmuch as B had sold to C every right that may accrue to him. Therefore, if the twelve months have passed, and A did not redeem the house, it remains in perpetuity in the hands of the final buyer, C.

Thus also, if A gives the house to B as a gift, and does not redeem it within the twelve months, it remains in the hands of the recipient of the gift, B.

5. If the year is a leap year, permanent possession does not take effect until the end of the year, as it is said, *Within the space of*

a full year (Lev. 25:30), implying the inclusion of the additional month of the leap year.

6. If a person sells two houses, one in the middle of the first month of Adar, and the other at the beginning of the second month of Adar, the rule is that for the house sold on the first day of second Adar the year of grace is completed on the first day of Adar of the following year, while for the house sold in the middle of first Adar the year is not completed until the middle of Adar of the following year, because the buyer had taken possession in the additional month.

7. If the last day of the twelve months has arrived, and the buyer cannot be found in order to redeem the house from him, the seller may deposit his money with the court, and then break down the door and enter his house. The buyer, whenever he appears, may come forth and take his money.

8. If A consecrates a house in a walled city, and B then redeems it from the Sanctuary, the rule is that once a year has passed from the time the house was redeemed from the Sanctuary, and A has not redeemed it, it remains in perpetuity in B's hand, inasmuch as it is not the Sanctuary that acquires permanent possession but the buyer, as it is said, *To him that bought it, throughout his generations* (Lev. 25:30).

9. If A sells a house in a walled city to B, and the year of Jubilee arrives within the year of grace, the house is not restored to A in the year of Jubilee, and remains in B's hand until A is willing to redeem it within the year of the sale, or until the year of grace has passed and B acquires permanent possession.

10. If one sells a house in a hamlet or in a city that is not surrounded by a proper wall, it may be redeemed on the basis of the same privilege as that which prevails in the law governing both fields and houses surrounded by a wall.

How so? If he wishes to redeem the house immediately, he may do so, the same as in the case of houses in a walled city. If twelve months have passed without his redeeming it, he may redeem it thereafter up to the year of Jubilee, the same as in the case of fields, and when he does redeem it, he must make an accounting with the seller and deduct what the latter had enjoyed. If the year of Jubilee arrives without the original owner redeeming it, it returns to him without any payment, the same as in the case of fields.

11. Whatsoever is inside the wall, such as gardens, bathhouses, and dovecotes, counts the same as houses, as it is said, *That is in the walled city* (Lev. 25:30).

Fields within the city, however, may be redeemed in the same way as fields outside of the city, as it is said, *Then the house that is in the walled city shall be made sure in perpetuity* (*ibid.*), implying a house as well as whatsoever is like a house, but not fields.

12. A house less than four cubits square cannot pass into the buyer's permanent possession, as do houses within a walled city, nor can any house in Jerusalem. A house built into the city wall is not regarded as a house within a walled city.

13. A city whose wall is formed by gardens or by the sea is not regarded as a city surrounded by a wall.

14. A place cannot be called a walled city unless it comprises three or more courtyards, each containing two or more houses, the wall having been erected first, and then the courtyards built within it. A place settled first and then encompassed by a wall, or one comprising less than three courtyards with at least two houses in each, is not a walled city, and the houses in it are regarded as the houses of a hamlet.

15. A proper encompassing wall is one that was in existence at the time of the conquest of the Land. How so? A city not surrounded by a wall at the time when Joshua conquered the Land, even if so surrounded now, is considered the same as a hamlet. Contrariwise, a city encompassed by a wall in the days of Joshua, even if not so encompassed now, is considered as if it were still walled.

When the children of Israel were exiled at the time of the destruction of the First Temple, the sanctity of cities walled in the days of Joshua ceased. When Ezra came up at the time of the second return, all the cities that were encircled by a wall at that time became sanctified, inasmuch as the coming up of the Israelites in the days of Ezra, which was the second coming up, was analogous to their coming up in the days of Joshua: just as at their coming up in the days of Joshua they commenced counting septennates and Jubilees, sanctified houses in walled cities, and became liable to tithes, so also did they do the same when they came up in the days of Ezra.

16. Similarly in the days to come, at the third coming up, when the Israelites shall reenter the Land, they will commence counting septennates and Jubilees and sanctifying houses in walled cities, and all places occupied by them will become liable to tithes, as it is said, *And the Lord thy God will bring thee into the land which thy fathers possessed, and thou shalt possess it* (Deut. 30:5), thus comparing your possessing of the Land to your fathers' possessing: just as your fathers' possessing involves the renewal of all these things, so does your own possessing.

CHAPTER XIII

1. Even though the Tribe of Levi was given no share in the Land, the children of Israel had already been commanded to give them cities to dwell therein, together with the adjacent open land. These cities were the six cities of refuge, and forty-two cities in addition to them. When in the days of the Messiah additional cities of refuge will be designated, all of them will go to the Levites.

2. The open space around these cities has already been defined in the Torah as a space 3000 cubits wide in every direction from the city wall outward, as it is said, *From the wall of the city and outward a thousand cubits round about* (Num. 35:4), and further on, *And ye shall measure without the city for the east*

side two thousand cubits, etc. (Num. 35:5). The first 1000 cubits is open space, and the 2000 cubits measured beyond the open space are for fields and vineyards.

3. Each city was given a burial ground outside of this boundary, inasmuch as their dead were not buried within the city limits, as it is said, *And their open land shall be for their cattle, and for their substance, and for all their beasts* (Num. 35:3), implying that the open space was given to them for use by the living and not for burial.

4. In the cities of the Levites, city space may not be converted into open space, or vice versa, nor open space into field, or vice versa, as it is said, *The fields of the open land about their cities may not be sold* (Lev. 25:34).

5. The Sages have learned by tradition that *may not be sold* (Lev. 25:34) means "may not be changed"—the field, the open space, and the city space, each one of the three must remain as it is forever after.

6. Similarly, in all the other cities of the Land of Israel one may not tear down his house in order to turn it into a garden, nor plant the site of a ruined structure in order to convert it into a garden, so that people would not cover the Land of Israel with ruins.

7. Priests or Levites who have sold a field out of the fields of their cities, or a house out of the houses of their walled cities, need not redeem it in the regular manner; rather, having sold the fields, even if close to the year of Jubilee, they may redeem them immediately. If they have consecrated the field, they may redeem it from the Sanctuary after the year of Jubilee. As for houses in walled cities, they may redeem them whenever they wish, even after the lapse of several years, as it is said, *The Levites shall have a perpetual right of redemption* (Lev. 25:32).

8. An Israelite who has inherited the estate of his Levitic maternal grandfather has the Levite's right of redemption, even though he himself is not a Levite, inasmuch as the cities or fields

of the Levites are subject to unlimited redemption, which is attached to these properties and not to their owners.

9. So also a Levite who has inherited his Israelite maternal grandfather's estate has the Israelite's, not the Levite's, right of redemption, since the Biblical rule, *The Levites shall have a perpetual right of redemption* (Lev. 25:32), refers only to the cities of the Levites.

10. The entire Tribe of Levi was enjoined to have no share in the Land of Canaan. They were likewise enjoined to seize no share in the spoils of war while the cities were being conquered, as it is said, *The priests, the Levites, even all the Tribe of Levi, shall have no portion nor inheritance with Israel* (Deut. 18:1), implying, *no portion* in the spoils *nor inheritance* in the Land. Similarly it is said of Aaron, *Thou shalt have no inheritance in their land, neither shalt thou have any portion among them* (Num. 18:20), *portion* referring to spoils of war. If a Levite or a priest seizes a share in the spoils, he is liable to a flogging, and if he takes a share in the Land, it must be taken away from him.

11. It would appear to me that this rule applies only to the Land which is subject to the covenant made with Abraham, Isaac, and Jacob, and which was inherited by their children and was divided among them. As for other lands conquered by one of the kings of Israel, the priests and the Levites have the same share in them and in their spoils as do all other Israelites.

12. Why was the Tribe of Levi granted no right to a share in the Land of Israel and in its spoils, together with his brothers? Because they were set apart to worship the Lord, to serve Him, and to teach His upright ways and His righteous judgments to the many, as it is said, *They shall teach Jacob Thine ordinances, and Israel Thy law* (Deut. 33:10). They were consequently set apart from the ways of the world: they may not wage war as do the rest of Israel, they have no share in the Land, and they may acquire nothing for themselves by physical force. They are rather the host of the Holy Name, as it is said, *Bless, Lord, his host*

(Deut. 33:11). It is He, blessed be He, who acquires for them, as it is said, *I am thy portion and thine inheritance* (Num. 18:20).

13. Not only the Tribe of Levi, but also each and every individual of those who come into the world, whose spirit moves him and whose knowledge gives him understanding to set himself apart in order to stand before the Lord, to serve Him, to worship Him, and to know Him, who walks upright as God had made him to do, and releases his neck from the yoke of the many speculations that the children of man are wont to pursue—such an individual is consecrated to the Holy of Holies, and his portion and inheritance shall be in the Lord forever and evermore. The Lord will grant him in this world whatsoever is sufficient for him, the same as He had granted to the priests and to the Levites. Thus indeed did David, upon whom be peace, say, *O Lord, the portion of mine inheritance and of my cup, Thou maintainest my lot* (Ps. 16:5).

LIST OF ABBREVIATIONS

Tractates of Mishnah, Tosefta, and Talmud

’Aḇ—*’Aḇoṯ*
‘Ar—*‘Ăraḵin*
AZ—*‘Aḇoḏah Zarah*
BB—*Baḇa Baṯra*
Beḵ—*Bĕḵoroṯ*
Ber—*Bĕraḵoṯ*
Beṣ—*Beṣah*
Bik—*Biḵḵurim*
BḲ—*Baḇa Ḳamma*
BM—*Baḇa Mĕṣi‘a*
Dem—*Dĕmay*
‘Eḏ—*‘Eḏuyoṯ*
‘Er—*‘Eruḇin*
Giṭ—*Giṭṭin*
Ḥaḡ—*Ḥăḡiḡah*
Ḥal—*Ḥallah*
Hor—*Horayoṯ*
Ḥul—*Ḥullin*
Kel—*Kelim*
Ker—*Kĕriṯoṯ*
Keṯ—*Kĕṯuḇoṯ*
Ḳid—*Ḳiddušin*
Kil—*Kil’ayim*
Ma—*Ma‘ăśĕroṯ*
Mak—*Maḵḵoṯ*
Maḵ—*Maḵširin*
Me—*Mĕ‘ilah*
Meḡ—*Mĕḡillah*
Men—*Mĕnaḥoṯ*
MḲ—*Mo‘eḏ Ḳaṭan*
MSh—*Ma‘ăśer Šeni*
Neḏ—*Nĕḏarim*
Neḡ—*Nĕḡa‘im*
Nid—*Niddah*
’Oh—*’Ŏhaloṯ*
‘Or—*‘Orlah*
Pa—*Parah*
Pe—*Pe’ah*
Pes—*Pĕsaḥim*
RH—*Roš haš-Šanah*
Sanh—*Sanheḏrin*
Shab—*Šabbaṯ*
Sheḇi—*Šĕḇi‘iṯ*
Sheḵ—*Šĕḵalim*
Soṭ—*Soṭah*
Suk—*Suḵḵah*
Ta—*Ta‘ăniṯ*
Tem—*Tĕmurah*
Ter—*Tĕrumoṯ*
Ṭoh—*Ṭohăroṯ*
ṬY—*Ṭeḇul Yom*
‘Uḵ—*‘Uḵṣin*
Yaḏ—*Yaḏayim*
Yeḇ—*Yĕḇamoṯ*
Zeḇ—*Zĕḇaḥim*

B. prefixed to the name of a tractate indicates a reference to the Babylonian Talmud; P. indicates a reference to the Palestinian (Jerusalemite) Talmud; and Tos a reference to the Tosefta (ed. Zuckermandl, Pasewalk, 1880; 2nd ed., Jerusalem, 1937). Otherwise the reference is to the tractate of the Mishnah.

Other Sources and Commentaries

Aḥai, Gaon—*Šĕ'eltot̲,* Jerusalem, 1953
ARN—'Ab̲ot̲ dĕ-Rabbi Nathan
Bertinoro—commentary on the Mishnah by R. Obadiah of Bertinoro (15th century)
Corcos—commentary on the Book of Agriculture by R. Joseph Corcos (d. after 1575)
DER—Derek̲ 'Ereṣ Rabbah
DEZ—Derek̲ 'Ereṣ Zuṭa
JE—Jerusalem Edition of the Code, published by the Kook Foundation; Book of Agriculture, with commentary by A. Karlin and H. Pardes, Jerusalem, 5722/1962
Klein, S.—*Sefer hay-yiššub̲,* Jerusalem, 1939
KM—*Kesef Mišneh* (commentary on the Code, by R. Joseph Caro [1488–1575])
Mĕlek̲et̲ Šĕlomoh—commentary on the Mishnah, by R. Solomon Adeni (b. 1567)
Midrash Tanḥuma
Midrash Tannaim
PhM—*Peruš ham-Mišnah* (Maimonides' commentary on the Mishnah)
RABD—*Haśśag̲ot̲* (critical notes on the Code, by R. Abraham ben David of Posquières [1125–1198])
Rashi—R. Solomon ben Isaac of Troyes (1040–1105), author of commentaries on the Bible and on the Babylonian Talmud
RDBZ—commentary on the Code by R. David ibn Abi Zimra (16th century)
RaN—R. Nissim ben Jacob of Kairouan (11th century), author of a commentary on some tractates of the Talmud
ŠEE—*Šulḥan 'Aruk̲, 'Eb̲en ha-'Ezer* (by R. Joseph Caro [1488–1575])
ShM—*Sefer ham-miṣwot̲,* by Maimonides
Sif—Sifra (on Leviticus) and Sifre (on Numbers and Deuteronomy)
Sif Zuṭa—Sifre Zuṭa
ŚK—*Śift̲e Kohen* (commentary on Šulḥan 'Aruk̲, by R. Shabbethai Kohen [1621–1662])
ŠYD—*Šulḥan 'Aruk̲ Yoreh De'ah* (by R. Joseph Caro [1488–1575])
ṬaZ—*Ṭure Zahab̲* (commentary on the Šulḥan 'Aruk̲, by R. David ben Samuel ha-Levi [d. 1667])
ṬEE—*Ṭur 'Eb̲en ha- 'Ezer* (by R. Jacob ben Asher [1269–1343])
Tosafot̲—critical and explanatory glosses on the Talmud by the successors of Rashi
YJS—Yale Judaica Series

NOTES

Cross-references consisting of four numbers indicate passages found in the Code outside the present volume, the numbers referring, respectively, to the Book, the Treatise, the Chapter, and the Section where the passage in question occurs.

TREATISE I: LAWS CONCERNING DIVERSE KINDS

Chapter I

1. B. Kiḏ 39a; Beḵ 54a.

2. B. Mak 21a; Kil 7:8; B. Men 70a.

"flogging for disobedience"—*makkat mardut,* literally "lashes of rebellion"; flogging prescribed by Rabbinic law, the number of lashes inflicted at the discretion of the court, in contradistinction to the Biblical lashes which are limited to no more than thirty-nine.

3. P. 'Or 3:7; Kil 8:1; P. Kil 8:1; B. AZ 64a; Ḳid 39a.

"diverse kinds"—*kil'ayim,* literally "two kinds." Here the term refers to species of seeds that may not be sown together. In Scripture, and as explained in this treatise, there are four types of diverse kinds that are forbidden: sowing a field or garden with diverse kinds of seeds; crossbreeding cattle of different species; sowing a vineyard with diverse kinds of plants; and wearing garments of diverse kinds of stuff (woolen and linen) mixed.

"in behalf of a heathen"—in the Land of Israel; the acquisition of land in the Land of Israel by a heathen does not exempt it from the Laws Concerning Diverse Kinds.

"but it is permitted to tell him"—the difference is that here the heathen himself does the sowing, whereas in the preceding case the Israelite does it for the heathen.

"a matter of tradition"—based on the Scriptural expression *thy field* (Lev. 19:19), which is interpreted as meaning a field in the Land of Israel. The other prohibitions of diverse kinds are not so limited by Scripture.

5. B. Ḳid 39a; Kil 1:7; P. Kil 1:7; B. Sanh 60a.

6. Tos. Kil 1:12; B. Ḳid 39a.

"to allow a heathen to intergraft for himself . . . trees"—heathens, too, are subject to the prohibition of grafting (XIV, v, x, 6; YJS, *3,* 236).

7. Kil 8:1; Tos Kil 2:15; P. Kil 1:4.

8. B. Neḏ 55a; Ber 37a; Kil 2:2, 5:5.

9. "mustard"—below, iii, 18 and II, iii, 11, it is stated that mustard is sown in single furrows.

Chapter II

1. Kil 2:1; P. Kil 2:1.

"one twenty-fourth"—this proportion was deemed sufficient to neutralize the admixture. In the case of food the criterion is whether the forbidden ingredient imparts a flavor to the permitted ingredient, and if the former constitutes more than one-sixtieth by volume, it is considered potent enough to do so. Here the determining factor is visibility, and if the proportion is one of twenty-four or less, the admixture is considered inconspicuous. There is also an additional distinction: in foods, the admixture is either intrinsically forbidden (e.g., ṭĕrefah or nĕbelah), or intrinsically permitted but forbidden as admixture (e.g., milk mixed with meat). Here all commixture is intrinsically permitted, and it is only the combination with other seed that is forbidden.

2. Kil 2:1. One sĕ'ah comprises six ḳab; hence the three added species together amount to one sĕ'ah.

3. Kil 2:2.

"garden seeds"—which by reason of their fineness can cover a larger area of sowing than grain or pulse.

"bet sĕ'ah"—a standard measure of cultivated area, 2,500 square cubits, required for sowing a sĕ'ah of wheat. In relation to our problem this works out thus: Since garden seed is so much finer and its produce takes up more space, even as little as ½ ḳab of it may suffice for a bet sĕ'ah. One twenty-fourth of this quantity, or 1/48 ḳab of garden seed mixed with a sĕ'ah of wheat, may accordingly be sufficient to render the mixture diverse kinds. See the next Section.

4. P. Kil 2:1.

"two sĕ'ah to a bet sĕ'ah"—unlike other garden seeds that require less per bet sĕ'ah.

5. P. Kil 2:2. 1/24 of 3 sĕ'ah (18 ḳab)=¾ ḳab.

6. P. Kil 2:1.

"or sowed"—intentionally, but without first intentionally mixing them together, the mixing having occurred accidentally.

7. B. MḲ 6a.

"the other (*'aḥer*) species"—so the *editio princeps* (Rome, 1480); later editions read "the one (*'eḥad*) species."

8. Kil 2:5.

9. *Ibid.;* B. BḲ 81a.

"is injurious to the original crop"—clear evidence that it was not the planter's intention to intermix them.

10. B. BḲ 81a; Kil 2:5. Bare ground is required for effective threshing.

"The same applies"—in the *editio princeps* Sec. 10 begins here, which is more logical; in later editions the Section begins with the preceding paragraph.

11. Kil 1:8–9; P. Kil 1:9.

"the stump"—cut even with the ground (cf. Loew, *Flora der Juden, 1,* 277). The sycamore was said to be very tenacious of life, and its timber very durable.

"a bunch . . . exposed to view"—which is not the way a vegetable would be grafted onto a tree, hence no apprehension of unlawful grafting is justified.

"even under a vine"—whose roots are soft and can be penetrated by the roots of other plants, and thus become involved in the law of diverse kinds.

12. Kil 2:5.

13. Kil 2:3; Tos. Kil 1; P. Kil 2:2.

"he must wait. . . . Then he must turn it over with a plow"—RABD interprets the Mishnah to mean "he must either wait . . . or turn it over."

"rotted"—literally "become wormy."

"customarily plowed before the rain"—a light plowing which does not go deep into the soil.

14. Kil 2:3; Tos Kil 1.

15. P. Sheḳ 1:1.

16–17. P. Sheḳ 1:2; B. MḲ 6a.

16. "the owners"—who failed to clear their fields.

17. "they did not wait"—until the next first of Adar; see above, Sec. 15.

Chapter III

2. Kil 1:4–6.

3. Kil 1:1–3.

4. P. Kil 1:5; Kil 1:4.

5. P. Kil 1:5.

"outward appearance"—see above, note to ii, 1.

6. *Ibid.;* Kil 1:3.

9. Kil 2:10; Tos Kil 2:6; P. Kil 1:9.

"beṯ roḇa'"—the area required to sow one-quarter of a ḳaḇ of seed, or approximately 104⅔ square cubits; here given less accurately as 104 1/25.

"middle . . . side"—i.e., if the boundary is not straight, there need not be the space of a *beṯ roḇa'* all along the boundary.

10. Kil 2:10, 3:7; Tos Kil 2:6; P. Kil 1:9.

"be it even pumpkin"—see Kil 3:7, where according to one opinion more space must be left in the case of pumpkin, because its leaves spread out farther. See below, Sec. 13.

11. Kil 3:7.

12. Kil 3:3.

"two fields"—the standard dimensions of a field are not stated here. Below (II, iii, 4) a "small field" is defined as measuring 50 x 2 cubits or less.

"a furrow no more than six handbreadths long"—and not one the length of the entire field. This is a sufficient indication that the two species are not mixed.

13. Kil 3:7; Tos Kil 2:3.

"one row"—these rules stress the difference between two fields adjacent to each other, and a row of one species adjoining a field or a row of another species. In the case of a single row it is obvious that one did not intend to plant diverse kinds together.

"as will be explained"—in Tos Kil 2:3.

14. Kil 2:11.

"as we have explained"—above, Sec. 13.

15. Kil 2:8.

"or a trench . . . deep"—missing in the *editio princeps,* perhaps owing to a copyist's error.

"and four handbreadths wide"—it is not clear whether this refers to the trench alone or to both the fence and the trench. Some commentators therefore say that this refers to the trench only. Those who make it refer also to the fence interpret it to mean the length of the fence, which must be four handbreadths long.

16–17. Kil 2:7.

16. "the end of B's field"—the prohibition, according to Scripture, applies in general to sowing heterogeneous seed with one and the same throw of the hand; otherwise it extends only to situations where it would appear to strangers that diverse kinds had been sown deliberately. In this case it is clear to everyone that there was no such intention, and that it is just a case of "here one field ends, and the other begins."

17. "sown it in order to waste it"—in which case the prohibition of diverse kinds does not apply.

"two fields"—so the manuscripts; the printed text has "two rows," which is less logical. The two words, *śaḏoṯ* and *śuroṯ,* look very much alike.

"unless one leaves an intervening space"—it is only in the case of flax that people would think he was experimenting, and such cases are not subject to the law of diverse kinds. With other species he lays himself open to the suspicion that he is deliberately violating the law, which makes the interval necessary.

18. Kil 2:5, 7.

"only a single furrow"—see above, note to i, 9.

"cause no harm"—the principle is that where this does cause harm there can be no suspicion of intent to sow diverse kinds, since the mustard must have been planted "in order to waste it," i.e., as an experiment. Where it causes no harm there are grounds for such suspicion.

"as we have explained"—above, Sec. 16.

Chapter IV

1. Kil 3:4; B. Shab 85b; Tosafoṯ *ad loc., s. v. 'eṯeḇeh.*

2. Kil 3:6.

"He must then skip an area of vegetables"—the idea is that he may plant vegetables between rows of pumpkins, so long as there is an interval of twelve cubits (not including the width of the furrow, for which there is no required standard) between one row of pumpkins and the other.

3. Kil 3:7; 2:10.

4. Kil 3:2.

5. Kil 3:5; P. Kil 3:4. Later authorities were hard pressed to explain this rule, and remarked upon the difficult and complicated nature of the whole subject of diverse kinds. The reason for this particular rule seems to be that by inclining the plants in different directions one clearly manifests his intention to observe the law and not to violate it, and it is intention which is, in matters of diverse kinds, of paramount importance.

6. Kil 2:6; P. Kil 2:4.

7. Kil 2:9.

"approximately ten cubits less a quarter"—PhM gives the approximate figure of 9⅞ cubits.

8. The explanation implies that a long bed is tantamount to a row, while a patch is tantamount to a field; in the case of a field one must leave a greater space.

9. Kil 3:2, 1; B. Shab 85a; Tosafoṯ *ad loc., s. v. wĕ-ḳim.*

"only a little at a time"—in small regular beds; vegetables occupying whole fields are sown broadcast irregularly.

"as we have explained"—above, i, 9.

10–11. B. Shab 85a.

12. Kil 3:1; B. Shab 85a.

"lest it should fill it"—because its leaves are large and the several species would be so intermixed as to seem to be diverse kinds.

13–14. Kil 3:2.

13. "species of seeds"—garden seeds customarily occupying an entire field (above, i. 9).

"as we have explained"—above, Sec. 9.

14. "as we have explained"—above, Sec. 12.

"is reduced"—so that it is less clearly visible than it was before.

15. B. Shab 85b.

16. "as we have explained"—above, Sec. 9.

"when they appear separated"—even without a space; see above, Sec. 5, and iii, 15, 18.

Chapter V

1. Kil 5:5.

"two species"—even one species is forbidden (below, Sec. 7), but flogging is incurred for two species (below, Sec. 2).

2. B. Ber 22a.

"in the Land of Israel"—outside of the Land of Israel diverse kinds are forbidden by Rabbinic law only; see the next Section.

"if he merely covers them with earth"—see above, i, 2.

"Likewise, if he sows," etc.—the point is made that vegetables are subject to the same rule as grain.

3. B. Men 15b; Ḳid 39a.

"only"—in addition to grain, pulse, and vegetables.

"become available"—produce of newly planted fruit plants may not be harvested in the first three years; the fourth year's produce or its equivalent must be consumed in Jerusalem; from the fifth year onward the fruit may be consumed (Lev. 19:23–25). Hemp, arum, etc., are thus held subject to the same rule, on the same Scriptural authority.

4. B. Ḳid 39a. Cf. above, i, 3.

"outside . . . of seeds?"—some editions omit this clause, as it is merely explanatory. Since the question refers to the concluding statement in the preceding Section, the explanation was deemed superfluous.

"the law is more severe"—hence the same pattern is followed outside the Land of Israel, although to a lesser degree.

5. B. AZ 63b f.; Tos Kil 2:16.

"diverse kinds"—in a vineyard.

"to reduce what is improper"—uprooting eliminates what is unlawful, while hoeing improves it.

6. Kil 5:8.

"needless to say"—since the prohibition of diverse kinds of trees applies only to grafting. See above, i, 6.

7. Tem 7:5; B. Ḳid 56b; Pes 26a; Tos Kil 4:10.

"is forfeit"—literally "is rendered holy." In this case the implication is that one may not derive any benefit from it.

8. Kil 5:5–6, 7:4–5; B. Yeḇ 83a; P. Kil 7:3.

"during the Sabbatical year"—when all produce is considered ownerless, and thus is not his to render it forfeit.

9. Tos Kil 3:12.

"is forbidden"—since the moment he notices the diverse kinds he automatically renders them unlawful to himself.

"as we have explained"—in the preceding Section.

10. Kil 7:6; P. Kil 7:4.

"usurper"—in Hebrew *'annas,* one who has seized property by force and violence and is holding it illegally.

"If the . . . owner's name cannot be traced"—PhM explains it to mean that the original owner has "sunk," i.e., hidden himself in terror of the usurper. Other commentators explain it literally, i.e., the field is no longer associated with its original owner but only with the present occupant.

11. Kil 7:7.

"and has trained them"—so that they took root.

"one must cut them away"—in Hebrew *yigdor,* which can also mean "build a fence around them"; PhM reads *yigror,* "drag them away." The Hebrew letters *daleṯ* and *reš* look very much alike and are often interchanged by error. The fence would prevent the vine from overhanging the grain.

12. Kil 7:6.

"must harvest the seed"—for being the rightful owner, he is obligated to allow no diverse kinds to remain. See above, Sec. 8. Another explanation is that he must do so for the sake of appearance, so that people would not be under the impression that the rightful owner allows diverse kinds to remain in his vineyard.

"even . . . during the intermediate days of a festival"—when as a rule only such work may be done as is necessary to prevent deterioration or loss; here this consideration does not apply, but the work is permitted in order to remove diverse kinds.

"one-third over their regular wages"—the Mishnah states only "one-third," and the Palestinian Talmud raises the question whether the reference is to one-third of the wages or one-third of the value of the harvest. PhM gives a concrete illustration: if the prevailing wage is three coins, he should offer four.

"he may proceed at leisure to harvest"—this corresponds to the statement in the Mishnah, "he may harvest it in his usual way, even if he has to continue after the festival," by which time the produce may have increased by an amount that would render it forfeit.

"both"—seed and grapes.

13. Kil 7:6; P. Kil 7:5.

14. Tos Kil 4:12; B. Men 15b.

"for benefit"—but not for consumption.

15. Kil 7:7.

16. Kil 7:2, 8.

"one may not sow"—for the sake of appearance.

"unperforated pot"—since the earth in the pot is not exposed to the soil of the vineyard.

"flogging for disobedience"—for disobeying an enactment of the Sages. See above, note to i, 2.

17. Kil 5:7; Sif Deut. 22:9; Tem 7:5; P. Kil 7:6.

"a grain field"—literally "a white field," sown with grain or the like, and not planted with shade-giving trees.

"all of it"—including the straw.

"they must be burned"—by that time they have dried enough to be easily burned.

18. Kil 5:8; B. Shab 144b.

19. Kil 5:8; Tos Kil 3:13.

"The cowpea . . . does not render forfeit"—contrary to the present text of Tos, *ibid*. Maimonides evidently had a variant text of it before him.

20. Tos Kil 3:15; P. Kil 5:7; B. Ber 36a.

"a tree"—note that the Hebrew term (*'ilan*) covers both trees and bushes.

21. Kil 5:6.

"this is allowed"—because his evident readiness to remove the vegetables shows that they are there without his knowledge.

"render the vineyard forfeit"—since he had knowingly allowed the offending vegetables to remain among the vines for a substantial period of time.

22. P. Kil 5:3.

23. P. Kil 7:2; Kil 7:8; Tos Kil 4:13.

24. B. Neḏ 57b.

"and thereafter the vineyard is uprooted"—had it not been uprooted, it would have been condemned as diverse kinds.

"two hundredfold"—the amount that usually neutralizes diverse kinds in a vineyard.

Chapter VI

1. Kil 5:5; P. Kil 5:5.

"retains them"—they become forfeit only after they have ripened by one two-hundredth. If he himself has planted them, however, they become forfeit immediately.

2. Kil 5:5.

"forty cubits"—thirty-two cubits, plus four cubits at each end of the diameter.

3. Kil 5:5, 3:7, 6:1.

"only one grapevine"—which is not accounted a vineyard.

4. B. Soṭ 43a; Men 15b.

"If the whole vineyard"—even if planted entirely with young shoots, it is accounted a vineyard.

4. B. Soṭ 43a; Men 15b.

"If the whole vineyard"—even if planted entirely with young shoots, it is accounted a full-fledged vineyard.

5. P. Kil 6:2, 7:2.

6. Kil 2:4.

"he may then sow the seeds"—this does not contradict Sec. 4, which rules that a whole vineyard consisting of young vines less than a handbreadth high renders other seeds forfeit, since by cropping the grown vines down to less than a handbreadth one manifests his intention not to plant diverse kinds.

7. Kil 7:1; B. BB 19b.

"pumpkin . . . earthenware pipe"—these are porous, and the vine roots may extend through them.

8. Kil 7:1; P. Kil 7:1.

"as will be explained"—below, vii, 1.

9–10. Kil 7:1.

9. "between them"—between one bent vine and the other, or between the row of bent vines and the next row of standing vines.

"four to eight cubits"—the customary distance between one row of vines and the other in a vineyard; cf. below, vii, 2.

10. "leave six handbreadths' space"—as is required for a single vine. See below, vii, 1.

11. Kil 6:9; P. Kil 6:4.

12. Kil 6:3, 5, 4; P. Kil 6:3.

13. "either to retain this *or* to turn back the shoots"—so the printed text; the correct reading, in conformance with the preceeding Section, seems to be "to retain. . . *and* to turn back."

14. Kil 6:8. The poles are not considered part of the trellis, unless one deliberately uses them for the same purpose of training the vines over them.

15–16. Kil 6:9.

Chapter VII

1. Kil 6:1, 4:5, 3:7; B. BB 18a.

2. P. Kil 4:3; B. Beṣ 14b; Yoma 87a; Kil 4:8, 5:2; P. Kil 4:6.

3. Kil 4:9; P. Kil 4:6.

"less than four cubits' space"—that is, the interval between the two outer rows is less than eight cubits wide.

"eight or more cubits"—in which case, as stated above. Sec. 2, there is no vineyard, but only separate rows of vines.

4. Kil 4:9; P. Kil 7:2.

"a vineyard that has gone to waste in the middle"—see below, Sec. 11.

5. Kil 4:7.

"less than ten handbreadths high"—the minimum height for an effective barrier is ten handbreadths (above, iii, 15).

6. Kil 6:2.

7. Kil 4:6.

8–9. Kil 5:1.

10. Tos Kil 3:3; P. Kil 5:1.

11. Kil 4:1, 7:3.

12. Kil 4:2, 3; B. 'Er 93a.

13. Kil 4:1, 7:3, 2; P. Kil 4:1.

14. Kil 4:3; Tos Kil 4:1.

15. Kil 2:8, 4:4.

16. Kil 4:4; B. 'Er 15b.

"an entrance"—which is *de jure* the same as a wall or a fence, and one is therefore permitted to sow immediately in front of it, a vine on one side of the boundary and seeds on the other, just as if the fence were actually standing between them.

"many breaches"—less than ten cubits high and more than three handbreadths wide.

17. Tos Kil 3:3; B. BḲ 100a; BB 2a; Tosafot to BḲ, *s. v. 'omer,* and to BB, *s. v. 'omer lo gĕḏor.*

"the partition"—with vines on one side of it and vegetables on the other, each belonging to a different owner. By failing to repair the partition, the owner of the vineyard exposes the vegetable owner's crop to forfeiture.

18. B. 'Er 93a.

19. B. 'Er 92. The small courtyard adjoins the large one and is fully open to it; hence it is the same as if it were part of the large one. Therefore, if the vines are in the large one, one may not sow seeds in the small one, because that would constitute diverse kinds. If, however, the vines are in the small courtyard, one may sow in the large one, inasmuch as its side which adjoins the opening of the small courtyard extends beyond either side of the small courtyard (i.e., has posts or strips there), thus making it a separate entity which is not affected by the seeds sown in the small courtyard, and is therefore not accounted as constituting diverse kinds.

20. Kil 5:3.

"as we have explained"—above, v, 8.

"bare patch"—see above, Sec. 11.

21. Kil 4:2, 3.

22. Kil 5:3; P. Kil 5:3.

"square . . . round"—the explanation of the difference is that if the mound is square, it is more easily recognizable, that is, more clearly distinct from the vineyard surrounding it.

23. Kil 5:4; P. Kil 5:3.

"more than three handbreadths by three"—so the *editio princeps.* Later editions add "up to four handbreadths," i.e., one may enlarge it to four handbreadths square and then sow in it.

"closed"—i.e., the ground is regarded as solid and not separated by the structure's walls. See commentary to P. *ad loc.*

24. Kil 5:4; Tos Kil 4:2.

"one is forbidden to sow within it"—see the next Section.

25. Tos Kil 4:2; Kil 7:3. The difference between a single vine and a vineyard in this respect is explained above, Sec. 1.

Chapter VIII

1. "There are some other rules"—see above, vi, 12.

2. Kil 6:1.

"and he must leave a space of four cubits from the arbor"—unlike the case of an adjoining single row of vines, where a space of six handbreadths is sufficient. See above, vii, 1.

3. Kil 6:1; P. Kil 6:1.

"From the base of the fence"—from what follows it is clear that this means sowing on the side of the fence. If, however, one sows on the side of the roots of the vines, he must measure from the roots. The emphasis on the fence is due to the fact that usually a fence constitutes a barrier, and seeds may be sown right alongside it (see above, vii, 15). When the wall itself serves as a trellis, it is not regarded as a barrier, because the whole space from the vines to the wall, including the wall, constitutes an arbor.

4. P. Kil 6:4; Kil 6:6.

5. Kil 6:6.
"as we have explained"—above, vi, 12.

6. Kil 6:6, 7:3; P. Kil 6:7.

7. 'Ed 2:4.
"enough room"—two cubits (P. Kil 6:1).
"inasmuch as it is surrounded by a fence"—if there is only a straight (not circular) or rectangualr fence, the law is different; see above, Sec. 2. Here the garden is surrounded on all sides by a wall, with the vines trained over it on the outside, and the question is whether one may sow vegetables inside the surrounding wall.

8. Kil 6:2.

9. Kil 6:7.
"next to each other"—the two walls form an angle with each other, and the arbor begins at the point of the angle but extends only over part of the walls; one may then sow in the space between the walls beyond the arbor's end, provided he has left the required space from the roots of the vines.

10. Kil 7:1. PhM limits this rule to cases where the actual root of the vine is not readily visible, in accordance with P. *ibid.*

11. P. Kil 5:6.
"We have already explained"—above, vi,11.
"uprooted . . . permitted"—the explanation given is that the roots of the vine are accounted the same as the vine, while the roots of the grain do not have the status of grain (RDBZ *ad loc.*).

12. B. 'Er 2b.
"spread wide"—literally "laughing."

13. B. Ḳid 39a; Shab 139a.

14. 'Or 3:9; B. Ḳid 38b f.; P. 'Or 3:7.
"forbidden for consumption"—but permitted for benefit.
"in the Laws Concerning Forbidden Foods"—V, II, x, 11–12 (YJS, *16,* 204).

Chapter IX

1. P. Kil 1:1; B. BḲ 54b; Sif Lev. 19:15; B. BM 91a.

2. B. BM 91a.
"mate it"—with an animal of a different species.

3. B. Ḥul 115a.
"in the Laws Concerning Forbidden Foods"—V, II, i, 13 (YJS, *16,* 155).

4. B. Beḳ 7a; Kil 1:6, 8:6.

5. Kil 8:5; B. BḲ 55a; Bik 2:11.
"ḳoy"—a kind of bearded antelope or wild sheep, concerning which the Sages were in doubt as to whether it was a domestic animal or a wild beast.

6. Kil 8:4; B. Ḥul 79a.

"he is [not] liable to a flogging"—the text has "he is liable to a flogging," which seems to contradict the principle just stated.

7. Kil 8:1–2; Tos Kil 5:5; Ḳid 1:9; Sif Deut. 22:10; B. BM 90b. Here the law is more stringent, prohibiting the joining not only of diverse kinds but also of clean with unclean (ox and ass).

"in every place"—even outside the Land of Israel.

"merely pairs them"—as a team.

8. Kil 8:6; B. BḲ 54b f.

"in the Laws Concerning Forbidden Foods"—V, 11, vi, 1 (YJS, *16,* 181).

"carp"—*šibbuṭ,* Arabic *shabbūṭ,* a large species of carp (which may reach fifty pounds in weight, according to the *Encyclopedia Britannica*). Jastrow (*A Dictionary of the Targumim, the Talmud Babli, and the Midrashic Literature,* New York/Berlin, 1926, p. 1556) has "mullet," which is too small a fish to fit here.

9. Kil 8:3, 6; Sif Deut. 22:10; B. BḲ 54b.

10. Kil 8:6; Sif Deut. 22:10; B. BḲ 54b.

11. B. Mak 22a; Tem 33a.

"disqualified consecrated property"—animals dedicated to the Sanctuary which have become unfit for sacrifice. These were redeemed, but retained some degree of consecration.

"on account of diverse kinds"—since such an ox is regarded as a combination of clean and unclean animals, and thus, all alone, comes under the rules of diverse kinds.

Chapter X

1. P. Kil 1:1; Kil 9:2; Shab 2:1; B. Shab 20b.

"*ḳalaḳ*"—according to Maimonides, this is a plant, but according to Loew (*3,* 458) it is not a plant. Uriah Feldman, *Ṣimḥe hat-TĕNaḴ* (Tel Aviv, 1956), p. 3, identifies this with the *Ferula communis,* but it is evident that this is not the same plant as the one Maimonides refers to. Most commentators say that it is coarse silk. According to Jastrow, this is cissaros blossom.

"the Salt Sea"—the usual term for the Dead Sea, but Maimonides appears to use it as a synonym of the Great, that is Mediterranean, Sea.

"silk"—which resembles linen, may not be mixed with kalaḳ, which resembles wool.

2. B. Beḳ 17a; Sif Lev. 19:19; Kil 9:9, 8; B. Nid 61a.

"The wool of a ewe sired by a goat"—whose texture is different from sheep's wool.

"Once wool is joined"—in some editions Sec. 2 begins here, which is perhaps more logical.

"spins them"—without first carding them.

3. Kil 9:8–10; Tos Kil 5:22; Sif Deut. 22:11.

"or even with twisted twine"—even though the prohibition mentions only wool and linen.

4. B. Yeḇ 4a; Men 39a.

"next"—Deut. 22:11 (clothing of diverse kinds) and Deut. 22:12 (fringes).

5. B. Nid 61a.

6. Kil 9:1; Tos Kil 5:12.

"are not threads of wool"—i.e., not whole threads composed entirely of wool, but rather single woolen fibers mixed with camel's hair to form each thread.

8. Kil 9:1.

9. Tos Kil 5:12.

10. *Ibid.* 5:13.

11. *Ibid.* 5:15.

"twist"—*yiḵroḵ;* some manuscripts and the *editio princeps* read "hackle (*yiṭrof*)." The two cords must be separate, so that each garment may be taken off individually.

12. Kil 8:1; B. Yeḇ 4b; P. Kil 9:6; Tos Kil 5:26.

13. B. Yoma 69a; Kil 9:2.

14. B. Beṣ 15a.

15. *Ibid.;* Kil 9:7.

"Footwear"—called *'ardal* in B. Beṣ 15a. According to Rashi, who quotes a Gaonic authority, this is an overshoe worn beneath the regular shoe, consisting of a goatskin sole with a heel lined with woolen cloth.

"with only a heel"—that is, consisting only of sole and heel (see preceding note). The text has "with no heel," which seems to be a scribal error.

16. Kil 9:6, 5; B. Yeḇ 4b; Shab 29b.

"in their usual fashion"—holding the garment in their laps as they sew.

17. P. Kil 9:2.

18. *Ibid.;* P. Kil 9:1.

"to escape customs dues"—in which case he is liable not for diverse kinds but for theft. See XI, III, v, 11 (YJS, *9,* 108). Clothing worn by a person was not subject to duty.

19. B. Beṣ 15a; Tos Kil 5:25; P. Kil 9:7.

20. B. Shab 57b.

21. B. Shab 54a–b.

22. Kil 9:3.

"cloths used to wrap Scrolls of Torah"—*u-miṭpĕḥoṭ sĕfarim;* so *editio princeps.* Later editions read the second word as *sapparim,* "cloths used by barbers" (likewise mentioned in Kil 9:3), and consequently inserted before this phrase an additional phrase, "cloth used to wrap a Scroll of Torah," which is tautological. Barber's cloths cover the body like a garment and are obviously subject to the law of diverse kinds, so that they really need not be mentioned separately.

23–25. Kil 9–10.

23. "of no value"—and is so minute that it might otherwise be nullified by the bulk of the garment (above, Sec. 7).

25. "even if only in order to carry out manure"—this does not contradict Sec. 16, because his intent is to "wear" it, at least to the extent of benefiting by the protection it gives him from the manure.

26. Tos Kil 5:26.

27. B. Pes 40b; Nid 61b.

"in the Laws Concerning Forbidden Intercourse"—V, 1, xviii, 17 (YJS, *16,* 118).

28. Kil 9:7.

29. B. Ber 19b.

"accost him and rip it off him"—other authorities forbid such action, since the companion is most probably unaware of his transgression. See R. Asher on B. Nid, Chap. 9, "Laws Concerning Garments of Diverse Kinds," Sec. 6.

"in the case of the return of a lost article"—where a Sage is not obligated to return a lost article of slight value if it is beneath his dignity to do so (XI, III, xi, 13; YJS, *9*, 130).

"involves money"—here one can waive his right to it and thus exempt the Sage from the obligation to return it.

"in the case of defilement by the dead"—where a priest, who is forbidden to defile himself by touching a corpse, may nevertheless do so if he finds an unclaimed body (XIV, IV, III, 8; YJS, *3*, 171–72). The whole verse reads, *He shall not make himself unclean for his father, or for his mother, for his brother, or for his sister, when they die.*

"Scripture says"—referring to the oral sentence of a priest or a judge.

30. Mak 3:8; B. Mak 21b.

31. Tos Mak 4:10.

"submitted to it willfully"—since B knew the garment was of diverse kinds, he should have removed it immediately. His willfulness is expressed also by his cooperation with A.

32. B. 'Ar 3b.

"as in the case of the fringes"—where the law of diverse kinds is set aside by the law of fringes. See above, Sec. 4, and VIII, II, viii, 1–2 (YJS, *12*, 69–70).

TREATISE II: LAWS CONCERNING GIFTS TO THE POOR

Chapter I

1. Pe 4:1; P. Pe 1:1; B. BḲ 94a; Sif Lev. *Ḳĕḏošim* 1:7; P. Ḥul 137b; P. Pe 1:4.

"the end of the field"—according to Rashi (on B. Shab 23b), this means the end of harvesting. But see below, ii, 12, where it is apparent that Maimonides refers to space and not to time.

"corner crop"—literally "corner" or "edge" (*pe'ah*).

2. Pe 4:1; P. Pe 4:1; B. BḲ 94a; Mak 16b.

"give it to the poor"—in this case it is not sufficient to *leave* it for the poor; he must *give* it to them.

3. B. Mak 16b.

"negative . . . positive commandment"—the obligation to leave corner crop for the poor consists of two commandments: a negative one, *Thou shalt not wholly reap the corner of thy field,* and a positive one, *Thou shalt leave them for the poor and for the stranger.* The negative commandment is thus transformed into the positive one, i.e., if one has failed to leave corner crop, he still must give it to the poor. In this case, the harvest having been destroyed, he has violated the negative commandment and made it impossible for himself to perform the positive commandment.

4. *Ibid.;* Pe 4:10.

"and binds"—that is, "or binds"; even if the ear drops out of his hand, it must be left to the poor.

5. Pe 7:3, 4.

"one sheaf"—below, v, 14, no more than two sheaves may be regarded as accidentally overlooked. More than two are presumed to have been meant to be picked up later.

6. Pe 6:7; Sif Deut. 24:19; P. Pe 4:6; B. Ḥul 131b.

"to all trees"—but not in all respects. Whereas in produce the rule applies only to species that are harvested all at once (see below, ii, 1), in trees this distinction does not apply (see below, Sec. 12).

7. B. Ḥul 131b; Tos Pe 2:13.

8. B. Ḥul 131b f.

"must be exacted from him"—since Scripture uses the term *thou shalt leave,* not "thou shalt give"; once the produce is left, the owner no longer has any jurisdiction over it, no matter how needy he may be.

9. Sif Lev. 23:22; Giṭ 5:8.

"a righteous proselyte"—the Scriptural term *ger,* translated *stranger,* originally meant a resident non-Israelite alien. Later the term came to mean specifically a proselyte, and this is the one mentioned in connection with gifts for the poor. A righteous proselyte is sincerely and wholeheartedly converted to Judaism; a resident alien, on the other hand, was only a partial proselyte, in that he was obligated to observe only some commandments and not all.

"the poor man's tithe"—so some manuscripts, in accordance with the Scriptural verse here quoted; the printed text has "the second tithe."

"son of the covenant"—a full-fledged Israelite.

10. B. Ḥul 134b.

"the corpus of these gifts is not holy"—in the case of heave offering, its corpus becomes inherently holy, whether the priest takes it or not. In the case of gifts for the poor, there is no such inherent holiness in the given produce; hence, if there are no poor people to take it, anyone else may do so.

11. Pe 8:1; B. BM 21b; P. Pe 8:1.

12. Pe 7:2. The idea is that a tree is not subject to the rule that it must be harvested all at once. See below, ii, 1, and above, note to Sec. 6.

13. P. Pe 8:1.

"with which the poor choose not to concern themselves"—choose not to gather them for some reason or other.

14. B. Ḳid 36b; Ḥul 137b.

15. Pe 1:1–2; B. Ḥul 137b; P. Pe 1:1; B. Neḏ 6b.

Chapter II

1. Pe 1:4; Sif Lev. *Ḳĕḏošim* 1:7.

"as it is said"—the Sifra explains that the term *harvest* refers to produce which has the enumerated five characteristics, and therefore the law of corner crop applies only to produce possessing these characteristics. See the next Section.

2. *Ibid.;* Pe 1:5; P. Pe 1:4; B. Shab 68a; Ber 40b; Pe 3:4; B. Nid 50a; Šĕ'iltoṯ, *Ḳĕḏošim* 98.

"they do not grow out of the soil like other produce of the soil"—while they do grow out of the soil, they are said not to draw nourishment out of it, as do other plants (B. Ber 40b).

3. Pe 1:6; B. Ḥul 135b.

4. Pe 2:7; P. Pe 2:7; Sif Lev. *Ḳĕḏošim* 1:6.

5. Pe 2:7; Tos Pe 1:9.

"the buyer must render corner crop for the whole field"—some editions omit "for the whole field." PhM explains that since corner crop is what remains after the field as a whole has been harvested, that remainder is ipso facto part of the second half of the harvest, and not subject to sale. Hence the buyer is aware at the outset that what he buys is subject to deduction of corner crop for both halves of the harvest. The same reasoning applies to the following cases.

"the treasurer"—in charge of the Sanctuary's property and funds.

6. "on either side of the vineyard"—the distinction between the two kinds of harvesting is the following: The rule is that if the owner does only some casual cutting, and not systematic harvesting, all over the vineyard, he is not subject to corner crop. If, however, he confines the cutting to one side of the vineyard, this constitutes the first step of regular harvesting, and he is subject to corner crop for the entire field.

"he is exempt from gleanings," etc.—because even though plucking is one of the modes of harvesting, his manner was casual and not indicative of intention to perform regular harvesting.

7. B. Men 71b; Pe 2:1.

"before the crop has ripened," etc.—this is not harvesting, since the yield is not yet usable.

8. Pe 4:7; Sif Lev. *Ḳĕḏošim* 1:6.

"and then redeems it while the corn is still standing"—so *editio princeps;* later editions omit this sentence.

9. Pe 4:6; P. Pe 4:4.

"even though forgotten sheaf," etc.—the law of forgotten sheaf takes effect at the time of the binding of the sheaves, but obligation for it arises once the harvesting has begun.

10. Tos Pe 3:1; P. Pe 2:7.

11. B. BḲ 94a; Pe 1:6; Tos Pe 1:1; B. Neḏ 6b. Tithe becomes due after all the work of harvesting has been completed, whereas gifts to the poor become due once harvesting has begun.

"the greater part of the harvest"—not all of the harvest, which cannot constitute gifts to the poor, since the owner reserves no part of it for himself, and therefore cannot act as giver.

"to the amount due from that field"—before deducting the amount of the tithe.

12. Tos Pe 1:5; P. Pe 4:2; B. Shab 23a; Pe 1:3.

"so that the poor would know the place"—and would be spared the humiliation of standing by to see where is would suit the owner to leave corner crop.

"suspicion"—of withholding the gifts due from him.

13. Tos Pe 2:6; B. Neḏ 6b.

"both are valid corner crop"—since no set amount is prescribed for corner crop, the presumption is that the owner meant to give more than one.

14. Tos Pe 2:7–8; Sif Lev. *Ḳĕḏošim* 1:9.

15. Pe 4:14; Sif Lev. *Ḳĕḏošim* 3:5.

"to divide them among themselves"—i.e., that the owner should give each one his share. See the next Section.

16. Pe 4:2; P. Pe 4:1.

17. Pe 4:5; P. Pe 4:3.

"the time of the afternoon prayer"—before sunset.

18. Pe 4:3; Tos Pe 2:1.

"throws it . . . or . . . falls . . . or spreads his cloak"—in the mistaken belief that he thus acquires title to all of the corner crop.

"the same applies to . . . sheaf"—some texts omit this sentence as self-evident.

19. Pe 4:9.

20. Tos Pe 2:2.

Chapter III

1. Pe 2:1; Sif Lev. *Ḳĕḏošim* 1:11; P. Pe 2:1; Tos Pe 1:8.

2. Pe 2:12; P. Pe 2:1, 2; Tos Pe 1:8.

"he cannot harvest both sides together"—i.e., while standing on one side of the water he cannot harvest the other side (PhM *ad loc.*).

"the channel is . . . fixed"—i.e., there is water in it both summer and winter. Note that a natural watercourse may be stagnant, while an artificial watercourse must be flowing constantly.

3. Pe 2:1; B. BB 99b.

"a public path that is less than sixteen cubits wide"—not that a lesser width converts a public road into a private path; it is still a public path, even though it may be an invalid divider.

4. Pe 2:1; P. Pe 2:1; Tos Pe 1:8.

"fallow land"—*editio princeps* reads "fallow field."

"before the crop is one-third grown"—if it is one-third grown, it is subject to corner crop, which would certainly make the partition a valid divider.

"three initial furrows"—the first three furrows made when one commences plowing. These are usually wider than the rest.

5. Tos Pe 1:8; P. Pe 2:1.

6. Pe 2:2.

7. Tos Pe 1:9; Pe 3:1; P. Pe 2:2.

"intermixed"—the terraces are situated on an incline, so that the end of a row of plants on a lower terrace is in line with the beginning of a row on an upper terrace.

"lift up the plow . . . and put it down"—the same criterion should apply to slanting terraces when the rows are not intermixed, since there too one must lift up the plow from the lower row and put it down on the upper row. Possibly the reason is that the ground is, theoretically at least, fertile and can be plowed, even if with difficulty, whereas a rock is forever barren.

8. Pe 3:1.

9. Pe 3:1.

10. Pe 3:4.

"even though the vegetables divide them"—there are two reasons why one would have had to render corner crop separately for each onion bed: a) they form separate beds; b) other vegetables separate them, and according to Sec. 4, above, this makes these beds separate fields. The reason why this is nevertheless considered one field is that, as in Sec. 8, above, it is obvious that it is in fact one field, and that the onions were sown in separate beds simply because of the intervening vegetables. In other words, one would not ordinarily plant the onions in such an unusual and laborious fashion unless one was compelled to do so by the presence of other plants.

11. Pe 3:2; P. Pe 3:1.

"whole fields"—and not separate beds.

12. Pe 3:1.

"When does this apply?"—referring to the rule that one corner crop may be rendered for both the dry and the green plants.

"the green ones in the middle"—the deciding factor is, does it look like one field or like several fields? If the green plants are in the middle, it looks like one field; if the green plants are on either side and the dry ones are in the middle, it does not look like one field.

13. Pe 3:3.

"he must leave corner crop separately"—here the distinction is not between green and dry but rather between produce for sale in the market place and produce for storage, or between one destination and the other.

14. Pe 2:5. The owner is presumed to have failed to render corner crop as he was harvesting, and must therefore render it out of produce already harvested.

15. Pe 2:5–6.

16. Pe 3:5; P. Pe 3:4

"Two brothers who have divided . . . become partners again"—in both cases prior to the harvest.

"need set aside nothing"—since corner crop is due from standing corn at the conclusion of the harvest. The first partner, who takes already harvested produce while the harvest is still incomplete, is thus analogous to the owner who harvests only half of his field and then sells the rest of the crop, in which case the buyer alone is obligated to render corner crop (above, ii, 5). The second partner, however, is responsible only for his own half (and not for the entire field, as is the buyer, *ibid.*), since he had never had any title to the first partner's half.

"for his partner's share"—not really. What is meant is that since they hold the field in partnership, whatever is given is given for both of them, and it thus can be construed as if each one were rendering corner crop for the other's share in the standing corn.

17. P. Pe 3:4. To simplify this case, it may be divided as follows: a) one-half of the field crop reaches one-third of its growth and thus becomes subject to corner crop; b) the owner then harvests one-half of this half, i.e., one-quarter of the entire field, which harvest thus becomes subject to corner crop; c) meanwhile the other half of the field too reaches one-third of its growth, and becomes likewise subject to corner crop. There are now three installments of the crop: 1) the first quarter, termed "initial"; the second quarter, termed "middle"; and the second half, termed "final." The middle crop has an affinity with both the initial and the final crops—with the initial, in that it has ripened simultaneously with it; with the final, in that it was harvested continuously with it. The initial crop, however, has no affinity with the final crop. Hence some MSS and printed editions add at the end, "but not out of the initial installment for the final installment."

18. Pe 3:5; B. Ḥul 138a; Pe 2:3; Sif Lev. *Ḳĕḏošim* 2:3.

19. Pe 2:3; Sif Lev. *Ḳĕḏošim* 2:3.

"high fence"—Sif Lev. defines it as ten handbreadths high, and so does PhM to Pe 2:3. Moreover, in regard to mixed seeds, the Mishnah (Kil 4:3) explicitly defines a fence as ten handbreadths high. Maimonides appears to have reasoned that here the statutory height of the fence is insufficient, and that branches and foliage also must be completely separated, which is why Pe 2:3 does not define the height of the fence again with respect to corner crop.

20. Pe 2:5.

"only one corner crop"—since they own jointly the whole of the tree.

21. Pe 2:4; P. Pe 2:3. Carobs are tall trees with copious branches and abundant foliage.

"either side"—the text has "on the first extreme side for the intervening trees and out of the intervening trees for the trees on the first side." The translation follows the text of the Palestinian Talmud, on which this is based, which reads *min ha-rašim* instead of *min ha-rišonim*.

22. Pe 2:4; P. Pe 2:4. The reason for this rule is not given, and it is presumably one of the "rules given to Moses on Mount Sinai."

23. Pe 3:3; P. Pe 3:3.

"thinned out"—Maimonides defines thus the term *meḏel* used in the Mishnah.

"We have already explained"—above, ii, 6.

"even if the harvested grapes were meant for the market"—see above, ii, 6. If he thins out the vines and sells these grapes in the market place, he is exempt from corner crop, as is stated immediately hereafter. Here, however, having done the thinning on one side only, his act is accounted not as thinning but as harvesting.

"to bring . . . to his house"—in preparation for pressing.

Chapter IV

1. Pe 4:10, 6:5.

"the back of the sickle . . . of the hand"—which is evidently pure accident caused by the shaking of the sickle or of the hand, and is therefore not subject to gleanings.

2. P. Pe 4:7; B. Ḥul 137a; Pe 4:10; Sif Lev. 19:9.

"is pricked by a thorn"—see above, note to Sec. 1.

3. Pe 5:2; Tos Pe 2:15.

4. P. Pe 5:2; Tos Pe 2:21.

"saves . . . saved"—for the owner.

"For it is," etc.—this refers not to the preceding sentence, but to Sec. 3, above, and supplies the reason for the rule stated there.

"remain among the stubble"—uncut ears are generally not gleanings (below, Sec. 9).

5. Pe 5:1; P. Pe 5:1; B. BM 105b.

"Four ḳaḇ . . . for each kor plot"—equal to 1/45 of the total yield, or a little over two percent, for each plot of ground required to plant a kor measure of seed.

6. Pe 5:1, 4:11; P. Pe 4:8.

"set down"—a term suggesting contact with the ground, hence the preceding rule.

7. P. Pe 4:1.

8. Tos Pe 2:20; P. Pe 4:3.

"to sprinkle"—and as a result the poor people cannot enter the field. Sprinkling settled the soil and prevented it from being blown away by the wind.

9. Pe 4:11.

10. Pe 5:2; P. Pe 4:2.

"two ears"—KM calls attention to the fact that if the second ear is the actual gleanings, the first ear, which is given to the poor, would constitute untithed produce that may not be eaten until the tithe is separated from it. Hence a third ear would be required, with the same declaration made over it, which would provide for all eventualities. Hence "two ears" is presumably to be understood as additional to the original ear mentioned first in this Section.

11. Pe 5:6; Tos Pe 5:1; B. BM 12a.

"a laborer"—the difference between a laborer who is paid monetary wages per diem and the sharecropper who contracts to give his labor for a share of the crop, is that the former is not permitted to stipulate that members of his family be allowed to pick up gleanings after him, since he would otherwise sell his labor at a lower price and thus deprive the poor of their share; in the sharecropper's case, the price is fixed, and he cannot make any stipulation, but his wife and children are allowed to glean after him. The poor will not consider this as depriving them of their share; they will rather be happy with this arrangement in the hope that the same privilege will be extended to them.

"sharecroppers, tenants"—the former pay the owner a share of the crop; the latter pay a fixed amount of produce or its monetary equivalent.

"purchasers"—who buy only the crop, and not the land, and thus do not come under the classification of owners.

"A laborer may," etc.—without previous stipulation.

12. Pe 5:6.

13. B. BM 12a; Tos Pe 3:1.

"as are not entitled"—who have two hundred zuz, or engage in business worth fifty zuz (below, ix, 13), and are thus not really poor.

"exclude them"—by peaceful means, such as persuasion.

14. B. Tem 25a.

15. Pe 6:5; 7:3.

16. Pe 7:3; P. Pe 6:3.

"it is not grape gleanings"—because the grapes did not become separated in the course of gathering the vintage, but rather accidentally.

"if . . . he finds even half a cluster," etc.—this is a penalty imposed for the protection of the poor, similar to the penalty imposed above, Secs. 6–7.

"Indeed, if a whole cluster," etc.—for the same reason. This sentence is omitted in some old editions.

17. Pe 7:4; Tos Pe 3:11; P. Pe 7:4.

"as an ear of corn," etc.—so *editio princeps;* other editions have instead "as an ordinary cluster."

"or vice versa"—this is presumably added in order to agree with the text of Tos and P.

18. Tos Pe 3:11; P. Pe 7:4.

19. Pe 4:1; 7:4; P. Pe 4:1.

"A single grape"—i.e., a defective cluster which has single grapes attached to the spine (RDBZ *ad loc.*).

20. Pe 7:4.

21. Pe 7:7; B. Ḥul 131a; Tos Pe 2:13.

"And from . . . clusters"—usually translated *And thou shalt not glean thy vineyard.*

22. Pe 7:7; P. Pe 7:7.

"take the defective clusters"—usually translated *glean it after thee.*

"a quarter"—the commentators differ as to whether a quarter of a log̅ or a quarter of a ḳab̲ (=one log̅) is meant; three clusters, when pressed, produce a quarter of wine.

23. Pe 7:8; P. Pe 7:7.

"they do not belong to the poor"—because consecrated property is exempt from the obligation of defective clusters.

"after the defective clusters have become recognizable"—at which moment they become the property of the poor, and the owner cannot consecrate what is not his.

"the value of their growth"—the value of their improvement from the moment of their dedication to the time of vintage.

24. Pe 7:5. The poor are entitled to defective clusters at the commencement of vintage; pruning is not vintage.

25. Tos Pe 3:12.

"for the vintage"—i.e., only the yield, and not the vineyard itself, is sold to the Israelite, entitling him to harvest the grapes, usually for a share of the vintage. He is then obligated for defective clusters, because it is the harvesting that gives rise to the obligation.

26. Tos Pe 3:14; P. Pe 7:4.

"he must give them to the poor"—since what belongs to the poor is exempt from heave offering.

"cut off together with the normal clusters"—see above, Sec. 20.

"but must designate it as heave offering for another crop"—such a mixture of normal and defective clusters belongs in toto to the Levite (above, Sec. 20), and he consequently may use these grapes for heave offering for another gift of produce that he may receive from someone else.

27. Tos Pe 1:10; P. Pe 3:3.

"fourth year's fruit"—see below, V, ix, 1.

"he is exempt"—in the case of all gifts, with the exception of defective clusters, the obligation to give them arises at the time of harvesting; here the harvesting has deviated from the usual, since the bulk of the crop would customarily be pressed into wine (cf. above, ii, 6).

"unless he leaves over"—the obligation would then fall upon the last produce to be harvested, from which the gifts would then be due.

Chapter V

1. Pe 5:7; P. Pe 5:6; Tos Pe 3:1; B. Soṭ 45a.

"seen by other persons"—who would be likely to call it to the owner's or laborers' attention.

"if forgotten"—by everyone; a hidden sheaf is usually meant to be recovered later.

2. B. BM 11a.

"did indeed forget it"—JE emends in both instances "and he then forgets it," meaning that the owner, after remembering it, forgets it. In the city the principle is that the sheaf in the field has the status of forgotten sheaf only if both the owner and the laborers have forgotten it in the end. In the field the principle is that the forgetting by the owner takes place in the beginning, and even if he subsequently reminds himself, this has no effect. The reason for this is that if the owner is in his field while the laborers forget the sheaf, the fact that he did not forget it preserves his ownership of it, and his subsequent forgetting cannot invalidate it. If he is in the city, however, his original remembering does not preserve his ownership, which is invalidated by his subsequent forgetting.

3. Pe 5:7; P. Pe 5:6; Pe 6:3, 2; Tos Pe 3:2; B. Soṭ 45a.

"stand . . . cover"—so as to hide the sheaf from the owner's sight and cause him to forget it.

"it is not accounted forgotten sheaf"—the fact that he has expressed his intention to convey it to the city indicates that he has not forgotten it.

"even if he leaves it next to the fence gate," etc.—which indicates only change in location, and not intention to make final disposition of it.

4. P. Pe 6:3; B. Soṭ 45a.

"the lower one is not accounted forgotten sheaf"—since it has been forgotten only because it was covered up by the upper sheaf. Cf. above, Sec. 3, the case of the sheaf hidden under a covering of straw.

5. B. Soṭ 45a.

6. P. Pe 5:2.

"If there is a sixth sheaf"—if there are six sheaves, and he pauses before the fourth but not before the fifth or sixth sheaf, it is reasonable to assume that he has not forgotten it, but has rather paused in order to divide them into two even rows of three sheaves each. In the case of five sheaves, this assumption would be unlikely.

7. Tos Pe 3:4; P. Pe 6:3.

"two sheaves"—variant reading: "the sheaves in a field."

"the surrounding sheaves"—which are regarded as hiding the forgotten sheaf (above, Sec. 3).

8. Pe 6:10, 11.

"they are buried"—the edible bulbs remain underground.

"if a blind man forgets," etc.—since he is sightless, he is presumed to exercise special care not to overlook any sheaves; hence it is not his blindness that has caused him to forget one or more sheaves.

"the larger plants only"—since it was not his intention to harvest the small plants, the law of forgotten sheaf cannot yet be applied to them.

9. Pe 6:10; P. Pe 6:9.

"before it is ripe"—but after it is one-third ripe, since otherwise the term "harvesting" would not apply (above, ii, 7).

10. Pe 6:4.

11. *Ibid.;* Tos Pe 3:4; P. Pe 6:3.

"The overlooked sheaf behind"—RDBZ points out that this is really not a sheaf but standing corn that they had left behind without cutting it off.

"proving that it was not really forgotten"—but had rather been left there to mark the east-west row, to be cut and made into sheaves later.

12. Pe 5:8.

14–15. Tos Pe 3:5; Pe 6:4.

14. "bundles"—bunches of produce smaller than sheaves.

16. Tos Pe 3:5; P. Pe 6:4.

"if there are two"—since Scripture mentions only two kinds of beneficiaries, we may assume that it means no more than two units of produce.

17. Pe 6:1; P. Pe 6:1.

18. Pe 6:6; P. Pe 6:5; B. BB 72b.

"two sĕ'ah"—equal to twelve ḳaḇ.

19. Pe 6:7; P. Pe 6:5.

"according to this estimate"—so corrected in recent editions; the text in earlier editions seems to be corrupt.

20. Pe 6:9.

21. Pe 6:8; P. Pe 6:7; Tos Pe 3:5, 6.

22. Pe 7:1.

"even if it is laden with fruit amounting to several sĕ'ah"—which amount, in the case of sheaves of corn, exempts them from the law of forgotten sheaf. See above, Secs. 17–18.

23. Pe 7:1; P. Pe 7:1.

"yielding particularly rich oil, or particularly lean oil"—the Hebrew terms, taken by Maimonides from the Mishnah, are *šafḳani* and *bešani,* literally "pouring" and "sparing, dry." There is a difference of opinion about the meaning of these terms. Some consider them descriptive adjectives, as translated here. Others regard them as derived from place names, *bešani* thus meaning from Beisan (Beth Shan, in Galilee). P. Pe 7:1 interprets the word as "shaming," i.e., one that puts other trees to shame by its rich oil, and thus practically synonymous with *šafḳani*.

24. P. Pe 7:1; Pe 7:2; Tos Pe 3:9.

"If . . . he does commence the harvest"—then the tree ceases to be marked in his mind; ordinary trees would usually be harvested first.

"as we have explained"—above, Sec. 22.

25. Pe 7:2.

"at that time"—presumably after Bar Kokhba's revolt, when Palestine was laid waste by the legions of Emperor Hadrian.

26. Pe 7:7; Tos Pe 3:16; P. Pe 6:5.

"cannot be reached . . . by the reaper's hand"—from the place further on, where the reaper finds himself when he realizes his omission.

"a vineyard"—where the vines are not trained, but trail on the ground.

"one or more vines"—i.e., one or two vines, since more than two are not forgotten sheaf (above, Sec. 16).

27. B. BḲ 28a.

"as will be explained"—below, III, ii, 9, 11, 12.

Chapter VI

1. "sixth"—the other five are: gleanings, forgotten sheaf, corner crop, grape gleanings, and defective clusters.

"yield"—literally "seed."

2. "great heave offering"—as distinguished from the lesser heave offering, or heave offering of the tithe (Num. 18:25–26; below, Sec. 6).

4. B. RH. 12b.

5. Sif Deut. 14:38; Yaḏ 4:3.

"the lands outside the Land of Israel"—for a definition of these lands see below, III, i, 7. The lands of Egypt, Ammon, Moab, and Shinar (Babylonia) are adjacent to the Land of Israel, and there was much traffic to and from them. On the applicability of the laws of heave offering and tithes to Egypt, Ammon, and Moab see below, III, i, 1. Shinar, on the other hand, was regarded as a distant land.

7. B. Neḏ 84b; P. Pe 8:4; Sif Deut, 26:12.

"enough . . . to satisfy his hunger"—for a day, i.e., two meals; see the next Section.

8. Pe 8:5; B. 'Er 82b, 29a; P. Pe 8:4.

9. Pe 8:6.

10. B. Neḏ 84b; Tos Pe 2:13; B. Ḥul 131a.

"has no optional right"—since Scripture says *thou . . . shalt lay it up* (Deut. 14:28), not "thou shalt give."

11. Pe 8:6; P. Pe 8:5.

12. Tos Pe 4:2; B. Ker 6b.

"in the field . . . in the house"—the distinction is that in the field, if the poor man does not get enough, he has nowhere else to turn for more, while in the house (*within thy gate*), if there is not enough, he can obtain more from another house.

13. B. Yeḇ 100a; Ḳid 32a; Tos MSh 4:7; P. Pe 5:4; B. Giṭ 12a; Ḥul 131a.

"give to the woman first"—because she is more sensitive, and unnecessary waiting would humiliate her further.

"may give him"—since the poor member of the pair may not take it himself.

14. Pe 5:5; B. Giṭ 12a; Ḥul 131a.

15. Pe 5:5.

"gained possession"—and may not take the poor man's tithe out of his own produce. In the former case the poor man has become in effect holder in tenancy of the entire field; in the latter case he is no different from a hired laborer.

16. Pe 5:6; P. Pe 5:5.

"The buyer"—if he, too, is poor.

17. Tos Pe 4:16; Sif Deut. 26:12.

"One may not . . . repay a debt"—the principle is that the poor man's gifts may be used for his household needs only.

"One may use it . . . to repay an act of charity"—the Yemenite MSS read "One may send it as an act of charity," that is, to another poor man.

"but the recipient must be informed"—that no repayment is expected.

"captives . . . wedding"—in the former case, a captive is not necessarily poor, and hence does not qualify; moreover, in both cases a reciprocal gift or other return is expected.

"as a matter of optional right to the town scholar"—i.e., he may give it to the scholar as a present but not as a charitable gift (which would humiliate the scholar).

Chapter VII

1–4. B. Ket̠ 62b f.

1. "the poor of Israel"—this is not meant to exclude the heathen poor; see below, Sec. 7.

5. B. BB 31b; Ket̠ 50a; P. Pe 1:1; B. BB 9a; Giṭ 7b.

"a person of evil eye"—i.e., an immoral miser.

6. B. BB 9a; P. Pe 8:6.

"hungry . . . naked"—the difference is explained in the Talmud: the hungry person is suffering intense pain which must be relieved immediately, while the person who needs clothing is, in the mild climate of the Near East, not a case of dire emergency.

7. B. Giṭ 61a, 59b; BB 9a; P. Pe 8:5.

"One is not obligated to give him a large gift"—Rashi and Tosafot̠, *ad loc.*, explain that this refers to the officials charged with dispensing alms out of public funds, who are thus required to provide the beggar with only a small contribution, for the reason that since he is going from door to door, he will in the end collect enough to provide for his needs.

8. Pe 8:7; Tos Pe 4:8; B. Ket̠ 67b.

"not less than one loaf of bread"—the minimum necessary to sustain life.

"We have already explained," etc.—III, II, i, 12–13; VII, i, 2–4 (YJS, *14*, 203, 411).

"three meals"—the prescribed number of meals on the Sabbath.

"If he is known"—and thus his standard of living is also known.

9. B. Ket̠ 67b.

"or a loan"—if he refuses to accept it even as a present.

10. B. BB 8b; B. Ket̠ 48a.

"in his presence"—but not in his absence and not on the assumption that had he been present, he would have given without coercion.

"even on the eve of the Sabbath"—when the householder may plead that he is busy preparing for the Sabbath.

11. B. Ta 24a; B. BB 8b.

12. B. BB 9b; BḲ 119a.

"bondsmen"—who have no property rights anyway.

13. Sif Deut. 15:17; B. BM 71a.

14. B. Meḡ 27a f.

"must contribute"—even though he is there only temporarily. See below, IX, 12, where three months' residence is required as a minimum.

"If a large group," etc.—the absence of one or two local contributors will not seriously affect the alms fund of the hometown; the absence of many (more than two, according to some authorities) will. Hence if several contributors are visiting another city and are there assessed for alms, they must pay the assessment, but when they are about to return home, the authorities of that city should refund the assessment, which upon the travelers' return should be handed over to the hometown authorities.

"to be distributed as he sees fit"—he will see to it that the poor of the other town do not suffer by it.

15. Tos BḲ 11:3.

"Give"—out of the amount which he is contributing.

Chapter VIII

1. B. RH 6a; ʿAr 6a.

"to change the selaʿ or to combine it with others for conversion into gold coin"—smaller coins were more readily disposable. If not used immediately, however, the selaʿ could be combined with other coins for conversion into larger silver coins or gold coins, which were more convenient to hold but less readily disposable.

2. B. Neḏ 7a.

"linking up"—an object or person with another object or person already subject to vow or oath (VI, 1, ii, 8; YJS, *15*, 10).

"like this one"—obviously "this one" has already been pledged to charity.

3. Men 13:4.

4. B. ʿAr 6a; BB 8b.

"for other collectors, not for themselves"—lest they should be suspected of realizing a personal profit from the exchange.

5. B. ʿAr 6a.

"in order to encourage others to contribute"—by telling them, quite truthfully, that there is at the moment no cash on hand for distribution to the poor.

"borrow"—and assume personally the obligation to repay.

6. B. ʿAr 6b; Tos Meḡ 2:3.

"secular reason"—not connected with any religious duty.

7. B. ʿAr 6a.

8. B. ʿAr 6a; P. Sheḳ 1:4.

"easily identified"—which might enable the heathen to point boastfully to his contribution or to its importance for the Temple.

"intended it for the Lord"—a "gift for the Lord" is the same as a gift for the Temple, which may not be accepted, as stated at the beginning of this Section.

9. B. Sanh 26b.

10. B. BB 8b; Sanh 73b.

"his very life is in jeopardy"—since his captors may kill him if he is not ransomed. A beggar's life, contrariwise, is not in jeopardy, since the community, however niggardly it may happen to be, will surely not let him starve or freeze to death.

11. B. BB 3b.

"and set them up"—*gĕḏarum;* perhaps read *gĕrarum,* "polished them," to parallel the following "planed them."

"make a new collection from the community"—but if that is not feasible, they must sell the synagogue (see ŠYD 252; and ŚK Sec. 1, ṬaZ Sec. 2, *ad loc.*). Even a Scroll of the Torah may be sold in order to redeem captives.

12. B. Giṭ 45a, 47a.

"fair value"—the price a person would fetch if offered for sale at the slave market.

"yoke"—the yoke or collar placed upon a captive's neck to prevent his escape.

13. B. Giṭ 46b, 47a.

"if their father dies in captivity"—according to the Talmud, in order to prevent the children from going astray in the ways of the heathens, of which there was not so much danger while the father was living. It would be more logical to say that they are not to be redeemed while their father is living, in order to teach him to be more prudent in his ways, which would not apply after his death.

14. B. Giṭ 37b, 47a.

"ritually immersed"—a bondsman owned by an Israelite must go through the ritual of conversion (V, I, xiii, 11; YJS, *16*, 89).

"out of spite"—the law makes a distinction between a person who eats forbidden food in order to appease his hunger and one who eats it as a mark of his contempt for the law.

15. Hor 3:7; B. Keṯ 67a; P. Hor 3:8.

"sin"—in the case of the woman, adultery; in the case of the man, pederasty.

16. Keṯ 6:5; B. Keṯ 67a f.; Tos Keṯ 6:8.

"her sense of shame"—at her unmarried estate. See Lieberman's note to Tos *ibid.*

"she should be given"—for her dowry. The Mishnah states the minimum as 50 (denar), which is interpreted as common denar worth 1/8 of a pure silver denar, or 6¼ pure silver denar in all.

"treasury"—literally "purse" (*ḳis*).

17. Hor 3:7; B. Hor 13a.

"profaned priest"—*ḥalal,* the child of a priest and a woman whom he is forbidden to marry (e.g., a divorcée).

"a person of unknown parentage . . . a foundling . . . a bastard"—the former two may possibly be of legitimate descent, whereas the bastard is assuredly illegitimate.

"a person of unknown parentage"—*šĕṯuḳi,* a person who knows his mother but not his father.

"a foundling"—*'ăsufi,* a person who knows neither his mother nor his father.

"has grown up . . . in the state of holiness"—and thus was never a heathen, while a proselyte was once a heathen.

"included among the accursed"—i.e., is the progeny of Canaan who was cursed by Noah, Gen. 9:25.

18. Hor 3:8; B. Hor 14a; P. Hor 3:7.

"teacher or father"—note that the teacher is mentioned first.

Chapter IX

1. B. Sanh 17b; P. Ḳid 4:5; B. BB 10b.

"alms fund"—literally "basket" (*ḳuppah*).

2. B. BB 8b.

"alms tray"—*tamḥuy,* used in modern Hebrew in the sense of "soup kitchen."

3. B. Sanh 17b.

4. B. Sanh 35a.

"On fast days"—i.e., on the eve of each fast day, when one must fortify himself for the ensuing day of total abstension from food and drink.

"If . . . the alms collectors merely delay . . . money or wheat"—the bread and the fruit (food requiring no preparation) having already fortified the poor for the following day of fasting, the money and wheat can wait until the termination of the fast, to provide the wherewithal to break the fast.

5–6. B. BB 8b.

5. "a demand for money may not," etc.—literally "an office conferring authority in money matters over the community may not be filled by," etc.

"money involved in a civil action"—the disposition of which is decided by a panel of three judges (XIV, 1, xi, 1; YJS, *3,* 31). The distributors of the alms fund must adjudicate the amount to be given to each poor man according to his individual need. The collection of the moneys for the alms fund, on the other hand, requires no adjudication, since the assessment is already set upon each citizen, hence two collectors are sufficient.

"since . . . it is not set"—each citizen contributing according to his own judgment. The Talmud gives the reason that "it is distributed as soon as it is collected," which Rashi explains to mean that if only two persons were sufficient for the solicitation and a third one had then to be found to complete the triumvirate of distributors, it would cause a delay and prolong the suffering of the needy.

6. "the poor of the whole world"—in other words, all comers.

7–11. *Ibid.* and f.

8. "gate . . . shop"—so long as they are in sight of one another and appear to be acting in concert.

9. "in the alms pouch"—so that people would not suspect him of misappropriating public alms money.

11. "convert the small coins into denar"—see above, note to viii, 1.

12. B. BB 8a; Tos Pe 4:9; P. BB 1:6.

13. Pe 8:7.

"two meals"—standard fare for one day. Since alms tray food is distributed every day, this is sufficient.

"fourteen meals"—standard fare for one week. Since alms fund money is distributed once a week, this is sufficient.

"or he who has fifty zuz . . . trade"—some texts omit this clause.

14. Pe 8:8; B. Keṯ 68a.

"he may not be compelled to sell his house and his furnishings"—since if he did not have a home and furnishings, these would have had to be provided for him at public expense (above, vii, 3–4).

"accept alms . . . ask for public assistance"—there is a distinction between accepting help from private individuals and applying for public assistance.

"eating and drinking vessels"—the Talmud explains that a person used to tableware of this quality would find low quality utensils repulsive or even loathsome. This does not apply to other household furnishings.

15. Pe 5:4.

16–17. B. BḲ 7a.

16. "houses"—i.e., rental houses, excluding his own home (above, Sec. 14).

18. Sheḳ 2:5.

19. Tos Pe 4.10.

"If a poor man contributes"—as is his moral duty (above, vii, 5), provided he can spare the money or the food without going hungry or naked himself.

"they too should be accepted"—the worn-out clothing may still be put to some use by another poor man; in any case his generous impulse should not be rebuffed.

Chapter X

1. B. BB 9a; Yeḇ 79a; Shab 139a.

"righteousness"—Hebrew *ṣĕḏaḵah,* the term used for both "righteousness" and "almsgiving, charity." This applies to the other Scriptural quotations cited further on.

"The throne of Israel"—the temporal and spiritual kingdom of Israel; cf. I Kings 8:25, 10:9.

2. B. Ta 9a; Prov. 28:27; B. Shab 151b; Beṣ 32b.

3. B. BB 10a.

"a base fellow"—literally "a Belial," a Biblical term signifying "uselessness, worthlessness," later understood as a synonym for Satan.

"Thou hearest the cry of the poor"—presumably a misquotation from memory of *He heareth the cry of the poor* (usually translated *of the afflicted;* Job 34:28).

"a covenant has been made"—by the Lord Himself.

4. "his face averted to the ground"—a mark of displeasure, of doing something reluctantly and unwillingly.

"loses his merit"—most of it, not all of it; cf. below, Sec. 14.

5. B. BB 9b; Prov. 13:7.

6. *"the work of righteousness"*—the word "work" (*ma'ăśeh*) in Hebrew implies (if read *mĕ'aśśeh*) also causing others to do the same thing.

"peace"—the second half of the verse, *and the effect of righteousness, quietness and confidence for ever,* is taken to refer to him who only gives alms himself. His reward will be quietness and confidence, evidently a lesser boon than peace.

7. B. Shab 63a.

"eight degrees"—these are the famous eight degrees of charity. Maimonides starts from the top; the popular version starts from the bottom.

8. B. BB 10b; Sheḳ 5:6.

"contributes directly to the alms fund"—and thus cannot know who will receive his contribution.

"Rabbi Hananiah ben Teradion"—who had a reputation for utter reliability. See Tosafoṯ to B. BB 10b, *s. v. 'ella',* and B. AZ 17b. His first name is also spelled Ḥanina.

9–10. B. BB 10b; Keṯ 67a.

15. B. BB 10b.

"in righteousness"—or *with righteousness,* i.e., with the alms given immediately before prayer.

16. B. Keṯ 50a; Tosafoṯ to B. Ḳid 32a, *s. v. 'oro.*

"for his grown sons and daughters"—though not legally obligated to do so. See IV, 1, xii, 14 (YJS, *19,* 76–77).

"provides maintenance for his father and mother"—though not legally obligated to do so unless they are destitute. See XIV, III, vi, 3 (YJS, *3*, 154–55).

"one's relative has precedence"—see above, vii, 13.

17. 'Aḇ 1:5; 2:7.

"bondsmen"—were thought to have low moral standards, the males being suspected of thievery, and the females of promiscuity.

18. B. Pes 112a; Yoma 35b; Giṭ 67b; Keṯ 105a; Ber 28a.

19. P. Pe 8:9.

TREATISE III: LAWS CONCERNING HEAVE OFFERING

Chapter I

1. Ḳid 1:9; Bik 2:3; Yaḏ 4:3.

"Land of Shinar"—i.e., Babylonia.

"it is adjacent to the Land of Israel"—actually it is not adjacent, but was considered so because there was much traffic between the two countries. Egypt, Ammon, and Moab, on the other hand, are literally adjacent to the Land of Israel.

2. B. AZ 2a.

"Israelite king"—some texts add "or judge."

"with the consent of the majority"—otherwise it is conquest by an individual, which does not affect the status of that land in relation to the commandments that apply in the Land of Israel.

3. B. Soṯ 44b; Sanh 2a; Sif. Deut. 'Eḳeḇ 11:24; Tosafoṯ to Giṭ 8a, *s. v. ḳibbuš*.

"Aram-Naharaim, Aram-Zobah"—these conquests of David are mentioned in 2 Sam. 8:3–6 and Ps. 60:2. The Biblical Aram-Naharaim, "Aram of the two rivers," refers to Mesopotamia. Aram-Zobah was a town and kingdom north of Canaan.

"Ahlab"—mentioned in Judg. 1:31, a town on what is today the Lebanese coast. The Biblical Ahlab, however, is listed as within the limits of the Land of Israel, whereas Maimonides' Ahlab is outside of it. Perhaps it refers to the modern Aleppo, called in Arabic Ḥalab.

"the High Court"—the Sanhedrin consisting of seventy-one judges, whose consent was necessary to wage war (XIV, V, v, 2; YJS, *3*, 217).

"Seven Nations"—which had inhabited Canaan when it was invaded by the children of Israel. They are enumerated in Deut. 7:1.

"Syria"—much more inclusive than modern Syria, and corresponding to what we term today the Fertile Crescent. It was roughly the same as the province of Syria in the Roman Empire, and the Seleucid Kingdom of the Hellenistic period. See below, Sec. 9.

4. B. Giṭ 8a is more explicit: "Our Rabbis have taught: In three respects Syria is in the same category as the Land of Israel, and in three others in the same category as foreign parts. Its earth is ritually unclean like that of foreign parts; selling a slave into Syria is like selling him into foreign parts, and a bill of divorcement brought from Syria is reckoned the same as one brought from foreign parts. On the other hand, in three other respects it is like the Land

of Israel: it is subject to the obligation of tithe and Sabbatical year like the Land of Israel, it is permsisible for an Israelite to enter it in a state of ritual purity, and a field bought in Syria is like one bought on the outskirts of Jerusalem."

5. B. 'Ar 32b; Yeḇ 16a; Ḥul 6b; P. Shebi̱ 6:1; Tos 'Oh 18:18.

"first consecration"—there is a controversy in the Talmud as to whether this consecration was permanent or temporary, valid only while the children of Israel occupied the Land. Maimonides sides with the opinion that the first consecration was temporary, and that the second consecration, which embraced less territory, was permanent. The areas made sacred by the first consecration but not reconsecrated by the second consecration were thus exempted from the "commandments which depend upon the land." Scribal law, however, retained for these areas the obligation to render heave offerings and poor man's tithe, in order to help the poor in the rest of the country, where these were not available during the Sabbatical year.

"Our Saintly Rabbi"—Rabbi Judah the Prince, compiler of the Mishnah and Patriarch of the Jewish community in Judea during the second half of the second century C. E.

"Beth-Shan"—is in the Jordan valley near the Sea of Kinnereth (Gennesaret), while Ashkelon is down south on the Mediterranean shore. Both are within the boundaries of the Land of Israel, but Rabbi Judah exempted them from tithes because they were not reconsecrated by those who had returned from the Babylonian exile. Obviously these rules had a very sound economic reason behind them.

6. "Shinar"—mentioned in the Bible (Gen. 10:10). See above, Sec. 1, and below, Sec. 9.

7. Giṭ 1:1; B. Giṭ 7b f.; Shab 15a; Shebi̱ 6:1; Tos Ter 2:12.

"Rekem"—Josh. 18:27. Targum Onkelos (Gen. 16:14) identifies it with Kadesh on the southern border of Palestine. Josephus (*Ant.* IV, 7, 1) identifies it with Petra and calls it Rekemos.

"the Great Sea"—the Mediterranean Sea.

"Acco"—the modern Acre.

"Chezib"—Gen. 38:5, perhaps the same as Achzib (Josh. 19:29) which is north of Acco. The passage is difficult because left of Chezib is the Mediterranean Sea, unless Maimonides refers to a thin strip of land near the coast which he considered part of the Land of Israel (see B. Giṯ 7b).

"Ummanom"—identified with Mt. Hor, Num. 34:7, which Targum Pseudo-Jonathan renders as Tauros Ummanim (variant: Manos), the modern Giaour Dagh. See below, Sec. 8.

"the Stream of Egypt"—identified with the modern Wādī al-'Arīsh, or (less likely) the Nile.

"This is its shape"—evidently a drawing of the cord followed this sentence.

8. Shebi̱ 6:1; P. Dem 1:3.

9. Giṭ 1:1.

"outward"—literally "below." The term is due to the belief that the Land of Israel was situated higher than all other lands, hence all other countries were considered below it. In modern Hebrew he who immigrates into Israel is termed *'oleh,* "he who has ascended," while he who leaves Israel is called *yored̲,* "he who has descended."

"Haran"—Gen. 11:31.

"Migbad . . . Ṣahar"—otherwise unidentified.

10. B. Giṭ 47a.

"it retains its status of holiness"—and is subject to all the obligations "dependent upon the land."

"individual conquest"—see above, Sec. 2.

"as will be explained"—below, Sec. 15.

11. B. Bek̲ 11b; Giṭ 47a.

"not 'the heathen's corn' "—the Hebrew word *dağan,* "corn," can be read also as the verbal noun *diggun* (as below, in Sec. 13), "completion of the growth of the produce."

"I am acting," etc.—i.e., the priest could not have demanded the heave offering of the tithe from the heathen, since he had no authority over him. Nevertheless the Israelite purchaser may not consume what is formally due to the priest, who therefore must purchase it from him. This does not apply to the great heave offering, which is not mentioned in Num. 18:26.

12. Ma 5:5.

"and must sell the other third to the latter"—here again the produce is subject to the tithe by Scribal law, but the priest is in no position to claim it, while the owner may not consume it. Hence this arrangement.

13–15. See above, Sec. 11.

13. "except by Scribal law"—enacted as a preventive measure against wealthy landowners who used to evade the statutory dues by fraudulent sales of their produce to heathens.

15. Ḥal 4:1; Ma 5:5. Cf. above, Sec. 10.

"as we have explained"—above, Sec. 4.

16. Ma 5:5.

17. Ḥal. 4:7; Tos Ter 2:11; B. AZ 15a.

"as we have explained"—above, Sec. 10.

18. B. Giṭ 47a.

19. P. Dem 6:8.

"bought it in the market place"—and it is therefore exempt from tithes. In P. the explanation given is that the tenant has stated expressly that the produce had already been tithed before the rental payment was sent to the landlord. Maimonides' copy of P. presumably did not contain this explanation, which is why he gives his own here. When the liability is doubtful, the owner is given the benefit of the doubt.

20. B. Ḥul 135b; P. Dem 6:6.

"untithed and profane produce are intermixed"—untithed produce may not be consumed by an Israelite commoner until the tithes have been set aside. Profane produce is produce from which all that is due to the priest and the Levite has already been separated, thus making it permissible for common consumption. Here these two kinds are considered intermixed, and not subject to the principle of retroactive differentiation, according to which what the heathen and the Israelite receive when the produce is finally divided had been so divided from the very beginning, which would have rendered the heathen's share exempt.

"The liability . . . is required only by Scribal law"—therefore, according to Sec. 21, the principle of retroactive differentiation should apply. RDBZ explains that while in this particular aspect the obligation is only Scribal, the basic principle that the yield of a field in the Land of Israel is subject to tithes has Biblical authority.

"as we have explained"—above, Sec. 13.

21. B. Beṣ 38a.

"the principle of retroactive differentiation"—i.e., that a condition existing now explicitly has existed before implicitly, in which case the share that the heathen receives now when the produce is divided would be regarded as having been his before the division.

22. Ḥal 2:1; P. Ḥal 2:1; Dem 6:11.

"dough-offering"—see Num. 15:20.

"had been designated"—while still outside the Land of Israel.

23. Ḥal 2:2.

"so long as the ship remains in contact with the Land of Israel"—the wooden ship is regarded as analogous to a perforated wooden pot, which cannot prevent the roots of the plant from feeding upon the ground underneath or adjacent to the pot.

24. Ma 3:10; B. Giṭ 22a; BB 27b.

"a mixture"—since the tree derives its sustenance from both lands and blends it in its trunk; see above, note to Sec. 20. If the roots are all in one land or the other, the decision goes with the land.

25. B. Giṭ 22a.

26. B. Pes 44a; Shab 68a; Keṯ 15a.

"When ye are come"—this verse refers to the dough offering, but Maimonides seems to have applied it also to the heave offering. The ordinance dealing with heave offering begins with *When ye come* (Num. 15:18).

"the third settlement"—the future final restoration of Zion.

Chapter II

1. Ma 1:1; B. Shab 68a; P. Ma 1:1.

"human food"—which excludes food fit only for animals.

"watched over"—which excludes ownerless crops.

"grows out of the soil"—the commentaries explain that this excludes truffles and mushrooms, which were thought to have no roots and to derive their sustenance from the air.

2. P. Ḥal 4:4; B. Nid 51a.

"If the courtyard does not serve to protect the produce"—the produce is then considered ownerless, which makes it exempt from all dues. This would be true, for example, if the courtyard is so open that it can be used as a public thoroughfare.

3. Ma 5:8; ʿUḳ 3:6.

"garden seeds"—the seeds themselves are not edible but the plants produced by them are.

4. Ma 4:6; P. Ma 4:5.

"they are not produce"—they are not ordinarily meant for human consumption at this stage of their growth.

5. Ma 4:5; P. Ma 4:5.

6. P. Ma 1:1; B. RH 12a; Ḥal 4:4; B. Zeḇ 76a.

7. P. Ḥal 4:4; Tos Shebi 2:8; B. 'Er 29a.

"even in places"—like Shinar, Egypt, etc. See above, i, 15.

"we have said"—above, i, 1.

"precautionary dues"—enacted in order to prevent nonpayment of dues in other cases where they are obligatory.

"even if they have some earth clinging to their roots"—and are therefore presumed to continue growing in the Land of Israel.

"for their greens"—and not for the sake of the ripened grain and beans, an uncommon occurrence.

8. Tos Ter 10:4; MSh 2:3; Ma 1:3.

"in the beginning"—when it is still soft.

9. Ḥal 1:3; P. Ḥal 1:3; P. Kil 6:2; B. Ber 40b; Soṭ 43b.

"he must separate . . . from them"—as a precautionary measure, since passers-by would have no way of knowing that this pile belongs to the poor man and not to the owner of the field.

10. Pe 4:9, 6:1; Ma 1:3; P. Ma 1:2; B. Ḳid 62b.

"are liable to heave offering and tithes"—inasmuch as the heathen owner in the Land of Israel is not subject to the law of gifts to the poor, these are no different from the rest of the produce of that field, which is liable to these dues once it is acquired by an Israelite (above, i, 10–11).

"unless he declares them ownerless property"—see the next Section.

11. Ter 1:5; Ma 1:1; Pe 4:9; P. Ma 1:1.

"has declared it so"—as above, Sec. 10, in order to release the indigent recipient from tithing.

"If . . . one plants an ownerless field"—the produce is liable to tithe, because while the field is ownerless, its produce is not.

12. P. Ma 1:1; Ma 5:8.

"standing corn . . . ears of corn"—the difference between the two cases is that standing corn is in a state in which it is not yet liable to heave offering; hence one's declaring it ownerless does not change its status. Therefore, when he repossesses it, he is liable to the priest's due. In the case of ears of corn, they are already liable to heave offering (cf. above, II, v, 27), and when he declares them ownerless and thus exempts them from heave offering, he changes their status. Therefore, when he repossesses them, they are exempt from heave offering.

"pungent . . . sharp"—literally "causing weeping . . . causing dripping" (*ba'al beḳi . . . riḳpah*); both terms are also interpreted as names of localities.

13. Ḥal 3:9.

"gleaners' olives"—literally "knockdown olives," left on the tree to be knocked down by the poor man, and exempt from tithes.

14–15. B. Shab 25a; Ter 11:10.

14. "whether it is clean or unclean"—the plural of *My heave offerings* implies more than one kind.

"by burning it"—not by feeding it to cattle, which is forbidden.

15. "benefit . . . by burning it"—for illumination or fuel; cf. the preceding Section.

16. Tos Ber 6:9.

"in the Laws concerning Benedictions"—II, v, xi, 3, 12.

17. Shebi 6:5, 6: P. Shebi 6:5, 6; Ḥal 4:5.

"even if it is unclean"—so that taking it to a foreign land would not defile it. Nevertheless one may not take it thither because of the insistence that it must be burned in the Land of Israel.

"it has been defiled by its sojourn in the land of the heathen"—the Sages have decreed that the lands of the heathens were to be considered intrinsically unclean, and that sacred food, such as heave offering, would become defiled if transported thither (B. Shab 14b). Modern historians attributed this restriction not so much to chauvinism as to economic policy. Evidently, because of the political situation, many of the inhabitants of the Land of Israel emigrated to foreign lands, and this decree was meant to discourage people from leaving the Land of Israel.

Chapter III

1. B. Ḥul 137b: P. Ter 4:2.

"the amount set by the Sages"—see the following Section.

"stands to be burned because of uncleanness"—i.e., becomes readily defiled and requires burning.

2. Ter 4:3; B. Ḥul 137b; Tos Ter 5:3, 8.

3. Tos Ter 5:6; P. Bik 3:1; P. Ter 1:1; B. Giṭ 52a.

"about which the priests are not particular"—because they are readily available at a low price.

"acorns"—*ḳĕlisin* (variant *ḳĕlisin*), variously interpreted as acorns, a species of figs (same as *ḳĕlusin*), a kind of pea or bean, or the fruit of the carob tree (Mishnah, tr. Danby, p. 65, n. 11). Cf. below, note to xi, 10.

"planted heave offering"—i.e., the priest, instead of consuming his due, has planted it. Although the produce that grows out of this planting remains forbidden to commoners by Scribal law (see below, xi, 21), it is nevertheless subject to tithe and heave offering.

"already taxed . . . not yet taxed"—from which heave offering has, respectively, already been and not yet been separated.

"defiled by accident"—if one defiles it deliberately, he is not permitted to offer the minimum, because he would then be rewarded for his transgression.

"sycamore figs"—regarded as of low quality.

"produce grown in an unperforated pot"—which is subject to tithe and heave offering by Scribal law only (cf. below, v, 15–16).

4. Ter 1:7; P. Ter 1:7; B. Ḥul 137b. According to the Talmud, a single grain of wheat discharges one's obligation for a whole heap (above, Sec. 1).

5. Ter 4:5; P. Ter 4:4; Ḥal 1:9; Tos Ter 5:8.

"his heave offering is valid"—since his intention was realized, for even though the amount he actually came up with is less than what he had intended, it still is included in it.

"his heave offering is invalid"—since the amount he actually came up with exceeds what he had intended and cannot be included in it.

6. B. Keṯ 99b; Ter 4:3.

"that supplement is subject to tithes"—because it has the status of untithed produce.

"If in setting aside heave offering"—with the intention of giving the proper amount, at least one-sixtieth.

"one comes up with one part out of sixty-one"—which is less than the minimum (above, Sec. 2).

"the total amount that he had in his mind"—i.e., the amount corresponding to his disposition (above, Sec. 2).

"this additional amount he may set aside by measure"—inasmuch as the first installment has already been separated as required, by conjectural estimate.

"from produce adjacent to the original produce"—as is required for all heave offering (below, Sec. 17).

7. Ter 4:1.

"only part of the heave offering"—fully intending to complete the process up to the minimum due.

"is considered not heave offering"—his partial action is in effect null and void; nevertheless, the separated produce must pay its due out of itself, and not out of other produce.

8. P. Ter 4:1; Ter 3:5; B. 'Er 37b.

"he may not assign one part's offering to the other"—for the heave offering must be set aside out of that same heap (see the preceding Section).

9. P. Ter 3:3.

"ends at the same point"—i.e., if he has assigned the offering of the first heap to the northern part of that heap, the same must apply to the second heap also. So the reading in most editions of the text (*ḳa-zeh*). The *editio princeps* reads *ba-zeh* (as does P), which would change the meaning of the rule as follows: "and yonder heap is to be included in this," i.e., the combined heave offering for both heaps is to be levied from the first heap, and from the same (for example, northern) part of the heap.

10. P. Ter 1:4; B. Giṭ 31a.

"even in this time"—when heave offering is obligatory only by Scribal law.

11. Ter 4:6; P. Ter 4:5. This rule refers to the heave offering of the tithe, for which the Torah prescribes an exact amount, namely one-tenth. Not so in the case of the great heave offering, for which the Torah prescribes no exact amount.

12. B. Giṭ 31a.

"An Israelite . . . may"—the regular procedure is for the Israelite to give the tithe to the Levite, who in his turn must give one-tenth of it to the priest. If the Israelite wishes, however, he may give to the priest both the great heave offering, which is about two percent of the total harvest, and the heave offering of the tithe, which is one percent (one-tenth of one-tenth) of the remainder. He may then give the remaining nine-tenths of the tithe to the Levite.

13. B. Ber 47a.

"while the corn is still in its ears"—the principle behind these rules is the following: The obligation to set aside the great heave offering takes effect once the produce has been threshed. In the first case the Levite receives the produce before it has become subject to this obligation; in the second case he receives it after it has become subject to it.

14. B. Beṣ 13a; Ter 10:6. By accepting the tithe in improper form, the Levite caused loss of revenue to the priest.

15. B. Giṭ 30a.

"You"—meaning the Levite.

"no apprehension need be felt"—when no measure or quantity is mentioned, it is assumed that all of it has already been taxed and it now belongs to the Levite; if a quantity or a measure is given, the assumption is that the produce has not yet been taxed, and therefore the Levite must first set aside the priest's share out of it.

16. Tos Ter 10:6; Ter 11:8.

"one-eighth of one-eighth"—or 1/64 (*ḳurṭub*) of a loḡ. Since loḡ is a liquid measure, what is meant is juice of produce. Such a minute quantity is of no appreciable value, except in the case of grape juice and oil (see further on).

17. Ḥal 1:9; B. 'Er 32a; Yeḇ 93a; BḲ 11b.

"he has said nothing"—because such sure to perish produce is regarded as virtually nonexistent.

18. Tos Ter 3:10; MSh 3:12.

"in two storerooms in the same house"—our text of Tos states that in this case heave offering must be set aside out of each storeroom individually. Maimonides presumably had a different reading in the Tos text before him.

19. P. Ter 2:1; Tos Ter 3:8, 9.

"another species"—*min 'aḥer,* so corrected by Prof. Saul Lieberman; the text of Tos reads "one species" (*min 'eḥaḏ*). In Hebrew the words are identical except for the last letter, and are easily confused.

"five heaps"—the last remaining heaps, the rest having been already threshed and removed to storage.

"as long as the ground of the threshing floor is intact"—i.e., so long as the ground of the threshing floor that connects them is uninterrupted and makes them as one. See Lieberman, *Tosefta ḳi-fĕšuṭah,* to Tos Ter 3:9.

20. Bik 2:5; P. Ter 2:1; B. Giṭ 30b.

"Disciples of the wise"—who are particularly scrupulous in the performance of their religious obligations.

21–22. Sif Zuṭa Num. 18:24.

21. "so that . . . it remains untaxed"—the Levite had in mind to use this batch of produce, received as tithe, as source of heave offering of the tithe for some other tithe produce in his possession. The batch itself thus remains untaxed, and later on, when used as heave offering for another batch, must be so used in its entirety.

22. "he has done nothing"—for once he has set aside the heave offering out of it, it is released from any further obligations. Hence (see below, v, 13) he may not set aside out of it heave offering of the tithe for other produce which is still under obligation for the priest's share.

"sacred portions"—the part of the tithe that is due to the priest. The principle behind these rules is that the priest's due may not be taken out of produce liable to it for produce that is exempt from it or vice versa (see Ter 1:5; and below, v, 12).

23. Ter 3:7, 6; B. Tem 4b.

24. Dem 5:2.

"should proceed as follows"—Maimonides' seemingly cumbersome procedure which preserves the sequence prescribed by the preceding Section is really simple The great heave offering is about two percent of the produce. The heave offering of the tithe due from the Levite, i.e., one-tenth of one-tenth, is one percent. The total is three percent, or approximately one out of 33⅓.

"and let it be profane produce"—temporarily, until the great heave offering is set aside, whereupon it, too, will become holy as the heave offering of the tithe.

"What is left of what I have set aside"—that is, the remaining two one-hundredths.

"the one hundred parts"—actually 98 parts (100 less 2 for the great heave offering).

"let it be adjacent," etc.—that is, let it be that part of the remaining 98 percent of the produce which is adjacent to the separated 2 percent.

Chapter IV

1. B. Ḳid 41b.

"which implies," etc.—the Talmudic rule is that whenever the word *gam* ("also") occurs in Scripture it implies something additional not mentioned previously or explicitly. In this case it indicates authorization of an agent. The implicit addition, however, must have some connection with the explicit subject; hence the qualification that the agent, like the principal, must be a son of the covenant.

2. Ter 1:1; B. Ned̲ 36b; P. Ned̲ 4:3.

"it is valid heave offering"—since A is, in effect, acting as B's agent.

"optional benefit"—A is doing a favor to B, at his own expense, and is therefore rewarded with the privilege, now denied to B, of choosing the particular priest who is to receive the offering.

3. B. Ḳid 52b; BM 22a; P. Ter 1:1.

"for the purpose of making it his own"—the problem then is whether this produce is to be considered A's property or stolen property. If it is A's property, he is permitted to set aside heave offering from it in behalf of B's produce. If it is not A's property, he may not set aside heave offering therefrom. The rest of the Section describes how it can be determined whether the produce gathered by A is his or not.

"to convey his protest"—i.e., B's remark is to be understood in the sense of "How dare you make free of my property!"

4. Ter 1:6, 2; B. Shab 23a.

"cannot speak"—and thus cannot pronounce the benediction.

"the naked"—must first cover his nakedness before reciting a benediction (cf. Deut. 23:15).

"they cannot direct their attention well enough to set aside the choice produce"—the rule is that the offering must be taken out of the choice produce (see below, v, 1). The blind and the intoxicated are incapable of distinguishing between poor and choice produce.

5. Ter 1:3.

"the age when his vows are valid"—in the case of a boy twelve years and one day, and in the case of a girl eleven years and one day, when they are capable of knowing to whom the vow is made. Heave offerings and vows belong to the same category, since they both depend on the spoken word.

"as we have explained in the Laws Concerning Vows"—VI, II, xi, 1–4 (YJS, *15,* 101–02).

6. B. Ḥul 12a; Giṭ 64a.

"without A's consent"—and therefore the offering is invalid.

7. Ter 4:4; B. Ket̲ 100a.

"his offering is valid"—since B can claim that he had changed his mind and had reclassified A as, respectively, generous or miserly. A himself, however, cannot plead a change of mind (above, iii, 6).

"even if he comes up with one-ninety-ninth"—this seems illogical, since the agent intended to set aside a greater amount, not a smaller one. Some texts therefore have "one-forty-ninth," which is of course nearer the amount due from the generous-minded.

8. B. Ḥul 136a; P. Giṭ 5:4; Ter 3:3; P. Ter 3:2.

"your"—in the plural (not "thy").

"depend on each other"—to set aside heave offering for the total amount of produce.

"if A has set aside the required amount"—thus demonstrating that his intent was to cover the entire crop. If he did not set aside the amount required for the entire crop, he has obviously meant to cover only his own share.

"B's heave offering . . . amount"—omitted in some editions, but required by the sense. See KM.

9. Ter 3:4; Giṭ 23b; P. Ter 3:2.

"If the agent has not deviated"—the Mishnah makes a more logical distinction: if A revokes the agency appointment before the agent has set aside the offering, it is invalid. Maimonides followed the Palestinian Talmud which quotes an opinion offering the distinction given here. RABD in his strictures, therefore, disagrees. See also KM.

10. Tos Ter 1:7, 9; B. Giṭ 22a; Ḳid 41b.

"the offering is valid"—since the tenant has title to a share of the produce in payment for his labor, and is thus part owner. Tos Ter 1:7 adds here, "But the tenant may not pay out the tithe except for his own share of the produce"; presumably this limitation applies to heave offering also, in the sense that the tenant may set aside the offering for the whole crop, but may actually hand to the priest only the offering for his own share and must leave the balance of the offering for disbursement by the owner. Maimonides omitted this ruling because he relied on the ruling he gave to that effect below in IV, vi, 17. See Prof. Saul Lieberman *ad loc.*

11. Tos Ter 1:6; P. Bik 1:2; B. BḲ 67a, 114a f.

"is valid"—since the rightful owner is presumed to have despaired of recovering his property, the seizor becomes subject to the obligation.

"if the rightful owner is in hot pursuit"—showing thereby that he has not yet despaired of recovering his property and that he therefore is still the real owner.

12. Tos Ter 1:6.

13. Ter 3:4.

"with the exception of those who tread the wine press"—Maimonides explains this exception as due to the fact that the owner trusts them to handle the wine properly. RABD explains it as due to the fact that inasmuch as the treaders could easily render the wine defiled immediately, the owner prefers not to have them wait for him, but to set aside heave offering before the wine is defiled.

14. Tos Ter 1:7.

"his offering is valid"—although he did not follow the sequence enjoined by the owner. Some manuscripts omit "and set aside heave offering therefrom," and RABD rightly challenges this rule as trivial, prefering the text of the Tosefta, which states that while a laborer may not set aside heave offering on his own authority, once the owner tells him to gather in the harvest, he implies thereby that the laborer should also set aside heave offering, since one is obligated to do so before gathering in the harvest. See Prof. Lierberman's comment *ad loc.*

15. Ter 3:9; B. Giṭ 23b; 'Ar 6a; Men 67a.

"holders of purses"—wealthy people may be greedy enough to resort to fraud, by assigning their produce to a heathen at the time when it becomes subject to heave offering, and thus rendering it exempt.

"with an Israelite's consent"—that the heathen's offering should go to the priest, the same as the Israelite's offering.

"it should be hidden away"—since its status is in doubt: if the heathen is sincere in his turn "toward heaven," his offering deserves to be valid; if he is not, his motives are highly suspect.

"land outside of the Land of Israel"—where even an Israelite is not obligated to set aside heave offering.

16. Ter 3:8; B. Beṣ 13b.

"reckoned"—literally *thought.*

17. ṬY 4:7; P. Ter 2:1; B. Neḏ 59a.

"regrets it"—i.e., changes his mind.

"consult a Sage"—if one makes a vow and then changes his mind, he must consult a Sage, who upon questioning him may absolve him, if he finds valid grounds for it (VI, II, iv, 5; YJS, *15,* 73).

18. ṬY 4:7; P. Ter 2:1.

"by saying"—thus stipulating a condition.

"but not from uncleanness"—the reason for the distinction is that breakage and spillage are common accidents, and therefore he must have had them in mind when he stipulated the condition; defilement, however, is not common, and he is therefore presumed not to have allowed for it.

"render forbidden"—Hebrew *mĕḏamma'.* If heave offering produce falls into one hundred times its bulk of profane produce, it is annulled, and a commoner may consume all of it. If heave offering produce falls into less than one hundred times its bulk of common produce, the whole amount becomes *mĕḏamma',* i.e., a mixture forbidden to a commoner, and is permitted, like whole heave offering, only to a priest.

19. Tos ṬY 4:8.

"even out of produce that is not adjacent"—above, iii, 20. Therefore, if one wishes to prevent breakage or spillage and the consequent defilement of the rest of the wine, he can achieve this by setting aside the offering from another container of wine where this apprehension does not exist.

20. P. Ter 4:8.

21. Tos Ter 3:6, 4; Ma 5:4.

"pressing vat"—so one manuscript (*gaṯ*). Other texts read "olive peat" (*gefeṯ*), which does not fit the context. However, ŠYD (331:50) reads at the beginning of this paragraph "olive oil (*šemen*) tank" instead of "wine (*yayin*) tank" (the two words are paleographically very similar), in which case "olive peat," i.e., a cake of pressed olive skins, would be appropriate.

Chapter V

1. Ter 2:4; B. Men 55a.

"fresh figs for dried figs"—fresh figs are preferred to dried figs.

"dried figs for fresh figs"—dried figs keep longer, and will not spoil before they reach the priest.

"even where there is no priest"—because the fresh figs will be eventually converted into dry figs and will keep.

2. Ter 2:5; B. Bek̲ 53b.

"In all places"—whether a priest is resident there or not.

"increase . . . increase—the repetition of the word is taken to indicate that no substitution is permitted.

3. Ter 2:6; B. Yeb̲ 89b.

"darnel"—an inedible weed usually found in fields of grain.

4. Ter 1:10; Sif Num. 18:27.

5. Tos Ter 3:11; B. Beṣ 13a.

"sifted"—the processing of the grain is now complete. Some texts read "gathered into a heap." See Prof. Lieberman's comments to Tos *ad loc.*

"out of the unparched ears"—since the ears are ready for parching, their preparation is regarded as complete.

6. Tos Ter 3:12, 13.

7. Ter 2:1; P. Ter 2:1.

"the hallowed part"—implying that the other part is not hallowed, having become unclean.

8. Ter 2:2; B. Pes 33a; Yeb̲ 90a; P. Ter 2:1.

"if he does so unwittingly, it is valid"—because according to the Torah he may do so even deliberately.

"what he has set aside retains its status of heave offering"—for the same reason.

"When does this apply?"—this refers back to the first sentence in the Section.

"the same as if he had acted willfully"—that is, what he has set aside is valid heave offering, but he must go back and set aside an offering once more.

9. Ter 1:5; Sif Num. 18:26; B. Ḳid 62a; Tos Ter 2:7.

"How so?" etc.—the answer seems to offer two contradictory rules. In the first case Maimonides says that if A sets aside heave offering out of detached produce for attached produce, his act is of no effect even if he adds, "after the latter shall have been detached." In the second case, if A adds "after the latter shall have been detached," and if the latter produce is then detached, his act is effective.

Various solutions have been proposed. One suggestion is that the same principle is involved in both cases, namely, if it is within A's power to detach the second bed, no act can be deemed lacking, and therefore A's act is valid. Accordingly, in the first case the attached produce is evidently not A's property, and he is therefore not empowered to detach it—an act is therefore deemed lacking. In the second case the attached produce is A's property, and it is therefore within his power to detach it (JE).

Another suggestion is that the first case must deal with the fruit of trees, the plucking of which is sometimes fraught with some difficulty, such as the necessity to climb the tree in order to gather the fruit. The second case deals with beds of vegetables, where it is in A's power to detach them easily (KM).

Neither of these two suggestions is supported by the text (R. Joseph Corcos).

The third suggestion seems to be the most plausible, but it requires the omission of a word, translated here "even if he adds," and is suggested by the fact that in our text this word is enclosed in parentheses. Accordingly, in the first case the implication is that if A sets aside heave offering out of plucked produce for unplucked produce, the offering is invalid even after the unplucked produce has been plucked, since A presumably has no authority to pluck it. In the second case A explicitly stipulates when he sets aside the heave offering that it is to become effective after the second bed has been plucked. Such a stipulation is immaterial if it is assumed that it is within his power to pluck the unplucked produce, and once it, too, is plucked, the offering becomes effective (KM).

10. Tos Ter 4:5; P. Ter 9:11.

11. Ter 1:5; Tos Ter 2:6; B. Bek̲ 53b; Sif Deut. 14:22; B. RH 12a, 14b; RH 1:1.

"year by year"—implying that each year's heave offering should cover the produce of that same year.

"before sunset . . . after sunset"—sunset marks the end of the preceding day in the Jewish calendar.

12. Ter 1:5; Tos Ter 2:9.

"nor from produce of the Land of Israel for produce of Syria"—this would seem to contradict the rule above, i, 4, that Syria is the same as the Land of Israel in regard to heave offering. Prof. Lieberman, in his commentary on Tos *ad loc.,* explains that what is meant here is finished produce imported from Syria and assumed to have come from the field of a heathen, which in Syria is exempt (above, i, 10).

13. Ter 1:5.

"from which no heave offering has yet been taken"—the statutory procedure is for the farmer to set aside heave offering first and then tithes. In this case he has given tithe to the Levite before giving heave offering to the priest. The priest therefore has a claim on the tithe for the great heave offering as well as for the heave offering of the tithe, and the owner may not set aside heave offering out of this produce for other produce before he has first regularized the former produce.

"second tithe or consecrated produce"—both are considered sacred property, out of which no heave offering need be set aside.

14. Dem 5:10, 11.

"If he does so nevertheless, his heave offering is valid"—this rule applies only to the second alternative, in which he has set aside the offering out of produce liable to it only by Scribal law: the priest may consume this offering even if the owner had not set aside out of the original produce the gifts due from it, since the obligation is only Scribal. Where the produce is subject to the gifts from it by Scriptural law, and one sets aside the offering therefrom for produce subject to it only by Scribal law, the offering is invalid, inasmuch as the priest may not consume it, since it is still liable to heave offering.

15. Dem 5:10; ʿUk̲ 2:10; B. Shab 95b; Men 70a; P. Ma 5:2.

"is deemed the same as the soil itself"—i.e., the plant in the pot is deemed the same as if it had grown in the underlying soil itself.

16. Dem 5:10; B. Yeḇ 89b; Ḳid 46b.

"it is valid heave offering"—because both are regarded as grown in the soil and are therefore subject to the same Scriptural ordinance.

"it is valid . . . but he must . . . set aside heave offering once more"—the standard commentaries to the Mishnah that is the source for this rule explain it thus: It is really invalid, because he has set aside heave offering out of produce that is exempt for produce that is liable, and hence he must set aside heave offering for the liable produce once more. It has been declared valid, however, since he had assigned it by name as a precaution, so that people should not make light of heave offering. See Bertinoro *ad loc.*

"the priest may not consume it"—since the produce in the unperforated pot is still subject to heave offering and tithes, which have not yet been set aside in its behalf, it is considered untithed produce.

17. Dem 5:11.

"he must . . . set aside heave offering once more"—because heave offering set aside out of doubtfully tithed produce cannot discharge the untithed produce from the duty of heave offering; the doubtfully tithed produce from which the heave offering has now been set aside may have been set aright by its former owner, and the present owner may thus be setting aside heave offering out of what is no longer liable for what is still liable.

"but it may not be consumed by the priest"—because if the doubtfully tithed produce had already been tithed, the offering would now be set aside out of what is still liable to it for what is no longer liable, in which case it is invalid heave offering and needs to be set aright.

18. Ter 1:4, 9; Sif Num. 18:27; P. Ter 1:2, 5; B. AZ 66a; Tos Ter 4:1, 3, 4; B. Men 54b; P. Ter 2:4.

"out of wheat ears," etc.—on the principle that heave offering may not be set aside out of produce still in the processing stage for produce in a completed state. See above, Sec. 4.

"out of oil-rich olives for pickling olives," etc.—on the principle that heave offering may not be set aside from a superior kind for an inferior kind, or vice versa. See above, Sec. 3.

"unless he calculates"—wheat when made into bread gains in weight and volume. Hence if one sets aside wheat for bread, the priest is the gainer, since he will get more. If one sets aside bread, however, the priest will be the loser, unless the owner calculates exactly the amount of wheat in the loaf of bread, and not its volume. The same applies to other kinds of produce. In some texts "unless" is omitted, evidently by error.

19. Ter 1:8.

"pressed olives . . . trodden grapes"—from which the oil and the juice, respectively, have not yet been drawn off. This is an intermediate stage between raw produce and processed produce, and thus requires a rule to prevent any doubt as to the proper procedure.

"but this is not so with the second heave offering"—which is prescribed by Scribal law only.

20. Ter 1:9.

"out of oil for olives"—the reason in all these cases is that when the offering was set aside, the preparation of the produce was thought to be complete, and a change of mind afterward cannot affect it.

21. Tos Ter 4:6, 8; B. BB 84b.

"One may not set aside," etc.—on the principle that one may not set aside heave offering out of inferior produce for superior produce, but may sometimes do so out of superior produce for inferior produce. See above, Sec. 3.

"If his intention was . . . it is invalid heave offering"—this sentence is omitted in some texts, apparently by error.

22. Ter 3:1; P. Ter 3:1; Tos Ter 4:6.

"one must . . . set aside heave offering once more"—because if the wine set aside as heave offering had turned into vinegar before being set aside, the wine for which it was set aside has not been regularized properly.

"punctured"—by a snake, and therefore potentially poisonous.

23. Ter 3:2.

"in proportion"—for example, if the larger offering amounts to two sĕ'ah, and the smaller one to one sĕ'ah, and both, amounting together to three sĕ'ah, fall into one hundred sĕ'ah of other produce, the proportion is one to one hundred, not two or three to one hundred (PhM). Since both offerings are subject to doubt, neither of them alone can affect the other produce; but if both of them fall into the same lot of other produce, they do affect it, since one of them must be valid.

24. Tos Ter 4:8; B. BB 96a.

"heave offering for other wine"—i.e., he has other wine in process of preparation, and as it matures and thus becomes liable to heave offering, he means to set aside the amount due from the one jar, and to continue to do so until the entire jarful becomes heave offering for all the other wine.

"for three days thereafter"—during that time the wine in the jar is presumed to have remained unchanged. Beyond that the matter becomes doubtful, and much depends on atmospheric conditions. Cf. the next Section.

25–26. Giṭ 3:8.

25. "At three times"—because during these times wine is particularly liable to turn sour. At other times the assumption is that wine will remain sweet.

26. "except out of adjacent produce"—above, iii, 17; here he has evidently set aside the retained produce for produce that is not adjacent.

"the assumption is that the retained produce is still in existence"—the situation is as follows; One has on hand several heaps of produce. He designates heap A to serve as collective heave offering for all the other heaps, B, C, D, etc. When he comes to heap B, he designates part of heap A as heave offering for it, and the same for heaps C, D, etc., until all of A becomes heave offering. Now the problem is that the while he was relying on A, it may have wasted away and is thus no longer available as heave offering for the other heaps. The rule therefore is that normally it may be assumed that heap A is still intact, and one may lawfully consume the other heaps. If, however, it is discovered that heap A has indeed wasted away, even though it cannot be determined exactly when this wastage has occurred, heaps B, C, etc. are presumed to have had no heave offering set aside in their behalf, and the offering must be set aside for them once more.

"there are grounds for apprehension"—that the other produce may not yet be lawfully consumed.

Chapter VI

1. Num. 18:19; Sif Lev. 5:1; B. Giṭ 12b; Tos Ter 10:8; B. Keṯ 24b f.

2. B. Keṯ 24b; Ḥul 130b; Ḥal 4:9.

"a priest of proven priestly descent"—who has witnesses to attest his priestly descent, or documentary proof of it.

"Priests by presumption"—who have no proof of their priestly lineage, except that they have always been accepted as of priestly descent.

"a priest who is a disciple of the wise"—and is not an unlearned person. This is quite discriminatory. The sources, however, indicate that all priests were in fact entitled to these offerings; see Ḥal 4:9; P. Ḥal 4:9. Maimonides' opinion of the unlearned is well known, but he did not allow this feeling to affect his legal decisions. It has therefore been suggested that this rule simply expresses a preference, meant to prevent defilement of heave offering by a poorly informed priest, as the text does indeed suggest.

"all unlearned persons are presumed to be unclean"—because the rules of ritual uncleanness are so minute and involved that unlearned persons have difficulty in adhering to them.

3. Nid 5:4; Keṯ 5:7; B. Keṯ 5:7b.

"of the breast and thigh"—of peace offering; see VIII, v, x, 4 (YJS, *12,* 202–03).

"she comes at that moment into her husband's possession"—cf. IV, 1, iii, 11 (YJS, *19,* 17–18).

4. B. Yeḇ 113a.

"even if her father had given her in marriage"—which is a binding marriage according to the law of the Torah, giving her the same rights as those of any other married woman. See IV, i, iii, 11 (YJS, *19,* 17).

"lest a deaf-mute priest should marry a deaf-mute commoner"—a binding marriage according to Scribal law only; inferred from IV, 1, iv, 9 (YJS, *19,* 25).

5. B. Sanh 83b; Yeḇ 70a; Yeḇ 9:4; B. Yeḇ 68a.

"A Hebrew bondsman"—who had sold himself into servitude and must be emancipated after six years (Exod. 21:2); he is therefore in fact a temporary servant and not a permanent bondsman (Lev. 25:39–40).

6. B. Sanh 83b; Tos Ter 6:4; Ter 6:1; P. Ter 7:1; B. Keṯ 30b.

"death at the hand of heaven"—not at the hand of a human court. This usually implies that he will die prematurely.

7. B. Yeḇ 68b; Yeḇ 10:1.

"Two matters"—the two instances when the daughter of a priest forfeits her priestly privilege.

"as we have explained . . . Intercourse"—V, 1, xviii, 1; xix, 1 (YJS, *16,* 113, 121).

"the breast and the thigh"—see above, note to Sec. 3. Once she is married to a commoner, she forfeits this right forever. If she is divorced or widowed, however, she regains her right to eat of heave offering (see the next two Sections).

8–9. B. Yeḇ 69a.

10. B. Yeḇ 68a.

"by a priest"—to whom she was lawfully married. See below, Sec. 12.

11. 'Ed̲ 3:6; B. Ket̲ 36b, 22a; B. Yeb̲ 59b.

"A captive woman"—is regarded as a harlot out of doubt, since the possibility exists that she had been raped by her captors. She must therefore cite some evidence that she had not been so molested, and in her case the rules of evidence have been considerably relaxed, one witness being sufficient, who may be even a child, a bondsman, or a relative, and two captive women may vouch for each other; see V, I, xviii, 17–18 (YJS, *16,* 117–18).

"who is to be believed"—by virtue of some evidence other than that of a witness.

"A woman who has lain with a beast"—even though she is liable to death for it (Lev. 18:23).

12. Yeb̲ 9:5; B. Yeb̲ 72a, 70a.

"and have no child"—Hebrew *wĕ-zera' 'en lah;* the Talmud suggests that *'en lah* might be read *'ayyen lah,* "inquire concerning her" whether she has any descendants, however distant.

13. Yeb̲ 9:15, 7:5; B. Yeb̲ 70a.

"disqualifies her"—from eating of heave offering.

"blemished"—and therefore disqualified from the priesthood.

"entitles his late mother's mother"—although the grandchild is disqualified, his grandmother is regarded as having priestly progeny, and in the one case she is qualified to eat of heave offering, while in the other case she is disqualified (see the next Section).

14–15. Yeb̲ 7:5, 6.

14. "If his maternal grandmother"—this is a continuation of the preceding Section.

"or even is not Israelite at all"—Hebrew *'eno;* perhaps read *me-'eno,* "or even was sired by a non-Israelite father."

16. *Ibid.;* B. Ḳid 68b.

"bondsmen have no genealogy"—i.e., the child is not regarded as the descendant of his natural father because his mother is a bondswoman. Hence his grandmother is not affected by the child's survival.

17–19. Ter 9:5.

17. "she may not eat . . . on account of her son by the Israelite"—because her remarriage to the Israelite has severed her close relationship to her first priestly husband, even though she has a son by the latter.

"If . . . her son by the Israelite . . . dies"—thus restoring her close relationship to her first priestly husband.

18. "but not of the breast and the thigh"—see above, Secs. 3, 7, 9. If she chooses not to return to her father's house, she still may eat of heave offering on account of her son by her priestly husband, and may also eat of the breast and the thigh.

Chapter VII

1. Yeb̲ 8:1; B. Sanh 83a; Yeb̲ 74b, 73b; Tos Ter 6:4.

"death at the hand of heaven"—and not at the hand of a human court.

"is therefore liable to a flogging"—at the hand of a human court; the rule is that those who incur the penalty of death at the hand of heaven must be flogged. See ShM, Negative Commandment 133.

"even though this is forbidden by a negative commandment"—the transgression of which is usually punishable by a flogging.

2. Ber 1:1; B. Ber 2a f.

"and it is clear"—it referring to the day; so interpreted by the Talmud, which quotes an Aramaic proverb, "When the sun is set, the day is cleared (i.e., gone)." The usual translation of this clause is, *he* (*the unclean priest*) *shall be clean.*

3. P. Bik 2:1; B. Yeb̲ 73b.

"this is forbidden by a positive commandment"—hence the transgressor is not liable to a flogging.

4. Nid 5:1–2.

"take firm hold"—to prevent the emission of the semen or to delay it until after the food is swallowed.

5. B. Nid 13b.

"a copper sheath"—for the genital organ. A copper sheath was used because it is not absorptive, and should there be a discharge, it would be visible.

6. B. Nid 14a.

"upon the bare hide of a camel"—without the customary saddle.

7. B. Nid 42a.

"make violent motions"—literally "turn herself over," in order to eject the semen and thus avoid conception; cf. IV, I, xiv, 5 (YJS, *19,* 88, 395).

"as shall be explained"—X, v, v, 11 (YJS, *8,* 272).

8. B. Bek̲ 27a.

"the root of whose sanction is Scribal authority"—according to the Torah, only produce of the Land of Israel is subject to the law of heave offering. See above, i, 1.

"whose uncleanness is issuing out of his own body"—see X, IV, i, 1; v, 1 (YJS, *8,* 207, 269–70).

"nocturnal pollution"—emission of semen while asleep.

"even if that day's sun has not yet set"—normally they have to wait until after sunset. See above, Sec. 2.

"from which it is impossible for us to be cleansed today"—because this involves the use of the ashes of the Red Heifer, which is nonexistent today. See Num. 19.

9. *Ibid.;* B. Pes 67b.

"a priest of proven genealogy"—in contradistinction to a priest by presumption; cf. above, vi, 2; and V, I, xx (YJS, *16,* 127–33).

10–11. Yeb̲ 8:1; B. Yeb̲ 70a, 72a; Shab 135a.

10. "who has had his prepuce drawn forward"—who has had a fold of skin drawn over the *glans penis* to disguise his circumcision.

"to be circumcised a second time"—before he may eat of the offering.

11. "ṭumṭum"—since his sexual organs are covered over.

12. Yeb̲ 8:1.

"their wives and bondservants"—while these priests themselves may not eat because of their personal blemish, they still retain their priestly status and confer the right to eat upon their dependents.

13. *Ibid.;* B. Yeb̲ 57a.

"their wives may not"—because they became profaned by marrying a priest so injured; see Deut. 23:2.

"If such a priest has had no sexual intercourse"—obviously in his case the marriage was contracted before the husband was injured and was consummated normally.

"if such a priest marries a proselyte's daughter, she too may eat of it"—since the man does not lose his status as a priest, and is merely not permitted to marry a daughter of Israel.

14. B. Yeḇ 56b, 72a, 83a.

"she may not eat"—because such a marriage makes the wife a profaned woman.

"a congenital eunuch"—is not forbidden to marry (IV 1, ii, 10; YJS, *19,* 25).

"but not their wives"—since their marriage is of doubtful validity, owing to the husband's indeterminate sex (IV, 1, ii, 11; YJS, *19,* 26).

15. Tos Yeḇ 9:2; Nid 5:3.

"these bondservants may not eat"—because deaf-mutes, imbeciles, and minors are incompetent to effect legal purchase (XII, 1, xxix, 1; YJS, *5,* 102).

16. P. Yeḇ 8:6.

"the same as do women"—who, if they have had intercourse with disqualified men, become themselves disqualified.

"no male may disqualify another male"—in this case, the assumption is that both hermaphrodites are acting as males, since if both are acting as females, there is no valid intercourse.

17. P. Ter 11:5; B. Yeḇ 81a.

18. Yeḇ 7:1–2; B. Yeḇ 66a.

"they may eat of heave offering"—in both cases they are in the priest's custody. In the case of his wife's mĕloḡ property, he is entitled to the usufruct thereof; in the case of her iron sheep property, it is actually his so long as the marriage endures; cf. IV, 1, xvi, 1–2 (YJS, *19,* 98–99).

"bondservants buy bondservants"—a bondsman's property belongs to his master.

19. Yeḇ 7:1–2. In this case neither the husband nor the wife is entitled to eat of heave offering; hence their bondservants may not eat of it either.

20. *Ibid.;* B. Yeḇ 85a; P. Yeḇ 9:4; Keṯ 11:7.

"a widow married to a High Priest," etc.—all these marriages are forbidden, but if contracted are valid, and require a geṭ for their dissolution.

"forbidden . . . by a negative commandment"—in which case the marriage is valid, but the wife is rendered a harlot. See V, 1, xviii, 2–3 (YJS, *16,* 113–14).

"secondary degree of consanguinity"—forbidden by Scribal law only. See IV, 1, i, 6 (YJS, *19,* 6).

21. Yeḇ 7:1; B. Yeḇ 57b.

"a priestly widow"—who is entitled to eat of heave offering.

"she is awaiting intercourse"—although she has not yet been profaned; therefore, if these women subsequently become separated from their husbands by divorce or death, prior to the consummation of the marriage, they regain their former status. See below in this Section.

22. B. Yeḇ 58b.

"her levirate tie to the profaned priest"—once she becomes a childless widow, she is bound by levirate tie to all the surviving brothers-in-law. Since one of them is a profaned priest, her tie to him deprives her of the right to eat of heave offering.

"verbal declaration"—Hebrew *ma'ămar,* a declaration of intention to marry made by the levir to his widowed sister-in-law; it alone does not complete the levirate marriage. See IV, III, ii, 1 (YJS, *19*, 271).

23. *Ibid.;* Yeḇ 6:4, 5.

"forbidden to him"—because the geṭ makes her forbidden to a priest.

"still tied to him"—the rule is that the levir must submit to the rite of drawing off the shoe (*ḥăliṣah,* Deut. 25:5 ff.), if he wishes to be released from the obligation of levirate marriage. If instead he gives a geṭ to his sister-in-law, he thereby renders her forbidden to him, since a priest may not marry a divorcée; at the same time he does not yet release her from her levirate tie to him. See IV, III, v, 1 (YJS, *19*, 294–95).

"Similarly . . . may eat"—here, too, her disqualification is by Scribal law only.

"marries a barren woman"—such a marriage, though discouraged, is valid.

Chapter VIII

1. Yeḇ 7:4.

"An unborn child"—literally "a fetus."

2. B. Yeḇ 67b; Nid 44a; Sif Lev. 22:13.

"by a priest . . . by an Israelite"—who had since died.

3. Yeḇ 7:5; B. Yeḇ 69b.

"she may have become pregnant"—which would disqualify her from eating of heave offering.

4. B. Yeḇ 67a; Nid 44a.

"who died"—without living issue.

5. B. Yeḇ 67b, 68a.

"under levirate tie to a priestly levir"—i.e., her husband died without issue, and she is now awaiting levirate marriage.

6. B. Yeḇ 56a, 54a.

"widowed sister-in-law"—who is an Israelite woman.

"as we have explained"—IV, III, ii, 3 (YJS, *19*, 272).

7. B. Yeḇ 54b, 68a.

"as we have explained"—above, vi, 3.

8. Ḳid 3:1.

"Behold, thou art consecrated unto me," etc.—in the first case the betrothal becomes effective thirty days hence; in the second case the formula "as of now, and after thirty days" is ambiguous: it may mean, "as of now, provided that I do not change my mind within thirty days"; or "and after thirty days" may be meant as an amendment of "as of now." Because it may be the first, which makes her betrothed immedately, she is forbidden to eat of heave offering as of that moment. See IV, I, vii, 10, 12 (YJS, *19*, 45).

9. B. Yeḇ 68a, 112b.

"enactment of the Sages"—see IV, I, iv, 9 (YJS, *19*, 25).

10. B. Yeḇ 56a.

11. B. Yeḇ 68a; Soṭ 26b; Yeḇ 7:4.

"is nine years and one day old"—see IV, I, xi, 3 (YJS, *19*, 68).

"as we have explained"—above, vi, 7.

"the statutory two hairs"—the tokens of puberty (IV, I, ii, 10–11; YJS, *19*, 10–11).

12. B. Yeb̲ 113a.

13. B. Yeb̲ 112b; Yeb̲ 7:5.

"The following persons," etc.—in each case the woman is a priest's daughter and the man is an Israelite, or vice versa. In the former case he could otherwise deprive her of the privilege; in the latter, he could otherwise confer it upon her. In neither case is there a valid marriage.

"as we have explained"—above, vi, 7.

14. Yeb̲ 7:5; B. Yeb̲ 69b.

"on account of her child"—who is accounted a priest in spite of the irregularity of his conception.

"on account of the fetus"—which is Israelite, and not priestly (above, Sec. 2).

15. Soṭ 1:3; B. Ned̲ 90b; Soṭ 3:6.

"attested by witnesses"—see IV, v, i, 1 (YJS, *19,* 343).

"until she has drunk of the water of bitterness"—and thus proven her innocence or guilt; see IV, v, iii (YJS, *19,* 353–59).

"If her husband dies"—see IV, v, ii, 7 (YJS, *19,* 350).

"neither drink nor receive their kĕt̲ubbah"—see IV, v, ii, 2 (YJS, *19,* 348–49).

"I have been defiled"—meaning "I have committed adultery." In this case no witnesses are available.

16. B. Ḳid 45b.

"minor . . . female"—two acts are involved here: *ḳiddušin,* betrothal, i.e., acceptance of the token of betrothal; and *niśśu'in,* nuptials, marking the completion of the marriage. In this case the father was party to the first act but not the second. During the second act he stood silent, which is usually interpreted as signifying consent, but here is interpreted as possibly signifying resentment. Since the bride is still a minor, her father's vocal protest could invalidate both acts.

Chapter IX

1. Giṭ 6:4; B. Giṭ 28a.

"A woman"—an Israelite woman married to a priest.

"receiving agent"—who has the same power as the woman herself; the moment the geṭ reaches his hand, it is the same as if it had reached her own hand, and she is validly divorced.

"doubtfully divorced may not eat of heave offering"—although her priestly husband is still obligated to provide her with food other than heave offering (IV, 1, xviii, 25; YJS, *19,* 120).

"is immediately thereafter forbidden"—not exactly, since the agent must be given enough time to execute his mission. The import of the statement is that the agent is presumed to have executed his commission without unnecessary delay.

"an agent to fetch"—in contradistinction to the receiving agent, the fetching agent's mission is limited to conveying the geṭ, and she is therefore considered divorced only when the writ reaches her own hand.

"is immediately forbidden"—since for all he knows his death may occur at any moment.

2. Giṭ 3:4; P. Giṭ 3:4; B. Giṭ 29a.

"sea voyage . . . caravan"—upon a calm sea and across a peaceful desert, respectively, where danger, while possible, is not imminent.

"led to execution"—in all these three cases death is probable but not absolutely certain: the man might have escaped from the conquered city, been cast ashore by the sea, or been released by bribed heathen court officials.

"condemned by an Israelite court"—which is not likely to pervert the law or be swayed by bribery.

3. B. Giṭ 28a; Yeḇ 15:5.

"in a state of imminent death"—the Hebrew is *goses,* the term for the final death throes.

4. Yeḇ 15:4, 6.

"the five women"—her mother-in-law, her mother-in-law's daughter, her co-wife, her husband's brother's wife, and her husband's daughter. See IV, II, xii, 16 (YJS, *19,* 251).

5. Giṭ 1:6; B. Giṭ 42b.

"in the Laws Concerning Bondservants"—XII, v, v, 3; viii, 13–17 (YJS, *5,* 265, 277–78).

6. B. BB 138a; Ḥul 39b. Cf. XII, II, iv, 3 (YJS, *5,* 121–22); the transfer is by way of gift.

"remains silent"—indicating consent to accept the gift.

"a change of mind"—which is invalid, since B's previous silence has established his consent, so that the estate has already become his property.

7. Ter 11:9; B. AZ 15a.

"he may feed it heave offering"—because the beast is still the property of the priest, and hiring it out has not altered its status.

8. Ter 11:9.

"accepts a cow"—the procedure is as follows: A conveys a cow to B at an agreed price, with the stipulation that B is to fatten her up and sell her, and that A and B are then to share in her increased value. The cow thus actually belongs to B, and if B is a priest, she may eat of heave offering, while if B is an Israelite, she may not. Such a transaction is lawful, and is not regarded as a usurious loan (cf. XIII, III, viii, 12; YJS, *2,* 108–09).

"he had set a value upon her for himself"—that is, he had actually bought her, although with the proviso that both are to share in her increased value.

9. Tos Ter 10:7.

"may store . . . vetch . . . in his dovecote"—the emphasis is on "store"; his intention is not to feed it to the doves, whom he is not obligated to feed anyway, since they are not regarded as domesticated fowl. Hence he is not responsible if the doves happen to eat of the sacred vetch, which is normally not used for human food (cf. above, ii, 2).

10. B. BM 47b.

"act of drawing"—the property to be acquired toward himself, thus signifying that he has legally taken possession of it. See XII, I, ii, 6 (YJS, *5,* 9–10).

"as will be explained"—XII, I, iii, 1 (YJS, *5,* 11).

Chapter X

1. Ter 6:1.

"the death penalty"—at the hand of heaven.

2. *Ibid.*; B. Nid 32a; Tos Ter 7:3.

"which includes the person who anoints himself"—according to the Talmud, this verse, seemingly superfluous, is intended to add something not included in the previous verse which mentions eating. It could not mean drinking, since drinking is included in eating. Hence it must refer to anointing. See B. Nid 32a.

3. Tos Ter 7:3.

"half a loaf"—the customary daily ration given to a bondservant or a member of the household.

4. Me 4:2; P. Dem 4:2, 1:2.

"liability to the death penalty"—KM calls attention to the fact that this cannot apply to doubtfully tithed produce, which is liable to heave offering by Scribal law only.

"the term 'heave offering' is used with reference to all of them"—the term "heave offering" (*tĕrumah*), or its derivative, is used in each case (dough offering, Num. 15:20; first fruits, Deut. 12:17).

"as will be explained"—below, v, v, 4.

5. Ter 7:1; Tos Ter 7:2; B. Pes 32a.

"he was forewarned . . . he was not forewarned"—some texts omit this part of the sentence.

6. B. Pes 32a.

7. Tos Ter 7:1; B. Ḥul 9b; P. Ter 6:1; B. Ber 37a.

"punctured . . . uncovered"—in both cases consumption is forbidden, because a snake may have bitten into one or drunk of the other, and thus poisoned it.

"wine and oil together . . . oil and vinegar together"—his individual custom being to drink these together, an unusual procedure.

"chews wheat grains"—unripe ones, which are ordinarily not eaten.

"swallows vinegar"—likewise an unusual way of drinking it neat.

8. B. Yoma 80b.

9. B. Yoma 81a.

"the former must pay"—swallowing is accounted the same as eating, even though no chewing has taken place.

"their value as firewood"—see above, Sec. 5.

10. Ter 6:3; P. Ter 6:2.

"the value of their meal"—the cost of their meal, had it been prepared out of profane produce, which is higher than the cost of the same meal prepared out of heave offering produce, since the latter usually sells for less than profane produce.

"inasmuch as one recoils in awe"—this is the reason for penalizing the employer or host and making him pay more than the real cost of the meal. He has made them eat something abhorrent to them, and it would not be right that he should be the gainer, i.e., he was supposed to feed them profane food which is more expensive, yet he fed them heave offering which is less expensive.

11. Ter 7:3; Tos Ter 7:1; B. Ber 35b.

"minor"—who is not subject to the law, hence is not liable to pay.

"bondservants"—have no property rights, hence cannot pay.

"heave offering imported from outside of the Land of Israel"—so classified only by Scribal enactment.

"a Nazirite"—is forbidden to drink wine, and his drinking of it is therefore not amenable to monetary compensation.

"drinks . . . oil . . . anoints himself with . . . wine"—this is abnormal usage. See above, Sec. 2.

12. Ter 7:2, 8; Giṭ 1:6; B. Pes 72b.

"disqualified from the priesthood"—by marrying a man who is disqualified from marrying into the priesthood. See above, vi, 7.

"must pay the principal thereof but not the additional one-fifth"—which is paid only by a person totally alien to the priesthood. Besides, a priest's daughter may yet qualify again to eat of heave offering on her return to her father's house after her husband's death (Lev. 22:13). Since the sanctity of priestly stock thus clings to her, she is not deemed totally estranged from heave offering. She must repay the principal, however, inasmuch as she has eaten what was not rightfully hers.

"a woman . . . a bondservant . . . a priest"—in all these cases, as in the preceding one, there is an ingredient that ties them to some extent to the priesthood.

"son of a divorcée," etc.—and is therefore a profaned priest forbidden to eat of heave offering.

"they are exempt"—because it is the same as if they had eaten it under duress.

13. Ter 8:2.

14. Ter 7:5.

15. Ter 6:1; B. BḲ 65b; Ter 7:1, 4; Ter 6:1.

"if the priest is willing to waive his right to it, he may not do so"—because the payment constitutes not restitution but expiation. See B. Ket̠ 30b, and Tosafot̠ *s. v. Zar*. The Mishnaic source says that the priest may waive his right, but this refers to the case when heave offering was eaten deliberately. See the next note.

"Whosoever pays the principal alone"—after having deliberately eaten of heave offering.

16. Ter 7:2.

"not yet acquired by a priest"—the Israelite farmer had already separated it but has not yet given it to a priest. Her marriage to a priest entitles her henceforth to eat of heave offering, which is why the payment is made to herself.

17. P. Ter 6:1.

"in either case"—whether the offering has already been given to a priest or not.

18. Ter 6:1. B. BM 54a; P. Ter 6:5; B. Ber 47b.

"from which heave offering has already been separated"—that is, heave offering of the tithe. See above, iii, 13.

"even if the redemption was made not according to the rule"—see below, v, iv, 9.

"new produce for old"—see above, v, 11.

19. Ter 6:6.

"must wait," etc.—in the meantime chate melons of the sixth year, having become hard, are anyway no longer fit to be eaten, while from those grown during the Sabbatical year no benefit whatsoever may be derived. Payment, which must be of the same kind, can therefore be made only with chate melons grown after the seventh year.

"as will be explained"—below, VII, vi, 10.

20. B. Yeḇ 90a; Ter 6:1.

"his payment is valid"—and the unclean profane produce becomes unclean heave offering, which the priest cannot use. The payee must therefore make a second restitution out of clean profane produce.

21. Tos Ter 7:5; P. Ter 6:1.

"a learned priest"—literally "[a priest who is] an associate scholar (*ḥaḇer*)."

"clean things may not be delivered to an unlearned person"—since he will surely defile them, which is forbidden.

22. P. Ter 6:1; Tos Ter 7:5.

"if one steals . . . if one inherits"—in both cases he is not permitted to eat of it. The difference is that in the former case the offering came into his possession illegitimately, while in the latter case he acquired it legitimately. Hence in the former case no benefit may accrue to him.

23. Ter 6:4.

"with the monetary equivalent"—he may pay it in money and not necessarily in kind.

24. *Ibid.;* B. Pes 32b.

"as will be explained"—XI, II, ii, 1 (YJS, *9,* 64).

25. "take hold"—referring to the preceding Section dealing with consecrated heave offering, where the law of consecrated things prevails over the law of heave offering, and the payee is exempt from double restitution.

"Laws Concerning Forbidden Intercourse and Forbidden Foods"—V, I, xvii, 9–10; II, xiv, 19 (YJS, *16,* 110, 230). The second prohibition can take hold only if it adds something new to the already existing first prohibition.

26. P. Ter 6:1; B. BM 54a; Pes 32a.

"worth five coins"—the one-fifth being reckoned of the grand total; today we would say that the penalty is one-forth of the value of the eaten produce.

Chapter XI

1. Sheḇi 8:2; Tos Ter 9:10; Shab 9:4; B. Yoma 76a; Ter 11:10; B. Shab 24b.

"wine or vinegar"—which are not normally used for anointing.

2. Ter 11:3, 2.

"the only exceptions are olives and grapes"—in Scripture (Num. 18:12–13) these are included in "fruits." The reasoning behind this rule is that since olives and grapes are for the most part converted respectively into oil and wine, these are their natural forms.

3. Ter 11:1.

"fig cakes," etc.—of heave offering.

"brine"—*muries,* the Latin term for salted water in which chopped fish or locusts are pickled. Occasionally wine is added to it.

"this spoils them"—after soaking in the brine the figs were squeezed out and thrown away, having served their purpose to sweeten the brine and become inedible.

"removes it from the category of food"—thus causing depreciation of heave offering, which is forbidden.

"wine, honey, pepper, and the like"—making a spicy beverage called *'inomilin* (Greek *oinomeli,* "honey-wine").

4. *Ibid.;* Tos Ter 9:4, 10:15.

"reduces it . . . spoils the vinegar"—heave offering must not suffer damage either by reduction of volume or by diminution of quality. Moreover, condensed wine is too potent.

"One may not mix grain with pulse"—because this reduces the value of the grain, pulse being cheaper.

"which become separated"—and thus can be easily identified and kept apart.

"Since the devastation of Judea"—when all produce became scarce, and the impoverished people could not be particular about mixing several species and grinding them together.

5. Ter 11:5.

"inasmuch as it is edible"—the residue contains coarse flour as well as bran, and may not therefore be wasted.

6. Tos Ter 10:11.

"One may not use heave offering oil"—the principle here is that heave offering oil may be used only for anointing the human body. If by chance other objects are anointed by coming in contact with the body, there is no objection.

"a new leather spread"—which absorbs much oil; see Lieberman's commentary to Tos *ad loc.* Marble is not absorbent, and some oil would run off.

7. Ḥal 1:9, 4:5; B. Ḥag̃ 18b; Tos Ter 10:11; B. Yeḇ 71a; P. Yeḇ 1:1.

"even if just fruit"—which usually does not require washing of hands (II, v, vi, 3).

"as will be explained"—X, v, viii, 8 (YJS, *8,* 282).

"he may rub it in with unclean hands"—because the oil has already become profaned.

"within seven days of his birth"—when he is still uncircumcised and thus should be forbidden to touch holy oil. See above, vii, 10.

8. Tos Ter 10:11; B. Ker 7a.

"take his daughter's Israelite son"—the most common occurrence of such intimate and loving care, according to Rashi *ad loc.;* the rule applies equally to any other unpriestly person.

"a commoner may massage him"—because the oil has already become profaned. The Hebrew verb *lĕ-mašḥo* means literally "to anoint," and KM emends it to *lĕ-mašmĕšo,* literally "to feel with one's hands."

9. Tos Ter 10:4.

"roll her hair"—since the fenugreek has already soaked into the priestly woman's hair, it has become profaned and permitted for use by commoners.

"Because fenugreek is not fit for human consumption"—and may therefore be used for other purposes without fear of wasting heave offering. See, however, below, xii, 7, and above, ii, 8, where it is stated that young fenugreek is used for human consumption, which is why it is subject to heave offering and tithe.

10. Ter 11:4; Tos Ter 10:2; P. Ter 11:4.

"inferior figs"—*ḳĕlisim,* translation suggested by the context. The Soncino Talmud and Danby's Mishnah render it "acorn." Feliks regards it as a species of figs. Rashi thought it was a kind of pea or bean; others thought it was the fruit of the carob tree. Cf. above, note to iii, 3.

"vegetable trimmings discarded by gardeners"—who trim away only those parts that are completely useless.

11. Tos Ter 10:2; Ter 11:5; P. Ter 11:4.

"Citron seeds are permitted"—because they are bitter and inedible.

"even if not gathered by a priest"—who has thus indicated that he did not care for them; nevertheless they are forbidden to commoners, since they may have some meat attached to them.

12. Ter 11:4; P. Ter 11:4; Tos Ter 10:4. New bran is soft and moist, and some of the edible wheat may adhere to it.

13. Tos Ter 10:3, 12; B. BB 97a; P. Ter 11:4.

"the third is permitted"—since there is little taste of wine in it.

"even the third filtering is forbidden"—because it is still undiluted raw wine.

14. Ter 11:6.

"tap it with his hand"—upon the ground, to wipe away any oil that still remains on the surface (PhM).

15. Ter 11:8.

"emptying"—from one jar to another.

"the oil residue is still heave offering"—and it cannot be said that one may disregard it after the third drop.

16–17. Tos Ter 10:9.

17. "may even dip," etc.—concern for the Israelite's personal safety overcomes the prohibition (which is only Scribal) to enjoy the benefit obtained from indirect consumption of sacred oil.

18. Tos Ter 10:9; P. Ter 11:5; B. Shab 23b.

"the Israelite may kindle"—since the priest too benefits from this light.

"in synagogues," etc.—since the prohibition of the use of unclean heave offering oil by an Israelite is only Scribal, an exception is made in such cases where the public welfare is involved.

"for the sick"—i.e., Israelite sick. Here only an individual's welfare is involved, hence the priest's permission is necessary.

19. Ter 11:10.

20. Ter 9:1; P. Ter 9:1; B. BB 93a; Tos Ter 8:1.

"he must plow it under"—without incurring liability for destroying heave offering.

"he must allow it to remain"—as a penalty, since the produce will assuredly be forbidden to him, and he will have to sell it at a lower price, seeing that heave offering produce fetches a lower price than profane produce, because the owner cannot use it himself.

"If it is already one-third grown"—having attained one-third of its growth, it is now fit for food, and were he to plow it under, it would appear as if he had destroyed heave offering.

"If it is flax"—which is inedible. The additional stringency in the case of flax is meant to prevent him from deriving benefit from the stalks, on the plea that only the seeds are heave offering and forbidden and not the stalks (cf. below, Sec. 22), whereas in fact the chief part of flax is precisely the stalk and not the seed.

21. P. Ter 6:1; Tos Ter 8:2; B. Shab 17b.

"is the same as profane produce"—i.e., is subject to gleaning, forgotten sheaf, corner crop, heave offering, and tithes (see below, Sec. 27).

"therefore"—since the outgrowth has in reality the same status as profane produce.

"with unclean hands"—without one washing his hands first.

"after a day's uncleanness"—cf. above, vii, 2.

22. Ter 9:4; B. Neḏ 59b; P. Ter 9:2; Tos Ter 8:2; Ma 5:8; B. BB 93a.

"in every respect"—and may even be consumed by commoners. See the preceding Section.

"the basic root"—i.e., the seed.

"heave offering imported"—whose sanctity is of Scribal sanction only.

"slightly intermixed"—but more than 1/100 by volume.

"supplementary heave offering"—referring to the case where one sets aside less than a minimum, and must then add to it; see above, iii, 6.

"heave offering flax"—whose seeds are likewise inedible.

23. Ter 9:7; Sif Lev. 11:30; P. Ter 9:3; Pes 34a.

"forbidden for consumption"—even by a priest.

24. Ter 9:7; P. Ter 9:3.

"is permitted for consumption"—by a priest, not by a commoner, since this is not a second outgrowth of the plant itself, as in Sec. 22, above.

25. P. Ter 9:3.

"because they have been cleansed"—cf. above, Sec. 23.

26. B. Men 70a; Ter 9:6.

"and if one smooths down the entire heap"—which makes it liable to tithe.

"has flown away from it"—has departed from it, since planting sometimes changes the status of produce.

"punishment by death"—at the hand of heaven.

27–28. Ter 9:2–3.

27. "is subject to gleanings," etc.—since according to the law of the Torah it has the status of a field of profane produce (above, Sec. 21).

"at the price of heave offering"—which is lower than the price of profane produce.

28. "baskets"—open baskets out of which the beasts can feed.

"muzzles the beasts"—which is forbidden (Deut. 25:4).

Chapter XII

1. Ter 8:1; B. Zeḇ 33b; Yeḇ 75a; Beḵ 34a, 27a; Shab 15a.

"nor may one cause"—indirectly. In the previous instance direct defilement is meant.

"defiled only in foreign lands"—by the famous decree which declared foreign lands to be unclean. One may now add to it a defilement of a higher category, one declared to be so in the Torah. This is permitted, inasmuch as in foreign lands heave offering is such by Scribal law only.

"as we have explained"—above, vii, 8.

2. Ter 8:8; B. Shab 15b.

"there are . . . some doubtful cases"—the word *šam,* "there," is emended by some to *šiššah,* "six," referring to the six doubtful cases mentioned in Toh 4:5.

"as will be explained"—X, v, xiii, 13 (YJS, *8,* 304).

3. Ter 8:8.

"neither move it"—to protect it from defilement; once a doubt has arisen, it no longer requires the protection due to the sacred nature of heave offering.

"nor uncover it"—to make it assuredly defiled.

4. Ter 8:9; B. Pes 20b.

"the lower part thereof is unclean"—the vat underneath the press itself contains profane, not sacred, wine.

"If one is able to save"—by using clean vessels and by quickly washing his hands.

"even though he thereby renders it unclean"—but at the same time saves the unclean profane wine from being mixed with unclean heave offering wine and thus rendered forbidden for consumption.

"as will be explained"—X, v, viii, 8 (YJS, *8,* 282).

5. B. Pes 20b, 21a; Ter 8:10, 11.

"subject to heave offering and . . . disqualified"—and the loss is then great.

"so that the proportion is one hundred and one to one"—in which case the wine from the jar is neutralized; see below, xiii, 1.

"he must act in haste"—and will act on impulse rather than adhere strictly to the rules.

6. Ter 8:11. Here, in contrast to the case in Sec. 5, no extreme haste is required to save the remaining loaves from defilement; hence one must observe the rules.

7. MSh 2:3–4; Ḥal 4:9.

"being unfit for human consumption"—see above, note to xi, 9.

"to an unlearned priest"—ordinarily heave offering may not be given to an unlearned priest (above, vi, 2).

8. B. Giṭ 16b.

"he is apt to consume it"—and will thereby defile it.

"covered with a tightly fitting lid"—which protects it from defilement.

"open to uncleannness"—Hebrew *mukšarin,* literally "are rendered fit or capable," and refers to the principle laid down in Lev. 11:34, 37, 38 that no food or produce of the soil can contract uncleanness unless at some time or another it had been moistened by water.

"lest his wife should shake it"—a menstruating woman defiles food not only by touching it but also by indirectly causing it to shift. See X, IV, viii, 2–3 (YJS, *8,* 232–33).

9. B. Giṭ 62a.

"One may not handle," etc.—an oil press operator working for an unlearned person should not in a clean state take part of the contents of the press and set it aside as heave offering, because people who see this will think that the rest of the olives is also clean, which is not so, because an unlearned person does not know how to keep things in a state of cleanness.

"one should tell him," etc.—the unlearned person is not careful about ritual cleanness but is careful about untithed produce. One should warn him therefore in a manner that will have an effect upon him, by telling him that the offering will revert to its untithed state (which is not true), rather than that it will become unclean (which is true, but will have no effect upon him).

10. B. AZ 55a; Shab 17a.

"One may not," etc.—because one thereby aids in the performance of a forbidden act.

"One may . . . carry jars"—because this act does not change anything, since the produce has already become unclean.

11. B. Pes 33b, 34b. To understand this Section, some basic rules governing uncleanness must be kept in mind. There are several gradations of uncleanness. The primary source is the "father of uncleanness"; a person who touches it becomes unclean in the first degree, and if he then touches some food, that food becomes unclean in the second degree. Profane food cannot go beyond this—if it is unclean in the second degree and touches other profane food, it does not defile it, but it does defile heave offering food, which consequently becomes unclean in the third degree. If this latter touches other heave offering food, it does not defile it, but it does defile sacred food of sacrifices, and there it stops. This Section deals with olives or grapes which have come in contact with something that was unclean in the first degree, and which have thus become unclean in the second degree. The juice enclosed within them is considered a separate body, and therefore does not become unclean, because profane food cannot go beyond the second degree of uncleanness. If they are heave offering, the juice pressed out of them will become unclean in the third degree, unless one takes the precaution of pressing them in amounts less than an egg's bulk, in which case no uncleanness results from contact. See X, v, xi, 2; VI, iv, 1 (YJS, *8*, 292, 344–45).

"one should press them"—emending *wĕ-soḥăṭan* to *soḥăṭan*.

"is merely a precaution"—since even an egg's bulk should be permissible, seeing that even if a drop of juice is pressed out of the egg's bulk, its volume becomes less than an egg's bulk.

"as will be explained"—*ibid.*

12. B. Pes 33b; Tem 7:5.

"to heat an oven or a stove"—such indirect benefit is permitted (above, ii, 14).

"should be buried"—that is, poured out and covered with earth, so that none of it would remain on the surface.

13. Ter 8:4, 6. The food may have been tasted by a poisonous reptile.

14. Tos Ter 7:13. The reason is the same as in Sec. 13.

15. Sif Zuṭa Num. 18:12; Yalḳuṭ Shim'oni 755; B. Ḥul 130b; Sif Num. 5:10; B. BM 6b; Neḏ 85a.

"this option is not accounted the same as money"—money unlawfully appropriated can be recovered in a court of law, whereas an option unlawfully forestalled cannot.

16. Tos Ter 10:16.

"stolen property"—heave offering must be set aside by the owner himself or his authorized agent. Hence any extraneous object found therein must have been put there with the owner's knowledge; it is not lost or abandoned property, and must be returned to its rightful owner. To retain it would constitute theft on the priest's part.

17. B. Ḥul 134b; Tos Ter 10:17.

"he would have been guilty of desecration of the Holy Name"—heave offerings are designated as holy things (Num. 18:19), hence desecration of them is blasphemy.

18. B. Beḳ 26a.

19. B. Ḳid 6b, 52b; Ḥul 133a; Suk 35a.

"this makes the priest the same as a helper"—i.e., it is a case of *quid pro quo:* the priest returns this heave offering with the understanding that he will be given heave offering from some other produce.

20. Tos Dem 5:16, 18; B. Beḵ 26b; 'Ar 8:7.

"watchman," etc.—all these employees are priests.

"gifts"—due to the priests out of cattle.

"son of my daughter," etc.—the payer's daughter or sister is married to a priest, thus making her son also a priest. See the next Section and note thereto.

21. P. Dem 6:2; Sheḇi 7:3.

"for my right of option"—i.e., for exercising my right to choose the recipient in your favor. The Israelite owner of the produce does not own the gifts due to the priest and the Levite, but has the option to choose the beneficiary, which evidently was marketable.

"to use heave offering as merchandise"—i.e., to trade with it in order to realize a profit.

22. B. Yeḇ 99b, 100a; Tos Ḥaḡ 1:2; Pe 4:7; Bik 2:6. To distribute heave offering publicly to such unsuitable persons would degrade and demean the sacred offering.

"spread his hands"—in priestly blessing, the priest spreading his hands over the congregation, with the four fingers divided in the middle to form a "V." Other texts read "spread his lap," to receive his share of the priestly gifts.

"separate kind of creature"—an abnormal creature.

"and testify that he is a priest"—and is thus fit to marry any woman of the children of Israel; yet being a bondservant, he may not in fact do so.

"uncircumcised . . . unclean"—who may not eat of heave offering themselves, but may confer the right to eat of it upon their wives and bondservants.

"lest she should be divorced"—meaning an Israelite woman married to a priest, who would lose that right if divorced.

"unlawful seclusion"—with any unrelated man, an easy thing to do in the open field around the threshing floor.

"unsuited to him"—a divorcée or any other woman who may not marry a priest.

"other holy things of the country"—literally "of the boundary." Some holy things must be consumed within the precincts of the Temple, for example, sacrificial offerings. Others may be consumed in "the country," outside the Temple, for example, heave offering, tithes, and dough offering. Here the latter are referred to. Heave offering of the tithe and dough offering are distributed not at the threshing floor but in the town.

"the only exceptions are the male," etc.—all three can rectify their disqualification: the mésalliance can be dissolved by divorce, the unclean person can go through the purification ritual, and the uncircumcised can undergo the required operation (unless he is a hemophiliac, in which case danger to life releases him from the commandment of circumcision).

Chapter XIII

1. Ter 4:7.

"inferior figs"—see above, note to xi, 10.

2. Ter 5:1, 6, 7.

"forbidden as heave offering"—*mĕdumma'*, a mixture of profane produce and heave offering produce, the latter in sufficient proportion to make the mixture forbidden to commoners in its entirety and permitted only to priests.

"at the price of heave offering"—which is lower than the price of profane produce, since commoners may not eat of it.

3. Ter 5:5, 6.

"in the ten sĕ'ah of the first mixture there is one sĕ'ah of heave offering"—$10 + 90 = 100$ sĕ'ah of mixture; since $1/10$ of this mixture is heave offering, it follows that 10 sĕ'ah of the mixture contains one sĕ'ah of heave offering.

4. P. Ter 5:3.

"that do not normally blend"—wheat kernels or flour particles, even of the same quality and indistinguishable from one another, yet do not blend to the same degree as does oil or wine, in which no separate particles are discernible.

5. Ter 5:7.

"one hundred . . . become mixed with one hundred"—this is Maimonides' interpretation of the Mishnaic statement, "until the heave offering exceeds the profane produce." RABD interprets it to mean, until 51 sĕ'ah of heave offering have fallen into 100 sĕ'ah of original profane produce—the amount taken up may have consisted entirely of profane produce, so that finally heave offering would have formed 51 percent of the 100 sĕ'ah of mixture.

6. Ter 5:8; Tos Ter 6:6.

7. P. Ter 5:3; Ter 5:9.

"How so?" etc.—fine quality wheat contains less coarse bran than inferior quality wheat. Once the flour is sifted, the ratio changes. Therefore, if one sĕ'ah of fine quality wheat falls into one hundred sĕ'ah of inferior quality wheat and they are ground together, the ratio of sifted flour will be not one to one hundred but rather one to less than one hundred; and vice versa in the reverse case.

8. Tos Ter 6:10.

"one need not," etc.—one need not clarify the entire 101 log̃ of the mixture, before drawing the one log̃ for the priest.

9. Tos Ter 6:7.

"because the water does not neutralize the wine"—water cannot combine with wine to neutralize other wine, and it is therefore the same as if the one log̃ of heave offering wine had fallen into ninety-nine log̃ of profane wine.

10. Ter 5:9.

"If . . . unwittingly"—if the additional profane produce was introduced into the mixture by accident.

11. B. Beḵ 27a.

"He thus annuls"—reading, with some manuscripts, *šeb-biṭṭel* for *šen-naṭal*. Each time he pours in one log̃ of heave offering wine, the greater part of the mixture is profane wine, and consequently annuls the heave offering wine.

12. Ter 9:5; Tos Ter 8:4.

13. Ter 7:5, 7; P. Ter 7:3.

"falls into one of them"—in this case both baskets are supposed to contain profane produce.

"according to the volume of the smaller basket"—in all cases of doubt the lesser penalty applies, on the plea that he who claims otherwise must bring proof of his claim; see above, x, 14. Therefore, if the smaller basket contains one sĕ'ah and the larger two sĕ'ah, and both fall into 100 sĕ'ah of profane produce, the mixture is considered profane, since the smaller basket has been neutralized by 101.

"cannot confer liability upon each other"—the source in P. has "proof" for "liability." If both baskets have been planted simultaneously, one must assume that one of them contained heave offering, whereas such an assumption cannot obtain if one basket is harvested before the other is planted, since either basket may have been profane.

14. Tos Ter 6:18; B. Yeḇ 82a; Pes 44a.

15–16. Ter 3:3.

15. "and all of the heap is rendered forbidden as heave offering"—between the sĕ'ah that fell in and the amount designated by the owner as heave offering for the heap, there is more than can be annulled by the heap itself.

"northernmost"—the southern half is obviously not included. The northern half is divided into two quarters, on the assumption that the owner had in mind the northern quarter of the northern half, and hence only that quarter is a mixture subject to heave offering.

Chapter XIV

1–2. Ter 4:9, 8; P. Ter 4:7.

1. "the white figs become heave offering"—the proportion in this case is 50:1, the black figs not being included; hence the single fig is not neutralized.

"If it is not known"—some of the figs are in this case presumed to have been eaten or lost in the meantime, so that counting the black and the white figs would not be conclusive.

3. Tos Ter 6:6.

4. Ter 4:11, 12; Tos Ter 5:12.

"into the mouth of a bin"—the bin is presumed to be full of profane produce, so that the heave offering remains in the mouth of the bin, and cannot fall any further into the body of the bin.

"are blended together"—see above, note to xiii, 4.

5. Ter 4:12; P. Ter 4:8.

6. P. Ter 4:8; Ter 4:11.

"How does one neutralize," etc.—the question assumes that while the two baskets may combine together, this affects only the possibility that the heave offering may have fallen into either one of the two baskets; it does not imply that the contents of the two baskets may combine to neutralize the heave offering.

7. Tos Ter 5:11; B. Beṣ 4a; Ter 4:10.

"Similarly"—referring to the previous Section which deals with one liṭra of heave offering figs pressed into a jug of profane figs and neutralized by the total of 100 jugs.

"if one presses over the mouth of a figcake"—this is a separate case involving two unknowns, which figcake, and which part of it.

"as if they were separated"—and not pressed together into one solid mass;

accordingly the doubt affects each fig in the vessel, even those at the bottom, which might belong to the liṭra that has fallen in. Hence all the vessels combine to neutralize that liṭra.

8–9. Ter 4:7; P. Ter 4:7.

9. "as we have explained"—above, Sec. 1.

"If . . . it falls in with fifty"—reading with one manuscript, *lĕ-tok ḥamiššim*. In accordance with the rule given in Sec. 8, if doubtful heave offering falls into profane produce, it renders it forbidden up to fifty times its volume. Therefore, if one part of heave offering falls into fifty parts of profane produce, and one part of the mixture is then lost, the status of the whole of the remainder is now in doubt for the reason mentioned, i.e., the part that was lost may have been heave offering or it may have been profane produce. The remainder therefore becomes permitted to commoners once one adds profane produce in an amount greater than the original amount, an act forbidden in cases where no doubt is involved. If one part of heave offering falls into fifty-one parts of profane produce, and one part of the mixture is then lost, there is no need to add profane produce, and the mixture becomes permitted to commoners, since it has become profane at the outset.

10. Ter 5:1.

"the whole thus becomes . . . heave offering"—and may not be eaten by a commoner.

"less the value of," etc.—for these belong to the priest in the first place.

11. Ter 5:1; P. Ter 5:1.

12. Ter 5:1.

"parched"—a process involving no liquid. Food does not become unclean unless it first becomes wet by contact with one of seven liquids. See X, VI, i, 1–3 (YJS, *8*, 333), and above, note to xii, 8.

"fruit juices"—other than the seven liquids.

13. Ter 5:2, 3.

"because the sĕ'ah that has fallen in is not the very same sĕ'ah that has been lifted out"—therefore it may be eaten without fear that it might be unclean.

14. Ter 4:1; B. Bek 22b, 23a.

15. Tos Ter 6:13; P. Ter 4:8.

"falls into one of two baskets"—and there is not enough in each basket to neutralize it.

"fell into the same basket"—so that the status of the other basket is unaffected.

16. Tos Ter 6:16; P. Ter 7:3.

"one may say that it must have fallen into the unclean basket"—which was already contaminated.

17–19. P. Ter 7:3.

17. "unclean profane produce"—variant reading, "clean profane produce."

"falls into one of them"—and it is not known which one.

"the profane produce must be eaten in cleanness"—because the assumption that the sĕ'ah of heave offering has fallen into the heave offering basket is meant only to prevent a loss, and not to permit the risk of making the sĕ'ah of heave offering unclean. The extra precaution involves no material loss.

18–19. These two Sections seem to be repetitious, with only a slight variation. Some manuscripts and old editions read them as one, omitting "in cleanness . . . parched."

"one sĕ'ah of unclean heave offering," etc.—the commentators ask, what advantage is there in saying that it must have fallen into the basket containing heave offering. Since the sĕ'ah is unclean, it causes damage whether it falls into the basket of clean heave offering or into the basket of unclean profane produce. They therefore suggest that the order of the Sections here should parallel the text of P. Ter, as follows: Secs. 17, 21, 20, 18, 19. This arrangement, suggested by R. Joseph Corcos, is more logical and obviates many of the strictures expressed against the text as it stands here.

20. P. Ter 7:3; Tos Ter 10:16.

21. P. Ter 7:3.

"Laws Concerning Forbidden Foods"—V, ii, xv, 1–6, 10–13 (YJS, *16,* 231–34).

Chapter XV

1. 'Or 3:7, 8.

"A sealed jar"—has substantial importance and distinctiveness, which makes it impossible to neutralize it so long as it remains sealed.

"spring open"—accidentally; one may not open them deliberately.

2. B. Zeḇ 74b.

"the Great Sea"—the Mediterranean Sea.

3. B. Zeḇ 74b; P. Ter 4:7.

"and one of them springs open"—thus losing its importance and distinctiveness.

"one one-hundredth"—actually $1/101$. The rule is a precautionary compromise, since strictly speaking, only two alternatives are possible here: either the opened jar is all heave offering, in which case all of it is forbidden to commoners, or it is all profane, in which case none of its contents is due to the priest.

"that is, one jar"—strictly speaking, the amount due out of 150 profane jars plus a jar of heave offering is $1/151$; hence the amount due out of the 100 open jars is $100/151$, or slightly less than $2/3$ of a jar. One jar is therefore a rounded figure and more convenient.

4. 'Or 2:4; Ter 10:2.

"we have already explained"—V, ii, xvi, 1 (YJS, *16,* 240).

5. Ter 10:12.

"even its yolk is forbidden"—let alone the white which envelopes the yolk. Although the eggs are cooked in their shells, the shells cannot prevent absorption of the spices.

6. Tos Ter 8:11; P. 'Or 2:3.

"falls into dough"—one hundred times its bulk.

"it is permitted"—because the presumption is that the leavening was caused not by the leaven, since it was lifted out immediately, but by the spontaneous action of the dough itself.

7. Ter 10:1.

"we have already explained"—V, ii, xv, 1–6 (YJS, *16,* 231–33); and above, xiii, 2.

8. "a dried onion"—that is, one with its outer shell dry and firm, not soft and juicy.

"other broths"—other than lentils. Cooked lentils were considered incapable of imparting or absorbing flavor (see the next paragraph).

9. Ter 10:10.

"the onions are permitted"—since the vegetables, being mild and not sharp, cannot impart any flavor of their own to the onions, which are sharp.

10–11. Ter 10:7.

12. 'Uḳ 3:4.

13. Ter 10:3; B. AZ 66b.

"places it"—to cool.

"wheaten bread"—is not absorbent, although it may acquire some of the wine's aroma; but aroma alone cannot affect the bread. See the next Section, and V, II, xv, 33 (YJS, *16*, 239).

14. Ter 10:4.

15. Ter 10:2.

16. Ter 10:5.

"the seed alone"—because the stalk, being inedible, does not partake of the holiness of heave offering (cf. above, ii, 1).

17. B. AZ 73b.

"we have already explained"—above, xiii, 9, where it is stated that water cannot join with profane wine to neutralize heave offering wine.

18. Tos Ter 8:14.

"rise to the surface"—the oil will then float on top of the wine and may be easily skimmed off.

19. B. Zeḇ 96b.

"it depends on whether the pot imparts a flavor"—see V, II, xv, 1–6 (YJS, *16*, 231–33). The earthenware pot soaks up the flavor of the food cooked in it previously.

20–21. 'Or 2:1; B. Me 15b.

20. "heave offering"—usually translated "gift."

"must be brought to the assigned place"—must be eaten in Jerusalem.

21. "is not liable to a flogging"—because no flogging may be inflicted whenever doubt is involved.

22. B. Shab 23a; Pes 73a.

"and must then recite the benediction"—see below, V, xi, 15.

"in the country"—that is outside the holy city of Jerusalem; cf. above, xii, 22, and note thereto.

TREATISE IV: LAWS CONCERNING TITHE

Chapter I

1. B. Ḥul 131a; Yeḇ 91a; RH 33a.

'the great heave offering"—in contradistinction to the heave offering of the tithe, which is set aside by the Levite.

"first tithe"—as distinguished from second tithe (below, Treatise V).

2. B. Yeḇ 85b, 86a, 91a; Sif Num. 18:31; Yeḇ 10:1.

"Wheresoever . . . Scripture speaks," etc.—e.g. Lev. 27:30; Deut, 26:13; Lev. 27:31; Deut. 14:25.

"if a Levite's daughter has been made captive"—and there is the apprehension of rape. In the case of a priest's daughter, she becomes disqualified from eating of heave offering. See above, III, vi, 7.

"has been penalized"—although theoretically she may eat of it. While her marriage is legitimate, this rule serves as an additional precaution, so that in cases like these the woman would be particularly careful.

3. "must set aside the first tithe"—referring to priests and Levites who have produce of their own and are theoretically not obligated to tithe. They must do so only in order to set aside the heave offering of the tithe, to which they too are liable.

"other heave offerings"—the term "heave offering" applies also to other gifts that must be set aside, such as dough offering and first fruits. See above, III, xv, 20.

"heave offerings and tithes"—JE (p. 257, note 30) suggests that the clause "and tithes" is the result of an incorrectly resolved abbreviation for "heave offering," since the second tithe and the poor man's tithe do not accrue to the priests themselves. Hence this clause should be crossed out.

"ye also"—since *ye* obviously (cf. the preceding verse 26) refers to the Levites, the addition implied by *also* must refer to the priests.

4. Sif Num. 18:28; Sif Zuṭa *ibid.;* B. Ḥul 131a; Yeḇ 86b.

"first tithe from the priests"—who have fields of their own that yield produce.

"Ezra has penalized," etc.—the penalty applies only to that generation of Levites. Thereafter the first tithe may be given either to a priest or to a Levite.

"did not go up"—cf. Ezra 8:15.

5. B. Ḥul 130b: Tos MSh 3:11. B. Sanh 83a; Sif Lev. *'Ĕmor* 6; B. Yeḇ 86a; Soṭ 48b; Beḵ 27a.

"they need not pay"—although these gifts are due to the priests, the priests have no claim to them until they are actually set aside. Hence no restitution is due to the priests if one has eaten of these gifts before they were actually set aside.

"Outside of the Land of Israel"—where the obligation of tithe and heave offering is based merely on Scribal authority.

6. Bik 2:5; P. Ter 2:4; Tos Ter 2:4.

"one need not tithe only from adjacent produce"—unlike heave offering, which may be given only from adjacent produce (see above, III, iii, 17).

7. P. Ter 1:2.

"one may not set aside from one for another"—as for example, from produce attached to the ground for produce detached (see above, III, v, 9).

"if he has so set it aside"—as for example, if he has set aside clean produce for unclean produce (see above, III, v, 7).

"whatsoever is exempt"—as for example, produce that is not used for human consumption (see above, III, ii, 1).

"whosoever is liable"—see above, III, iv, 2.

8. P. Ma 2:1. The reason for the difference between the two cases is that in the first case there is the presumption that an agent who on his own initiative undertakes a commission is likely to execute it; in the second case the agent is commissioned by the principal, and therefore there can be no such presumption. See above, III, iv, 6.

9. Sif Deut. 14:22. P. Ma 1:1, 3; B. Ḥul 25b.

"carobs"—below, V, i, 3; Maimonides explains that this refers not to carobs in general, which are edible, but only to a species grown in Zalmonah (southeast of Haifa; S. Klein, *Sefer hay-yišuḇ* [Jerusalem, 1939], p. 130).

"[most]"—required here as below, V, i, 3.

10. P. 'Or 1:2; Ma 3:8.

"since the yield of a fig tree standing in a courtyard"—Maimonides equates courtyard with house. RABD, in his strictures, argues that a courtyard is in this respect the same as a field.

11. Tos Ma 3:9; Ma 5:2.

"Onions rooted"—plucked onions stored indoors and spontaneously rooted in the earthen floor.

12. B. Pes 6b; P. Ma 1:1.

"and are . . . liable to tithe"—i.e., the person who stole them, or received them as a gift from the owner, must set aside the tithe from these remainders.

"and are . . . exempt"—because in this case they constitute ownerless property, which is exempt from tithe.

13. Ter 2:4, 6; Sif Num. 18:30.

"unto you"—quoted from verse 27 preceding; in verse 30 the reading is *unto the Levites.*

"set aside by the Levites"—i.e., the heave offering of the tithe.

14. 'Ab̲ 1:16; Ter 4:1; B. Ḳid 51a.

"by measurement, weight, or count"—here Maimonides suggests that these three methods are equal in rank. In the case of heave offering, however, he grades count, measurement, and weight in ascending order of merit. See above, III, iii, 11.

"his tithes are blemished . . . His own produce, however, is duly regularized"—since the tithe requires a set amount, one-tenth. Whatever one gives above that amount does not become tithe but remains untithed produce, which may not be eaten. The rest of his produce, however, having already been duly tithed, is not affected thereby.

15. Ter 4:1.

"from this same part"—so that it should not appear as if he were tithing from unregularized produce for regularized produce.

16. B. Pes 7b; P. Dem 5:2; Tos Ber 6:14.

Chapter II

1. B. BM 88a. This rule is not affected by a subsequent change of mind.

2. B. Bek̲ 11b. RABD, however, thought that the purchaser is exempt in any case, according to the Torah.

3. Ma 1:1; P. Ma 1:3, 2; Sheb̲i 4:9; Ma 1:2.

4. Ma 1:4, 1.

5. Ma 1:2–4; P. Ma 1:2; Tos Ma 1:2; B. AZ 14a; P. Sheb̲i 4:7.

"Pomegranates . . . seed"—reading *pered̲* (cf. *pĕrid̲ah* at the end of this Section; so *editio princeps*) for *peraḥ,* "blossom."

"bitter almonds," etc.—cf. above, i, 9.

"one-ninth of the amount of oil"—one sĕ'ah of ripe olives produces three log̃ of oil; 1/9 of this amount is 1/3 of a log̃.

"as we have explained"—above, Sec. 4.

6. Ma 5:3.

"waste"—pressed cakes of skins and seeds which remain after the extraction of the oil or juice.

"for the purpose," etc.—the purpose must be expressly stipulated; if it is not, the waste is assumed to be intended for its customary use as fuel, and may be lawfully sold.

7. Ma 5:6; B. Pes 42b; P. Ma 1:3; B. BB 96b.

"as we have explained"—above, III, iv, 21.

"If he decants less than four measures"—the excess over three is regarded as too small to be taken into account.

8. Pe 4:8. Consecrated produce is exempt from tithing only after the Temple treasurer has taken actual possession of it.

Chapter III

1. Ma 1:2, 5; B. Kiḍ 62a; P. Kiḍ 3:1.

2. Ma 1:5.

3. B. Beṣ 35a.

"courtyard"—including the house. The *editio princeps* reads here "house," but cf. below, iv, 2, where "house" is mentioned as additional to the six things. Sec. 4, next, however, speaks of the house, and not of the courtyard.

4. B. BM 88a; Ma 2:4; P. Ma 2:3; Tos Ma 2:2; Ma 1:5.

"The only exception is a basket of figs"—the exception was made because of the basket (*ḳalḳalah,* basket of choice produce meant to be eaten by the owner) and not because of the figs. The basket shows that the owner will not give them any further preparation, such as leaving them to dry or offering them for sale, but will use them forthwith for his own consumption. See *Mĕleḳeṯ Šĕlomoh* on Ma 2:4.

5. Tos Ma 3:5; B. Beṣ 13a.

"by little children or by laborers"—the difference between them and the owner is that when the owner brings these into his house, his intention is to give them no further preparation, and the liability to tithing is thus established. No such intent can be imputed to a child or to a hired laborer, since the former is incompetent, while the latter cannot make decisions for the owner.

"to make dough . . . to husk them"—again, in the first case the preparation of the grain is not finished, while in the second case it is.

6. B. Pes 9a.

7. P. Ma 2:1.

"without B's knowledge . . . the produce becomes liable"—because usually the owner is satisfied to have someone else perform this chore for him. The produce therefore becomes liable.

8. Ma 1:5; Tos Ma 1:6.

"one alongside the other"—watermelons were not piled one on top of the other for fear of splitting them open.

"one by one"—without the intention of processing the entire crop.

"One may . . . set aside heave offering"—in accordance with the rule above, III, v, 5.

9. Ma 1:5.

10. Ma 1:4.

"basket"—*ḳalḳalah;* see above, note to Sec. 4.

11. P. Ma 1:4.

12–13. Tos Ma 1:6; Ma 3:4; P. Ma 1:4.

14. Ma 1:7; B. BM 92b.

"from the upper wine press"—i.e., from the wine that has not yet reached the vat.

"from the spout"—formed at the mouth of the press, through which the wine flows into the vat. The wine which is still in the upper press (where the grapes are pressed) or in the spout is not yet completely ready for use.

15. Ma 1:7.

"pressing basket"—a basket made of rope, in which olives were placed for pressing.

"beam"—or stone.

"pressing boards"—against which the beam or stone is pressing. At this stage the processing of the oil is not yet complete; once the oil has reached the pressing trough, the processing is finished.

"a vessel that is away from the fire"—literally "a second vessel." The pot on the fire, with food cooking in it, was called "the first vessel"; when this food was cooked, it was poured off into another vessel, called "the second vessel."

"will boil"—which would render it liable to tithing (above, Sec. 3).

16. Ma 1:8.

"the jar . . . splits"—this clause is inserted here from the Mishnah. Once the container is breached, the contents are not regarded as a single entity, and the lower layer of figs is considered to be completely processed; hence the entire batch of figs becomes liable to tithing (above, Sec. 4).

17. Tos Ma 2:12.

"on the drying pad"—normally their processing is considered completed when they are gathered into a heap, but not if one takes them to the drying pad, in which case the work is considered completed only when they are sufficiently dried.

18. Ma 3:4.

"one may fetch some of them down"—until all of them are up on the rooftop for drying, their preparation is not considered complete.

19. B. Beṣ 13a, 12b; P. Ter 1, end; Ma 4:5, 1; P. Ma 1:1.

"hastens," etc.—the owner has tithed his produce before threshing it, and has given the tithe to the Levite. The latter may not eat of it until he has first set aside the heave offering of the tithe for the priest.

"while the produce is still in its ears"—when its preparation has not yet been completed.

"some of the wine"—from the press or vat.

"cold broth"—if the broth is boiling, the wine is not exempt, for boiling over fire carries liability to tithing (above, Sec. 3).

"the same as a small vat"—which makes it a regular, not casual, meal.

"over his flesh"—to anoint his skin.

"into his hand"—it is then the same as if he had put it into a container.

20. Pe 1:6; Tos Ma 2:20. B. Beṣ 13a.

"completed, even if . . . not yet . . . designated"—as for example, if one has planned to take the produce into his house after completing its preparation. See above, iii, 2.

"even inside one's house"—where it becomes liable to tithing.

"tying them into bundles"—which completes their processing.

21. Ma 3:4.

"single plucked berries"—so Jastrow for the Hebrew *ḳĕṣiṣoṯ*. Soncino translation of the Mishnah has "figs partly dried." When fully dried they are called *gĕrogĕroṯ*, and when pressed into a round cake, *dĕḇelah*.

"even beside a field"—and there is every reason to believe that they are from that same field and are therefore subject to this rule. They are nevertheless exempt, because the presumption is that the owner does not care about them, and they are therefore ownerless. Since they have thus become ownerless before their preparation has been completed (the owner has not gathered them up), they are exempt from tithing.

"the presumption is that their preparation has been completed"—and therefore their subsequent ownerless status does not affect their liability to tithing. See the preceding note.

22. Tos Ma 2:17; P. Ma 3:3. The principles involved in these rules are the following:

1. Produce which has become ownerless before the completion of its preparation is exempt from tithe.
2. Produce becomes ownerless if in our judgment the owner is not interested in processing it further.
3. One may not eat a casual meal out of produce the preparation of which has already been completed, and which has therefore become liable to tithe.
4. Small sheaves found in a private domain are liable to tithe, those found in a public domain are not. Large sheaves found in a public domain are liable, because the presumption is that the owner has put them there temporarily, assuming that because of their size and weight they will not be removed. A covered basket of produce is liable to tithe, because its preparation has been completed; etc.

"as heave offerings"—rather only as heave offering of the tithe, not as great heave offering, since such produce is exempt from the latter (hence Tos reads "heave offering" in the singular).

"no apprehension"—that it may have been already tithed.

"treating it as produce assuredly not yet tithed"—because it may be regarded as certain that heave offering and tithe have not yet been set aside.

23. Ma 5:1.

24. Ma 3:4; B. BM 21b.

"olives . . . carobs . . . figs"—the difference is that olives and carobs, being dry, do not become soiled when they fall, and therefore one may assume that they had dropped off from the tree underneath which they were found. Figs, however, being moist, become soiled immediately, and therefore one cannot tell where they had come from.

25. Pe 4:5; Ḥal 3:4; P. Ma 1:1; Tos Ma 10:29; B. Men 66a.

"meal offerings"—offerings made of dough, in contradistinction to animal sacrifices. See Lev. 2.

Chapter IV

1. B. BM 88a; P. Ma 2:1; B. Shab 7a; Ber 35a.

"brings it into his house"—for other means of establishing liability to tithing see above, iii, 3.

"outer enclosures"—enclosed areas at the back of the house, whose entrances do not serve as the regular entrances to the house.

2. P. Ma 4:1.

"It would seem to me"—Maimonides' personal opinion, based on a passage in P. Ma, which refers to means other than bringing into the house.

"as we have explained on the ground of tradition"—in the preceding Section.

"that we have enumerated"—above, iii, 3. These six factors are based on tradition.

"as we have explained"—above, iii, 3.

3. P. Ma 3:3; Ma 3:6.

"less than four cubits square"—such a tiny structure is not really a house.

"such a roof"—is not really a roof in the common sense of the word.

4. Ma 3:7.

"handmill . . . poultry"—which ordinarily characterize a permanent dwelling.

"potters' huts"—described below, Sec. 10.

"during the Festival of Tabernacles"—if used after the festival, such a booth would render produce liable to tithing, because it could no longer be regarded as a temporary shelter for a particular purpose.

5. P. Ma 3:7; Ma 3:7.

"a schoolhouse and a house of study"—the former is devoted to the study of Scripture, the latter to the study of Talmud, and considered more advanced.

6. Tos Ma 2:20.

"a dwelling"—for the precentor or beadle, respectively.

7. P. Ma 3:3; B. BM 88a.

8. B. BM 88a; Ma 3:5; B. Nid 47b.

9. Ma 3:6.

"gateway"—near the entrance of the courtyard, usually used as shelter for a watchman.

10. Ma 3:5.

"the potter's hut"—cf. above, Sec. 4. The inner hut served as the potter's dwelling, the outer hut was his shop.

"A merchant's shop"—while the merchant does not dwell in his shop, he spends all of his working day there, and it is thus almost his second dwelling place.

11. Ma 2:3; Tos Ma 2:1.

"is conveying"—even if for sale, since the owner is determined to sell the produce only at his final destination.

"his return journey"—in case he changes his mind and decides to carry the produce back to the place of origin.

12. Ma 2:3; B. BB 22a. The night's lodging makes the house the peddler's temporary home.

13. Tos Ma 2:8–10; Ma 3:1; P. *ibid.*

"in a courtyard that exempts them from tithing"—a courtyard that does not conform to the specifications enumerated above, Sec. 8.

"the courtyard of his own house"—*beṯo;* the text reads "his companion's (*ḥăḇero*) house," which seems to be a scribal error. This courtyard is presumed to conform to the specifications which render the produce in it liable to tithing.

14. Tos Ma 2:20; P. Ma 3:4; B. Suk 8b.

"hoed . . . seeded . . . planted"—the principle behind these rulings is that if the courtyard loses its status as a courtyard and becomes a garden, it comes under the rules governing a garden, and not those governing a courtyard. Thus, if one hoes the greater part of the courtyard, he is likely to hoe the rest of it also, and it therefore ceases to be a courtyard. If, however, he seeds or plants part of it, the assumption is warranted that he intends the rest to remain as a courtyard, and that the planting is but a temporary project (JE).

15. Ma 3:8; B. BB 89b; P. MSh 3:8.

16. Ma 3:10; P. Ma 3:4.

17. Ma 3:9; Ter 8:3.

"slice it"—reading *sofĕṯo,* as in the Mishnah; the text reads *ḵofĕṯo,* "bend it," which does not seem to fit the context. See also below, v, 3, and note thereto.

"passes"—deliberately, not forgetfully.

18. Ma 3:9.

"and eat it"—without tithing.

Chapter V

1. Ma 3:1; B. BB 88a.

"for his own consumption"—if for other purposes, see the next Section.

"as we have explained"—above, iii, 3 and iv, 2.

"even if he has not yet taken possession"—literally "drawn." It is the act of drawing the purchased object toward himself which confers possession upon the purchaser, not the payment of the price. See XII, 1, iii, 1 (YJS, *5,* 11).

"even if he has made up his mind"—a mental resolve alone is not a legally binding act.

"may then return"—the seller consenting thereto.

2. Ma 5:1.

"attached to the ground"—the sale fixes liability to tithe only in the case of plucked produce, not produce attached to the ground.

"or is detached . . . to be sent"—the Mishnah reads, "or picks it to be sent," which seemingly does not involve a purchase.

"to be sent to one's companion"—purchase for use as a gift does not fix liability in the case of attached produce.

3. Ma 2:5, 6; P. *ibid.*

"twenty figs"—for the other fruits that follow two of each kind are mentioned. Hence some later editions have changed the number here also to "two." Hebrew letters have numerical values, and the letters for "two" (*beṯ*) and "twenty" (*ḵaf*) are very similar, so that they are frequently confused with each other. The original text may have used the letter *beṯ* for "two," and a copyist may have misread it as a *ḵaf.* A subsequent copyist may have then spelled it out as "twenty." But cf. below, Sec. 5, which has "twenty (*ḵaf*) figs."

"slice it"—reading *sofeṯ* for *ḵofeṯ;* see above, iv, 17, and note thereto.

4. Ma 2:8. Exchange, like sale, renders detached produce liable to tithe, provided that its preparation has been completed. Fresh detached fruit is considered ready for consumption, while fruit meant to be dried first is considered not yet completely processed.

"as we have explained"—above, iii, 3.

5. P. Ma 2:1; Tos Ma 2:6.

"I will fill my belly"—the fact that A did not state the exact amount of B's figs that he was going to eat indicates that he gave his own figs to B as a gift. Hence the following statement that a gift does not designate produce for tithing.

"so that if one of them had eaten more than one fig at a time," etc.—the text as it stands seems to say, "but if one of them eats more than one fig at a time, he is liable." The translation given here follows the commentaries, which connect this sentence with the preceding statement.

"A gift . . . does not designate," etc.—in the printed editions this statement begins Sec. 6. For the reason why it belongs here see above, the first note to this Section.

6. Ma 2:1; P. Ma 2:1, 3:2.

"an unlearned person"—who is not reliable in regard to tithes. The explanation for this rule is as follows: If the produce is donated at a time when it is liable to tithe, the recipient must set aside the tithe out of doubt, for the unlearned owner may or may not have tithed it. If the produce is donated before it becomes liable, and becomes so liable in the recipient's hand, the latter must set aside the tithe, because in that case the produce is assuredly not yet tithed.

"If the majority of the people," etc.—the owner is in that case regarded as one of the majority, and therefore his produce is assuredly not yet tithed.

"and carry them into your houses"—this might be an indication that he has already tithed, hence their doubtful status.

"a large one"—which certainly cannot all be eaten outside the house.

"not . . . eaten raw"—and must be brought into the house to be cooked.

"eating in the market place"—was regarded as the mark of an ill-mannered and disreputable person unfit to serve as a witness in court (XIV, II, xi, 5; YJS, *3*, 105–06).

7. Ma 2:1.

"A may eat"—an occasional meal outdoors, because it has not yet become liable to tithe.

8. Ma 2:2. The situation here is as follows: Several persons are sitting in a gateway or a shop; evidently some are owners and some are visitors. An unlearned person tells them to help themselves to some of his produce. The rule then is as follows: The visitors may eat an occasional meal out of it, on the principle that a house that is not one's own does not designate produce that he brings into it for tithing; the owners of the gateway or the shop may not eat of it until they are certain that the produce has been tithed, for in regard to the owners it is as if the donor had told them to take the produce into their own houses, which in fact they are, and which designates the produce for tithing.

"we have already explained"—above, iv, 11.

"when that produce passes through it"—see above iv, 3.

9. Ma 2:7; B. BM 89a, 87b, 92b.

"according to the Torah"—see Deut. 23:25–26 (cf. 25:4).

"of the detached produce"—once the produce is processed, the Scriptural rule no longer applies.

10. Ma 3:3; P. *ibid.*

"to weed"—pluck the small olives, leaving the large ones on the tree.

"several olives at a time"—this constitutes a miniature threshing floor, and renders the produce liable to tithe.

11. Ma 3:3, 2:7; B. BM 89b.

"the salt designates them for tithing"—see above, iii, 3.

12. Ma 3:8; B. BM 91b.

"poor quality figs . . . fine quality figs"—*lĕḇasin* and *bĕnoṯ šeḇa'*, respectively. See Löw, *Flora, 1,* 241.

13. Ma 3:2; B. BM 83a; P. Ma 3:2.

"one may not pay off a debt," etc.—because paying off a debt out of produce is equivalent to selling it, and one may not sell untithed produce. See below, vi, 9.

14–15. Ma 4:1.

15. "does . . . not . . . render it designated for tithing"—since such artificial ripening is not equivalent to natural ripening.

16. P. Ma 1:7, 4:4.

"hot dish," etc.—all these come under the category of "fire," which is one of the six things that designate produce for tithing (above, iii, 3).

"even if it is still in the wine press"—and thus its preparation is not yet complete.

17. P. Ma 4:1; Tos Ma 1:9; B. Shab 144b; BB 97b.

"in the field"—in the house it becomes liable automatically, even without grating; in the field grating makes its preparation complete; but not if poured into a tray, because the tray has food in it, and the wine is merely an adjunct to the food.

18. Ma 4:1, 3; P. Ma 4:1. Salting designates produce for tithing (above, iii, 3).

19. "that he must then set aside a second heave offering"—for example, if he has set aside heave offering from produce liable to it by Scribal law only instead of from produce liable by Scriptural law (above, III, v, 14). Generally heave offering designates produce for tithing (above, iii, 3).

20. P. Ma 2:2; B. Beṣ 34b.

"Sabbath eve"—the onset of the Sabbath renders produce the preparation of which has been completed liable to tithe (above, iii, 3).

21. Ma 4:2.

22. P. Ma 4:3; Ma 4:2; Ter 8:3.

"is intended to be eaten only on the Sabbath"—because of its extra fine quality.

"he may not eat of it"—at all, even before the Sabbath, and even an occasional meal.

"until he has tithed it"—before the Sabbath, since one may not tithe on the Sabbath.

Chapter VI

1. Ma 1:8.

"level the surface"—the surface of a cake of dried figs or raisins was leveled by daubing it with the juice of figs or grapes, respectively. One was permitted to do so with the juice of untithed figs or grapes because such a small amount of juice was considered of no consequence. See R. Joseph Corcos *ad loc.*

2. B. Shab 26a.

"kindle"—for illumination or for fuel.

"nor, needless to say, on the Sabbath"—that is, for use as Sabbath light kindled before the onset of the Sabbath; the use of untithed oil for such a purpose is certainly forbidden, because on the Sabbath, unlike the weekdays, one may not regularize such oil by setting aside what is due to the priest and to the Levite.

"implying"—by the plural *offerings.*

3. Tos Dem 1:18; P. Dem 1:3; B. MḲ 2b; P. Ma 5:1; Ma 5:1, 2.

"cover . . . with earth"—which is tantamount to planting.

"to promote their further growth"—which does not complete the processing of their produce.

"get their seeds"—which completes the processing of their produce.

4. B. Neḏ 58b, 57b.

"the whole of it is liable to tithe"—including the original liṭra which itself was tithe, because it becomes part of the new yield.

"out of some other produce"—i.e., out of the produce of the same year, since when it was planted no heave offering of the tithe was set aside out of it.

"in proportion"—that is, heave offering of the tithe equal to 1/10 of the original liṭra.

"regularizes"—i.e., sets aside the tithe and the heave offering of the tithe due from the liṭra before planting it.

"the entire yield"—if the new yield totals ten liṭra, the entire ten liṭra are regarded as untithed and must be tithed, even though the original liṭra has already been tithed before.

5. B. Men 69b.

"one-third grown"—before that they are not liable to tithe (above, ii, 3–5).

"analogous to onions"—where the rule is that the entire amount must be tithed. See the preceding Section.

"onions are not customarily grown out of seeds"—but only out of bulblets. Cf. Löw, *Flora, 2,* 127; *Encyclopedia Britannica,* under "Onion."

6. P. Dem 1:3, 5:8, 3:3; Ter 9:6.

"If the seed can still be picked up"—i.e., if it has not yet struck root.

"he must be penalized"—for transgressing the rule in Sec. 3, above.

"The same rule"—i.e., that the outgrowth is forbidden.

"in case of need"—as when some tithed produce has become mixed with other produce, and can be set aright only by setting aside heave offering and tithe for it out of another similar lot of produce. If the owner does not have such a similar lot, he must sell the mixture to someone who has.

7–8. Tos MSh 3:7; P. Dem 7:1.

8. "being thus penalized"—although generally the burden of proof in sales is on the purchaser. Cf. below, xii, 18.

9. Dem 3:1.

"tantamount to selling it"—which is forbidden (above, Sec. 6).

10. Tos MSh 3:8.

"the government"—literally "the king's house."

"in satisfaction of a debt"—which he may not pay with untithed produce (above, Sec. 9).

11. Dem 5:8, 6:1.

"leases . . . rents"—there are two kinds of agricultural rental: in one, here termed lease, the lessee (*mĕḳabbel*) pays a portion of the yield; in the other, termed hire, the lessee (*ḥoḳer*) pays a fixed amount of produce regardless of the actual yield (see below, Sec. 13). In the first case the tenant may pay the owner his share untithed, since it is the owner's property at the outset. In the second case it depends on whether the tenant pays the owner out of the yield of that same field or of another field. If of that same field, he must set aside the heave offering out of the rental, because the produce becomes liable to heave offering while on the threshing floor and still in the tenant's possession; he need not set aside tithe, which the owner himself must do. If the tenant pays with the produce of another field or of another kind, he must first tithe the rental as well, for this is the same as if he paid a debt with his own produce, and he is therefore bound to tithe it before it leaves his possession.

12. Dem 6:2; P. Dem 6:2; B. BM 101a. These rules, based on the Mishnah and best understood in the context of the time, were intended to make it unprofitable for a Jew to lease or rent a field from a heathen, particularly a field originally confiscated from another Jew. This might induce the heathen to sell his field outright to a Jew, rather than leave it uncultivated and producing no rental income. In either lease or rental the heathen owner is presumed to be unwilling to cultivate the field himself.

13. BM 9:6; Tos Dem 6:2.

"a hirer"—*śoḳer;* since the hire is paid in money, the hirer is obviously responsible for all dues required out of the yield, which is all his own.

14. Dem 6:8.

"A may say"—and need tithe his own share only, because the principle of retroactive choice would apply to him. See below, note to vii, 1.

"the same as if they sold untithed produce"—in which case A is liable for both shares if B does not set aside tithe.

15. Pe 1:6; P. Pe 6:5.

"heave offering and tithes"—heave offering due to the priest, and tithes due to the Levite (who must set aside from it heave offering of the tithe, due to the priest).

"must be collected from the purchasers"—since priests and Levites also are liable to them (above, i, 3).

"This penalty"—were it not for this penalty, they could have retained the heave offering or the tithes for their own use.

"If . . . they have bought . . . before its preparation is completed"—here there is no ground for apprehension that they had bought the produce in order to deprive other priests and Levites of their due, because of the work entailed in completing the processing of the produce.

16. Tos Dem 7:13. The reason for the difference in the two cases is this: In the first case it is assumed that the seller will certainly not give away anything that he is entitled to; in the second case, inasmuch as the preparation of the produce has been completed, it is logical to expect that the seller would explicity stipulate that he wishes to retain the priest's or the Levite's due. Since he had not so stipulated, he has obviously renounced his right (see R. Joseph Corcos, *ad loc.*).

17. Dem 6:3, 4; Tos Dem 7:2; P. MSh 5:5.

"leased"—for a share in the yield (see above, note to Sec. 11).

"is due from the owner of the field"—the assumption is that the lease was made with this condition in mind.

"the other gifts"—such as second tithe or poor man's tithe.

18. Dem 6:5.

"has leased olives"—i.e., has contracted with the owner to receive his crop of olives, press the oil from them, and share it with him.

"because oil is an important item"—and therefore the assumption is that if any other arrangement had been intended, it would have been expressly stipulated. Olive oil is the most important product of the olive crop.

19. B. BB 63a; P. Dem 6:2; Tos Dem 7:14.

"that the tithes thereof are to accrue to me in perpetuity"—see above, i, 4. The clause "in perpetuity" is not found in B. BB 63a; from what follows here it appears that "in perpetuity" is equivalent to "during my lifetime."

"is yours"—literally "is before you."

20. Tos Dem 7:15.

"are to accrue to me"—meaning "are to be paid by me to some priest other than the lessor, as I may choose." The lessee thus acquires ownership of the priestly gifts in so far as he has the optional right of choosing the priest to whom he wishes to give these gifts.

"a priest cannot make a priest"—variant reading, "a priest cannot make an Israelite a priest" (RABD). The priest may cede his constitutional privilege to a nonpriest for a limited time but not for a lifetime, since in the latter case the nonpriest would in effect become a priest, which is patently illegal, since priesthood is hereditary and not appointive.

"for himself"—since an Israelite may eat of the first tithe (above, i, 2); even though in this case he is applying it toward the discharge of the Levite's indebtedness, and thus, as it were, handing it to the Levite who in turn hands it to him. Nevertheless, it is forbidden for the reason given.

21. B. Bek̲ 11a; Ḳid 58b.

"priestly gifts due to be set aside," etc.—the heir is an Israelite whose mother was a priest's daughter. He is therefore entitled to his inherited priestly gifts, even though his priestly grandfather was himself heir to his own Israelite maternal grandfather. The reason for this rule is that for a while on their journey these gifts were in the hand of a priest who had the right to dispose of them as he wished. In this case the Israelite heir, though he may not eat of the heave offering, may sell it to a priest; the tithe he may consume himself, even though the Israelite forebear should have paid the tithe to a Levite but for some reason failed to do so.

22. Tos Dem 7:11–12; P. Dem 6:1; B. Shab 23a. The purpose of these rules, too, is to discourage contracts with persons likely to act illegally.

"the season for tithing"—see above, ii, 5.

"to an unlearned person"—the majority of such persons do tithe, but because of the minority who do not the following precautionary rule has been enacted.

"need not worry"—he is not obligated in this case to tithe before the season for tithing, out of fear that the lessee might eat of the crop before it is tithed, because lessors are not responsible for lessees who violate the law by eating of the produce before it has been divided between them and the lessors.

Chapter VII

1. B. Giṭ 25a; Beṣ 38a.

"with the thought"—*'ad̲,* literally "until," is presumably a scribal misinterpretation of the abbreviation *'d̲ = 'al da'aṯ,* "with the thought (or intention)."

"as if it had been selected in the beginning"—this is the well-known principle of *bĕrerah,* literally "choice," meaning the selection retroactively of one object rather than another, although designated by a term equally applicable to either object. In this case what is left at the end should have been regarded as having been selected in the beginning, but since an obligation prescribed in the Torah is involved, the principle of choice does not apply.

2. Tos Dem 8:14.

"the liquid therein is intermixed"—and therefore what has been designated as tithe is intermixed with the rest of the liquid. This cannot apply to a bin which holds solid produce.

3. P. Dem 1:4.

"redeem it"—with money, whereupon the second tithe becomes profane produce (below, v, iv, 1).

4. Giṭ 3:8.

"so that it may serve as the source of tithes"—the situation here is as follows: The owner of a large amount of produce sets aside a certain quantity (batch A) with the idea that each time he eats out of the rest (batch B), he will designate part of batch A as the required tithe. He may continue to do so until the whole of batch A has been used up as tithes for batch B.

"not out of certainty"—because while he was eating out of batch B, batch A may have been still in existence.

5–6. Giṭ 3:7.

5. "gifts due to them"—reading *me-ḥelḳan* (as below, Sec. 7) for *me-ḥelḳo* ("due to him").

"may have died"—in which case the creditor must first secure the heirs' permission to proceed.

"may have grown rich"—the Gemara (B. Giṭ 30b) remarks sadly, "It is a common thing for people to die; it is not a common thing for the poor to grow rich."

6. "the first or second tithe"—variant reading, "the first tithe or the poor man's tithe."

"through a third party"—who then surrenders the tithe to the creditor. The third party must likewise be a priest, Levite, or poor man, as the case may be.

"may sell the heave offering"—since he himself is not allowed to eat of it.

"he may eat of them himself"—above, i, 2.

7. B. Giṭ 30a.

"accounted . . . as . . . interest"—the Talmud explains it thus: Inasmuch as when the creditor has no produce he need not give anything to the priest, etc., when he has something and gives less than all of it, this is not accounted as charging interest.

"The Sabbatical year cannot cancel such a debt"—since the creditor cannot claim anything from the debtor, repayment being restricted to produce set aside as tithe.

"retract"—the debtor who holds the money is given the advantage.

"despaired of recovering"—as when the creditor's field has been stricken by blight, and there is no produce to be tithed.

"he may not set aside tithe"—out of next year's crop.

"provided that the debtor has left them some land"—for only in that case are the heirs responsible for the testator's debt (cf. XIII, III, xi, 8–9; YJS, 2, 119).

"in court"—according to the Talmud, the whole Tribe of Levi became responsible for a priest's or Levite's debt (otherwise people would be reluctant to make loans to them).

"If the poor debtor has grown rich"—see above, note to Sec. 5.

8. B. BM 49a.

"a kor of tithe"—the amount must be specified, otherwise the amount of heave offering of the tithe cannot be determined.

"voice his complaint"—that the Israelite did not act properly. The Levite thus has no legal remedy, since the Israelite is entitled to give the tithe to any Levite he wishes, and is guilty only of breaking his promise to give it to the first Levite —a moral, not legal, offense.

9. Ter 4:2; Tos Ter 5:2. In this case the owner pays his dues to the Levite and to the poor man first, and then may reserve a proportional share for himself only if the two sĕ'ah are still in existence. If the Levite and the poor man have already consumed part of what they had received, the owner of the produce may reserve for himself a quantity to correspond to the quantity that still remains with the Levite and the poor man, because to allow the full amount (eight sĕ'ah) would be possible only by applying the principle of retroactive choice, which cannot be done when it involves a Scriptural rule. See Giṭ 3:8.

10. Pe 5:5.

"even if the exchange is made with his consent"—the principle here is that what the poor man has is exempt from tithe, while what the owner of the produce has is liable. The exchange does not alter the status of either vis-à-vis liability to tithe.

11–13. Dem 7:6; Tos Dem 8:15.

12. "his declaration is effective"—*ḳara' šem,* literally "he has pronounced (their exact) name." Having declared that the tithes are already located inside these baskets, he may not change his mind and replace them by tithes taken from some other batch of produce.

13. "for both baskets out of one of them"—and not out of one for the other, otherwise he would face the difficulty described above, Sec. 11, that is, of setting aside tithe out of produce already tithed for produce not yet tithed.

"out of the small basket for the large one"—i.e., he should set aside one-tenth of the produce in the small basket as tithe for a portion of the contents of the large basket that is equal to the total amount of produce in the small basket. Out of the untithed residue in the large basket he should then set aside tithe for that untithed residue, as well as for the produce in the small basket. In this manner he would at no time tithe out of produce already tithed for produce not yet tithed.

Chapter VIII

1. PhM to Dem 7:7.

"regularized"—Maimonides makes a distinction between produce from which only the two priestly heave offerings have been set aside, which is then permitted for consumption by a commoner, and produce from which all prescribed dues have been set aside.

2. Ḥal 3:9; Tos Dem 5:12; Dem 7:7. To understand the next several sections properly the following must be kept in mind:

a) The great heave offering given to the priest out of an Israelite's crop equals on the average 2 percent of the produce and may be eaten only by a priest.
b) The heave offering of the tithe which the Levite must give to the priest out of his share of the tithe equals 10 percent of that share or 1 percent of the original crop from which the tithe was set aside.
c) In the case of a mixture of tithed and untithed produce, one may not simply take out of it a certain amount of tithe corresponding to what is due from the untithed portion of the mixture, because if one does so, he might pick the tithe out of produce already tithed, which would mean that he has set aside tithe out of produce exempt from tithe (since it has already been tithed) for produce not yet tithed, which is forbidden.
d) A commoner and a Levite may not eat of heave offering produce, or produce mixed with heave offering, but may eat of tithe or produce mixed with tithe.

In Sec. 2–6 we have the following cases:
a) One hundred sĕ'ah of produce from which the great heave offering, but not the heave offering of the tithe, has been set aside, have become mixed with one hundred sĕ'ah of regularized profane produce (Secs. 2–3).
b) One hundred sĕ'ah of tithe have become mixed with one hundred sĕ'ah of regularized produce (Secs. 4–5).
c) Untithed produce has become mixed with an equal or unequal amount of tithe, from which the heave offering of the tithe has not yet been set aside (Sec. 6).

"of the same species"—if it is of a different species, the proportion can be determined by the flavor. If the forbidden produce imparts a distinct flavor, the mixture is forbidden. If there is no change in flavor, the mixture is permitted.

3. Dem 7:7; Tos Dem 5:12; P. BM 10:1. The case here is as follows: An Israelite has one hundred sĕ'ah of produce, from which the heave offering of the tithe has not yet been set aside, and which has become mixed with one hundred sĕ'ah of tithed profane produce. The problem is, how is he to set aside the heave offering of the tithe, forbidden to him, that is due out of this mixture. Simply to take out the required one sĕ'ah at random will not do, since it may happen to come out of already tithed produce, and he would then give out of the exempt for the liable, which is forbidden. He must therefore take 101 sĕ'ah out of the mixture and declare that in this 101 sĕ'ah there is 1 sĕ'ah of heave offering of the tithe; he may then eat of the remaining 99 sĕ'ah, since no matter what has actually taken place, he may eat of these 99 sĕ'ah. For if we assume that in the 101 sĕ'ah there are 100 sĕ'ah of untithed produce and 1 sĕ'ah of regularized profane produce, the remaining 99 sĕ'ah are surely profane produce and may be eaten. If we assume, on the other hand, that the 99 sĕ'ah are produce from which the heave offering of the tithe has not yet been set aside, it would mean that of the 101 sĕ'ah, 1 sĕ'ah is like the 99, and since the owner has declared that one sĕ'ah is heave offering of the tithe, that 1 sĕ'ah may serve this purpose, inasmuch as it is just as liable as the 99, and therefore the owner has set aside out of liable produce for liable produce, which is proper.

"Similarly" etc.—in this case both heave offerings, the great and that of the tithe, have not yet been set aside. Hence the owner must set aside three sĕ'ah, two for the great heave offering and one for the heave offering of the tithe.

"Why does he forfeit this one sĕ'ah?"—the answer given by Maimonides, that it is a matter of identification, is not shared by the standard commentaries on the Mishnah. They say that this is rather the only way in which one can avoid giving heave offering out of the exempt for the liable or vice-versa.

4–6. Dem 7:7.

4. "the mixture is rendered forbidden"—if he leaves it as it is; but of course he can remedy this condition, as described in the next Section.

5. "If one hundred sĕ'ah of tithe have become mixed with one hundred sĕ'ah of regularized profane produce"—in this case one must take 10 sĕ'ah out of the 100 sĕ'ah of the tithe for the heave offering of the tithe. He may not take it out of the profane produce, because that would be taking out of the exempt for the liable. If he takes 110 sĕ'ah, he is safe, for the remaining 90 are then assuredly profane produce, and one may eat of them under any circumstances; or if they are tithe, he has already set aside 10 sĕ'ah for them out of the 110 of which 100 are certainly tithe.

6. "untithed produce mixed with unregularized tithe"—in both cases what makes the mixture forbidden is the heave offering of the tithe, because the great heave offering has already been set aside. The 101 sĕ'ah of tithe now set aside make the remaining 99 sĕ'ah untithed produce, on the principle explained above, note to Sec. 3. It is understood that while the 1 sĕ'ah is heave offering of the tithe for the remaining 99, one is still liable for 10 sĕ'ah of heave offering of the tithe for the 100 sĕ'ah of tithe.

"which are tithe"—i.e., 100 are tithe, and the additional 1 sĕ'ah is heave offering of the tithe for the 99 sĕ'ah.

"If the . . . untithed produce exceeds," etc.—as for example, 100 sĕ'ah untithed produce and 90 sĕ'ah tithed produce; one must set aside 10 sĕ'ah out of the tithe as heave offering of the tithe, 1 sĕ'ah for the 100 sĕ'ah, and 9 for the 90, and he is then in the clear, since even if the 10 sĕ'ah had come from the untithed produce, they would still serve as the heave offering of the tithe for both the 100 untithed sĕ'ah (1 sĕ'ah) and the 90 sĕ'ah of tithe (9 sĕ'ah).

"after deducting the price of the two heave offerings"—which he must give to a priest and may not sell.

7–9. Dem 7:8. A square of one hundred jars is meant, each side of which could be regarded as an outer side. The Hebrew terminology here vacillates from "jug" to "jar" and back, but presumably jars are meant.

7. "a diagonal line"—these two jars can thus be regarded as components of all four sides.

Chapter IX

1. Tos Soṭ 13:9–10; MSh 8:15; B. Yoma 9a; Soṭ 48a; Dem 4:3; B. Shab 13a.

"Johanan the High Priest"—John Hyrcanus, who reigned over Judea 135–104 B.C.E.

"Simeon the Just"—see 'Aḇ 1:2, and Josephus (Loeb Classical Library), 7, 732–36 (and further references given there).

"scrupulous about setting aside great heave offering"—knowing as they did how severe the law concerning it was, and also because it did not amount to much.

"the unlearned"—meaning only a minority of the unlearned; "the majority of the unlearned do render tithes" (B. Shab 13a).

"were lenient"—thinking that untithed produce, even though it contained the heave offering of the tithe, did not involve the death penalty.

"dĕmay"—interpreted as a composite of two words, *da may,* "what is this?", indicating doubtful status.

2. B. Soṭ 48a; Yeb̲ 86a.

"involves the death penalty'—above, i, 5.

"is to be consumed by the owner of the crop"—below, v, ii, 1.

3. Dem 4:3.

"in order to designate the second tithe"—i.e., to remind the owner that the poor man's tithe in that year is a substitute for the second tithe, so as to keep him mindful of the latter; he is liable for it in its time, although he is not liable for the poor man's tithe.

4. B. Shab 23a; Dem 1:4. Cf. III, x, iii, 5 (YJS, *14,* 464–65).

"while he is naked"—cf. above, III, iv, 4.

5: Dem 5:1.

"the rest of the tithe"—i.e., 9⁄100, making together with the first 1⁄100 one-tenth of the total for the first tithe.

"at the outset"—but if one does set it aside before the tithe, it is nevertheless valid. Cf. above, III, iii, 23–24.

6. Dem 4:1; P. Dem 4:1.

"from a baker"—who is an unlearned person, hence his loaves have the status of doubtfully tithed produce.

7. Dem 7:1, 5:1; P. Dem 4:1. A, an unlearned person, invites B, a scholar, to share his meals during the Sabbath, when B cannot go to A's house beforehand and rectify the food. Generally a scholar should at the outset decline such an invitation (below, x, 1).

8. Dem 7:5.

"he would not be permitted to tithe . . . during the Sabbath"—see V, I, xxiii, 14 (YJS, *14,* 152–53). The prohibition covers not only the actual setting aside of the tithe but also the mere declaration of it, as in this case.

9. P. Dem 7:1, 5.

"may be moved"—if no such verbal stipulation is made, the untithed produce may not be moved (V, I, xxv, 19; YJS, *14,* 167).

10. Dem 7:2. This refers back to Sec. 7, above, where the invitation specified eating and not drinking.

11. Dem 7:3; Tos Dem 8:6. Here the invitation specifies drinking and not eating.

"a laborer"—his share of his employer's meal is part of his wages, and the meal is a weekday meal, not a Sabbath meal; therefore he can rectify the produce.

12. Dem 2:4; B. Yoma 9a.

"were not obligated"—because they were compelled by the officials of the government to sell at a set price and to give good weight, so that their profit was small; the buyer therefore had to set aside the second tithe.

"If . . . he sells them to a bread shop"—i.e., sells them wholesale to a retailer, in which case he is allowed a larger profit.

13. Dem 6:7. In this case the total wine in the vat amounts to 400 log̃ composed of 200 log̃ from A's doubtfully tithed grapes and 200 log̃ from B's assuredly

tithed grapes. Each log̃ of wine is therefore half doubtfully tithed and half assuredly tithed. Since B's share is 200 log̃, half of it, or 100 log̃, is consequently regarded as doubtfully tithed, and B must set aside the tithes due out of them (1 log̃ as heave offering of the tithe, and 10—or rather 9—log̃ as second tithe). The same applies, in proportion, if B's share is ⅓ or ¼ of the total of 400 log̃.

Chapter X

1. Dem 2:2; P. Dem 2:2; Tos Dem 2:2.

"everything he eats"—whether his own food or that of others who are not reliable.

"sells"—of his own crops.

"buys"—from someone who is unreliable, for resale to others.

"eschew the hospitality"—lest he should be tempted to eat of untithed produce.

"he is to be believed"—i.e., he has established a presumption of trustworthiness.

2. B. Beḵ 30b; P. Ma 3:1; B. Pes 4b, 9a.

"to have been regularized"—on the principle that an associate scholar would not let go of anything that has not been regularized (B. Pes 9a).

3. Tos Dem 2:16–19; B. Beḵ 30b; PhM to Dem 4:4.

4. Tos Dem 3:9.

5–7. Tos Dem 3:9; P. Dem 2:2.

5. "May a curse befall him"—a pious woman ought not to marry an impious man in the first place, and he may eventually subvert her too.

6. "on stipulations"—see above, ix, 7–11.

7. "even if that person is attending the same feast"—because that person is not his responsibility.

"his father had stipulated in his behalf"—as described above, ix, 10.

8. Dem 6:12; P. Dem 6:8; Tos Dem 8:1; B. ʿEr 37b.

"to an associate scholar"—who is going to market to purchase vegetables for himself.

"without explicit specification"—as explained further on in this Section.

"is exempt from tithing it"—even though the actual purchase is made by the scholar, it is accounted the same as if made directly by the unlearned person, so long as the original coin is used in payment.

"If he exchanges this mĕʿah"—once the original coin is exchanged for another coin or coins, the purchase is regarded as made by the scholar, hence he is responsible for tithes.

"even if one . . . with one hundred"—the one bunch is not annulled or neutralized by the one hundred bunches, because there is a way to render it permitted, and anything that can eventually become permitted cannot be annulled by any admixture of similar things.

9. Tos Dem 8:1.

10. Tos Ma 2:5; B. ʿEr 32a.

"for yourself"—so in the sources: the text here has "for me," presumably meaning "in my behalf, as a gift from me to you." A gift, unlike a sale, does not designate produce for tithing (above, v, 5–6).

"otherwise"—if he wishes to eat a regular, not an occasional, meal (above iii, 1).

"for him"—meaning the unlearned person, as a gift.

"the latter may eat"—at the unlearned person's invitation, as a guest.

"Even though," etc.—since an associate scholar is presumed to observe the law meticulously, and the rule is that one must set aside sacred gifts out of produce that is adjacent, how can we assume that he has set aside tithe out of produce in another place? The answer is that here he has presumably acted leniently in order to prevent a more grievous trespass, namely that the unlearned person would eat untithed produce.

"that is not adjacent"—which is forbidden (above, III, iii, 20).

11–12. Dem 3:1; P. Dem 3:1.

11. "It is permitted"—in order to encourage charity and hospitality.

12. "without further ado"—without inquiring whether the contributions have been tithed, and whether the contributors are learned or unlearned.

13. Dem 3:1.

"to a sick unlearned person"—if the patient's illness is not serious. If it is serious, any treatment whatsoever is permitted.

Chapter XI

1. Dem 2:3.

2. Dem 2:4–5.

"they usually add to the exact measure"—i.e., are generous in measuring out their merchandise, and their profit is smaller than it would be otherwise. Hence the Sages did not wish to burden them further and have obligated the buyer or the recipient to set aside the tithe. In the case of a gift, where no profit at all is involved, they applied the same rule in order not to complicate matters.

3–4. Dem 2:5.

4. "in the lump"—i.e., without exact measure.

"may not be sold"—but must be tithed first, because their bulk is usually below the minimum wholesale quantity specified in the preceding Section.

5. P. Dem 2:5.

"a retailer"—in which case the buyer must do the tithing. The court has thus ruled that the two may not share the tithe equally.

6. Dem 6:9; Tos Dem 6:8.

"may say," etc.—the difference between the two cases is that in the first case, where both shares are of the same species, the principle of retrospective designation applies. In the second case, where two species are involved, this principle cannot apply, and the transaction thus amounts in effect to a sale. Cf. above, vi, 14.

7–8. Dem 3:2; Tos Dem 4:2; P. Dem 3:2.

7. "vegetables"—that are doubtfully tithed. Cf. above, iii, 20.

8. "the act of drawing"—which is the statutory way of transferring title. See above, v, i, and note thereto.

9. Maḵ 2:10.

"he is exempt"—because the vegetables have not yet been designated for tithing, and since they have become ownerless, the finder is exempt (above, iii, 20–22).

"is accounted doubtfully tithed"—hence if it is destined for the market, it becomes liable as soon as its processing is completed.

10. Dem 3:3; Tos Dem 4:3.

"a stumbling block"—to the people in his house, who might eat of it unwittingly.

"to prevent it from going to waste"—without any intention of keeping it for himself, in which case he may keep it and is not liable to tithe, because one is not liable for produce that is not his. The assumption is that the original owner meant to sell the produce in the market place.

11. Tos Dem 4:4; P. Ter 11:4.

"found in the garden"—only the completely wilted leaves are trimmed. See above, III, xi, 10.

"found in the house"—where, in contradistinction to the garden, even the slightly wilted leaves are trimmed. In the garden, it is the gardener who does the trimming, and trims only that which has deteriorated completely; in the house, it is the householder himself who does the trimming, discarding even slightly wilted leaves.

"everywhere"—i.e., whether trimmed by the gardener in the garden or by the householder in the house, or found in the refuse heap.

12. Dem 3:5–6; B. Ḥul 6b; P. Dem 3:6; Tos Dem 4:31–32.

"he must first tithe"—the hostess might rationalize that she is committing no trespass in exchanging the superior produce given to her for her own inferior produce, because of the work she puts into its preparation. Although her act is wrong, the lodger is nevertheless obligated to tithe his produce before he hands it to her, thus preventing her from eating food that is forbidden to her. Normally a person cannot be held responsible for others who are cheaters, but here, inasmuch as he personally hands the produce to her, he has the responsibility of preventing her from eating untithed produce.

"mother-in-law . . . woman neighbor"—who is not likely to take unfair advantage of a son-in-law or a neighbor, respectively.

"If he does not hand these to her"—she might use her own spices or leaven, which might be doubtfully tithed, without realizing her trespass (cf. below, xiii, 18).

13. Dem 3:4; Tos Dem 4:27. The wheat is presumed to have been tithed.

14. Tos Dem 3:9; B. Ber 52b; Pes 110b.

"merely goes in and out"—and is not constantly supervising the clerk's actions.

15. Dem 3:4.

"as we have explained"—above, III, i, 11.

"are subject to doubt"—in which case he need not give them to the priest or the Levite, since the burden of proof is upon them.

Chapter XII

1–2. Dem 4:1; Tos Dem 5:1; P. Dem 5:1.

1. "is overtaken by the Sabbath"—on which day, as stated here, one may not

tithe (III, I, xxiii, 9; YJS, *14,* 151); but not after the Sabbath, when this prohibition does not apply.

"or by a festival"—when one is likewise forbidden to tithe (III, IV, iv, 26; YJS, *14,* 292).

"some other untrustworthy person"—other than the seller. The presumption is that the third person would not perjure himself in a matter in which he has no personal interest, Read *'aḥer* (as in the Mishnah) for *'eḥad*.

"the awe of the Sabbath"—its exceeding sacredness impresses even the unlearned, and their conduct on that day is more ethical.

2. "the two days of the festival as observed in the Dispersion"—the Biblical festivals are celebrated in the Land of Israel for one day, and in the Dispersion for two days. During these two days, if the purchaser was told on the first day that the produce had been tithed, he may eat of it on the second day also.

3. Dem 4:2; P. Dem 4:4.

"B may inquire of A . . . and may then eat"—in order to prevent ill feeling between them.

"has vowed to enjoy no benefit"—this is a more pressing invitation, refusal of which is bound to produce even greater resentment on the part of A.

4. Dem 4:1; Tos Dem 5:1.

"the fear of produce mixed with heave offering, etc."—i.e., although they are suspect in regard to heave offering of the tithe, they are not so when that offering is mixed with other produce.

5. Tos MSh 3:16; P. MSh 4:3.

"because the second tithe remains his"—so that there is no reason for him to lie about it. The first tithe is due to the Levite, and he might lie to retain it.

"no heave offering and tithes"—that is, heave offering of the tithe and first and second tithes; so *editio princeps*. Later editions read "offerings," including the great heave offering.

"from which heave offering and tithes"—so read, meaning heave offering of the tithe and the first and second tithes; the text reads "heave offering of the tithe." Cf. the preceding note.

"he must redeem all of that produce"—he is not to be believed when he says that he had also set aside the heave offering of the tithe and the first tithe, and must redeem them; he must also redeem what he said he had set aside as second tithe, although he is to be believed with regard to the latter.

6. Dem 4:5; Tos Dem 5:1; B. Beḵ 36a f.

"from someone who sets aside tithes"—the Mishnah reads "from someone who is trustworthy (PhM: in matters of ritual cleanness) and who sets aside tithes." A person particular as to clean and unclean in general would certainly be particular about tithes.

"Buy it from C"—B, who is himself not trustworthy, will not bother to look for a trustworthy seller unless A insists that he buy the food from a specific person, so that he can be checked. In other cases, if checked, B might claim that to him the seller seemed trustworthy.

7. Dem 4:6; Tos Dem 5:5; P. Dem 4:5; B. Ḳid 20a.

"Who is deemed trustworthy here?"—see above, note to Sec. 6.

"C is such a one"—a person is not likely to lie for the benefit of someone else.

"old wine"—which is superior to new wine, and hence more expensive.

"both B and C may be believed"—the Sages were inclined to leniency when such a policy helped in their struggle for a livelihood, and at the same time made things easier for a person in a strange town.

8–9. Tos Dem 5:5.

10. Dem 4:7.

"in collusion"—though seemingly in an innocent one, since the first seller cannot profit from his lie. However, as Rashi suggests, the two may have conspired to reverse their roles later, so that one might sell his merchandise in one city, and the other in another city.

11. Dem 6:11; P. Dem 6:11.

"Syria"—see above, III, i, 3–4. Syria occupied an intermediate position between the Land of Israel and the lands outside of it.

"the purchaser must tithe it"—as doubtfully tithed produce (below, xiii, 4).

"the mouth that forbids is the mouth that permits"—had the seller wished to deceive, he would have kept silent, and people would have assumed that the produce was grown in Syria and was therefore exempt. Since it is he who has established his own liability by saying that the produce had come from the Land of Israel, he must be believed also when he says that it has already been tithed.

"from my own field"—here in Syria, where he must tithe his produce by Scribal law (above, III, i, 4).

"If it is known that the seller has land in Syria"—and therefore the principle that "the mouth that forbids is the mouth that permits" does not apply.

12. Pe 8:2; P. Pe 8:2.

"gleanings . . . forgotten sheaves . . . corner crop"—which are exempt from tithes (above, III, ii, 9).

"within the same day"—based on Ruth 2:17–18, *So she gleaned in the field until even . . . and she . . . went into the city.*

13–14. Pe 8:3–4.

13. "flour . . . bread"—it is not usual to give such processed products to the poor, on account of the additional trouble and expense involved.

14. All these rules refer to the form in which various kinds of produce were customarily given to the poor; produce in an unusual form made the poor man's claim suspect.

"It was pressed from olive gleanings"—which were usually too few to yield an appreciable amount of oil.

"he may likewise say"—cooking must obviously follow the gift of the vegetables; hence if he is to be believed in regard to the gift, he must perforce be believed in regard to his own cooking.

15. Pe 8:2.

"he is to be believed at all times"—unlike the poor man, who is to be believed only at certain times (above, Sec. 12).

"just as an Israelite"—who is an unlearned person (above, IX, 1), as is the aforementioned Levite.

16. B. Bek 29b, 30b; AZ 2:7.

"All these rules"—permitting the purchase of produce from an unlearned person, provided that it is treated by the purchaser as doubtfully tithed.

"with oil"—from which heave offering may not have been set aside.

"would lose all of it"—because the authorities would forbid everyone to purchase anything at all from him.

17. B. Beḵ 35a; BM 5b; Tos Dem 5:2.

"needless to say an unlearned person"—who is not suspected personally, but whose produce is subject to doubt because of the status of unlearned persons in general.

"as we have explained"—above, Sec. 10. RABD rejects such testimony under any circumstances.

18. Tos Ter 2:1; B. Giṭ 54b.

"a firstling"—which must be given to the priest.

"wine of libation"—used in idolatrous worship, and therefore forbidden.

"even if he is an associate scholar"—in this case belief or disbelief depends not upon whether A is an associate scholar or an unlearned person, but rather on the fact that it is not within his power to do anything about it, since the merchandise is no longer in his possession. A's motive may be laudable (he may wish to repossess the merchandise in order to regularize it) or fraudulent (the merchandise may be perfectly regular, and he may wish to repossess it for more profitable resale)—B has no way to determine which alternative is true, and may therefore choose either one.

Chapter XIII

1. Dem 1:1; Tos Dem 1:1; B. Ber 40b, 36a. Such plants were not customarily cultivated and grew wild.

"guarded crop"—i.e., meant to be harvested, and thus not ownerless property.

2. P. Dem 1:1; Tos Dem 1:2–3.

"in the valley"—i.e., in an open field next to a fenced and guarded garden.

3. Dem 1:3; P. Dem 1:3.

"who had come up from Babylonia"—see above, III, i, 8.

"in which it is found"—and was not imported from the Land of Israel.

4. Dem 2:1; P. Dem 2:1.

"needless to say"—the land occupied by those who had come up from Egypt is subject to a stricter rule than Syria.

"large cumin"—perhaps read "straight-shaped (*hay-yašar* instead of *hag-gadol*) cumin." P. states that Cyprian cumin was crooked in shape.

"all of the Land of Israel"—whether the part occupied by the returnees from Babylonia or the part occupied by those who had come up from Egypt. See above, III, i ,5.

5. P. Dem 1:3; Dem 2:1. Tyre is outside of the Land of Israel, and a caravan of asses is presumed to have come from far away.

"as we have explained"—in the preceding Section, referring to pure white rice.

6–7. Dem 1:3.

6. "a single ass"—a single ass is assumed to have come from an area nearby.

7. "storehouse keepers . . . ass drivers"—storehouses usually contain produce from nearby fields, while ass drivers bring produce from distant places.

"Sidon is nearer to the Land of Israel than Tyre"—actually Sidon is north of Tyre and therefore farther away from the Land of Israel. R. David Ḳimḥi (12th–13th century), in his commentary on Josh. 11:8, *unto Great Sidon,* suggests that there may have been a lesser Sidon south of Tyre.

8–9. P. Dem 2:1; Tos Dem 4:20.

8. "buys . . . from Israelites"—if the heathen merchant buys his produce from heathens, he is exempt from tithes (above, III, i, 11).

10–11. Tos Dem 1:9.

12. P. Dem 2:1; Tos Dem 4:11.

"not by appearance"—excepting conspicuous color, as in the case of pure white rice (above, Sec. 4).

13. Tos Dem 4:11.

14. P. Dem 1:3; B. Shab 79a; Tos Dem 1:19, 25; Bik 3:10; Tos Bik 3:5; P. Dem 4:1.

"were excluded from the decree"—the reason for the exclusion of certain kinds of produce from the decree is that they are of inferior quality and unlikely to be used for human food, and their owner consequently would not mind losing them, which makes them analogous to ownerless property that is exempt from tithing. Inasmuch as they are of no account, the assumption is that the unlearned owner did not set aside tithes out of them. Furthermore, since the whole institution of doubtfully tithed produce is based on a doubt, the Sages have been lenient in cases like these.

"poultices . . . plasters"—the two terms (the former Greek, the latter Hebrew) are really synonymous.

"mixed with heave offering"—the unlearned person is not suspected of causing others to eat of such forbidden produce.

"residue of meal offering"—see Lev. 2:3; the residue reverts to the priest.

"addition to first fruits"—added by the owner above the required amount.

15. B. Ḥul 7b; P. Ḥul 1:3.

"he may neither sell it to a heathen"—because when he bought it, it became doubtfully tithed produce, and one may not sell such produce to a heathen (above, vi, 6).

"nor feed it to cattle"—his own cattle, for the same reason (above, iii, 20).

"not even to other persons' cattle"—for a fee, so that he would derive some benefit from untithed produce, which is the same as selling it.

16. Dem 1:3, 4; Shab 9:4.

"between his fingers"—to protect them against the friction of the woolen thread.

"anointing is the same as drinking"—this seemingly contradicts the preceding rule that perfumed oil, obviously meant for anointing, is exempt. The answer is that perfumed oil is unfit for use as food, hence the Sages treated it leniently. Weaver's oil, on the other hand, while of inferior quality, is still fit for food.

17. Tos Dem 1:22–23; P. Dem 1:3.

"permitted to use doubtfully tithed oil"—the Sages were lenient in cases of illness.

"an Israelite may sit on that board"—the little oil that is left on the board has lost its status of doubtfully tithed or even untithed produce, since the heathen has already made use of it. Cf. above, III, xi, 6–8.

18. Tos Dem 1:22–23; P. Dem 1:3; B. Ḥul 6a.

"If . . . wine . . . or pulse"—is bought doubtfully tithed and is then mixed, one has a distinct doubtfully tithed product and is liable for it. In the other case the doubtfully tithed ingredient is bought ready-mixed. Even in the latter instance, however, an exception is made for an ingredient that retains its sharp and easily recognized taste.

19. Tos Dem 1:28, 4:5; P. Dem 1:3.

"as if they were doubtfully tithed"—heave offering of the tithe and second tithe that have been set aside in this case, while unnecessary, still violate no law. In the second case, where one treats doubtfully tithed produce as if it were assuredly tithed, he goes against the rule that if the great heave offering has already been set aside, he may not set it aside a second time. Hence, if he has done so, his act has no validity.

20. Tos Dem 4:6. This rule involves the principle of *ḳabu'a,* fixed object: a prohibited thing which has a fixed place among several permitted things, is deemed not a minority against the majority, but rather the equal of the majority. In this case, therefore, in regard to produce bought from one of the townspeople among whom the person who sells doubtfully tithed produce has his place fixed and determined, the possibility of its having been purchased from that exceptional person is regarded not as infinitely small but rather as equal ("half and half") to that of its having been purchased from one of the majority of sellers. Hence the buyer must satisfy the requirements of both alternatives (cf. V, II, viii, 11 [YJS, *16,* 193]).

"the second tithe only"—see above, ix, 2.

21. Tos Dem 4:7.

"one of them was lost"—and it is not known which one.

Chapter XIV

1. Dem 5:6; Tos Dem 5:11; P. Dem 5:5. Cf, above, xi, 2.

"we have . . . explained"—above, i, 6.

2. Tos Dem 5:10.

"and heaped it up in front of him"—coming from several growers, they must have become intermixed, and it would be impossible to tell which is tithed and which is untithed.

3–4. Dem 5:7.

3. "from different towns"—both purchases were made from the same householder though in different towns, but there is no cause for apprehension that in one of these towns the householder may have been reselling someone else's produce, since a householder is not normally engaged in merchandising other people's produce.

5. Dem 5:3.

"freshly baked . . . stale"—literally "hot . . . cold."

"I say"—introducing Maimonides' own explanation of the reason for the rule.

6. Dem 5:4.

"bread shop . . . monopolist"—the former buys bread from one baker, the latter from several bakers.

"for the baker who sells his bread to the bread shop"—variant reading, "to the monopolist," which seems to be an error. The monopolist held exclusive rights to the retail sales of bread in a particular locality.

7. P. Dem 5:4.

"one of the monopolists must have bought . . . from two bakers"—since these bakeries are "fixed objects," the rule does not go after the majority. See above, note to xiii, 20.

8. Dem 5:5; Tos Dem 5:9.

"he must tithe every piece"—because if he should tithe out of one piece for the others, he might tithe out of the liable for the exempt, or vice versa.

"he may lump them together"—the commentators interpret this in this fashion: If the gifts are large, the givers are generous, and have most probably already set aside the tithes. If the gifts are small, it is because the givers are not generous, and most likely did not set aside tithes (KM, RDBZ).

9. Tos Dem 5:8.

"the bread"—i.e., a whole loaf, which is considered a large gift.

10. Tos Dem 5:7.

"have left some pieces"—these pieces would generally be rather small, and the recipient is therefore liable for each piece separately. See above, Sec. 8.

TREATISE V: LAWS CONCERNING SECOND TITHE AND FOURTH YEAR'S FRUIT

Chapter I

1. "As we have explained"—above, II, vi, 4; IV, ix, 3.

2. RH 1:1; B. RH 12a, 15b; Tos Shebi 2:13.

"Grain and pulse"—for vegetables see below, Sec. 4.

3. B. RH 16b; Sif Deut. 14:22.

"as of the following year"—i.e., according to the time when the carobs are gathered, and not according to the season of tithing. Thus, when the tithing is ordained by Biblical law, it is governed by the arrival of the season of tithing—whether the fifteenth of Shebat or after (see above, IV, ii, 5)—regardless of the time when the fruit is actually harvested; when the tithing is ordained by Scribal law, the rule follows the time of the actual harvesting.

"fully formed"—see above, IV, ii, 5.

4. B. RH 13b, 12a.

"gathered on New Year's Day"—obviously by a heathen laborer, since an Israelite may not work at harvesting on a festival day.

5. B. Ḳid 3a; RH 15a.

"is considered the same as a vegetable"—the Talmud explains that the citron tree, like vegetables, requires "a multitude of water" (i.e., artificial irrigation).

"and Sabbatical year"—see below, VII, iv, 12.

6. B. RH 15a. Cf. above, IV, ii, 5.

7. Ma 4:6; P. Ma 4:5; B. Ber 36a.

"trees and seed plants"—the deciding factor being the formation of fruit for trees and the time of harvesting for seed plants.

"after the fifteenth"—the text has "before the fifteenth," which seems to be a scribal error.

8. B. RH 3b.

"even if rooted before the New Year"—according to others, it is the time when they take root that is decisive.

"cowpea"—literally "Egyptian bean."

"one may collect the whole crop into one heap"—literally "one may heap up the threshing floor in the middle of it."

9. Shebi 2:9.

"seedless onions"—literally "castrated onions." Unlike other species of onions that are kept in the soil in order to yield seed, these remain seedless.

"cut off from water"—a sign that they are fully grown.

10. Shebi 2:5; Tos Shebi 2:4, 5; B. RH 13b.

"one may collect the whole crop," etc.—above, Sec. 8.

11. Mak 2:11; Yad 4:3.

12. "whose heave offering . . . is valid"—see above, III, iv, 4.

"is invalid in any case"—see above, III, iv, 2.

13. P. Ma 2:4.

"designated for tithing"—see above, IV, iii, 3.

14. B. Tem 21b; P. Pe 7:5.

"involves bringing it to Jerusalem"—literally, "bringing it to the appointed place"; see below, ii, 1.

"the same as the firstling of cattle"—which may not be brought to Jerusalem from outside of the Land of Israel. See IX, III, i, 5 (YJS, *4,* 63–64).

"It would appear to me," etc.—the Rabbis made second tithe obligatory in these countries because they were close to the Land of Israel, and thus second tithe served another purpose, namely to benefit the poor.

Chapter II

1. B. Zeb 112b; Bik 2:3; Sif Num. 18:21, Deut. 14:23; B. Mak 19a; Zeb 5:8.

"by its owner"—even though it is holy (below, iii, 17), it need not be given to the priest or to the Levite.

2. P. MSh 4:1; B. 'Ar 29a; Šĕ'iltot, Ḳĕdošim, 100.

"the practice of piety"—to go beyond the requirement of the law at present, when the Temple is no longer in existence.

"full value"—see below, iv, 1.

"consecrated things"—which may be redeemed at the present time with one pĕruṭah, to be cast into the Dead Sea (VI, IV, viii, 10; YJS, *15,* 209).

"One must thereupon cast the pĕruṭah into the Great Sea"—so that he would derive no benefit from it. The Great Sea usually designates the Mediterranean, but perhaps here, as in the note above, the Dead Sea is meant.

3. "the redemption of the fourth year's fruit"—see below, ix, 1–2.

"as we have explained"—V. II, x, 17 (YJS, *16,* 205–06).

4. Tos MSh 3:13; P. MSh 3:3; Pes 7:3; P. Pes 7:3.

"so also may one not redeem it there"—because one may only eat it, but not redeem it, there. And since at present one may not eat it either, it must be left to rot, once it has been brought into the city.

5. Tos Ter 7:3; Mak 3:3; B. Mak 18a; Ker 4b.

"as it is said, *Thou mayest not*"—transgression of a negative commandment is punishable by flogging.

"for each of these three varieties"—mentioned in the Biblical verse, corn, wine, and oil.

6–7. B. Mak 19b.

8. B. Mak 19b; P. MSh 1:3; MSh 3:5.

"even if he holds it suspended from his staff"—so that bringing it into the city would not entail his going outside the city wall.

9. MSh 3:5–6; Dem 1:2; P. MSh 3:3; B. Ḥul 68b; BM 53b.

"And thou shalt eat before the Lord thy God"—the implication is that one must eat it in Jerusalem under any circumstances.

"one-fifth of which is worth less than one pĕruṭah"—i.e., the value of the whole tithe is less than five pĕruṭah, in which case one is obligated to bring it into Jerusalem by Scribal law only. Nevertheless, if he does bring it into Jerusalem, he may not remove it.

"Redemption money of second tithe"—if one resides far from Jerusalem, he may redeem second tithe with money, and carry the money to Jerusalem.

"may be brought into, and out of, Jerusalem"—for the Scribal decree applies only to produce.

10. MSh 3:6; P. MSh 3:3.

"fully processed"—see above, IV, iii, 8.

"Nor may it be redeemed outside Jerusalem"—even though, when the produce was in Jerusalem, second tithe had not yet been set aside, nevertheless it is considered the same as if it had already been in Jerusalem, on the principle that the priests' and the Levites' dues present in an unseparated state are regarded the same as if they had already been set aside (B. Mak 20a). Hence they come under the rule governing second tithe that has already entered within the walls of Jerusalem.

11. MSh 3:6.

"may be redeemed outside Jerusalem"—because, if the processing of the produce has not yet been completed, the principle mentioned in the note to Sec. 10 above, does not apply.

12. MSh 3:9; Tos MSh 2:16; P. MSh 3:9.

"If the uncleanness was caused by a secondary source of defilement"—there are degrees of sources of defilement. Primary defilement results when a person comes in contact with a dead body. Anything he touches is thereafter affected by a secondary defilement, or defiled in the second degree. If tithe touches such a secondarily defiled object, it becomes unclean only by Scribal enactment. Hence there are two reasons to require that the redemption money of that tithe be consumed in Jerusalem: (a) when the produce entered Jerusalem it was clean; (b) according to Biblical law, the produce is still clean.

13. P. MSh 3:9.

"When does this apply?"—that tithe defiled by a secondary source of defilement outside Jerusalem, which has entered Jerusalem, may be redeemed and the redemption money may be consumed outside Jerusalem.

"with the stipulation"—evidently one was permitted to stipulate that his own entrance into Jerusalem was not to give the tithe the status that such entrance normally bestows.

"and may not leave"—but must be redeemed, and the redemption money must be consumed in Jerusalem.

14. MSh 2:3, 4; P. MSh 2:3.

"it is only in this stage that it is fit for human consumption"—while it is still green and soft; later, after it hardens, it is fit only for fodder; cf. above, III, ii, 8; xii, 7.

"vetch"—is usually not consumed by humans at all except in an emergency (cf. above, III, ii, 2). Therefore, if it is kneaded into dough to make a substitute for bread, the law is inclined to leniency.

15. MSh 3:7; P. MSh 3:7; B. Mak 12b; Ma 3:10.

16. MSh 3:7; Pes 7:12; B. Zeḇ 56a.

"houses adjacent to the wall"—each house was evidently integral with the wall, with parts of it protruding inward and other parts outward.

Chapter III

1. B. Yeḇ 73b; Mak 19b; Mak 3:2; B. Pes 24a.

"prior to its redemption"—after its redemption it is no longer holy.

2. B. Yeḇ 73b.

"put away"—*bi'arti,* literally "burned up," hence "have utterly removed."

3. P. MSh 3:3, 5:12.

"We have already explained"—above, ii, 8.

"even in Jerusalem"—where clean second tithe may not be redeemed (above, ii, 8).

"as will be explained"—below, Secs. 10, 11.

"has become unclean while untithed"—and one would think that under these circumstances it would be exempt from second tithe.

4. B. Yeḇ 74a; Neḡ 14:3.

"even if the sun of that day has not yet set"—although in the case of heave offering the unclean person must wait until after sunset (above, III, vii, 2).

5. Bik 2:12.

"mourning"—the mourning period consists of the interval between death and burial, followed by a length of time thereafter. The former part is termed *'aninuṯ* (grieving), and the latter *'aḇeluṯ* (mourning proper); cf. YJS, *17,* 34, 102. Obviously here the former is meant.

6. B. Zeḇ 99b; Pes 92a; Ber 16b; Sif Lev. Šĕmini 2:11.

7. B. Zeḇ 101a.

8. MSh 3:3; P. MSh 3:3; P. Dem 1:2, 2:4; Dem 1:2.

"One may not give," etc.—all the items mentioned here have holiness attached to them, and therefore may not be eaten in a state of uncleanness.

"provided that one eats a corresponding amount"—the Sages have been generally lenient with respect to doubtfully tithed produce, inasmuch as its status is doubtful. At the same time, in order to discourage giving it to an unlearned person, they required that the giver eat a corresponding amount of produce in Jerusalem.

9. Dem 1:2; P. Dem 1:2. A clear distinction is made between second tithe of assuredly tithed produce and second tithe of doubtfully tithed produce. In the first case, no matter how difficult the road, the law insists that all of it, or its full monetary equivalent, must be carried to Jerusalem; in the second case the law is lenient and permits a small amount of it to be thrown away.

10. MSh 2:1, 1:7, 5:12; P. MSh 2:1; B. Yoma 76b; Shab 9:4.

"for the dead"—food, drink, and ointment are absorbed by the living body; vessels, clothing, and attending servants are not. Hence they are regarded as analogous to the coffin and the shroud which may not be bought with second tithe redemption money.

"for an unclaimed corpse"—if a strange corpse is found and no relatives come forth to claim it, everyone in that locality is under religious obligation to give the deceased a decent burial.

11. MSh 2:1; P. MSh 2:1; Ter 13:3; Tos Shebi 7:2.

12. Tos Ter 10:12; P. MSh 2:1; B. BB 97a.

"Whatever a commoner may eat of heave offering"—see above, III, xi, 10–13.

13–14. MSh 2:1.

13. "proportionately"—see below, Sec. 15.

14. "accrues to the second tithe"—and the dough must be redeemed at the price of baked bread, without deduction for the cost of baking, etc.

"the appreciation is recognizable"—by the flavor.

15. "four and one-quarter zuz"—the Venice edition reads "four less one-quarter," and so do some commentators. The latter reading seems to be correct according to the respective proportions. The value of the ingredients is three zuz for the wine and one zuz for the honey and spices. The appreciation, one zuz, if divided proportionately, would entitle the wine to ¾ of a zuz, and the honey and spices to ¼, with the result that the appreciated wine would be worth 3¾ zuz.

16. Tos MSh 1:3; B. Shab 61a.

"for it is not meant to be used as a medicament"—but only for eating, drinking, and anointing (above, Sec. 10).

17. B. Ḳid 54b ff.; Tos MSh 1:1.

"sacred property"—*mamon gaboah,* literally "money of the Exalted One."

"may not be acquired as a gift"—the phraseology ("acquired," instead of "given") is intentional: the donor may make a gift of it to the donee (below, end of Sec. 18), but the latter does not thereby acquire title to it; like a guest, the donee merely partakes of the host's food, drink, or ointment, without becoming the legal owner of it.

18. Tos MSh 1:1; MSh 1:1.

"money"—Tos and *editio princeps* read "profane produce."

"exchange it"—since exchange is analogous to sale, both being commercial transactions.

19. Tos MSh 1:1; MSh 1:1.

"may not be used as a counterweight"—to weigh some other second tithe produce, or even gold coins, on the scales of the balance. Since second tithe is holy, this is considered a desecration.

"lest one should"—so *editio princeps* and most old editions; other editions add "not," which seems to be an error.

20. P. MSh 1:1; MSh 1:1; Tos MSh 1:1; Tos BM 4:1.

"may not be used as counterweight"—there are two reasons why second tithe may not be used as counterweight: (a) it is a desecration of a holy thing; (b) the tithe may become a standard counterweight, which is bound to lead to deception, since the weight of produce diminishes with the passage of time. In the following cases at least one of these two reasons applies.

"nor sold"—for money-changing or for smelting.

"to advertise himself"—to show that the money-changer has a large operating capital and thereby attract more customers.

"to prevent the coins from tarnishing"—coins kept in circulation do not tarnish.

21. Tos Shebi 7:9; B. BB 145b. Even though the owner may consume the second tithe, it is holy, and he may not use it for his personal advantage.

"voluntary charitable gifts"—see above, II, vi, 17, and notes thereto.

"but he must inform the recipient"—that the repayment consists of second tithe, so that he would take care of it properly.

22. MSh 3:1; B. AZ 62b; P. MSh 3:1.

23. MSh 2:2.

"in Jerusalem"—outside of Jerusalem such use of second tithe oil is forbidden (above, ii, 5).

24. Tos BḲ 7:21; B. BḲ 76a.

"We have already explained"—above, Sec. 17.

"We say"—the *editio princeps* reads "I say" (Maimonides' own opinion). In either case Maimonides explains that inasmuch as we maintain that second tithe is sacred property, it follows that if it is stolen or is obtained by robbery, the guilty ones do not have to pay the fines imposed upon those who steal or seize profane produce. Obviously those who disagree and insist upon these fines in the case of second tithe must be of the opinion that second tithe is not sacred property.

"he who steals"—see XI, II, i, 3–4; ii, 1 (YJS, *9*, 60, 64).

"he who obtains . . . by robbery"—see XI, III, i, 3; vii, 1–2 (YJS, *9*, 90, 114–15). Obviously both thief and robber are exempt only from the additional fine, and must restore the stolen property to its rightful owner in any case.

25. Tos MSh 5:22; B. Men 82a.

"he must give to the Temple what is due to it," etc.—that is, the Temple receives the redemption money, to be used for its upkeep, while the produce now assumes the status of unencumbered second tithe and must be treated accordingly.

Chapter IV

1. MSh 4:7; B. 'Er 46b; Tem 26b.

"He who wishes to redeem"—as prescribed in the Torah (Deut. 14:24–26) for one who resides far from Jerusalem, and would find it burdensome to carry the produce itself thither.

2. Tos Shebi 7:3; P. MSh 4:1; MSh 4:3; B. Suk 41a; BM 44b; BB 83b.

"by substituting other produce"—so Tos Shebi 7:3, and less directly P. MSh 4:3. B. Suk 41a, on the other hand, states, "Second tithe may be rendered profane by substituting money only." This contradiction is resolved by making the latter statement exclude substitution of living creatures when produce is not available.

3. P. Dem 1:4.

"need not recite a benediction"—where doubt is involved, a benediction would be improper.

4. P. MSh 4:1; Tos MSh 4:11.

"in order not to cast contempt upon sacred property"—an instance of touching sensitivity in regard to holy things. It is as if it were sacrilegious to say, "How much is this sacred produce worth?", especially since the price of second tithe produce is lower than the price of profane produce.

5. Dem 1:2; P. MSh 2:3.

"both moneys are silver"—so that there is no advantage in the exchange.

"or the first money is silver and the second copper"—copper coins had greater currency in the market place, hence he who received copper coins in exchange for silver was the gainer. On the other hand, silver coins have the advantage of not tarnishing as quickly as copper coins. See below, Sec. 7.

6. MSh 1:5; P. MSh 1:9; B. Suk 40b.

"lest he should breed herd upon herd out of them"—instead of eating them in Jerusalem as required. The assumption is that the animals have some bodily defect which disqualifies them for sacrificial purposes; if they are physically perfect, they must be brought to Jerusalem for sacrifice as peace offering and for consumption (see below, vii, 12).

7. MSh 2:6.

"In case of emergency," etc.—copper coins circulate more easily.

8. Dem 1:2.

"of doubtfully tithed produce"—concerning which the Sages have been lenient even in the absence of any emergency.

"silver for silver"—by substituting silver money for the original silver redemption money; the same applies to what follows.

"silver for copper"—the *editio princeps* adds here "and copper for silver," which too is forbidden in second tithe of assuredly tithed produce (see above, Sec. 5).

9. Sif Deut. 14:25; B. Suk 4a; BM 54a, 47b; Tos MSh 1:4; P. MSh 4:1.

"bind up the silver"—the usual translation is *the money,* but the Rabbinic sources take the Biblical term literally as coined silver.

"the required one-fifth"—see below, v, 1; the additional fifth is required only when one redeems his own second tithe produce.

"plain bullion"—the original term is *'ăsimon,* metal that unlike a coin has no engraving upon it. In modern Hebrew the word signifies a slug or token used for phone calls or as fare in public transportation.

"worth less than one pĕruṭah"—since a pĕruṭah is the smallest copper coin, any silver worth less than that would be a chip of metal rather than a regular coin.

10. Tos MSh 1:6; P. MSh 1:2; MSh 1:2.

"if still current in their name"—that is, if the old coinage has not been recalled because of the present government's hostility to the former king (*damnatio memoriae;* see *Oxford Classical Dictionary, s. v.*), and is still officially permitted to circulate as legal tender.

11. Tos MSh 1:6; MSh 1:2; P. MSh 1:2; B. BḲ 98a; Sif Deut. 14:25.

"since it is regarded as still in his possession"—it is not so at the moment, but it is within his power to get it back into his possession.

12. Tos MSh 1:6; B. BḲ 115b.

"he may redeem with it"—the major commentaries suggest that the rule applies only *ex post facto,* i.e., if he has already made the verbal declaration, it is not invalidated by the likelihood of his being robbed of the money; if he makes the declaration after noticing the robber, it is null and void. Cf. above, III, iii, 17, end.

13. Giṭ 3:3.

"to redeem second tithe produce therewith"—he sets aside money beforehand to be used for such redemption whenever the need for it should arise.

14. Tos MSh 1:6; P. MSh 1:1; B. BḲ 97b.

"located in Tiberias"—RABD disagrees and says that according to the Talmudic sources, the deciding factor is whether the money used for redemption is current where it is at the moment.

15. Tos MSh 4:12–13; B. 'Er 37b; P. Dem 7:5.

16. Tos MSh 4:12–13.

17. Tos MSh 3:17; P. Dem 7:5; Ter 3:5.

18. B. BM 52b, 57a; MSh 4:2; P. MSh 4:3; Tos MSh 3:1–3.

"lower price . . . higher price"—wholesale and retail, respectively.

"the money"—used to replace the original redemption money.

"at the rate," etc.—when the money-changer changes a large coin for small coins, he pays out less than the fair value of the large coin, the difference being his fee. When he changes small coins for a large one, he charges a fee above the fair value of the large coin.

20. MSh 4:12; Tos MSh 3:5–6; B. Sanh 14b; P. MSh 4:10.

"has begun"—as the deterioration proceeds the value diminishes, and since the exact degree of deterioration cannot be easily determined, it is difficult to make an exact appraisal of market value at a particular moment.

"The owner should be compelled"—since if the owner himself redeems it, he must add one-fifth (below, v, 1).

"in the case of consecrated property"—where the owner is compelled to bid first only when the law of Jubilee is in abeyance (VI, IV, V, I; YJS, *15,* 188–89).

21. MSh 4:1; P. MSh 4:1; Tos MSh 3:1. Cf. the next Section.

"at the price in the place where it is redeemed"—even if he has incurred some expense in bringing it there.

"it might have been sold at that price"—while it was in the place where the prevailing price was low.

22. MSh 4:1.

"at the price prevailing in the city," etc.—which is higher. The law obligates him to convey his second tithe to Jerusalem at his own expense. How he routes it on its way to Jerusalem—directly to Jerusalem or via the nearest city, via a high price place or a low price place—does not matter, and he must bear the cost of conveying. If he chooses to redeem the tithe in a high price place, he has only himself to blame for the higher cost of redemption.

Chapter V

1. MSh 4:3; B. BḲ 69b; Zeḇ 6a; BM 53b; Sif Lev. Bĕ-ḥuḳḳoṯay 12:9; Sif Num. 5:7.

"as a gift while the original produce was still untithed"—the second tithe itself, being holy, may not be given as a gift (above, iii, 17).

"as we have explained"—above, iii, 17.

"one-fifth"—of the grand total, or as we would say, 25 percent of the original amount (100 percent of the original value plus 25 percent totaling 125 percent).

2. B. Ḳid 24a; Men 82a.

"Similarly," etc.—this goes back to the rule in Sec. 1, above, requiring the payment of the additional one-fifth at redemption.

3. B. BM 54b.

"without adding"—the penalty applies only to the original principal value.

4. B. BM 53b f.; P. MSh 4:2–3; Dem 1:2.

"amounts to less than one pĕruṭah"—meaning that the entire second tithe is worth less than four pĕruṭah.

"Similarly," etc.—this paragraph seems to have no connection with the preceding. Perhaps it should follow Sec. 5, where it logically belongs.

"if the value . . . is not known"—and one-fifth thereof may be less than one pĕruṭah. Otherwise it must be appraised by three merchants (above, iv, 20).

"any second tithe that one redeems with money other than his own"—variant reading (supported by P.): "of which neither the tithe itself nor the redemption money are his own." This would agree with the rule in Sec. 1, above, that only when one redeems his own second tithe is he liable to the one-fifth.

"doubtfully tithed produce"—even in the case of assuredly tithed produce the nonpayment of the one-fifth does not invalidate the redemption (below, Sec. 12). When the tithing is doubtful, and especially in view of the fact that even unlearned persons are presumed to set aside tithes, the Sages have been lenient to the buyer of doubtfully tithed produce by requiring him to pay the principal only and not the added one-fifth (see Bertinoro's commentary to Dem 1:2).

5. B. BM 52b, 53b.

"in lieu of the first money"—that is, the redemption money is to cover both batches.

"to apportion his money with exactitude"—it is possible therefore that the redemption money of the first batch is sufficient to cover the redemption of the second batch as well.

6. P. MSh 4:2.

"does not assume the status of tithe"—meaning that it does not become holy and consequently need not be used in the prescribed manner, but rather remains profane and may be spent by the owner in any way he wishes. In the case of consecrated property the rule is that if one redeems it at more than its fair value, the excess value also becomes sacred. The Talmud explains this difference by stating that a person is more generous with the Sanctuary than with himself. In the case of second tithe, since it is consumed by the owner, he would wish to redeem it for its fair value and no more (P. MSh 4:2; VI, IV, vii, 11; YJS, *15*, 203).

7. MSh 4:3. The owner is obligated to make the first bid (above, iv, 20).

"he has added to the principal"—thus increasing the basic value of the second tithe, even though this increase may be less than the one-fifth that the owner would have had to add, because the other bidder is exempt from the one-fifth. See Bertinoro to MSh 4:3.

8. MSh 4:4; Tos MSh 4:3; P. MSh 4:3; B. Giṭ 65a.

"subtlety"—"artifice used to make something permitted constitutes subtlety (*ha'ăramah*); artifice used for any other purpose constitutes fraud (*mirmah*)" (PhM to Tem 1:1). Note that the owner's wife is not mentioned here, because a gift to her becomes her own property, with the husband entitled only to its usufruct.

"Hebrew bondsman"—who unlike a heathen bondsman is not the perpetual property of his owner, but is only indentured to serve him for a limited period of time, and is entitled to act independently.

"He may not, however, say to him," etc.—because in that case the bondsman would be acting as an agent for the owner, which is the same as if the owner himself were performing the redemption.

"Redeem for me with your money"—according to the commentators, this means that the owner first gave him the money as a gift and then asked him to use it for the redemption of the owner's tithe. A variant reading, however, has, "Redeem your tithe with your money," suggesting that he has been given both money and tithe.

9. MSh 4:4; B. Giṭ 65a.

"an unperforated pot"—see above, III, v, 14–16.

"can only be a minor"—and goes free the moment she reaches puberty (XII, v, iv, 5; YJS, *5*, 260).

"cannot acquire title for other persons"—but only for himself (XII, II, iv, 7; YJS, *5*, 123).

10. MSh 4:5; Tos MSh 4:3; P. MSh 4:5.

"A similar subtlety"—the advantage of this subtlety is that the recipient is less likely to abscond with the produce than with the money.

"while the original produce is still untithed"—if the produce has already been tithed, he may not give the second tithe as a gift, since it is no longer his (above, iii, 17).

11. Tos MSh 4:7; P. Pe 5:4; B. Ḳid 32a, 24a; P. MSh 4:4; 'Er 7:6; Ḳid 1:3.

"brings in"—as her dowry or her inheritance.

"as we have explained"—above, iii, 17.

12. B. BM 54a.

"for himself"—and is thus liable to the additional one-fifth.

13. MSh 2:7; B. BM 45a.

"to carry"—to Jerusalem.

"If he does it for himself"—meaning "even if he does it for himself," which is more logical. Such an exchange cannot be regarded as a second redemption of the copper or silver coins in lieu of the gold denars.

14. MSh 2:8–9.

"If one changes"—the exchange is understood to be a matter of emergency, since otherwise silver may not be changed into copper (above, iv, 7).

"all into copper coins"—which is more easily spent than, but is inferior to, silver.

15. MSh 2:8; P. MSh 4:8.

"is worth less than one denar"—and is not easily rendered profane by itself.

Chapter VI

1. MSh 2:5; P. MSh 2:5; Tos MSh 2:4.

"here and there . . . indiscriminately by the handful"—in the former case it is at least possible that only tithe money has been picked up; in the latter case the two moneys are bound to be intermixed.

"if . . . profane money"—or a mixture of tithe and profane moneys.

2. MSh 2:6; Tos MSh 2:5.

"the better of the two sela' "—which might have been the original tithe sela' as easily as the other less perfect sela', and here the decision is in favor of the tithe.

3. MSh 4:12; Tos MSh 5:7, 5.

"one mina"—one hundred zuz.

"it is considered profane money"—the presumption is that what was removed was tithe money, and what was found was profane money.

"the surplus is profane money"—the presumption is that the father knew there were two hundred zuz, but mentioned only one hundred, since he was concerned that the son should treat the tithe money with proper reverence.

"If he has deposited"—in this case the father knew how much he had deposited, and any variation from it indicates that what he found was not what he had deposited.

4. Tos MSh 5:5.

"the largest"—the presumption is that the father was referring to the largest bag.

5. Tos MSh 5:8; Giṭ 7:1; P. Giṭ 7:1; B. Giṭ 71a.

"as is done in the case of divorce"—see IV, II, ii, 16 (YJS, *19*, 176).

6. Tos MSh 5:9; P. MSh 4:9; B. Sanh 30a.

7. Tos MSh 5:10–11.

"it is profane money"—and he presumably wished merely to scare them so that they would not squander the money foolishly.

"a ruse"—to prevent them from appropriating the money.

"if he meant it literally"—that the money is holy, so that they would know it, in case he should forget to advise them or die.

8. MSh 4:11; Tos MSh 5:2.

"a vessel"—an earthenware vessel of slight intrinsic value, in which case the contents are more important, and the designation is therefore presumed to refer to the contents and not to the vessel itself.

"second tithe"—although this initial letter might refer to either the first or the second tithe, in this case it must refer to the second tithe, because the first tithe may be eaten without any restrictions, and the owner is therefore presumed to have inscribed the identifying letter in order to inform people that here there are restrictions.

"metallic vessel"—such a vessel is valuable and as important as its contents, and the designation must refer to it as well.

"in times of danger"—when people were afraid to reveal the nature of the vessel's contents, lest they should be persecuted for observing the commandments.

9. MSh 4:9; P. MSh 4:11; B. Pes 7a; BM 26b; Sheḳ 7:2.

"even if it includes gold denars"—it is not usual for people to mix them with lesser coinage, hence one might think that this is second tithe money.

"the streets of Jerusalem are swept"—and if this money had been of second tithe brought by pilgrims for the festivals, it would have been swept away.

"When does this apply?"—this goes back to the first paragraph of this Section.

10. Sheḳ 7:2; B. Men 82a; P. Sheḳ 7:2.

"the Chamber"—in the Temple, reserved for moneys used for purchasing animals to be offered up on the altar.

11. Tos MSh 2:11; B. Pes 7a.

"If most people," etc.—even if the amount of second tithe money deposited in the chest is usually less than the amount of profane money.

"if . . . half and half . . . profane"—RABD prefers the variant reading "second tithe," which is more logical. See Prof. Lieberman's comment *ad loc.*, p. 738. Some commentators explain that it is profane money when the users are divided half and half, but only if the amount of profane money deposited in the chest is usually larger than the amount of second tithe money.

12–13. MSh 2:11.

12. "some produce"—the presumption here is that this is extraneous produce, and not produce spilled from one of the piles. Hence, it is presumed to belong to the nearer pile rather than to the larger pile (JE).

"wash his hands"—above, III, xi, 7.

"until after sunset"—above, III, vii, 2.

"is forbidden to a mourner"—above, iii, 5.

"must be brought to Jerusalem"—above, ii, 1.

13. "the stricter of the rules governing the two kinds"—note the variation from "the stricter rules" used in Sec. 12. Here the stricter rules are on one side only, namely that of assuredly tithed produce, whose second tithe may not be consumed by a person in mourning.

14. Bik 2:2.

"intermixed with profane produce"—outside of Jerusalem and in a proportion which is insufficient for one component to neutralize the other. If the profane component is large enough to neutralize the holy component, the mixture becomes profane, but must be consumed in a state of cleanness, since the holy component cannot be eaten outside Jerusalem, and thus is in the category of a substance that will eventually be permitted (see the note below). There are two factors which one would think might put the produce into this category: (a) it is possible to bring the produce to Jerusalem, where one may consume it; (b) one can redeem the holy component. But these factors are not taken into consideration, the first because the bringing to Jerusalem entails much effort, and the second because redemption is an elective procedure, and not a compulsory one. Therefore these factors cannot affect the situation in this case.

"of the same species"—reading *bĕ-minan* for *ḳam-minyan* (*editio princeps* has *bam-minyan*). Cf. below, VI, iv, 15.

"being in Jerusalem"—where it must be eaten and may not be redeemed.

"a substance that can eventually be made permitted"—which cannot be neutralized. See V, II, xv, 12 (YJS, *16,* 234). This is the Talmudic principle with respect to the neutralization of a forbidden substance intermixed with a permitted substance. Though normally a certain proportion of the latter is sufficient to neutralize the former, it cannot do so if the former is destined to become permitted eventually without recourse to neutralization.

15. Bik 2:2; Ter 9:2.

"after it has entered Jerusalem"—when it becomes subject to the rule governing a substance that can eventually be made permitted, and therefore cannot be neutralized.

16. B. BM 53a.

"enclosure"—see above, ii, 9.

"is worth less than one pĕruṭah"—when the second tithe is required by Scribal enactment only (above, ii, 9).

"as we have explained"—V, II, 15 (YJS, *16,* 231 ff.).

Chapter VII

1. MSh 3:10; 1:5.

"primary defilement"—the carcass of an animal or a creeping creature, or a living person who had come in contact with a dead human body. See above, note to III, xii, 11.

2. B. Zeḇ 48b; P. MSh 3:8.

"This stringency"—in that the original second tithe produce may be redeemed outside Jerusalem, whereas produce bought with the redemption money may not.

"secondary defilement"—an object rendered unclean by contact with a primary defilement. See above, note to Sec. 1.

"enjoined by Scribal law only"—when an object touches a secondary defilement, the object is clean according to the Torah but unclean according to Scribal law.

3. B. 'Er 27b.

"grow out of the earth"—or, in the case of domestic animals, feed on the growth of the earth.

4. MSh 1:5; 'Er 3:1; Sif Lev. Way-yiḳra' 1:7; P. MSh 1:3; B. 'Er 27a.

"truffles, or mushrooms"—which were thought to receive their sustenance not from the soil but from the air.

"cannot reach Jerusalem"—because it is perishable.

"are not similar to cattle and sheep"—which, though not growing out of the earth, feed on the earth's growth, are not attached to the soil, and do not spoil in transit to Jerusalem.

5. Tos MSh 1:10, 2:18.

"honey"—was thought to be regurgitated nectar of flowers and not the product of the bee's body (V, II, iii, 3; YJS, *16,* 164).

6. MSh 1:3; B. 'Er 27b; Ḥul 26a.

"Grape skin wine"—an inferior wine or vinegar made by steeping husks and stalks of pressed grapes in water. The fact that it turns sour proves that it contains some wine essence, and is not mere water.

7. B. BB 96b f.

8. Tos MSh 1:13–14; 'Uḳ 3:7; P. Shebi 4:6.

"sweetened"—by pickling in wine or vinegar, which neutralizes their bitter taste.

"Terminal palm buds"—soft enough to be edible, even though they would eventually harden into wood.

9. Tos MSh 1:13–14; Midrash Tannaim to Deut. 14:26; B. Nid 51b.

10. B. Nid 51b; 'Er 27b; Tos MSh 1:10; 2:18.

"milk sauce"—*ḳamaḳ,* a Persian sauce of sour milk, bread crusts, etc., also called *ḳutaḥ.*

"fish brine"—*muryas,* Latin *muries,* brine containing hashed fish and sometimes wine.

"must be included in the price of the oil"—theoretically one should pay only for the oil (above, Sec. 4), and the other ingredients should be included free of charge, but actually one should set a higher price on the oil so as to pay for the other ingredients also.

11. MSh 3:2; P. MSh 3:2.

"would reduce the number of persons"—hence is to the disadvantage of everyone.

"have shed their uncleanness at sunset"—literally "priests for whom the sun has already set," that is, unclean priests who have not only purified themselves by immersion but have also waited past sunset of that day (above, III, vii, 2).

"is permitted to a mourner"—X, v, xii, 15 (YJS, *8,* 300).

"who has immersed himself"—even though the sun has not yet set (above, iii, 4).

"is forbidden to a mourner"—above, iii, 5.

12. Sif Deut. 14:26; MSh 3:2; P. MSh 1:2; B. 'Er 28a.

"may be consumed by commoners"—and thus the number of persons who may eat them is not reduced thereby.

"avoid sacrificing them"—when an animal is sacrificed on the altar, certain parts of it must be given to the priests. If the animal is bought for profane meat, there is no such obligation.

"are not fit for peace offerings"—hence the altar is not deprived of its due.

13. P. MSh 3:2.

"must clear it out"—hence the purchase is a waste of money.

"as will be explained"—below, VII, vii, 1.

14–15. MSh 1:5.

14. "water, salt,"—all these may not be bought with second tithe money (above, Sec. 4).

15. "If one buys produce outside of Jerusalem," etc.—even though this is forbidden (above, iv, 6).

"unwittingly"—unaware that the money is second tithe redemption money.

"to return the money"—and take back the produce, since the sale is null and void, seeing that the buyer has paid money that was not his to spend, although he thought it was his.

"deliberately"—in which case the sale is valid, since the buyer knew that the money was not his. He is therefore penalized by having to carry the produce itself to Jerusalem, since he is forbidden to reconvert it once more into money (above, Sec. 1).

"to rot"—above, ii, 4.

16–17. B. Ḳid 55b f.; MSh 1:6.

16. "buried together with its hide"—so that he would derive no benefit whatsoever from it.

17. "an unclean animal"—an animal which is unfit for consumption and may not be sacrificed on the altar in the Temple.

"burnt offerings"—which no one may eat (Lev. 1:3–9).

"sin offerings, or guilt offerings"—which only the priests may eat (Zeḇ 5:3, 5).

18. MSh 1:4; B. Men 82a; P. MSh 3:2.

"a wild beast"—a nondomestic animal which is permitted for food but not for a sacrificial offering (Lev. 1:2).

"for human consumption"—literally "for flesh of desire," referring to food eaten to satisfy one's hunger and not brought first as an offering (Deut. 12:20). Such a purchase is forbidden (above, Sec. 12).

"and a blemish befalls it"—so that it becomes unfit for a sacrificial offering.

"departs from it"—but its sanctity as a blemished sacrificial animal, which must be redeemed before it may be eaten, remains.

"he must redeem it"—not as second tithe but as a blemished sacrificial animal. The redemption money does not therefore become second tithe.

"he must add one-fifth"—as is required in the redemption of blemished consecrated animals (VI, IV, vii, 4; YJS, *15*, 201).

19. B. Men 82a.

"the money does not become peace offering"—and may be used for any other purpose.

"sacred property"—see above, iii, 17.

20. Tos MSh 3:9; P. MSh 1:1.

"as if it were profane produce"—i.e., outside Jerusalem.

"let him cry out to heaven"—reading, with Tos, P., and several manuscripts, *yiṣ 'ak̠ lĕ-šamayim;* the printed text reads *yaṣa' yĕd̠e šamayim*. The former reading means that he can do nothing but cry out to heaven and throw himself upon heaven's mercy, since the consumed sacred produce is no longer in existence and cannot be redeemed. The latter reading means that he has satisfied the requirements of heaven, which is obviously not the case.

Chapter VIII

1. MSh 1:3; P. MSh 1:1.

"a wild beast for human consumption"—see above, note to vii, 18.

"is not particular"—about the hide, and values only the beast's flesh.

"the hide becomes profane property"—and the buyer need not redeem it and spend its monetary equivalent in Jerusalem.

"does not become profane"—and comes under the rule governing things which, though bought for second tithe money, cannot be brought to Jerusalem for consumption. See above, vii, 17.

2. MSh 1:3.

"are . . . sold sealed"—the explanation is that in the case of sealed jugs, the jugs themselves are regarded by the seller as unimportant and go with the wine. If the wine becomes profane, so does the jug. In the case of unsealed jugs, however, the wine is sold by itself, and the jugs are not included with it.

"strict with himself"—and unseals the jugs, not because he wants to sell only the wine and charge extra for the jug, but because he wishes to make sure that the price he charges corresponds exactly to the amount of wine in the jug. The opening therefore does not affect the jug; it is the same as if it were sealed, and consequently it becoms profane.

3. MSh 1:4.

"unsealed or sealed"—what counts is local custom, and whether the seller is or is not a merchant. If the local custom is to sell unsealed, it is evidence that the jug itself is not included in the sale. The same is true if the seller is a merchant.

4. MSh 1:3; P. MSh 1:2.

"profane"—the principle behind the rule is that if the container is of insignificant value, it does not become second tithe.

"if the dates are not pressed"—the containers are analogous to the vessel of figs or grapes in the preceding Section.

5. MSh 3:12; Tos MSh 2:18, 1:10; P. Giṭ 3:8; P. MSh 3:10; B. Ket̠ 108b.

"they do not assume the status of second tithe"—because they are not included with the wine, having been assigned as temporary containers of it.

"without any assignment"—without designating the wine as second tithe wine, and the vessels as temporary containers.

"after sealing"—once the vessel is sealed, unsealing it would expose the wine to spoilage. The vessel thus becomes one with the wine, and therefore likewise acquires the status of second tithe.

"one quarter-log̠ of profane wine"—Prof. Saul Lieberman (to MSh 2:18, p. 748/110) explains this as follows; It was the custom to reserve a certain amount of one's produce in order to set aside therefrom the second tithe for other produce as the need may arise. After setting aside the tithe for all of one's produce, one might have one-quarter of a log̃ left over which would not become second tithe. It is assumed that the vessel is then filled to the brim with second tithe wine.

"oil, vinegar, fish brine, or honey"—which do not become second tithe, but are listed here by association with other places (Lieberman, *ibid.*, note to 107).

"it does not become second tithe"—because in each case the contents do not require the container, which is therefore not included with them.

6. MSh 3:11, 4:8.

"A gazelle"—for human consumption only (above, vii, 12).

"as any other produce that has been rendered unclean"—above, ii, 8; vii, 1.

"twenty mĕʿah worth . . . five mĕʿah worth"—the new equivalents, depreciated and appreciated, respectively, of the remaining half of the denar.

7. MSh 4:6; B. Ḳid 54b.

"takes possession"—literally "draws it" (see XII, 1, iii, 1; YJS, *5*, 11).

"only one selaʿ "—the produce becomes the property of the purchaser as soon as he takes possession of it by "drawing," but it retains its sanctity as second tithe until the price is paid to the seller. Therefore the increase in its monetary value becomes second tithe money, and the purchaser must redeem it at its new price of two selaʿ, which becomes second tithe and must be spent in Jerusalem.

"Then he shall give the money"—the Biblical text reads, *Then he shall add the fifth part of the money;* the inaccuracy is probably due to quotation from memory.

"the handing over of the money constitutes transfer of title"—according to Scriptural law. In the case of movables, Scribal law regards "drawing" as legal transfer of title (see XII, 1, iii, 1; YJS, *5*, 11).

8. MSh 4:6; Tos MSh 4:14.

"only one selaʿ "—because as far as second tithe is concerned, it is the payment of money asked at the time when the produce is worth one selaʿ that counts. On the other hand, as far as legal ownership is concerned, it is the formal act of "drawing" that counts, and at the time of "drawing" the produce was worth two selaʿ. Hence the purchaser must pay only one selaʿ out of second tithe money but must add another selaʿ out of profane money.

"If the seller is an unlearned person"—who is presumed to be careless about doubtfully tithed produce anyway.

"If the purchaser pays the seller," etc.—in this case the purchaser paid the money first and then gained legal ownership by "drawing." Hence, as far as second tithe is concerned, the transaction was completed with the payment of the one selaʿ. Legally the seller could withdraw from the deal after the payment of the one selaʿ, unless the buyer is willing to add another selaʿ; but if he does withdraw, he is subject to the oath prescribed for those who do not abide by their pledged word, even if legally he could break it. Hence the rule is that the produce, now worth two selaʿ, becomes second tithe, and as for the second selaʿ, the two parties must settle the matter between themselves, either by agreement or through litigation.

9. Tos MSh 4:14. See note to the preceding Section. In this case the loss accrues to the second tithe money, and not to one of the parties as before.

10–11. MSh 3:4. These rules do not conflict with the rule above, iv, 6, where both money and produce are located outside Jerusalem.

12. Tos MSh 3:3.

"and is in need of spending it"—for goods that cannot become second tithe (see above, iii, 10), while B has produce which he wishes to consume and not to sell. The solution given provides relief for A without causing any loss to B: the produce becomes second tithe, but since B is located in Jerusalem, he may consume it right there so long as he is ritually clean.

13. Tos MSh 3:3; P. MSh 3:3; ṬY 4:5; Tos MSh 4:9.

"only to an associate scholar"—and not to an unlearned person (above, iii, 8). Once A makes his declaration, B's produce becomes nominally A's property, which he then hands over to B to be consumed by the latter.

"to redeem doubtfully tithed produce"—in which case the status of A's money is doubtful, and the restriction (above, iii, 8) does not apply.

"to render second tithe"—either the produce itself or its redemption money.

"that these too might be second tithe"—one may not render one second tithe profane in lieu of another second tithe, or in other words, one sacred property cannot render profane another sacred property.

14. MSh 4:8.

"in order to go on consuming . . . against it"—the denar is redemption money for second tithe produce, and one renders it profane not all at once but piecemeal, against consecutive small amounts of other money or produce, as they are consumed. The rule here is that the denar becomes profane not piecemeal but only when the balance of it comes down to less than one pĕruṭah (cf. above, iv, 13).

"the statutory one-fifth"—above, v, 1. This is not required of doubtfully tithed produce (above, v, 4).

"four-fifths of one pĕruṭah"—the additional $\frac{1}{5}$ ($\frac{1}{5}$ of $\frac{4}{5} + \frac{1}{5} = \frac{1}{5}$) plus $\frac{4}{5}$ would equal the statutory minimum of one pĕruṭah, the balance being required to be below that minimum.

15. MSh 2:12; P. MSh 2:10. This Section can be understood better when compared with its source in the Mishnah: "If some of one's sons are clean and some unclean, he must lay down a sela' and say, 'This sela' is to be rendered profane in lieu of what the clean ones are drinking.' The clean ones and the unclean ones may then drink out of the same jug." Maimonides expands this rule to include any group of people consisting of both clean and unclean persons, but gives away his Mishnaic source when he abruptly changes the verb from the plural to the singular ("a person must . . . say"), which seems to suggest a third party interested in the proceedings. In the Mishnah the third party is the father, while here there is actually no third party, since the person who makes the declaration is one of the diners. Either Maimonides regarded the Mishnaic passage as a particular illustration of a general rule as stated here, not restricted to father and sons as in the Mishnah, or else he had a variant Mishnaic text before him, or perhaps the text here needs to be corrected.

Chapter IX

1. Pa 1:1; B. Ber 35a; Sif Num. 5:10; MSh 5:2; Sif Lev. Ḳĕḏošim 3:9; P. Pe 7:5; P. MSh 5:2.

"there is no second tithe in Syria"—above, i, 14.

"except fourth year's fruit"—therefore the aforecited verse, even though worded in general terms, must refer specifically to it.

2. Sif *ibid.;* MSh 5:3; P. 'Or 1:9; Tos MSh 5:19; P. MSh 5:2; Ter 1:5; Bik 2:4; B. Ḳid 54b.

"he must add one-fifth"—above, v, 1.

"produce"—usually translated *increase.*

"while it is still attached to the ground"—because it is difficult to assess its value for redemption purposes. See Lieberman *ad loc.*

"like second tithe"—above, iii, 17.

"it may not be acquired as a gift"—for one may not give away what is not his.

3. Tos MSh 5:19; Ter 11:3.

"grapes . . . olives"—are treated differently from other fruits, because when they are converted into a fluid, they are more easily assessed for redemption purposes, whereas other fruits are assessed more easily while in their natural form.

4. MSh 5:3; Pe 7:6. Cf. above, II, i, 7. Since the entire fourth year's yield is sacred property, it is not subject to sacred gifts.

5. MSh 5:2; Tos MSh 5:14–16; B. Beṣ 5b.

"grapes . . . must be brought up"—and may not be redeemed outside Jerusalem.

"to decorate the streets of Jerusalem"—usually grapes were first converted into wine and then brought to Jerusalem, thus there was a scarcity of grapes there. The owner was also permitted to redeem the grapes in Jerusalem by selling them to local merchants who would display them in their shops, while the owner consumed the redemption money. Normally such a sale was forbidden (above, iii, 17), but it was permitted in this case in order to decorate the streets of Jerusalem.

"close to the city wall"—and not inside the wall, as heretofore.

6. MSh 5:4.

"on condition," etc.—the estimated price should equal the full value of the fruit less the cost of production and transportation.

"During the Sabbatical year"—if the fourth year of the new plantation happens to coincide with a Sabbatical year.

7. MSh 5:5; P. MSh 5:1; B. BḲ 69a.

"within its years of *'orlah"*—the first three years, when the fruit is called 'orlah (literally "foreskin") and is entirely forbidden (see Lev. 19:23–25).

"is more stringent"—more so than the interdict of fourth year's fruit, which is forbidden for consumption only.

"Scrupulously pious persons"—who would not rely solely on the marking with clods of earth.

"as we have explained"—above, Sec. 2.

8. RH 1:1; Tos 'Or 1:1, 4; Sif Lev. Ḳĕḏošim 3:3; B. RH 9b, 10b.

"From the time of planting," etc.—if the planting takes root not less than thirty days before the end of the year, that year counts as the full first year, and the following year counts as the second year, etc.

"How much time do all trees need"—so manuscripts; the printed text has, "How much do all trees need generally."

9. B. RH 9b, 10b; P. RH 1:2.

"Nevertheless"—although the three years are over by the first of Tishri of the fourth year, one must wait until the fifteenth of Shebat of that year.

"or as fourth year's fruit"—reading *'o bi-rěḇa'i* for *'ella' ḇi-rěḇi'iṯ* of the printed text, which makes poor sense.

"New Year's Day for trees"—above, i, 2.

10. P. RH 1:2; P. Bik 2:4.

"on the fifteenth of Ab"—44 days before New Year's Day of the 11th year, which counts as a full year.

"a leap year"—which has an extra (13th) month (second Adar).

11. P. RH 1:2.

"on the sixteenth"—which leaves less than forty-four days of that year.

"from New Year's Day of the fourteenth year to year-end"—in this case one need not wait until the fifteenth of Shebat of the fifteenth year.

12. P. 'Or 1:2; Sif Lev. 19:23.

"from day to day"—from the day of planting to the same day three years and one year, respectively, hence.

"some statements ascribed to the Geonim"—see V, II, x, 15, 18 (YJS, *16*, 205–06); for the location of these statements see the notes to that Section, pp. 365–66.

13. 'Or 1:7, 8.

"Leaves," etc.—all these cannot be termed fruit.

"east wind"—blowing from the desert, hot and very dry.

"in all cases"—i.e., either as 'orlah or as fourth year's fruit. It is forbidden because even in this premature state it is regarded as within the term fruit.

Chapter X

1. P. 'Or 1:1.

"as it is said," etc.—the implication of the verse is that the fourth year is simply a continuation of the first three years.

2. 'Or 1:1; P. 'Or 1:1; Sif Lev. Ḳĕḏošim 3:3.

"and thereafter for food"—having changed his mind upon the expiration of the three years, when the law of 'orlah can no longer apply retroactively.

3. 'Or 1:1; P. 'Or 1:1; Ma 4:6; Tos 'Or 1:1.

4. 'Or 1:2; P. 'Or 1:2; B. Pes 22b; Sif Lev. Ḳĕḏošim 3:4; Tos 'Or 1:2.

"outside of the Land of Israel," etc.—the reason for this leniency is that even in regard to the Land of Israel there is a difference of opinion in this matter. Hence the stricter rule is applied to the Land of Israel and the more lenient rule to lands outside of the Land of Israel.

5. 'Or 1:2; Tos 'Or 1:2–3.

"in a public domain"—but for the planter's own benefit only.

"or in a ship"—when the ship is tied up at the shore, any living plant on board ship is regarded as connected to the soil underneath the ship (above, III, i, 23).

6–7. 'Or 1:1.

6. "craggy"—barren and unsuited to fruit trees.

7. "lulaḇ"—used during the Feast of Tabernacles; see Lev. 23:40.

"and then consecrates it"—consecration cannot invalidate an already existing liability to 'orlah.

8–9. P. 'Or 1:2.

8. A living tree, even if planted in an unperforated pot, may draw nourishment

from the underlying land through its roots, since the substance of the pot is porous. A mere seed cannot do so.

9. "implying, all trees"—including those planted by heathens; otherwise *shall come into the Land* would have been superfluous, since the Israelites themselves obviously could not possibly have planted trees in the Land prior to their arrival there. Hence this clause is taken as implying that the rule covers trees planted by heathens as well.

10. P. 'Or 1:2; Ter 3:9. Cf. below, Sec. 14, and above, Sec. 1.

11. Sif Lev. Ḳĕḏošim 3:4; 'Or 1:3.

12. 'Or 1:4, 3.

"wraps"—to assure the uniformity of the weaving.

"as it was"—without filling in with additional earth.

13. P. Sheḇi 1:8.

14. B. RH 9b; 'Or 1:5.

"bends . . . into the ground"—so that it would take root and grow into an independent new tree.

15. 'Or 1:5.

"one two-hundredth"—this is the minimum increase which renders produce holy (above, I, v, 21–22).

16. B. Neḏ 57b, Soṭ 43b.

"daughter-tree"—less than three years old.

"two-hundredfold"—which normally would neutralize forbidden 'orlah fruit (cf. above, note to Sec. 15).

"primary growth"—the fruit that was on the daughter tree from the beginning is primary, and was originally forbidden.

17–18. 'Or 1:5.

19. Tos 'Or 1:4; B. BM 119a; BB 82a; Soṭ 43b.

"it looks like a one-year-old plant"—and people who see the owner eating the fruit thereof might think that 'orlah fruit is permitted.

"two plants facing two others with a fifth . . . extending like a tail"—which constitutes a small vineyard (above, I, vii, 7). And what about two, three, and four plants? The idea is that when the entire vineyard is less than one handbreadth in height the matter becomes common knowledge; and this excludes even what is termed a small vineyard, i.e., two plants facing two others with a fifth extending like a tail.

"becomes common knowledge"—literally "has a voice," and those who see the owner eating the yield will know that it is not 'orlah.

20–21. 'Or 1:9; Tos 'Or 1:5; B. AZ 48b.

20. "as we have explained"—V, II, x, 9 (YJS, *16*, 204).

"what grows out of it is permitted"—on the principle of "two causes" stated in the following Section.

21. "what grows out of 'orlah fruit"—like the 'orlah nut in the preceding Section.

Chapter XI

1. Deut. 26:12–13; MSh 5:6, 10; B. Meḡ 20b; Soṭ 32a.

"confession of the tithe"—although it covers all the sacred gifts out of the yield of the land (see below, Sec. 12), it is called confession of the tithe, because basically it applies to the tithe (see below, Sec. 14).

2. *"all the tithe of thine increase"*—which includes the poor man's tithe, due in the third and sixth years of each septennate (above, II, vi, 4).

3. MSh 5:10; P. MSh 5:5, 3.

"In the afternoon"—literally "at the time of the afternoon prayer," the final hours of Passover.

4. Meg̃ 2:5; MSh 5:7; P. Ta 1:4.

"daytime . . . any part of daytime"—although it is preferable to do it in the afternoon, at the time of the afternoon prayer (above, Sec. 3).

"or is not in existence"—RABD regards the confession as obligatory only while the Temple is in existence.

5. Soṭ 7:1.

"jointly"—one person reciting the formula of confession, and the others signifying their participation in it by listening to him.

6. Ta 1:4.

"The commandment"—the thrust of the Hebrew word (*miṣwah*) is that it is the meritorious and preferred way.

7. MSh 5:10, 6.

8. MSh 5:6; Dem 1:2.

"of doubtfully tithed produce"—cf. above, IV, ix, 2.

"wherever they are"—although normally they should be brought up to Jerusalem (below, VI, ii, 1); once the time of putting away has come, one must dispose of them.

9. MSh 5:6.

"considered the same as put away"—there is a difference of opinion among the commentators about the reason for this rule. According to one view, it is that such comestibles deteriorate quickly if left to stand and will soon go to waste anyway. According to another view, the produce subject to putting away becomes neutralized when allowed to stand.

10. MSh 5:8.

"produce which . . . its season of tithing"—from which the hallowed gifts have not yet been set aside.

11. MSh 5:9.

"by dint of land"—on which the movables are located. If the title to the land is transferred simultaneously, the new owner automatically acquires title to the movables, even if the latter are not paid for and are in fact a bonus or a gift (XII, I, iii, 8; YJS, *5*, 13).

"symbolic barter"—transfer of title by symbolic exchange of goods (see XII, I, v, 5; YJS, *5*, 21).

"heave offerings"—omitted in the *editio princeps,* since they are not expressly mentioned in Deut. 26:12–13.

"must be given"—Deut. 26:12–13.

12. MSh 5:10.

"refers to second tithe"—cf. above, iii, 4, where it is called holy on the basis of Lev. 27:30.

"likewise called holy"—cf. above, ix, 1, where it is called holy on the basis of Num. 5:10.

"and also"—the word *also* is seemingly superfluous, and is therefore interpreted as implying other gifts not specified in this verse.

"gleanings, forgotten sheaf, and corner crop"—concerning which Scripture speaks not of giving but of leaving, Lev. 23:22; cf. above, II, i, 10.

13. MSh 5:11.
"in their proper sequence"—see above, III, iii, 23.
"During the time"—of Ezra (above, IV, i, 4).
14. P. Bik 2:2.
"the first hallowed thing to be set aside"—above, III, iii, 23.
15. MSh 5:11, 12.
"in my mourning"—above, iii, 5.
"being unclean"—above, iii, 1.
"for the dead"—above, iii, 10.
"to other mourners"—who might unknowingly eat of them while still mourning.
"chosen House"—the Temple in Jerusalem.
16. MSh 5:13.
17. MSh 5:14.
"given us"—meaning native born Israelites.
"cities with open land around them"—see Lev. 25:34.

TREATISE VI: LAWS CONCERNING FIRST FRUITS AND OTHER GIFTS TO THE PRIESTHOOD

Chapter I

1. Tos Ḥal 2:7; B. Ḥul 133b, 132b; BḲ 110b; Sif Num. 18:20.
"all are explicitly stated in the Torah"—see Num. 18:8–20.
"a covenant was made with Aaron"—see Num. 18:19.
"has no share in the priesthood"—i.e., is not allowed to participate in the service of the Temple.
2. "that is endowed with sanctity"—see below, Sec. 6, for examples of gifts that are hallowed and those that are not.
3. Tos Ḥal 2:7.
4. Sif Num. 18:20.
"sin offering"—see Lev. 4.
"guilt offering"—see Lev. 5:14–19, 7:1–10.
"congregational peace offerings"—see VIII, v, i, 4; x, 3 (YJS, *12*, 163, 202).
"the residue of the 'omer"—see VIII, VI, vii, 12 (YJS, *12*, 278).
"the two loaves"—Lev. 23:17.
"the shewbread"—Exod. 25:30; Lev. 24:5–9; VIII, VI, iv, 10–14 (YJS, *12*, 266–67).
"the leper's log̃ of oil"—see Lev. 14:10.
5. "the breast and the thigh"—Lev. 7:32–34.
"thank offering"—see Lev. 7:12–15.
"the Nazirite's ram"—Num. 6:14–17.
"the firstling of clean cattle"—Deut. 15:19.
"the first fruits"—see below, iii, 3.
6. "heave offering"—Num. 18:8–12.
"heave offering of the tithe"—Num. 18:26–29.
"dough offering"—Num. 15:19–21.
"first of the fleece"—Deut. 18:4.

"the field of possession"—see Lev. 27:16 ff., and VI, IV, iv, 1 (YJS, *15*, 183).

"accrue . . . only in the Land of Israel"—because if they are taken to a foreign land, they are thereby rendered unclean, which is forbidden. See above, III, ii, 17.

7. "The lay gifts"—the shoulder, the two cheeks, and the maw of every clean animal slaughtered for food; see Deut. 18:3, and below, ix, 1.

"the redemption money of the first-born son"—Num. 18:15–16, and below, Chap. xi.

"the firstling of an ass"—see Exod. 34:20, and below, xii, 2.

"things illegitimately seized from a proselyte"—if the proselyte dies without heirs to whom the stolen article might be returned, the article, plus the penalty (⅕ of its value), reverts to the priests. See XI, III, viii, 5. (YJS, *9*, 118).

"devoted things"—things devoted to the priests; see Lev. 27:28, and VI, IV, vi, 4 (YJS, *15*, 194).

8. "directly from the Temple"—i.e., these are not given to the priests by the sacrificing Israelites, but each priest receives his share directly from the Temple.

"burnt offerings"—see Lev. 1:3–17.

"all accrue to the priests"—see VIII, v, v, 19 (YJS, *12*, 183).

9. "as will be explained"—VIII, v, v, 3; ix, 1; x, 3; xii, 3 (YJS, *12*, 178–79, 197, 202, 211).

10. B. Men 84b; Bik 3:12; P. Ḥal 4:11; B. Ḥul 132a; Beḵ 4a.

"the priests of the division on duty"—literally "men of the watch." A watch (*mišmar*) was a division of priests whose turn it was to conduct the Temple service. The priestly class was divided into twenty-four such divisions. See below, iii, 1.

"as will be explained"—VIII, v, v, 19–20 (YJS *12*, 183); VI, IV, iv, 24; vi, 5 (YJS, *15*, 187, 194); XI, III, viii, 5 (YJS, *9*, 118).

11. "just as the first of the corn is given to females as well as to males"—above, III, vi, 1; and below, x, 17.

12. Tos Ḥal 2:7.

15. B. Ḥul 132a.

"if the peace offering . . . with it"—omitted in the *editio princeps*.

16. "Laws Concerning Offerings"—VIII, v, x, 3–5 (YJS, 12, 202–03).

"Laws Concerning Robbery"—XI, III, viii, 5 (YJS, *9*, 118).

"Laws Concerning Valuations"—VI, IV, iv, 1–2; vi, 4 (YJS, *15*, 183, 194).

"Laws Concerning Heave Offerings"—above, III, ii, 1.

Chapter II

1. Sheḵ 8:8; Tos Sheḵ 3:24; Sif Deut. 26:4; Kel 1:6; B. Ḥul 136a; Bik 1:10; Ḥal 4:11; Tos Bik 1:7; P. Shebi 6:6.

"Sihon and Og"—see Num. 21:21–29, 33–35.

"Ammon, Moab, and Babylonia, however"—omitted in the *editio princeps*, apparently by error.

2. Bik 1:3, 10; Deut. 8:8; P. Bik 1:3.

"mentioned"—in Deut. 8:8, where the seventh species listed is honey; P. interprets it as *pars pro toto* (date juice for dates).

3. Bik 1:3; P. Bik 1:3, 10; B. Men 84b.

"spoiled and worm-eaten"—*surot u-mĕrukkabot; editio princeps* reads, with P., *sukot u-mĕnukkabot,* "anointed with oil and punctured (to hasten ripening)," which seems to fit the context better; cf, below, VII, i, 5.

"dusted and fumigated"—to protect them from damage by insects.

4. Ter 11:3; B. Ḥul 120b; P. Ter 11:3; Ḥal 4:11.

"the liquid of olives and grapes"—since most of the olive and grape crops was used for oil and wine.

"implying . . . thereof"—omitted in the *editio princeps.*

5. Bik 3:3.

"dried figs and raisins"—fresh figs and grapes would spoil in transit.

6. Bik 1:3, 6; Ḥal 4:10–11.

"there"—in the Temple court (cf. below, iv, 10).

7. Ter 1:5; B. Men 84b; Yeb 73b; P. Bik 2:5; 1:8.

"fifteenth day of Shebat"—which is the New Year's Day for trees (above, III, v, 11).

8. B. Ḥul 136a.

9. B. Men 85a f.

10. Bik 1:12; P. Bik 1:1. Such encroachments are illegal at the outset.

11. P. Bik 1:1; B. BB 27b; BḲ 82a.

"it was on this condition"—i.e., that one is obligated to bring first fruits under such circumstances.

"that Joshua had distributed the land"—see XI, 1, v, 3 (YJS, *9,* 19–20), where this particular condition is not among the ten listed. According to B. BḲ 81b–82a, this (eleventh) condition was not acknowledged by some of the Sages.

12. Bik 2:3; P. Bik 1:2.

"Tenants, lessees"—a tenant works a field for a share of the crop; a lessee works a field against an agreed rental in money or in kind. See above, IV, vi, 13.

"even if the owners have despaired"—in the case of movables this would confer ownership upon the actual possessors. In the case of land, however, title remains forever with the rightful owners. See XI, III, viii, 14 (YJS, *11,* 120).

13. Tos Bik 1:2; P. Bik 1:8; Bik 1:6, 10; B. BB 81a.

"he does own land"—the purchase of three or more trees implies the right to pile the produce upon the ground, and is therefore regarded as conferring the right to use the ground, if not actually to own it. In the case of only one tree, purchase of it affects the land only to the extent that the tree may draw its nourishment from the ground. In the case of two trees, the matter is in doubt (below, iv, 4).

14. P. Bik 1:8; B. BB 81a.

15. Giṭ 4:9.

"as we have explained"—above, III, i, 10.

16. B. Bik 1:2.

"asherah"—originally a post or pillar, here obviously a living tree, or a grove of trees, attached to an altar dedicated to a pagan deity.

"hallowed things of the Temple"—see below, iii, 1; first fruits may not therefore come from a thing that was formerly the object of idolatrous worship. See RDBZ *ad loc.*

17. Pe 1:1; P. Bik 3:1; Pe 2:3; Bik 2:3.

18. Bik 3:11, 10, 9.

"seven species"—above, Sec. 2.

19. Bik 3:1, 2:4–5; P. Bik 3:1, 2:1; P. Pe 1:5; B. Yeb̲ 73a; Me 10a; Men 74a.
"that are bound to go to waste"—since, being unclean, they may not be used.
"as he would with unclean heave offering"—above, III, ii, 14–15.

20. Bik 1:89.
"to the Temple Mount"—more exactly, to the Temple court (cf. below, iv, 10).

21. P. Bik 1:6.

Chapter III

1. Bik 3:12, 2:1; Ḥal 1:9; B. Mak 18b f.
"priests of the division on duty"—see above, note to i, 10.
"We have already explained"—above, III, xv, 20.

2. B. Mak 19a.
"within the wall"—of Jerusalem.

3. B. Yeb̲ 73b; Mak 17a; P. Bik 1:7; Mak 3:3.
"as we have explained"—above, III, xv, 20.
"the altar of"—these words are not in the Biblical text.

4. B. Ḥul 68b; Mak 17a: P. Bik 17.
"made confession"—see below, Sec. 10.
"out of their precinct"—outside of Jerusalem, where they should be eaten. The point is that unlike some other holy things, the removal of first fruits from their precinct does not in itself make them unfit; cf. VIII, v, xi, 6 (YJS, *12*, 210).

5. Bik 2:2; B. Yeb̲ 73b.
"a mourner"—see above, V, iii, 5–6.
"prescribed place"—i.e., Jerusalem.
"the same as a clean Israelite," etc.—above, v, iii, 1.

6. B. Pes 36b; Yeb̲ 73b.

7. Tos Bik 2:8; Sif Deut. 26:20.
"barley at the bottom," etc.—the reason for this arrangement is that the more perishable the species, the nearer to the top it should be.
"at the top of the container"—Tos has "on top of everything" (reading *ḳol* for *ḳĕli*).
"else"—reading, with Tos, *'aḥer* for *'eḥad̲*.

8. Bik 3:8. B. BḲ 92a; MḲ 27a; P. Bik 1:10.
"metallic vessel . . . basket made of willow"—it is obvious that a metallic vessel is intrinsically valuable and should therefore be returned to the owner, while a wicker basket goes with its contents, and the priest should retain it. The Talmud (B. BḲ 92a) adds the sarcastic comment that the adage "Poverty follows the poor" has its source in this rule: the rich man, who can afford to bring first fruits in a metallic vessel, retains it, while the poor man, who can afford only a basket, must surrender it to the priest. However, since a holy gift honors the giver, the poor man may be said to receive the greater honor by surrendering the container as well.

9. Bik 3:5; P. Bik 3:4; B. Men 58a.

10. Bik 3:6; Soṭ 7:2; B. Soṭ 33a.

11. Bik 3:7.
"unable to read"—or unable to understand Hebrew.

"The Sages consequently ruled"—the purpose was not to prevent loss of gifts to the priests but to spare the feelings of the people who could not read or understand Hebrew.

12. Tos Bik 2:10; Bik 3:4, 6; 2:4; P. Bik 3:3; B. Men 46b, 58a; Suk 47b.

"and a psalm"—omitted in the *editio princeps,* but obviously necessary (cf. Sec. 13).

13. Bik 3:4.

14. Bik 3:4; P. Bik 3:3; Sif Deut. 16:7.

Chapter IV

1. Bik 1:1; B. Men 61b.

"offering, psalm, waving, and overnight stay in Jerusalem"—see above, iii, 12–14.

2. Bik 1:5.

"ṭumṭum"—a person whose sex cannot be determined with certainty.

"and therefore cannot say," etc.—because the Land of Israel was distributed to the heads of families, who were all male (Num. 26:55). In the case of guardians, bondsmen, and agents, they cannot say it because they are not the owners of the particular plot of land and of its yield.

"guardian"—of orphaned children.

3. Bik 1:4; P. Bik 1:4; Tos Bik 1:2.

"the Lord's oath was given first to Abraham"—even though it was later repeated to Isaac and to Jacob also.

"open land cities"—see Num. 35:1–5.

4. Bik 1:6; B. BB 81b.

"there is doubt"—see above, ii, 13: if A buys one tree he does not acquire the land under it; if he buys three trees, it is certain that he acquires the land as well; hence if he buys two trees, there is doubt as to whether he has acquired the land under them or not.

"What then should he do?"—the problem is this: if A owns the ground under his two trees, the produce is first fruits and accrues to the priests, so that A cannot consecrate what is not his, and is exempt from heave offering and tithes (above, III, iii, 23). On the other hand, if A does not own the ground, the produce is not first fruits, is subject to heave offering and tithes, and A may consecrate it. The solution offered takes account of both possibilities.

"to the repair of the Temple"—in which case they may be redeemed, in contradistinction to things consecrated for the altar, which may not be redeemed.

"by the hand of an agent"—who is not required to recite the confession (above, Sec. 2).

5. Bik 1:7; P. Bik 1:6.

6–7. B. Giṭ 47b.

6. *"and unto thy house"*—which includes the wife, the husband having the usufruct of her properties; after her death he comes into actual possession of them.

7. "When the Jubilee law is in force"—and the land must be returned to the original owner on the arrival of the Jubilee year. See Lev. 25:10–13.

"A's mind is not yet set"—to repossess the field, hence it is as if B owns the land as well as the yield. A's second sale is evidence that he had counted on repossessing the field.

8. P. Bik 1:6; B. Giṭ 47b.

9–10. Bik 1:8.

9. "the original first fruits are so liable"—the same as heave offering; see above, iii, 5, and III, vi, 6; Lev. 22:14.

10. "scatter them"—the fruit is thrown out of the basket. What is to be done then? The commentators differ as to whether the basket should or should not accrue to the priest, and whether the scattered fruit should be burned or allowed to rot.

11. Bik 1:9; P. Bik 1:9; Tos Bik 1:4.

12. Bik 1:6. Once the tree is dead, the owner's right to the ground from which it drew its nourishment lapses (cf. above ii, 13).

13. Bik 1:6; Sif Deut. 26:1; B. Pes 36b; P. Bik 1:11; Mak 18b; Giṭ 47b.

"up to Hanukkah"—after Hanukkah no first fruits may be brought in (above, ii, 6).

14. Bik 2:1; Ḥal 1:9; B. Ḥul 131a; Bik 3:12.

"the principal and the additional one-fifth"—paid by a commoner who has mistakenly eaten of heave offering (above, III, x, 15).

"gifts out of cattle"—i.e., the shoulder, the two cheeks, and the maw. See below, ix, 1.

"he may buy a Scroll of the Torah"—which, unlike the aforementioned purchases, may not be resold (II, III, x, 2).

15. Bik 2:1; P. Bik 2:1.

"must be consumed"—by the priests.

"a forbidden thing which will eventually become permitted"—according to this principle an object that is bound to become permitted does not become neutralized even when mixed with permitted objects in the required proportion. In the case of first fruits the required proportion is one hundred and one to one but even such a mixture is forbidden in Jerusalem, because it is bound to become permitted to the priests once the required procedure has taken place. The same rule applies to second tithe (above, v, vi, 14–15).

16. Bik 3:2; Tos Bik 2:8; P. Bik 3:2.

"post"—*"ma'ămaḏ,"* literally "stand," the name of a select group of Israelite commoners from each outlying district. There were twenty-four of them, corresponding to the twenty-four divisions ("watches") of priests, and each one served a week in turn. Some would go up to the Temple to attend the sacrificial offerings, while others would assemble in their district seat to conduct prayers during the day, at times corresponding to the fixed times of the sacrifices in the Temple. See B. Ta 26a, and VIII, II, vi (YJS, *12*, 62–64).

17. Bik 3:3, 4; B. Ḳid 31a; P. Bik 3:2.

Chapter V

1. Sif Num. 15:20; Ḥal 1:9; P. Ḥal 1:10.

"as it is said," etc.—the words *cake* and *gift* represent the Hebrew technical terms for two gifts due to the priest, *ḥallah* (cake), which is discussed here, and *tĕrumah* (gift), which is the subject of Treatise III, above. This latter term is sometimes applied to dough offering also. See below, Sec. 14.

"This *first* has no set amount"—cf. above, III, iii, 1.

"unless he has left some of it over"—*of the first* obviously implies part of the first, not all of it.

2. Ḥal 2:7; P. Ḥal 2:7; Tos Ḥal 1:7; Sif Num. 15:24.

3. Ḥal 2:7; P. Ḥal 2:7.

4. Ḥal 2:7.

"one forty-eighth"—in order to reduce the amount of waste, since unclean dough offering may not be eaten.

"the same as unclean heave offering"—above, III, xi, 1.

5. Sif Num. 15:19; B. Keṯ 25a.

"and in the days of Ezra"—when the children of Israel returned from captivity, they were liable only according to Scribal law, since only a segment of the whole nation of Israel had come back.

"in regard to heave offering"—above, III, i, 26.

6. Ḥal 2:1; Sif. Num. 15:22.

"is exempt"—RABD thought that it, too, is liable, by Scribal law.

7. B. Beḵ 27a; Ḥal 4:10; P. Ḥal 4:12.

"as heave offering"—above, III, ii, 17.

"and first fruits"—above, ii, 1.

8. Ḥal 4:8, 7; P. Ḥal 4:8; P. Bik 27a. For the explanation of the geographic terms and historical references see above, III, i, 2–9, and notes thereto. All three territories are regarded as more or less integral parts of the Land of Israel.

9. "the prevalence of uncleanness"—since the prescribed procedure of purification from such uncleanness (Num. 19) is no longer possible.

"authorized by the Torah"—and therefore must be burned (above, Sec. 4).

"From Chezib up to Amanah"—which is territory occupied by those who came up from Egypt but not by those who came up from Babylonia. See the preceding Section.

10. B. Beḵ 27a.

"even . . . dead bodies"—whose uncleanness is primary.

"a priestly minor"—who is not subject to such internal kinds of uncleanness.

"in Syria"—which, as far as dough offering is concerned, is regarded as part of the Land of Israel (above, Sec. 8).

"he need not set aside a second offering"—having openly consumed his first offering, he has done his duty of preserving the practice (above, Sec. 7).

11. Ḥal 4:8.

"even if it is not yet sunset"—normally one must wait until after sunset (above, III, vii, 2).

"contact with a dead body"—see note to preceding Section.

12. P. Ber 9:3; Ḥal 2:3; Ter 1:6; B. Ber 24a.

"her private parts"—literally "her face." A seated male's private parts are not concealed from view.

13. B. Beḵ 27a; Ḥal 4:8; P. Ḥal 4:10; B. Beṣ. 9a.

"and her like"—those suffering from uncleanness due to similar internal causes (above, Sec. 10).

"not against touching it"—in the case of domestic dough offering, she may not touch it either.

"as we have explained"—above, Secs. 10–11.

"equal in amount"—let alone if the amount of the admixture is greater.

"unlearned priest"—who cannot be trusted with clean dough, lest he should render it unclean.

"a priest assisting," etc.—see above, III, xii, 18.

14. B. Me 15b; 'Or 2:1; Ḥal 1:9, 2:8.

"as is the rule for heave offering"—see above, III, iii, 17, and v, 7.

15. "one may not set it aside"—for example, above, III, iv, 2.

"from one batch for another"—for example, above, III, v, 4.

"Whosoever may not eat"—for example, above, III, vi, 5.

16. P. Ḥal 4:3.

"Blind and intoxicated persons"—who may not at the outset set aside heave offering (above, III, iv, 4).

Chapter VI

1. Dem. 5:1, 3: P. Dem. 2:4.

"freshly baked . . . stale"—literally "hot . . . cold."

2. Ḥal 1:1, 4: P. Ḥal 1:1.

"only bread made out of these is called bread"—P. explains it thus: the term "bread" occurs in Scripture in connection with Passover and dough offering. In the case of Passover, it certainly refers to these five species, for only these can make leavened or unleavened bread; the same must therefore be true in the case of dough offering.

3. Ḥal 1:3, P. Ḥal 1:4; B. Men 66a.

"produce not yet one-third grown"—is not liable to heave offering because it is not yet produce, but is liable to dough offering because it can be made into dough.

"even though exempt from heave offering"—see above, III, ii, 9, 11; because even at the material time, i.e., when it was piled up, it was not liable to tithe.

"while still in its ears"—see above, III, iii, 13.

"it still contains the part which is great heave offering"—which has not yet been set aside, and from which the Levite is exempt in this case. See above, III, iii, 13.

"which have been redeemed"—and have thus become profane produce.

"the residue of the 'omer, of the two loaves, and of the shewbread"—see above, notes to i, 4.

"once all these residues have been redeemed"—they are profane, may be eaten by commoners, and are therefore subject to dough offering.

4. Tos Ḥal 1:5; Shebi 9:9; B. Bek 12b: Nid 47a; Ḥal 1:4.

"in Jerusalem"—where it may be eaten; outside of Jerusalem, where it may not be eaten, it is exempt.

"assuredly mixed"—in sufficient proportion to make it assuredly forbidden, is exempt, as are all consecrated foods.

5. Ḥal 1:5; Tos Ḥal 1:6; B. Pes 38b.

"Loaves of thank offering"—see Lev. 7:12 ff.

"the Nazirite's wafers"—forming part of the sacrifice brought by him when his period of Naziriteship is completed. See Num. 6:15.

6. B. Ḥul 135b; Tos Ḥal 1:4.

"Dough held in partnership"—even if each partner's share is too small to be subject to dough offering, and only the total is large enough.

"public consumption"—the owner does not declare the dough ownerless, but rather invites the public to partake of it.

7. Tos Ter 4:10; P. Dem 5:1.

"what he has done is done"—although the prescribed procedure is to set aside heave offering and tithe first. See above, III, iii, 23.

8. Sif Zuṭa Num. 15:19; Ḥal 1:8, 3:5; P. Ḥal 1:9; B. Men 67a.

"to feed . . . to cattle or to wild beasts"—Scripture says *your dough* (Num. 15:20), implying for human consumption and not that of cattle.

"The dough of a heathen is exempt"—since it is not the property of an Israelite; it is only *your dough* that is liable to dough offering.

9. Ḥal 3:5.

"the minimum"—see below, Sec. 15.

10. Tos Ter. 4:13.

"assigned it to the heathen"—that is, had him knead it.

11. Ḥal 3:7, 10. The proportion of each ingredient is immaterial—the criterion is taste.

"rice flour"—which is not liable to dough offering (above, Sec. 2).

12. Ḥal 2:2, 1:6; P. Ḥal 2:3; B. Pes. 37b: Ber 42a, 37b.

13. Ḥal 2:5. Lighter dough is used for baking bread; dough for other purposes is heavier.

14. B. Ber 37b.

"milk sauce"—Hebrew *kutaḥ,* which consisted of bread crusts and sour milk, well salted (see above, v, vii, 10)

"thin cakes"—the regular shape of bread in the Near East.

"shingles"—pieces of dough laid end to end over each other like shingles.

15. B. 'Er 83b; Ḥal 2:6, 1:1; B. Pes 109a; Men 39b, 41a.

" 'omer"—Maimonides makes the 'omer equal $10 \times 10 \times 3.11 = 311$ cubic fingerbreadths. If a fingerbreadth equals 2 cm., the result would be $20 \times 20 \times 6.22 = 2488$ cubic cm. By weight it would equal $86\frac{2}{3}$ sela', or 1.728 kilograms.

"forty-three and one-fifth . . . eggs"—1 log̱ = 6 eggs, and 1 ḳaḇ = 24 eggs; hence $1\frac{4}{5}$ ḳaḇ = 1 'omer = $43\frac{1}{5}$ eggs.

"Egyptian zuz"—meaning dirhams.

16. P. Ḥal 3:1; Ḥal 2:4, 4:4; B. Pes 109a; Sif Num. 15:9. The rule obviously does not apply to indigent persons who can afford only a little dough at a time.

"The oven . . . cannot combine"—because so long as the dough is inside it, being baked, it cannot be termed bread.

17. Ḥal 2:4; Tos Toh 1:2; B. Pes 109a.

"bite into one another"—and cannot be separated without one gouging a piece out of the other.

18. Ḥal 2:6; B. Shab 76b; P. Ḥal 2:6.

"bran . . . coarse bran"—the former is counted as flour, but the latter is not.

19–20. Ḥal 1:7.

19. In both cases the portions distributed are less than the minimum liable to dough offering. The difference between them is that in the first case it is a baker who distributes the leaven, and who, as the text says, is certain to use the unsold leaven in making bread for sale, and therefore the leaven is regarded as if originally intended for that purpose. In the second case the distributor is an ordinary householder, who is not likely to dispose of the remaining dough, if any, in the same manner.

20. "it is exempt"—even though the total amount of leaven is large enough to

be subject to dough offering; the reason is that it is taken for granted that each one of the women expects her share to be kneaded separately.

Chapter VII

1–2. Ḥal 4:1, P. Ḥal 4:1.

1. "bite"—see above, note to vi, 17.

2. "if his batches of dough"—made of the same species of grain.

3. Ḥal 4:2.

4. Ḥal 4:4; P. Ḥal 1:1.

"new grain"—grain harvested in the current year.

"old grain"—grain harvested the year before.

"set aside heave offering out of new produce for old"—which is forbidden (above, III, v, 11).

"Nor should one set aside," etc.—the situation here is as follows: batches A (new grain) and B (old grain) are separately less than the required minimum but jointly equal to it. Batch C (new grain or old grain) is added and combined with A or B, and dough offering is set aside from this triple combination, provided that the new or old ingredient, as the case may be, amounts now to the required minimum. See below, Sec. 6.

"When does this apply?"—referring back to the statement in Sec. 3 that dough of wheat and dough of emmer do combine together.

"as we have explained"—above, vi, 15.

5. Ḥal 4:3; Tos Ḥal 2:4; P. Ḥal 4:3.

6. Ḥal 4:3; P. Ḥal 4:3; Tos Ḥal 2:4. The principle in these rules is that while batch C in the middle cannot combine with either A or B in order to constitute the prescribed minimum, it does not form a barrier between them to prevent them from combining together, provided that it itself is subject to dough offering on its own account.

7. Ḥal 4:5.

8–10. P. Ḥal 3:5.

8. "inasmuch as it was not liable"—for had it amounted to the minimum while they were heathens, it would have been exempt, and once exempt it remains permanently exempt.

9. "with the intent aforethought to divide it"—see above, vi. 19.

10. "the Israelite's dough is liable"—inasmuch as the Israelite's share had not become exempt, while the heathen's share was originally exempt. RABD contends that the rule is the reverse, i.e., the Israelite's share is exempt, because originally he had the required minimum only in partnership with the heathen, while the proselyte's share is liable, because at the moment when it became liable the required minimum was on hand, and the previous exemption of it is not material, since at that time the heathen was not liable to dough offering.

11. Ḥal 3:8; B. AZ 56a.

"next to the dough"—the text reads "into the dough," which does not fit well with "adjacent dough" mentioned further on; cf. "next to the clean dough" in the following Section.

"out of adjacent dough"—that is why he must first put it next to the dough that has already been regularized, rather than just set aside dough offering out of the new batch wherever it may be.

"even the smallest amount . . . renders a mixture forbidden"—since the mixture can be regularized and thus become permitted, the unregularized ingredient cannot be annulled, in accordance with the principle governing "forbidden things which will eventually become permitted." See above, note to iv, 15.

12. B. Soṭ 30b; Ḥal 2:8.

"as will be explained"—X, v, xi, 2, 15. (YJS, *8,* 292, 295 f.).

"It is permitted to cause profane produce . . . to become unclean"—indirectly, not directly. If the produce has been regularized, one may do so directly. See X, vi, xvi, 9 (YJS, *8,* 393).

"and draw . . . over"—*u-mošeḵ;* presumably read *u-mĕnaššeḵ,* "and cause an egg-sized piece . . . to bite into the clean dough."

"egg-sized piece"—which is the size that makes unclean food capable of rendering other food unclean ('Or 2:4). Here this is permitted, because there is no transfer of uncleanness.

"in order . . . adjacent dough"—this explains the reason for causing a piece of the unclean dough to "bite" the clean dough.

13. Ḥal 4:6.

"even if these latter become unclean"—there are two minuses here: (a) the offering is set aside out of dough that is not adjacent; (b) it is set aside out of clean dough for unclean. Both of these are normally forbidden, but are permitted here, because there is doubt as to whether the several batches have or have not been regularized.

Chapter VIII

1. Ḥal 2:5; B. Ḳid 46b.

"out of flour"—before it is kneaded into dough.

"as something obtained by robbery"—since the priest is not entitled to it until after the flour has been kneaded into dough. He must therefore return it to the owner.

"an 'omer"—the statutory minimum of flour (above, vi, 15).

2. Ḥal 3:1; P. Pes 3:3; P. Ḥal 3:1. Strictly speaking, dough offering does not become due until the entire batch of flour is kneaded, but in order to avoid possible uncleanness the law permits setting aside the offering out of the first dough kneaded, while the rest of the batch has not yet assumed the form of dough. See below, Sec. 4.

"This is to be"—meaning this handful of flour already mixed with water.

3. Sif Num. Šĕlaḥ 110.

"it is still valid"—literally, "it does not matter." That is to say, although the rule is that one must set aside dough offering immediately, the moment it becomes due, and he has transgressed by delaying, it does not matter, and the offering is still valid.

"as we have explained"—above, vi, 16.

4. Ḥal 3:1; B. Men 70a.

"When does the mixture become liable to dough offering?"—the difference between this question and the question asked above, in Sec. 2, is that in the latter it is a matter of procedure, from a practical point of view: when is the best time to set aside dough offering? Here the question is, when does the liability arise, which entails a number of sanctions and prohibitions.

"rolled . . . becomes one cohesive mass"—rolled means completely homogenized into a uniform cohesive batter. Barley flour does not form as homogeneous a dough as wheaten flour; hence the difference in the required consistency.

5. Ḥal 3:1; P. Ḥal 1:1; B. Mak 13a.

"liable to the death penalty"—at the hand of heaven and not of human courts; hence he who so transgresses is punishable by the scripturally enjoined flogging.

6. Ḥal 3:2, 3.

"it is exempt"—because of the rule that in the case of dough mixed with heave offering, if there is not enough dough to neutralize the heave offering, i.e., less than one hundred times the heave offering's bulk, it is exempt from dough offering (above, vi, 4). The commentators say that this rule applies only to post-Temple times, when the law of heave offering is in force by Scribal enactment only. In Temple times it was not exempt, because then one was liable to dough offering by Scriptural authority, and according to Scriptural law heave offering becomes neutralized by a greater quantity of the admixture; therefore scripturally a lesser mixture is profane and consequently liable. Hence we must conclude that the rule here deals with post-Temple times, when the liability is only Scribal and may be overridden by another Scribal enactment.

"if after"—and thus already liable to dough offering.

7. Ḥal 3:3.

8. Ḥal 3:5.

9. Ḥal 3:6; P. Ḥal 3:5; B. Ḥul 134a.

"involves the penalty of death"—hence the severity of the rule even in case of doubt.

"not liable to the additional one-fifth"—on the principle that the burden of proof is upon the claimant.

10. Ḥal 3:2; Tos Ḥal 1:11; B. Nid 37a; P. Ḥal 3:2.

"suspicion of uncleanness"—if there is no uncleanness involved, one must be careful not to render dough offering unclean. Since here there is suspicion of uncleanness, one need not be so careful. See below, Sec. 11.

"burn its dough offering"—as is the rule with unclean dough offering (above v, 4).

"would have rendered profane produce unclean on the authority of the Torah"—for an explanation of the degrees of uncleanness see above, note to v, ii, 12.

"neither eaten nor burned"—one should wait until it becomes assuredly unclean, and then burn it (above, III, xii, 2).

11. Ḥal 2:3; P. Ḥal 2:3; B. Ḥul 122a.

"and a body of water"—wherein he might purify his person, and his vessels.

"make the dough in uncleanness", etc.—and give it as such to the priest, who must use it only for fuel (above, v, 4).

12. Tos Dem 3:1; B. Giṭ 62a.

"dough offering for an unlearned person"—who would not be careful to guard it against uncleanness and might defile it, which is forbidden.

"such vessels do not contract uncleanness"—see X, VII, i, 6 (YJS, *8,* 398).

"but should be told," etc.—this obviously is not true, but is told in order to keep him from defiling the dough offering.

"Why . . . allow him to do so?"—even with this precaution he might still defile the dough offering, in which case the transgression would be ultimately chargeable to the scholar.

"to safeguard the professional kneader's livelihood"—if he were forbidden to serve unlearned persons, his circle of potential customers would be so restricted as to jeopardize his livelihood.

13. Shebi 5:9; B. Giṭ 61a: B AZ 55b.

"once the latter has poured water into the dough"—because at that moment the dough becomes susceptible to uncleanness. See above, note to III, xii, 8.

14. P. Ḥal 4:7; Tos Ḥal 2:6. Cf. above, vi, 1.

"out of doubt"—on the ground that the baker may not be trustworthy.

"just as all Israelites . . . heave offering"—see above, IV, ix, 1.

15. P. Ḥal 4:7; Tos Ḥal 1:8; B. AZ 39b.

"baker . . . householder"—who are unlearned. Both the baker and the householder may be trusted to set aside dough offering when the dough is made for their own use. When the baker makes bread for sale, he may not be so scrupulous about not causing other people to transgress. When the householder sells bread, or gives it to his guest, the assumption is that it is part of that which he had prepared for his own use, and the buyer or guest is therefore not required to entertain any doubt and need not separate dough offering before eating of the bread.

Chapter IX

1. Ḥul 10:1; B. Ḥul 130b, 136a; Ḥal 4:9; P. Ḥal 4:11.

2. Ḥul 10:1; B. Bek 14a, 15a; Bek 2:23.

"suffered a permanent blemish prior to their consecration"—their consecration was therefore temporary, and they were destined at the outset for redemption, whereupon they became fit for profane food.

"they are exempt"—because even after their redemption they retain a residual holiness, seeing that when they were consecrated they were not destined for redemption.

3. B. Bek 18a f.; B. Ḥul 132a; Bek 2:6.

"the same as if the priest had acquired it first, and then returned it to its owner"—the Mishnah rules that the other animal must be left to pasture until it suffers a blemish, and there is therefore a difference of opinion as to whether the owner is liable to the gifts, the same as in the case of an unblemished animal. According to one opinion, the second animal is deemed unconsecrated in every respect and therefore subject to the gifts. The other opinion is that the animal is exempt, because it is the same as if the priest had acquired it and then given it (B. Bek 18a, sold it) to its owner. It is therefore exempt, because if the priest takes beast B as a substitute for beast A, beast A is exempt (Bek 7:8). Here the reasoning given in the case of the doubtful firstling, that the priest may claim the gifts because they are his in either case, cannot apply, because here the priest can plead only one alternative—if the particular beast is not a firstling, the gifts are due to him; he cannot plead that if it is a firstling, all of it belongs to him, since beast A is regarded as returned or sold to its original owner.

4. B. Ḥul 132a; Ḥul 10:3.

"it comes under the principle," etc.—and the priest has to furnish proof that the particular animal is not the one disqualified, and is therefore liable.

"only one of them is exempt"—since one of the animals is surely disqualified, and therefore exempt (above, Sec. 2).

5. B. Ḥul 132a; Bik 2:9. All the animals listed here are clean.

"offspring of a sheep and a goat"—both being domestic animals, though of different species.

"the *koy*"—a creature (a kind of bearded antelope), concerning which the Sages were undecided as to whether it is to be classified as a domestic animal or a wild beast.

"If a male gazelle is mated with a she-goat"—and the offspring is thus half domestic and half wild animal.

"If a he-goat is mated with a hind," etc.—although in this case too the offspring is half domestic and half wild animal, nevertheless it is exempt: in the preceding case the offspring is liable to half the gifts because the verse just cited takes the dam into consideration, although there is doubt as to whether the sire also should or should not be taken into consideration. In this case there is no such Scriptural proof, however indirect, and the offspring is therefore exempt, because should the priest claim his gifts on the ground that the young was sired by a domestic sire, the burden of proof is upon him to show that the wild dam should not be taken into consideration as well.

6. Tos Ḥul 9:1.

"for Israelite consumption"—literally "for human consumption." All the commentators exceept Corcos omit this phrase, and so do the Schulsinger edition and the Tosefta. Perhaps the correct reading is "Whether one slaughters an animal for human consumption, or for dog food," etc.

"it is liable to the gifts"—because it becomes liable the moment it is slaughtered; the purpose of the slaughtering does not matter.

7. B. Ḥul 136a.

"*from them that offer*"—which they may eat but not use in trade. Here, however, it is a matter of exchanging one kind of food for another (Corcos).

8. B. Beḵ 12b; Ḥul 132b, 131a.

"need not return them"—as is the rule in all such doubtful cases (Corcos).

9. B. Ḥul 132a.

"When does this apply?"—that priests and Levites are exempt.

"two or three weeks' grace"—the criterion is that he must slaughter three times, which ordinarily would cover a period of two or three weeks. Once he does it three times, and sells the meat, the presumption arises that he is engaged in a commercial undertaking. If he sets up a regular butcher shop, this presumption arises immediately.

10. Ḥul 10:3; B. Ḥul 131a.

"he is exempt"—because it is the owner who is liable and not the slaughterer.

"so that the gifts might be deposited by him with the share of the priest"—in other words, so that the gifts, marked with the Israelite partner's sign, might be clearly distinguishable from the priest's share, and thus would make it clear that the Israelite partner has fulfilled his obligation.

11. Ḥul 10:3; B. Ḥul 134a.

"having said . . . with the exception," etc.—the difference between "with the exception" and "on condition" is that in the former case the gifts are not part of the deal, hence it does not affect them; in the latter case they are part of the deal, and the condition is invalid, because according to the Torah the Israelite has the option of giving the gifts to any priest he may choose, and this rule may not be circumvented.

"Having reserved no partnership"—since he has stipulated that the gifts are to be entirely his, an invalid stipulation.

12. B. Ḥul 133b.

"inasmuch as the part that is liable," etc.—even though the rest of the animal belongs to the priest. In the preceding cases the part that is liable is the priest's share, not the Israelite's, hence it is exempt.

13. Ḥul 10:4; B. Ḥul 134a; cf. above, Sec. 10.

14. B. Ḥul 132b f., 130b, 134a.

"are distinct"—from the rest of the animal's body.

"no specific claimant"—since the Israelite is entitled to give the gifts to any priest he may choose.

"amenable to being seized by robbery"—things in the possession of the unlawful owner, acquired by seizure or by illegal purchase, are regarded as his property, although the aggrieved party may sue for recovery in a court of law. The transgression itself is liable to punishment at the hand of heaven.

15. Ḥul 10:3.

"the gifts"—i.e., in this case, the maw.

"by weight," etc.—in the former case the transaction implicitly excludes the maw, for it is common knowledge that the maw accrues to the priest, hence it cannot be included in the sale. In the latter case, the maw is included in the weight sold, and since it was at the outset not the seller's to sell, the total price must be correspondingly reduced.

16. Tos Ḥul 9:2; B. Ḥul 134b.

"In a place where there is no priest"—if the place has a resident priest, B must hand the gifts to him immediately.

"to avoid any loss to the priest"—by keeping the gifts and allowing them to spoil.

"any priest"—in some other place.

17. B. Ḥul 132b; Midrash Tannaim Jud. 18:3.

"those of an ox"—which are extra large.

18. Ḥul 10:4; B. Ḥul 134b.

"From the knee joint up to the shoulder socket"—in anatomic terms, from the carpus to the scapula; it thus includes two bones, the radius and the humerus.

"From the joint of the jaw to the knob of the windpipe"—i.e., to the tip of the thyroid cartilege, thus including the whole of the lower jaw and the tongue.

19. B. Ḥul 134b.

20. B. Ḥul 131b f.; Yeb̲ 74b; Bek̲ 47a; Tos Ḥul 9:1.

"there is no holiness in them"—they are in every respect profane comestibles (above, i, 7). Their only contact with holiness is when they are handed over to the priest—once this duty is fulfilled, they are no different from any other profane food.

"A profaned woman"—a woman who marries a priestly man forbidden to her (e.g., a divorcée who marries a priest), and the female issue of such a marriage. See above, III, vi, 7; and V, I, xix, I (YJS, *16*, 121).

21. B. Ḥul 133a, 132b. The assignor is evidently highly esteemed by everyone in the community.

22. B. Ḥul 133a, 132b; Zeb̲ 10: 7; B. Zeb̲ 90b.

"By reason of the anointing"—so AV: JV translates *for a consecrated portion.* The Talmud takes the word to refer to royalty, because kings were anointed; hence the implication is that these gifts, given to the priests in recognition of their exalted position, should be consumed in a manner characteristic of royalty. To be sure, the Biblical context deals with hallowed sacrifices, not profane ones, but it is here extended to cover the latter as well.

Chapter X

1. Ḥul 11:1; B. Ḥul 131b, 137b, 135b.

"the first of the corn"—referring to tithe and heave offering which are binding both in Temple times and after the destruction of the Temple. See above, III, 1, 1.

2. B. Ḥul 137a.

3. Tos Ḥul 10:3; B. Ḥul 137a.

"All consecrated animals"—see above, ix, 2, where this principle is applied to the gifts, and notes thereto, for an explanation of it.

4. Ḥul 11:1; B. Ḥul 137a.

5. Deut. 18:1; Sif Deut. 18:1. See above, i, 6–7, and notes thereto.

6. Ḥul 11:2; B. Ḥul 137a.

"and then dyes it"—in colors other than bleached white, thus making a change in it; he is therefore exempt, on the principle of alteration (Hebrew *šinnuy*): if A effects a change in an object belonging to B, A acquires title to the object.

"bleaches it"—bleaching is not considered a change, since it merely removes the original color from the wool without adding another color to replace it.

7. Tos Ḥul 10:1; B. Ḥul 136b.

"ḳoy"—see above, note to ix, 5.

"ṭĕrefah animal"—an animal torn by a wild beast or suffering from a serious organic disease, whose flesh may not be eaten.

8. Tos Ḥul 10:7; B. Ḥul 136b.

"until he gives it to the priest"—i.e., until he gives him an equal amount of wool to replace the lost wool; cf. above, ix, 14, where the gifts (cheeks, shoulder, and maw), if lost, are not repayable to the priest. The reason is obvious: there are no duplicates of these organs in that animal's carcass, whereas the lost wool and the extant wool, both from the same animal, are identical in nature. If all the wool is lost, it is likewise not repayable (KM).

9. Ḥul 11:2; B. Ḥul 138a, 136a.

10. Ḥul 11:2; B. Ḥul 138a.

"one may not sell the gifts due to the priesthood"—hence when he sold the fleece to A, he did not include these in A's share.

"If B does not retain any sheep"—the presumption obviously does not apply. The same is true if B's share of fleece is too small to furnish the required minimum of the priest's due.

11. Ḥul 11:2.

12. B. Ḥul 134a f.; cf. above, ix, 13.

13. Ḥul 11:2; B. Ḥul 137b.

"twelve sela' "—according to the Mishnah, "enough to make a small garment." See below, Sec. 16.

14. B. Ḥul 135a.

15. Tos Ḥul 11:35, 10:7; B. Ḥul 128a, 136a.

"they all combine together"—inasmuch as during the shearing he had the required minimum number of sheep.

"new fleece"—the current year's.

"old"—the preceding year's.

"they do not combine together"—since at the time of the shearing he did not have the required number.

16. Ḥul 11:2; B. Ḥul 137b f.

"unbleached"—literally "dirty, unwashed."

17. Sif Deut. 18:5; B. Nid 51a.

"out of cattle"—see above, ix, 20.

Chapter XI

1–2. Ḳid 1:7; B. Ḳid 29a.

2. "when he grows up"—when he reaches his majority, i.e., the age of thirteen years and one day.

3. "he must redeem himself first"—for his personal obligation has priority, since the grandfather had failed to redeem the father.

4. Ḳid 1:7; B. Ḳid 29a; Tos Beḵ 6:10.

"and only thereafter"—implying that the first duty has priority, even though it is postponable, whereas the second duty is not.

5. B. Pes 121b.

"and then the benediction . . . 'who has kept us alive' "—in II, v, xi, 10 Maimonides rules that this benediction is to be pronounced only over commandments performed in one's own behalf and not in behalf of another person. The answer is that since the redemption of the first-born is expressly imposed upon the father, he is in a sense performing it in his own behalf as well as in behalf of his son. Besides, the latter is of too tender an age to perform it himself, unless it is postponed until his majority.

6. B. Ḳid 37a; B. Beḵ 8:7, B. Ḳid 22b.

"the same as other shekel payments"—payable in coin or in equivalent goods, with the exception of the annual tax of one half-shekel. See Beḵ 8:7.

"writs"—of indebtedness to the redeemer.

7. Beḵ 8:8; B. Beḵ 51b; Ḳid 8a.

"but the son is not . . . redeemed"—the assumption is that this money is owed for some other reason, since the writ does not specify that it covers the redemption.

"the son is . . . redeemed"—because the transaction specifies the five sela' represented by the vessel.

"to ten priests"—what counts is the giver and not the receivers.

8. Beḵ 8:8; B. Ḳid 6b: Beḵ 52b.

"on condition that it be refunded"—this seemingly contradicts the rule stated at the beginning of this Section, that if the father gives the redemption money to the priest with the understanding that it is to be refunded, the son is not redeemed. The commentators (RDBZ, Corcos) explain that there is a difference between giving with the unspoken understanding that there is to be no gift at all, and giving as an explicitly temporary gift. The latter is a valid act, inasmuch as at the moment of giving it constitutes a valid transfer, and the donee has consented beforehand to the refund.

9. Bek̠ 1:1; B. Bek̠ 4a.

"the Levites had exempted," etc.—cf. Lev. 3:44–45: *And the Lord spoke unto Moses saying: Take the Levites instead of all the first-born among the children of Israel . . . and the Levites shall be Mine.*

10. B. Bek̠ 47a: Ḥul 132a; Bek̠ 8:1.

"That openeth the womb in Israel"—Maimonides follows the Mishnah in paraphrasing Exod. 13:2 (*Whatsoever openeth the womb among the children of Israel*) or Num. 3:12 (*That openeth the womb among the children of Israel*).

11. B. Ḥul 132a. The difference here between the daughter of a Levite and the daughter of a priest is that the former's illicit relationship does not impair her Levitical status, whereas the latter's illicit relationship nullifies her priestly status and makes her a profaned woman (above, ix, 20; III, vi, 7). The child of such an illicit union assumes the status of the mother and not of the father, and is considered legitimate (V, I, xv, 3; YJS, *16,* 57).

12. B. Bek̠ 47b; Giṭ 59b.

"has fathered a profaned son"—for example, when a priest marries a divorcée, his son does not have the status of a priest, although the father retains his priestly status, and is therefore entitled to the redemption money when due, and the son is thereby redeemed. Should the priest die before redemption becomes obligatory, the son must redeem himself, since the obligation is now his own and not his late father's, and must pay the redemption money to some other priest.

"within thirty days"—see below, Sec. 17.

13. Bek̠ 8:1; B. Ḥul 134a.

"That openeth the womb in Israel"—see above, note to Sec. 10.

14. Bek̠ 8:1.

"unclean by reason of parturition"—see Lev. 12.

"on the fortieth day"—or earlier, while the embryo is not yet fully formed (V, I, x, 1, 8; YJS, *16,* 67–70).

15. Bek̠ 8:1; Nid 3:5. The dead fetus is fully formed.

16. Bek̠ 8:2; B. Bek̠ 47b.

"out of its mother's side"—presumably by way of Caesarean section; cf. J. Preuss, *Biblisch-talmudische Medizin, 3. Aufl.* (Berlin, 1923), p. 492 ff.

17. B. Bek̠ 12b; Bek̠ 8:3; B. BḲ 11a.

18. B. Bek̠ 49a.

"as of now"—when the child is not yet subject to redemption.

"after thirty days"—i.e., after the child has become subject to redemption.

19. B. Ḥul 134a; Bek̠ 8:3.

"proof . . . from his father"—that he had redeemed him before the lapse of the thirty days in the manner described above, Sec. 18. Normally a father would not pay out the redemption money until it is actually due, on the 31st day.

20. Bek̠ 8:3.

"the priest has no claim"—since the female child, who is exempt, may have been first-born.

"even though it is not known which one is first-born"—the father must pay nevertheless, because the first-born is surely male.

21–25. Bek̠ 8:4.

21. "ten sela'"—five for each child, since each is first-born to his mother (above, Sec. 1).

23. Here too the principle applies that the burden of proof is upon the priest claimant.

24. "the father is exempt"—since the dead infant may belong to the primipara wife.

25. See below, note to Sec. 30.

26. Bek̲ 8:5; Tos Bek̲ 8:3.

"writ of authorization"—transferring title to the five sela' from father A to father B. B's claim is then beyond contest: if B's child is dead, B is entitled to a refund; if B's child is living, A is entitled to a refund, which he has already assigned to B.

27. "both fathers are exempt"—since each one can claim that he is the father of the female child.

"waiting three months"—to establish the child's paternity (IV, II, xi, 18: YJS, *19*, 244–45).

"both husbands are exempt"—since each one can claim that he is not the father.

28. Bek̲ 8:5. Here the chances of a male child being or not being first-born are equal, and the burden of proof is therefore upon the priest claimant.

29. Bek̲ 8:6.

"two male children"—one of them is assuredly first-born, even though it is not known which one; hence the husband of the primipara is in any case liable to redemption payment.

30. B. Bek̲ 49a.

"two doubts are involved"—there are five possibilities here: a) the primipara wife has given birth to only one male child; b) she has given birth to two male children; c) she has given birth first to a male child and then to a female child; d) she has given birth only to a female child; e) she has given birth first to a female child and then to a male child. In three of the five possibilities there is liability to redemption, and in the other two there is exemption. The former outweigh the latter in favor of liability, and therefore the husband of the primipara wife must pay the redemption money. The same rule seems applicable, in Sec. 25, where Maimonides declares the husband exempt; hence RABD reverses the verdict to liable in Sec. 25 and exempt in Sec. 30.

Chapter XII

1. Bek̲ 1:7; B. Ḳid 37a.

"each person"—man or woman, unlike the redemption of the first-born son, which lies upon the father only (above, xi, 1).

"in all places"—in the Land of Israel as well as in other lands.

"at all times"—while the Temple is in existance as well as when it is not.

2. Ḥal 4:9; B. Bek̲ 5b.

"as it is said," etc.—the first part of the verse states that all firstlings, both human and animal, belong to the priest (*shall be thine,* i.e., Aaron's). The second part states that firstlings of man and of unclean beasts must be redeemed. Since in the case of human first-born the redemption money belongs to the priest, it follows that the same applies to the firstling of an unclean beast, since both are mentioned here together.

"and the firstling . . . redeem"—in some editions this part of the verse is

annexed to Sec. 3. Logically it belongs to Sec. 2, and is so printed in the *editio princeps*.

4. B. Beḵ 9b, 10b; Ḳid 2:9; Beḵ 1:6; P. Ḳid 2:9.

"Priests may not . . . be trusted in this matter"—Rashi explains that a priest might very well say, "The redeeming lamb is mine to begin with, so why bother with the ceremony of redemption?" Yet one may not disregard a scripturally enjoined duty under any circumstances.

5. Beḵ 1:6; B. Beḵ 11b.

"is not answerable"—although it is still in his possession; once he has set the lamb aside, it is accounted as if it were already in the priest's possession.

6. B. Beḵ 13a.

"drawback"—one should not at the outset delay, but delay in any case does not cancel the obligation.

7. Beḵ 1:7; B. Beḵ 10b.

"from its back with a butcher's hatchet"—the proper instrument to be used in such a case, and one that causes instantaneous death.

"a cane"—Hebrew *maḳḳel;* the Talmud has "a sickle" (*maggal*).

8. Beḵ 1:5; B. Beḵ 12a.

"a *ṭĕrefah* animal"—see above, note to x, 7.

"a *ḳoy*"—see above, note to ix, 5.

9. B. Beḵ 12a.

"have become unfit for the altar"—and have been redeemed. Such an animal, even after redemption, retains some sanctity, in that it may not be used for work or shearing.

10. Beḵ 1.4; B. Beḵ 12a, 11a.

11. B. Beḵ 12b, 11a.

"as little as one denar"—1 sela' = 4 denar, hence 1 denar = 1/40 of 10 sela'.

"may be redeemed with their equivalent in money"—see VI, IV, vii, 8 (YJS, *15*, 202).

12–13. B. Beḵ 11a.

12. "three zuz"—that is, 3/4 of a sela' (i zuz = 1 denar; see note to Sec. 11, above).

13. "it is considered redeemed"—because the obligation does not rest exclusively upon the owner; anyone else may redeem it, just as anyone may redeem a consecrated animal. Other authorities disagree with this view.

14. Beḵ 1:1; B. Beḵ 4; cf. above, xi, 9.

15. Beḵ 1:1; B. Beḵ 3a f.

"embryo"—which is a firstling.

"although he is not permitted to do so"—because he thereby causes a loss to the priest.

"would make the animal blemished"—and unfit for sacrifice. See VIII, III, vii, 2, 9 (YJS, *12*, 110, 112).

16. B. Ḥul 134a.

17. Tos Ter 4:13; B. Men 67a.

18. Beḵ 1:2; B. Beḵ 6a. A she-ass is akin to a cow, in that both are subject to the law of the firstling. A she-ass is also akin to a mare, since both are unclean animals.

19–21. Beḵ 1:3.

19. "he must give one lamb to the priest"—because one of the two males is assuredly a firstling.

"The lamb . . . belongs to the owner"—there is doubt here as to whether the male was born before the female or vice versa; by setting aside a lamb for redemption, the owner releases the animal from the prohibition which affects the firstling, in case the male was born first. He is not required, however, to give the lamb to the priest, since the latter's claim is purely a claim to a debt as laid down in Scripture, the lamb itself possessing no sanctity and being like the ass which it redeems. Consequently the priest is in the position of a claimant who must prove his claim with evidence, or in this case prove that the male was born prior to the female.

20. "one lamb"—because the male, or one of the two males, is certainly a firstling.

21. "the priest receives nothing"—because it is possible that in both cases the firstling was a female.

22. Bek̲ 3:1.

23–24. B. Bek̲ 11a.

23. "set aside for a doubtful firstling of an ass"—and therefore remaining his property and not payable to the priest.

24. "inherited . . . from his own Israelite maternal grandfather"—and thus subject to redemption.

"these lambs remain his"—since if his late grandfather had redeemed them with lambs, he would have retained the latter (above, Sec. 4).

TREATISE VII: LAWS CONCERNING THE SABBATICAL YEAR AND THE YEAR OF THE JUBILEE

Chapter I

1. P. Shebi̲ 1:3.

"or tending the trees"—this is included in *Then shall the land keep a Sabbath,* which is followed by *Thou shalt neither sow thy field, nor prune thy vineyard* (Lev. 25:4), quoted below.

"and elsewhere"—this additional verse is more explicit.

"as it is said, *Thou shalt neither sow,*" etc.—the preceding verses are positive commandments, while this one is negative, thus making the transgressor responsible for violation of both a positive and a negative commandment.

2–3. B. MḲ 3a.

2. "liable to a flogging"—for those not liable see the following Section.

"reaping or gathering the vintage"—reaping refers to fields and gathering refers to vineyards.

3. "Why then did Scripture enumerate them separately?"—in Lev. 25:4–5.

"secondary . . . primary"—sowing and reaping are primary prohibitions, while pruning and gathering are secondary, but all four involve liability to a Scriptural flogging. All other such prohibitions involve liability to Scribal flogging. See the following Section.

4. B. MḲ 3a; Giṭ 53b; P. Shebi̲ 4:4.

5–6. P. Shebi̲ 4:4; Tos Shebi̲ 3:19; Shebi̲ 2:2.

5. All these cases involve improvement of the land, of the plant, or of the produce.

"sprinkle a treetop with powder"—to fertilize the tree.

"oil unripe figs, or pierce them"—to hasten ripening.

"tie plants together"—to make them grow straight up.

6. "light a fire in a thicket of reeds," etc.—either to clear the ground for cultivated crops or to make the reeds grow better for human use. See Prof. Lieberman's comment to Tos.

"except on sandy ground"—where, since no growth is possible, plowing is of no benefit to the soil.

"test seeds"—by digging up one seed to see whether it is growing.

"in a pot filled with manure"—obviously an unperforated pot; a perforated pot is regarded as part of the soil underneath it, and would be forbidden. It is not customary to plant seeds in whole manure.

"store aloe on the roof top"—where it might take root spontaneously, since this is not planting.

7. P. Shebi 4:4; B. Ḥul 77b f.; MḲ 3a; Tos Shebi 1:10.

"with red paint"—according to the Talmud, this was done to a tree that cast its fruit because of excessive fertility, so that the people would notice it and pray for its recovery (B. Ḥul 77b f.).

"load it with rocks"—to weaken it, so that it would not cast its fruit because of excessive fertility (B. *ibid.*).

"hoe the soil under grapevines"—which confers no benefit upon the vines.

"to plug the gaps"—in the branches, and cause them to grow straight, without interfering with each other, and thus prevent damage to the tree.

8. MḲ 1:1, 3; B. MḲ 2a, 4a.

"exceedingly thirsty"—and therefore must be irrigated, because rain water is not sufficient to keep plants growing.

"beṯ sĕ'ah"—a plot of ground planted with one sĕ'ah of seed. If there are more than ten trees to a beṯ sĕ'ah, and they are not watered, they will deteriorate. In both cases permission to irrigate is given because irrigation is not actually the same as tilling the soil. While this is normally forbidden by Scribal law, an exception was made here, in view of the probable serious loss. See below, Sec. 10.

9. MḲ 1:3; B. MḲ 3a f.

"but may not water the whole field"—using the trees as a pretext.

"grain field"—literally "white soil," a plot of open ground planted with grain or vegetables, with only a few trees on it.

"so that the trees . . . would not deteriorate"—see notes to the preceding Section.

"or dig a canal"—and this is not regarded as preparing the soil for planting.

"the connecting channels"—leading from the canal to the grooves around the vines.

10. B. MḲ 3a.

"salty"—meaning barren; cf. Jer. 17:6.

"the two primary acts"—sowing and reaping.

"the two derivatives thereof"—pruning and gathering the vintage.

"as we have explained"—above, Sec. 2–3.

11. B. Sanh 26b; P. Sanh 3:5.

"food stores"—so variant reading, *mĕzonoṯ;* the text has "camps," *maḥănoṯ;* cf. RDBZ *ad loc.*

"the Sages permitted"—in order to save lives, on the principle that where there is danger to life prohibitions are waived.

12. Ter 2:3; B. Giṭ 54a.

13. Shebi 4:2–3.

"or lets cattle manure it"—reading *diyyerah,* "confines the cattle successively in each part of the field, in order to manure it," as in the Mishnah.

"Nor may someone else rent the field"—because of this imposed penalty.

"his son may seed it"—because the penalty applies only to the father.

14. Shebi 4:2.

"the Sages did not hold him liable to a fine"—because the field is still not ready for planting, and requires plowing or digging.

15. Kil 1:9; Shebi 5:2, 2:10; P. Shebi 5:2. Burying for storage is permitted, provided that care is taken to prevent rooting and growth.

"exposed"—if the leaves are not exposed, the vegetables will strike root and grow.

"four ḳab," etc.—less than that resembles planting the bulbs.

"stir the soil of a rice field"—after wetting it with water. Otherwise the roots will dry out and die.

"but not to trim the rice plants"—which is not vital for their survival.

16. Shebi 4:1; P. Shebi 4:1.

"only large pieces"—by selecting large pieces, A makes it obvious that his intention is to use them for building or similar purposes and not to clear his field, which is forbidden.

"in B's field"—where it is unlikely that A would do it in order to clear that field.

"as a favor to B"—with the understanding that in return B would clear A's field, thus actually making it ready for planting.

17. P. Shebi 4:1.

"his intention"—to feed the cattle or to heat the oven, respectively, and not to clear the ground for sowing.

18. Shebi 4:4. When trees are uprooted, the earth is turned over, an act preparatory to planting, particularly in a piece of ground large enough to hold three or more trees.

19. Shebi 4:5.

"cover the cut"—so that it would retain its sap and not dry out.

"cultivation"—which is forbidden in the Sabbatical year.

20. Shebi 4:6. ,

21. Shebi 4:5; Tos Shebi 3:15.

"virgin sycamore"—never touched by an axe before; cf. the next Section.

22. Shebi 4:5; B. BB 80b. Leaving a stump under ten handbreadths high constitutes cultivation; cf. the preceding Section.

Chapter II

1. Shebi 3:1–3; Tos Shebi 2:14; P. Shebi 3:1. Cf. above, i, 4.

"make . . . a dunghill"—for use next year.

"after the moist dung has hardened"—literally "after the sweet has hardened"; "the sweet" is also interpreted as meaning a euphemistically sweet (that is, bitter) herb, specifically the colocynth vine; cf. below, vii, 17, and note thereto.

"If one wishes to add . . . he may do so"—because the larger the heap the more evident that it is meant for storage and not for application to the soil.

"If he already has a little dung pile"—from before the Sabbatical year.

"more than three dunghills"—in which case it may be construed as manuring.

2–3. Sheḇi 3:3; P. Sheḇi 3:2.

2. "he may do so"—without fear of arousing suspicion that his intention is to manure the field.

3. "there is no prescribed minimum"—since such depositories are unfit for planting, there is no suspicion that the intention is to manure the soil.

"several"—more than three (cf. above, Sec. 1).

4. Sheḇi 3:4; Tos Sheḇi 2:15.

"in the manner of all those who collect manure"—as described in the three preceding Sections.

"build another fold next to it"—i.e., attach the other three sides of the dismantled fold to the outer face of the remaining side, thus forming a new fold.

5. Sheḇi 3:4; P. Sheḇi 3:3; Tos Sheḇi 2:15.

6. Sheḇi 3:5; Tos Sheḇi 3:1.

"twenty-seven stones"—what the Mishnah states simply Maimonides makes complicated. The Mishnah says, "three layers of stones three cubits long, three cubits wide, and three cubits high, together making twenty-seven stones." Thus each stone must be one cubit square by one cubit high, which makes it evident that the stones were quarried for building purposes.

7. Sheḇi 3:6; Tos Sheḇi 3:16.

"because he needs them"—for building or for fencing, and not merely to clear the field for planting, which would not justify such a large expenditure of time and labor.

"until one handbreadth's height is left"—so that the ground underneath the fence is still not accessible for planting.

8. P. Sheḇi 3:6, 2; Sheḇi 3:6.

"When does this apply?"—that he must leave one handbreadth of the fence next to the ground.

"or when one commences removing the stones during the Sabbatical year"—the *editio princeps* reads instead "by commencing to remove the stones during the Sabbatical year," so that both factors must be involved: (a) he intends to improve his field; (b) he starts work during the Sabbatical year.

"under contract"—and thus has an interest in the outcome and should logically be subject to the same rule as the owner.

9–10. Sheḇi 3:7.

9. "each . . . a load for two men"—and thus fit for building purposes. Only two such heavy stones are required here, since the stones are scattered and had not been artificially joined to form a fence.

10. "those touching the ground"—for everyone to see that his intention is to build and not to plant.

"he may remove all of them"—because the ground underneath them is still unfit for planting even after the removal of the stones, so long as he does not remove the rock or straw as well.

11. Sheḇi 3:5; P. Sheḇi 3:7.

"by stretching his hand"—because it is evident that he is removing the stone in order to use it in erecting the embankment. If he has to go farther into the field or down into the ravine, he might be suspected of preparing the ground for cultivation. The embankment is erected to prevent run-off of rainwater from the field.

12. Sheb̲i 3:9.

"a lessee . . . from his leased field"—RABD understood the term *ḳabbĕlan* (lessee) as signifying here a contractor hired to erect the embankment, and interpreted the whole sentence as meaning "a contractor may remove stones . . . whether from the field in which he has contracted to erect the embankment or," etc., which seems more logical.

"may remove stones . . . from any place"—because it is assumed that since the field is not his, he is not interested in permanently improving it.

"even small ones"—so *editio princeps;* later editions read "even large ones," evidently an error.

13. P. Sheb̲i 3:8.

"if it does not trip up the public"—that is, is not a public thoroughfare; in which case repair of the fence may be intended to allow planting on the earthen portion of it.

"it is permitted to build it up"—the principle is that if people get the impression that what one does is not for the purpose of preparing the ground for cultivation, it is permitted. In this case the breach is part of a public thoroughfare, and planting it would be futile.

14. P. Sheb̲i 3:8; Sheb̲i 3:10.

"his companion's field . . . the public domain"—if one builds a fence between his own field and his neighbor's field, he does it in order to cultivate his own field right up to the fence; on the other hand, it is unusual for one to sow on soil closely bordering on a public thoroughfare, hence he would not be suspected on infringing upon the rules of the Sabbatical year.

"in the manner of all those who pile up manure"—see above, Sec. 1.

Chapter III

1. B. MḲ 3b f.; Sheb̲i 1:1, 2:1.

"a rule revealed to Moses on Mount Sinai"—a formula denoting an ancient oral tradition, not contained in the written Torah but of equal authority with it. See Pe 2:6.

"improving it"—which is forbidden.

"Pentecost . . . Passover"—all work performed thereafter being considered as intended to benefit the Sabbatical year's crop.

"When the Temple is not in existence", etc.—there are thus three kinds of limitation:

a) That of the Torah, which permits work in the fields in the sixth year up to New Year's Day of the Sabbatical year;

b) That of the rule revealed to Moses on Mount Sinai, which permits such work during the sixth year up to thirty days before New Year's Day of the Sabbatical year;

c) That of the precautionary enactment of the Sages, which forbids plowing a cornfield after Passover, and a field planted with trees after Pentecost, of the sixth year.

2. Sheb̲i 1:2, 3, 5; P. Sheb̲i 1:3; B. BB 83a.

"the same as fig trees"—the fruit of the fig tree is large and the yield abundant, hence this large criterion.

"for yoked oxen to pass through"—which is a space four cubits wide. If it is narrower, the trees will be eventually uprooted by the passing oxen.

3. Sheḇi 1:2–4.

"a clearance," etc.—comprising the space occupied by the fruit picker and his basket both under the tree and outside of it; so Albeck's commentary on the Mishnah. Beyond that area the orchard is considered the same as a cornfield.

4. Sheḇi 1:4; P. Sheḇi 1:4.

5. Sheḇi 1:4, 6; B. MḲ 3b.

"saplings"—which, being yet tender, would perish if not aided by plowing.

6. Sheḇi 1:6.

"in one row"—close to each other, and not scattered all over the beṭ sĕ'ah.

7–8. Sheḇi 1:8.

9. Sheḇi 2:3, 2; B. MḲ 3a.

"melons and pumpkins"—which would deteriorate if not hoed.

"as we have said"—above, Sec. 1.

"strip"—cut off excess leaves, to lighten the burden of the tree.

"fumigate"—to keep away insects.

"trim"—cut off twigs, leaving only the stem.

"prune"—when shoots abound, they are thinned to accelerate and strengthen growth.

"oil"—with rancid oil, to ward off vermin.

"wrap"—with rags for protection against heat or cold.

"clip them"—cut off the tops; some translate "cover them with ashes."

"build shelters"—as protection against the sun's heat, heavy downpour, and the like.

10. "may not be oiled or pierced"—because those who see one do so will think that the figs are of the Sabbatical year.

"steps"—their purpose is variously given as to facilitate drawing water from the ravine to irrigate the field; or to plant crops on the level surface of the steps, thus making at least part of the ravine productive.

11. Sheḇi 2:6; B. RH 10b; P. Sheḇi 2:6.

"At the present time, too"—after the destruction of the Temple.

Chapter IV

1. P. Sheḇi 8:6; B. Men 5b; Sheḇi 4:2; Sif Lev., Bĕ-har 1:5.

"thou shalt not reap"—the prohibition thus applies to harvesting in the regular fashion, not to the consumption of the spontaneous produce itself.

"harvests in preparation for the cultivation of the land"—harvests the spontaneous growth merely in order to improve the soil for future cultivated crops.

"as we have explained"—above, i, 1–2.

2. Sheḇi 9:1.

3. B. Ta 19b; Sheḇi 9:1; P. Sheḇi 9:1.

4. P. BB 5:1; P. Sheḇi 9:7.

"precautionary measure"—for appearance's sake, lest people who see one eating them should think that he is doing something forbidden; cf. above, iii, 11.

"to improve it for plowing"—it is better for the field to remain as it is, hence people will not think that one is to plant it presently.

"render his vineyard forbidden"—people who see him will not think that he is about to plant the vines, because this would render all of the field forbidden (above, I, vi, 1).

5. Tos Shebi 5:23.

"may not be consumed"—for the same reason mentioned above, Sec. 2, since transgressors might secretly plant them toward the end of the Sabbatical year.

"nor be plucked by hand"—the usual procedure with aftergrowth, which comes up haphazardly, and not in regular rows.

"plow them under," etc.—the prohibition covers only human consumption, and not utilization as fertilizer or fodder.

6. P. Dem. 2:1; Ter 9:4.

"From Hanukkah onward"—from three months into the eighth year onward, when the growth has probably taken place after the expiry of the Sabbatical year.

"their yield is permitted"—on the principle that the yield of two sources, one permitted and the other forbidden, is permitted. The seed being of the Sabbatical year and therefore forbidden, and the soil in which it was planted being of the eighth year and therefore permitted, the yield for which both are jointly responsible is permitted.

7. Tos Shebi 4:13, 17; Shebi 6:4; P. Shebi 8:3.

"If they have grown," etc.—the criterion behind these rules is whether or not people who see this produce might think that it was picked in the Sabbatical year. If it grew as much in the eighth year as produce grown entirely in that year, people are not likely to suspect its origin.

"buy it"—from a seller not known to be observant of the pertinent rules.

"It is permitted to buy vegetables"—because most of them were imported from outside of the Land of Israel, where the law of the Sabbatical year does not apply.

8. Shebi 5:5; P. Shebi 5:5.

"arum"—what is special about arum is that the plant grows for three years, and there is the possibility that the arum of the early part of the eighth year may have grown not in the Sabbatical year but in the sixth. See Kil 2:5.

9. RH 1:1; Tos Shebi 2:13; P. Ma 5:2; B. RH 13b.

"the season of tithing"—see above, IV, ii, 5.

10. This Section goes back to the concluding sentence of the preceding Section, "it is accounted produce of the Sabbatical year."

"partaking of the sanctity of the Sabbatical year"—and subject to the rules given below in chapters v–vi.

11. Shebi 2:7.

12. B. RH 13b, 15b.

13. B. RH 12b.

14–15. P. Shebi 2:8.

15. "while the leaves are permitted"—because in the case of vegetables the determining factor is the time of their gathering (above, Sec. 12).

16. Shebi 5:1.

"inasmuch as it takes them three years to complete their ripening"—and they are thus unfit for eating until the ninth year. But since they have become liable to tithing during the Sabbatical year, they are subject to the law governing that year.

17. Shebi 2:9.

"deprived of irrigation"—and now regarded the same as produce of naturally watered fields dependent on rain, and subject to the same rule.

18. Sheḇi 2:10; Tos Sheḇi 2:11–12.

"arum"—see above, note to Sec. 8.

"artichoke"—*ḳindes,* read *ḳinaras* (Greek *ḳynara*). Like arum, it remains in the ground for several years, but brings forth new leaves each year.

"he must . . . clip its leaves"—for appearance's sake, since the artichoke root is hidden from sight, and an observer would not know that the leaves have sprouted from a permitted root.

19. Sheḇi 5:4.

"metal mattocks"—which are otherwise regularly used in the cultivation of the soil.

20. Sheḇi 6:3; P. Sheḇi 6:3; B. Neḏ 59b. The onions have been detached from the ground prior to the onset of the Sabbatical year; the tender green leaves are regarded as the produce of the bulb itself.

21. B. Neḏ 59b, 58b; P. Sheḇi 6:3.

"by the soil"—as it is said, *Then shall the land ḳeep a Sabbath unto the Lord* (Lev. 25:2).

"its neutralization"—out text has *nĕṭilaṯah,* "the lifting (of its interdict)." The Talmudic source has *bĕṭilaṯah,* "its neutralization."

"through the soil"—since the onion was replanted in the eighth year, when the soil was no longer under the Scriptural interdict.

22. Sif Lev. *Bĕ-har* 1:3.

"in the same manner as in other years"—see above, Sec. 1.

23. Sheḇi 8:6; P. Sheḇi 8:6.

"process olives in a regular olive press"—the regular processing comprised three stages: (a) the olives were hand-crushed, placed in a loosely woven basket, and pressed for first oil; (b) the pressing beam was then applied to the caked olives for second oil; (c) the cake was then ground in a mill and returned to the press for the final oil. The procedures described here depart from this regular process, and the miniature press was not used in it at all.

"or in a small press"—so the Mishnah. In the *editio princeps* the phrase is omitted both here and in the next sentence; in later editions it appears only in the next sentence, where it is out of place, and is probably due to a scribe's eye wandering from one line to another.

24. Exod. 23:11; P. Pe 6:1; Sheḇi 5:7.

"everywhere"—whether in the field or in the owner's house.

"five jugs of oil"—the estimated amount needed for one's own use.

"if he brings in more"—claiming that he needs more for his household.

25. B. Ḳid 36b; P. Sheḇi 6:1. Scripture does not tie the Sabbatical year in any way to the Temple.

26. Sheḇi 6:1. For the explanation of the geographical terms see above, III, i, 7–9, and notes thereto.

27. Sheḇi 6:2; Yaḏ 4:3; P. Sheḇi 6:2.

"according to Scribal law"—see above, III, i, 1.

"Ammon, Moab, Egypt, and Shinar"—the fear that people might settle there did not apply here as it did in the case of Syria, since these lands were regarded as entirely foreign.

28. P. Sheḇi 6:1; Sheḇi 9:2.

"than the parts . . . occupied by those who came up from Egypt"—above, Sec. 26.

29. B. Sanh 91a.
"at Israelite transgressors"—above, Sec. 2.

30. Tos Sheḇi 4:8.

Chapter V

1. Sheḇi 8:2; Sif Lev. Bĕ-har 1:10; P. Sheḇi 7:2, 8:2.
"shall . . . be"—the complete sentence reads: And for thy cattle, and for the beasts that are in thy land, shall all the increase thereof be for food (Lev. 25:7).

2. Sheḇi 8:2; Tos Sheḇi 6:1.
"as is the rule for heave offering and second tithe"—above, III, xi, 1; v, iii, 11.

3. P. Sheḇi 8:2, 1.
"as is the rule for heave offering and second tithe"—Tos Ter 9:10; and above, v, iii, 11.

4. Sheḇi 8:7; B. Zeḇ 76a; P. Sheḇi 8:7.
"lest he should render the latter unfit"—if the heave offering oil becomes unclean, the entire mixture, including the vegetables, will have to be burned (above, III, xii, 1). Another interpretation: if the time for the disposal of Sabbatical year's produce supervenes, the heave offering oil also will have to be disposed of (RDBZ).

5. Tos Sheḇi 5:20; Sif Lev. Bĕ-har 1:7.

6. Sheḇi 8:2; P. Sheḇi 8:2.
"One may not perfume the oil"—since this makes it inedible; cf. above, III, xi, 3.
"nor anoint himself . . . in the bathhouse"—the oil is holy, and should not be carried into the bathhouse which is far from holy. If the anointing is done outside, the oil is absorbed by the skin by the time one enters the bathhouse.

7. Tos Sheḇi 6:10–12.
"seal an oven"—seal the cracks in the oven. All these are examples of using the oil for purposes other than anointing one's body.
"he may rub it in"—since the oil, once it touches the skin, loses its holiness (cf. above, III, xi, 7).

8. Tos Sheḇi 6:14–15; P. Sheḇi 7:3.
"burning fire"—used for warming one's self, and not for light.

9. P. Sheḇi 8:1.
"does not devolve upon dyes for animal use"—the sanctity of the Sabbatical year limits the use of food and feed in certain ways to prevent their waste, which does not apply here, since anointing and dyeing are restricted to humans.

10. Tos Sheḇi 5:5, 6:25; P. Sheḇi 7:1; B. BḲ 102a.
"Laundering substances"—which are inedible.
"one may launder with them"—because this is their normal use.
"as it is said," etc.—the same verse is used in the preceding case to include, and in this case to exclude, by omitting the last Hebrew word (for food) in the former and by citing it in the latter. The real reason behind it is not this her-

meneutical artifice but good logic: in the first case it is the normal way of using the substance, in the second case it is not.

"spray"—as a perfume.

11. Sheḇi 8:1; P. Sheḇi 8:1.

"anything not used . . . for human food"—but only for animal feed.

"but not for beast"—since Scripture specifies, *And for thy cattle . . . for food* (Lev. 25:7), not for medicinal application.

"wood"—which is not subject to the law of the Sabbatical year (below, Sec. 17).

12. Tos Sheḇi 5:19; P. Sheḇi 7:3.

"One may sell," etc.—see above, Sec. 8. The principle is that the restrictions appertaining to the Sabbatical year's produce are not removed by exchange or sale; rather the restrictions now cover both the original article and the one received in exchange. Cf. below, vi, 1.

13. Sheḇi 6:5; Tos Sheḇi 5:20; Sif Lev. *Bě-har* 1:6; P. Dem 3:1.

"hired man"—hired just for day-work, and returning to his own home for the night.

"Boarders"—variant translation: "billeted troops."

14. Tos Sheḇi 5:22.

"She may, however, maintain herself," etc.—in practical terms both procedures amount to the same thing, but in legal terms the situation is not the same. In the former case, the court would be transferring a portion of the absent husband's property to the wife in satisfaction of her claim to maintenance—a forbidden act during the Sabbatical year because it is considered as payment of a debt. In the latter case the wife, of her own initiative, consumes a portion of her husband's crop, to which she is lawfully entitled while the marriage endures.

15. Sif Lev. *Bě-har* 1:10.

"until it has reached the season of tithing"—for only then is it called produce (above, iv, ii, 3).

16. Sheḇi 4:7-9.

"one-third of a loḡ"—which is ⅑ of their full yield of 3 loḡ per sě'ah (cf. above, iv, ii, 5).

17-18. Sheḇi 4:10; B. Pes 52b.

18. "From what time on . . . to cut down trees," etc.—this explains the phrase "before any fruit appears on them" in the preceding Section.

"globules"—one of the early stages in the development of the grape.

"date berries"—small unripe dates.

"this is waste of fruit"—which is forbidden in the Sabbatical year.

"stony dates"—reading, with R. Joseph Corcos, *šiṣin,* "prickly twigs bearing hard inedible dates." The printed text has *šey-yaṣiṣ,* "blossom," which does not fit the context.

19. Sheḇi 9:7; Tos Sheḇi 6:16.

"olive or grape refuse"—skins and seeds pressed into cakes after the extraction of oil and juice, respectively; they are not considered edible and hence are not subject to the law of the Sabbatical year.

20. Sheḇi 8:11.

"heated"—ex post facto, since one may not do so deliberately (above, Sec. 19). Once the straw has been burned, however, one may utilize the heat.

"even if he pays a fee for it"—which makes it appear as if he was trading in produce of the Sabbatical year.

"other things"—other produce fit for human or animal consumption.

21. Tos Shebi 6:16; P. Shebi 4:6; B. Ber 36a; cf. above, III, xi, 10–11.
22. Tos Shebi 6:7.
23. Tos Shebi 5:18–19.

"One may not", etc.—because they are fit for animal feed (above, Sec. 19).

"destroyed"—and no longer existent, hence no longer fit for feed.

"An oven heated," etc.—cf. above, note to Sec. 20. On the analogy of the bath, it would seem that if one nevertheless utilizes the heat to bake bread, he may eat the bread, since baked bread and hot bath water are both products of unlawfully burned fuel.

"Once the second rain has fallen"—when the new crop is coming up, and the old crop loses its Sabbatical sanctity.

Chapter VI

1. Shebi 7:3; B.AZ 62a; Bek 12b; Suk 39a. Scripture specifies *for food* (Lev. 25:6), not for trade.

"to sell a small amount"—this is not called trading.

"the sanctity of the Sabbatical year"—see above, v, 10–11.

"the produce sold retains its sanctity"—see above, v, 8.

2. Shebi 7:3. The decisive factor is one's original intention.

"gather"—*lakah* here has evidently its original meaning "to take," and not its derived meaning "to buy." What is meant, of course, is vegetables growing spontaneously in an uncultivated field (above, iv, 4).

"with dye"—cf. above, v, 21.

"or daughter's son"—*ben bitto*, presumably a scribal error for *ben beto*, "a male member of his household."

3–4. Shebi 8:3.

3. "In the lump"—by a rough estimate of its market value.

4. "tied for use in the house" etc.—there was a difference in the method of tying produce into bundles for home consumption and for sale in the market place; in the latter case care was taken to adhere to exact size and number of ingredients.

5. P. Shebi 8:3; Tos Shebi 4:18.

"the same as the produce of the Land of Israel"—for appearance's sake, lest a person who sees one selling it should think that he is selling Israelite produce, which is forbidden.

6. Sif Lev. *Bĕ-har* 3:3; Shebi 8:7; cf. VI, IV, vi, 4 (YJS, *15*, 194).

7. Sif Lev. *Bĕ-har* 3:3; B. AZ 54b.

"clear it out"—see below, vii, 1.

"the same as other Sabbatical year's produce"—see above, v, 11–12.

8. B. Suk 40b, 41a.

"first produce"—the original Sabbatical year's produce.

"second produce"—other produce received in exchange for the first produce, or for the money realized therefrom.

9. Tos Shebi 7:8; B. Suk 41a.

"lest one should keep them"—which is forbidden (below, vii, 1, 7).

"with slaughtered animals"—where there obviously can be no such apprehension.

10. Tos Shebi 7:9; Shebi 8:8.

"discharge a debt"—for this is the same as trading.

"wedding gift"—presented by the groom's friends and considered obligatory (see XII, II, vii, 1; YJS, *5*, 133).

"repay a favor"—also considered obligatory.

"a promise"—equivalent to a debt.

"he must consume its equivalent"—for the money realized from Sabbatical year's produce must be used only for purchasing other food, which must be consumed within that year (above, v, 1).

"bird offerings"—see Lev. 15:14–15, 29–30; 12:6, 8.

"One may not oil vessels or hides"—since Sabbatical year's oil may be used only for personal anointing (cf. above, v, 7).

11. Shebi 8:5.

"to give him a drink"—which is a matter not of trade but of common courtesy, hence the tip given him is a free gift and not payment for skilled service.

12. Shebi 8:4. The difference between the two cases is that in the first case the money is ostensibly given as a gift, prior to the request for work; in the second case the money is conditioned on the performance of a service.

"vegetables"—of the Sabbatical year.

"is not penalized"—as he is in the next Section.

13–14. B. AZ 62a.

13. "enough for one's needs"—see above, iv, 24.

14. "the loaf is accounted the same as Sabbatical year's produce"—because it is acquired in exchange for Sabbatical year's produce; in the second case it is not an exchange but the payment of a debt. A variant view, however, regards this transaction as an exchange of free gifts, since the clause "for it" is not in our text of the Mishnah, and therefore exempt.

15. Shebi 4:2, 9:9; 'Ed 5:1.

"by favor"—of the owner, the consumer thus being morally indebted to the owner. Nevertheless, since the consumer may eat of the produce either way, when he eats of it by favor, it is not accounted a commercial transaction on a quid pro quo basis; cf. above, iv, 24. As a matter of decency, however, one should obtain the owner's permission before helping one's self to his produce, even though it is ownerless during the Sabbatical year.

Chapter VII

1–2. Sif Lev *Bĕ-har* 1:8; B. Pes 52b; Ta 6b; Nid 51b; P. Shebi 9:2; Shebi 9:2.

1. "wild beast"—which is not under man's control, and whose rate of consumption is determined by its own craving.

3. Shebi 9:8.

"If one has a large quantity of produce"—on the last day before removal, more than he himself can consume.

"Both poor and rich are forbidden to eat of it"—our text of the Mishnah reads, "The poor may eat of it . . . but not the rich, so R. Judah. R. Jose says, both poor and rich may eat of it after its removal." Maimonides' text of the last sentence must have read "may *not* eat of it."

4–5. Shebi 9:4.

4. "in the fields, gardens, and orchards"—which are accessible to wild beasts, since their produce is ownerless in the Sabbatical year and may not be guarded by watchmen.

6. Shebi 9:5; P. Shebi 9:5; Sif Lev. *Bĕ-har* 3:5, 1:18; B. Pes 52a.

"If he has begun eating out of that jar," etc.—the entire contents, even the other two species whose time for removal has not yet arrived, are liable to removal (RDBZ).

7–8. Shebi 7:1; cf. above, vi, 6.

9–10. Shebi 9:2–3.

9. "to the sea"—reading *'ad* for *'im*.

"three areas"—hill country, plains, and valleys.

11. Shebi 9:3; P. Shebi 9:3; B. Pes 53a.

"Zoar"—near Jericho, where dates were the last to be exhausted.

12. Pes 4:2; B. Pes 52b.

"may not be carried from place to place"—one need not bring it back to the Land of Israel and clear it out there, as is one opinion cited in the Talmud.

13–14. Shebi 7:1–2.

13. "inasmuch as its roots remain underground"—and only that which is exhausted aboveground in the field is forbidden in the Sabbatical year. Only the roots of these two plants are used to make dyestuff.

"until New Year's Day"—and thereafter as well.

14. "so long as it is not used for wood"—had one intended to use it for wood, it would have been exempt from the laws of the Sabbatical year (above, v, 21).

"until New Year's Day"—as well as thereafter.

15. Shebi 7:3.

"husks and blossoms"—are subject to the law of the Sabbatical year because they are used as dyestuffs, but are not liable to removal because the roots of the plant remain underground. See above, Sec. 13.

"Sprouts of service and carob trees"—are liable to removal because they are edible and eventually drop off.

"Sprouts of the terebinth", etc.—are subject to the law of the Sabbatical year because they are edible, but are not liable to removal because they remain on the trees and do not drop off.

"Their leaves . . . are liable"—because they do drop off from the branches.

16. Tos Shebi 5:12; Shebi 7:5, 9:6.

17. Shebi 9:6.

"until the ground moisture is dried up"—literally "until the sweet is dried up"; "the sweet" is also interpreted as meaning a euphemistically sweet (that is, bitter) herb, specifically the colocynth vine; cf. above, ii, 1, and note thereto.

"second rainfall"—usually in the month of Heshvan, the second month of the year in the Jewish calendar, about October/November in the western calendar.

18. Shebi 9:7. See note to preceding Section.

19. Shebi 7:6; B. Nid 8a.

"are subject to the law of the Sabbatical year . . . is not subject"—the principle is that if the produce is considered to be fruit, it is subject to the law of the Sabbatical year and is liable to removal; if not, it is exempt.

20. B. Nid 8a.

"is accounted the same as fruit"—since such trees bear no other fruit.

21. Shebi 7:7; P. Shebi 7:8. The principle is that the ingredient which has the power to impart a flavor determines the status of the entire mixture. The older ingredient has the power to impart a flavor to the younger ingredient, but not vice versa. Hence the rose of the Sabbatical year can impart a flavor to the oil of the eighth year, but not to the oil of the sixth year; the latter is therefore permitted for consumption.

22. Shebi 7:7.

"carobs"—are an exception to the rule above (note to Sec. 21), for they have the capacity to impart a flavor even when they are the younger ingredient.

Chapter VIII

1. Shebi 5:9.

"Just as it is forbidden to cultivate"—above, i, 1.

"so is it forbidden to encourage transgressors"—in PhM *ad loc.* Maimonides bases this interdict on Lev. 19:14, *Thou shalt not . . . put a stumbling block before the blind,* and explains that the transgressor is blinded by greed inspired by his Inclination to evil, hence he should be helped as far as possible to open his eyes to righteousness.

2–3. Shebi 5:6; P. Shebi 5:6; B. AZ 15b.

3. "which is permitted"—above, iv, 1, 24.

4. P. Shebi 5:6; Shebi 5:6.

5. Shebi 5:7.

"five oil jugs or fifteen wine jugs"—the maximum allowed to be collected out of the ownerless produce of the Sabbatical year (above, iv, 24).

"Outside of the Land of Israel"—where the law of the Sabbatical year does not apply (above, iv, 25).

6. Shebi 5:8; Tos Shebi 4:5; B. AZ 15b. The principle governing these rules is that where there is a possibility that the implement or beast will be used in a way that does not violate the laws of the Sabbatical year, the sale or loan is permitted, on the assumption that B will choose to use it in a lawful manner; cf. above, Sec. 2.

"unless he stipulates," etc.—since it is too much to expect that B would let the entire year's yield of the trees go to waste, A should eliminate such a stumbling block of temptation.

"a threshing floor"—containing Sabbatical year's produce, which may not be exactly measured for sale (above, vi, 3).

"laborers"—who might perform forbidden work, and A should not aid B in paying their wages therefor.

7. Shebi 5:9; Giṭ 5:9. The Mishnah gives the reason for this rule as "in the interests of peace"; but to help actively and directly in the violation of the law is still forbidden.

8. Shebi 5:9, 4:3; B. Giṭ 62a; Tos 'Uḳ 3:16; P. Shebi 4:3.

"in taking honey out of the beehive"—which comes under the heading of reaping (III, 1, xxi, 6; YJS, *14,* 131).

"cannot be penalized for plowing"—if an Israelite does it, he is liable to a fine (above, i, 13).

9. Shebi 6:2; P. Shebi 6:2.

"already cut, but not . . . still attached"—the reason for this distinction is as follows: The Sages permitted the processing of detached produce, because Syria was not regarded as part of the Land of Israel and hence was not subject to the rules governing the latter, but forbade the processing to produce still attached to the ground, in order to prevent people from leaving the Land of Israel and settling in Syria. See above, iv, 27.

10–11. B. Suk 39a.

10. "Just as it is forbidden"—above, vi, 1.

"to retain it"—beyond the time of removal (above, vii, 3).

11. "lulab̲ . . . citron"—the citron is a fruit and hence may not be used as merchandise. The lulab̲, being a palm branch, is not a fruit and is therefore not subject to the laws of the Sabbatical year. Both are required for the ritual of the Festival of Tabernacles (III, vi, vii–viii; YJS, *14*, 396–408).

12. B. Suk 39a; Shebi̲ 9:1.

"When does this apply?"—referring back to Sec. 10, above.

13. Ma 5:1; cf. above, iii, ii, 12.

"dyer's rocket bulb"—some translate "Richpah onion," to correspond to the other items here which are called after the place where they grow.

14. Bek̲ 4:8; P. Ma 5:6.

"To whom does this apply?"—that it is permitted to buy from him enough for three meals (above, Sec. 12).

"even if it is combed"—when by right it should no longer be considered produce; it is forbidden, however, because it may still contain some seeds, which is not the case after it is spun and twisted.

15. Bek̲ 4:10; B. Bek̲ 30a.

"tithes"—from the context it is evident that only the second tithe is meant here, since the first tithe does not involve conveyance of the produce to a certain place (i.e., to Jerusalem; above, v, ii, 1).

"is not redeemable . . . is redeemable"—above, vi, 6; v, iv, 1.

16. Bek̲ 4:10; B. Bek̲ 30a.

"unclean food," etc.—see V, v, viii, 10 (YJS, *8*, 283).

17. B. Bek̲ 35a.

"may . . . act as judge or witness"—the Mishnah (Bek̲ 4:10) explicitly rules that such a person may not do so, hence the stricture of RABD who follows the Mishnah. Maimonides, however, follows here the dissenting opinion given in B. Bek̲ 35a. See KM to iv, xii, 17.

18. B. Sanh 26a; Tos Ter 6:3.

"for they might say," etc.—the reason is Maimonides' own conjecture.

"as the law requires"—above, iii, xiii, 1–2. When the one sĕ'ah of heave offering produce is neutralized in one hundred sĕ'ah of Sabbatical year's produce, the fact that priests are suspect concerning Sabbatical year's produce is overlooked, because of their right to the heave offering produce. When the one sĕ'ah of heave offering is not neutralized, the whole mixture is forbidden to the priest, because of the fact that priests are suspect concerning Sabbatical year's produce, and the usual procedure of selling the entire mixture, minus the price of one sĕ'ah, to the priest cannot be followed.

19. Tos Shebi̲ 5:8; Shab 9:9; P. Shab 9:5.

"fatteners of animals"—or "druggists" (*paṭṭamin*).

"and need not be apprehensive," etc.—since aftergrowths are forbidden by Scribal law only (above, iv, 2), leniency prevails in case of doubt.

20. Tos Dem 3:17; P. Dem 3:1; Pe 8:2; B. AZ 62b.

"may be suspected only," etc.—they are suspect not about the yield of the field itself but only about things received for it by exchange or by sale, because of the erroneous impression that these do not become subject to the laws of the Sabbatical year. The assumption is that a person performing a meritorious act of charity would not vitiate it by deliberately doing something which he knows to be unlawful.

Chapter IX

1. Sif Deut. *Rĕ'eh* 112.
"passed over"—see below, Sec. 4.
2. B. Giṭ 36a; Ḳid 38b.
"the law of the Jubilee year"—see Lev. 25:10–13.
"when the land is restored to its original owners without any payment"—and hence this also is a form of cancellation of a debt.
3. B. Giṭ 36a.
4. Sif Deut. 31:10; B. 'Ar 28b; Sanh 32a.
"the nights of New Year's Day of the eighth year"—i.e., the nights of the 1st and 2nd of Tishri of the eighth year.
5. Sheḇi 10:2.
"has been intercalated"—i.e., has been declared by the court in Jerusalem to have thirty days instead of twenty-nine. Accordingly the day when the meat was distributed among the purchasers was actually the 30th of Elul, the last day of the Sabbatical year, hence the debts are forfeit. RABD disagrees and claims that this case comes under the category of store credit, which cannot be cancelled (below, Sec. 11); had the butcher known that the day was not New Year's Day, he would have extended no credit and would have demanded immediate cash payment.
6. Sheḇi 10:1; B. Giṭ 37a; Sheḇi 7:8.
"even one secured by a bond"—and it cannot be said that in such a case the debt is considered as in fact already collected. When a specific field has been assigned as security, the debt is ordinarily considered as already collected, just as if the field were a pledge. Cf. XIII, III, vii, 4 (YJS, 2, 103).
"removes the debtor's obligation to swear an oath"—an oath is obligatory upon the debtor who admits only part of the debt (see XIII, III, xiv, 4; YJS, 2, 129).
7. Tos Sheḇi 8:6; B. BB 42b.
"bailees, partners, and their like"—where the relationship of the parties is more like that of pledgeor to pledgee than that of debtor to creditor.
"he must pay it"—after the expiration of the Sabbatical year, and is not released by it from the obligation to pay.
8. P. Sheḇi 10:1.
"testify against B"—and B is consequently obligated to pay.
"the Sabbatical year does not cancel B's debt"—because A could not demand payment during the Sabbatical year. Had B admitted his indebtedness during the Sabbatical year, the debt would have been canceled. That year, however, cannot cancel a debt which is not actively claimed.

9. B. Mak 3b; P. Shebi 10:1.

"eventually"—ten years hence.

"during that year"—i.e., three years before due date.

"that A should not demand payment"—in order that the command *He shall not exact it* should not apply; the debt is canceled nevertheless, since the debtor must in conscience still feel obligated to pay.

10. B. Mak 3b.

"the stipulated condition is valid"—the Sabbatical year affects a debt in one respect only—it cancels the obligation to repay it. Any other obligation, even one not to claim cancellation, is not affected at all. Hence upon the expiry of the Sabbatical year B assumes a new indebtedness, which he is fully bound to repay; cf. IV, 1, vi, 9 (YJS, *19*, 38–9).

11. Shebi 10:1; Tos Shebi 8:3. Normally a purchase is payable for immediately, and wages are payable when due. Delayed payment of accumulated purchases or wages becomes in fact a loan by the payee to the payer.

12. Shebi 10:2; B. Giṭ 18a. A fine is a penalty rather than a debt.

"violator"—see Deut. 22:29.

"seducer"—see Exod. 22:16.

"defamer"—see Deut. 22:19.

13. B. Giṭ 18a; Tos Shebi 8:4.

"impaired it"—by receiving partial payment of it; the balance due assumes the status of a loan against the husband.

14. B. Giṭ 18a; Shebi 10:2; B. Shebi 44b; Tos Shebi 8:5.

"secured by a pledge"—the pledge becoming the temporary property of the lender, so that it is as if repayment had already been made for the duration of the debt.

15. Shebi 10:2; Sif Deut. 15:3; P. Shebi 10:2; B. Sanh 6b.

"already collected"—the situation has changed: there is now no longer a private debt but rather a court order to pay, and court orders are not subject to cancellation by outside factors.

16. Shebi 10:3; B. Giṭ 36b, 36a, 37a.

"prosbul"—or *prozbul*. The Talmud explains the term as *pĕros buli u-buṭi*, "a boon to the rich and to the poor," but it probably is the Greek *pros boulē*, "(declaration) before the council" (see L. Blau, in *Festschrift zum 50-jährigen Bestehen der Franz-Joseph Landesrabbinerschule in Budapest* [Budapest, 1927], pp. 96 ff.).

"at the present time"—RABD disagrees and maintains that Hillel's enactment applies at all times.

17. B. Giṭ 36b.

18. Shebi 10:4. As given by Maimonides, the form implies a court of two judges; three judges are actually required.

"at any time I please"—even after the expiration of the Sabbatical year.

"or witnesses"—if the judges are sitting in another place, and the document is to be forwarded to them there.

19–20. Shebi 10:6; B. Giṭ 37a.

19. "real estate"—which cannot be moved or hidden.

"lends"—which gives the debtor the quasi-proprietary right to use the land.

"land held as pledge"—either the debtor's land held as a pledge by a third person, or a third person's land held as a pledge by the debtor.

20. "their guardian's property"—provided that the guardian borrows the money for the orphans' benefit; cf. below, Sec. 24.

21–23. Sheḇi 10:5.

23. Adar precedes Nisan, and Iyar follows it. If the creditor predates the prosbul, it is to his disadvantage and to the advantage of the debtor, and therefore permitted. If he postdates it, the result is the reverse, and therefore forbidden.

24. B. Giṭ 37b; Keṯ 9:9; cf. below, Sec. 26.

"since the time of peril"—the persecutions under Hadrian that followed the revolt of Bar Kokhba (132–135 C.E.), when all Jewish religious practices were forbidden on the penalty of death, and it was hazardous to preserve a prosbul.

"Orphans need no prosbul"—to collect their claims, since the court is considered as their guardian; it is as if they had deposited their claims with the court, in which case the Sabbatical year cannot cancel them. See above, Sec. 15.

25. "he would have been believed"—on the principle of *miggo,* according to which a person's claim is allowed on the ground that had he wished to tell a lie, he could have offered a stronger claim without fear of contradiction.

26. B. Giṭ 37b.

"a loan"—subject to cancellation in the Sabbatical year.

"store credit"—see above, Sec. 11.

"the presumption is," etc.—see above, note to Sec. 25. The claim of a lost prosbul would have decided the matter forthwith in the creditor's favor, yet he chose the weaker argument of store credit not formalized into a regular loan. The presumption is that the latter is the truth, and that the creditor's conscience made him prefer the weaker truth to the more effective lie.

27. B. Giṭ 37a.

"and may be set aside by a mere verbal declaration"—RABD remarks, "Maimonides is contradicting himself here," without any further elucidation. Possibly he refers to the fact that the declaration is ineffective unless made, directly or indirectly, before the judges of the court (above, Sec. 18); this being so, the status of the parties, whether scholars or unlearned persons, is immaterial. RDBZ conjectures that Maimonides meant scholars may make the declaration before their colleagues, and need not forward it to the court.

28. Sheḇi 10:9, 8; B. Giṭ 37b.

29. B. Giṭ 37b.

30. Sheḇi 10:3; Sif Deut. 15:9.

Chapter X

1. B. RH 8b: Sif Lev. *Bĕ-har* 2:1.

"the Great Court"—in Jerusalem.

2. B. 'Ar 12b; Sif. Lev. *Bĕ-har* 1:2; B. AZ 9a; Ḳid 40b.

"from the year 2503"—that is from the end of that year, beginning the count with the year 2504.

"in the second year after creation"—creation is said to have begun in the final week of the last month, Elul, of year 1, Adam having been created on the following first day of Tishri, New Year's Day of year 2. Year 1 thus consisted actually of only five days.

3. B. ʽAr 13a, 11b, 32b; Yoma 9a.

"seventeen Jubilees"—not exactly, for, as Maimonides says further on, of the 17th Jubilee period only thirty-six years had passed.

"this count . . . ceased"—because the children of Israel went into exile, and the law of Jubilee applies only when the Land is inhabited by all of them. From Sec. 8, below, it would appear that the law ceased to operate earlier, at the time the Tribes of Reuben and Gad and half of the Tribe of Manasseh were exiled, but the count was continued nevertheless for the reason given at the end of this Section, i.e., for the sake of the Sabbatical years.

4. "the year in which the Second Temple was destroyed"—Maimonides evidently refers to the year following the destruction of the Temple, starting with the month of Tishri, about two months after the actual destruction which occurred on the 9th of Ab of the preceding year. That year was the first year of a septennate. The Second Temple stood for 420 years, but the count was resumed only in the seventh year after its erection (see the preceding Section), leaving 414 years, or 8 complete Jubilees plus 2 septennates; hence the Second Temple was destroyed in a Sabbatical year.

"the present year"—in which Maimonides is writing (1107 of the Second Temple = 1487 Sel. = 4936 A.M. = 1176 C.E.).

"one thousand one hundred and seven"—making twenty-two Jubilees and one septennate, the last year thus being a Sabbatical year.

"the Seleucid Era"—literally "Era of Documents," 312 B.C.E., the year of the establishment of the Seleucid empire, which is still used for the dating of documents by some oriental Jewish communities.

5. "which is traditional"—the Geonim based this reckoning on tradition, and did not deduce it from B. AZ 9b, here referred to.

6. "the year following a Sabbatical year"—1107 divided by 7 makes 158 septennates plus one year, which is the first year of the 159th septennate.

"tithes"—the second tithe, which is set aside in the first, second, fourth, and fifth years of the septennate, and the poor man's tithe, which is set aside in the third and sixth years (above, II, vi, 4).

"it is proper to rely upon them"—see, however, A.I. Zaslansky in *hat-Torah wĕham-mĕḏinah,* 9/10 (1958/59), 318, for some evidence to the contrary.

7. B. Neḏ 61a.

8. B. ʽAr 32b; Sif Lev. *Bĕ-har* 2:3; B. RH 9b; Ḳid 38b.

9. B. ʽAr 29a; Giṭ 36b.

"the law of the Hebrew bondsman"—see Exod. 21:2–6; Lev. 25:39–41.

"of houses in walled cities"—see Lev. 25:29–30.

"of the devoted field"—see Lev. 27:22–24.

"of the field of possession"—see Lev. 25:25–28.

"resident aliens"—who in order to acquire limited citizenship in the Land of Israel renounce idolatry.

"both only according to Scribal law"—a variant view holds that the Sabbatical year in the Land of Israel is authorized by Scriptural law. The reading in the MSS, however, applies "only according to Scribal law" to both.

"as we have explained"—above, IX, 3.

10. B. RH 9b, 30a; Sif Lev. *Bĕ-har* 2:5: RH 3:5, 4:9.

"tenth of Tishri"—which is the Day of Atonement.

"the Great Court . . . each individual"—based on the Scriptural pronouns used in the verse cited, *shalt thou* (singular) and *shall ye* (plural).

"the same as on New Year's Day"—as described in RH 4:9; cf. III, vi, iii, 1–4 (YJS, *14*, 375–76).

11. B. RH 33b; RH 3:5; B. RH 30a.

"in which the new month had been sanctified"—i.e., was declared to have begun, on the basis of the testimony of witnesses who had seen the new moon.

12. B. RH 30a; Sif. Lev. *Bĕ-har* 1:5.

13. B. RH 9b; Giṭ 36b.

"termed release of the land"—the Hebrew term for "release of the land" may also mean "rules of the Sabbatical year affecting land." Hence the definition here.

14. B. RH 8b; Sif Lev. *Bĕ-har* 2:1.

"wreaths"—customarily worn at banquets.

15. Sif Lev. *Bĕ-har* 3:2.

16. Sif Lev. *Bĕ-har* 3:6, 2:1; B. 'Ar 28b.

"as we have explained"—above, ix, 4.

Chapter XI

1. ARN 35; Midrash Tanḥuma, Lev. *Bĕ-har,* 1.

"have transgressed"—but are not liable to a flogging, since their act of sale and purchase, respectively, is null and void, and no flogging may be inflicted when there is no material act of malfeasance.

2. B. BM 79a. A term sale is by definition temporary and implies eventual repossession by the original owner, hence it is exempt from the Jubilee.

3. Sif Lev. *Bĕ-har* 5:1; Tos 'Ar 5:6.

"are validly sold"—because this prohibition is deduced indirectly from the aforecited verse, and is not explicitly stated therein; hence transgression of it cannot invalidate an otherwise legal act of sale.

4. See below, Chap. xii.

6. 'Ar 9:1; B. 'Ar 30a; Sif Lev. *Bĕ-har* 3:10.

7. Tos 'Ar 5:1; B. BM 79a, 109b.

"both buyer and seller are forbidden to enjoy it"—the seller, because the buyer is entitled to its benefit; and the buyer, because such benefit necessarily contributes to the destruction of the tree itself, which belongs to the seller. Usufruct may not be enjoyed if it results in the loss of the principal.

8. B. BM 109a.

9. 'Ar 9:1; Sif Lev. *Bĕ-har* 3:10; B. 'Ar 29b, 18b.

"as we have explained"—above, x, 8.

"two full years"—literally "two years, from time to time," the technical term for "full time," to the day and hour.

10. 'Ar 9:1; Sif Lev. *Bĕ-har* 3:10; B. BM 106a.

"years of the crops"—the plural *years* implies more than one, hence at least two.

11. 'Ar 9:1; B. 'Ar 29b.

"but does not plant it"—thus deliberately depriving himself of the second crop.

12. B. 'Ar 30b.

"for another year after the Jubilee"—even though the field returns in the meantime to its original owner, the buyer is still entitled to another year's crops.

13–14. B. ʿAr 14b.

13. "it must be restored"—Scripture says, *And he shall return unto his field of possession* (Lev. 25:27), without specifying whether the ground is fertile or not; cf. the next Section.

14. Cf. note to preceding Section.

"sell fruit trees"—exclusive of the ground underneath them.

15. Sif Lev. *Bĕ-har* 5:4.

16. Sif Lev. *Bĕ-har* 5:3; ʿAr 9:2; Tos ʿAr 5.

17. ʿAr 9:2; Sif Lev. *Bĕ-har* 5:2; B. ʿAr 30b.

18. ʿAr 9:2; Sif Lev. *Bĕ-har* 5:1.

19. Beḳ 8:10: B. Beḳ 52b.

"which is meant to cover gifts as well"—according to Rashi, *ad loc.*, this verse would be superfluous otherwise.

20. B. Beḳ 52b; Beḳ 8:10.

"is not rendered void"—the division of the estate is not annulled, but the heirs must exchange their shares among themselves (KM).

"has taken his widowed sister-in-law in levirate marriage"—and has thus inherited his deceased brother's share.

21. B. Beḳ 52b; BB 139b.

"need not restore it"—to the wife's kinsmen.

"is based only on Scribal law"—see IV, I, xii, 3 (YJS, *19*, 74).

"not to disgrace them"—by having strangers buried in their family plot, while they themselves must seek graves elsewhere.

"he is obligated to bury her"—see IV, I, xii, 2 (YJS, *19*, 74).

Chapter XII

1. ʿAr 9:3; Sif Lev. *Bĕ-har* 4:2.

"in a city surrounded by a wall"—see Lev. 25:29.

"twelve months"—as distinct from two years in the case of a field (above, xi, 9).

"and may not deduct"—as he may do in the case of a field of possession (above, xi, 4–5).

2. B. Ḳid 21a; ʿAr 31a; cf. above, xi, 17–18.

3. ʿAr 9:3; B. ʿAr 31b.

4. ʿAr 9:4; B. ʿAr 31b.

5. ʿAr 9:3.

"the additional month"—the thirteenth month of a leap year, second Adar, the year of grace thus comprising thirteen months.

6. B. ʿAr 31b. This is one of the peculiar cases involving the additional month of the Jewish leap year, in which the calendar year ends for the person who has purchased later. In this case the first sale was made on the fifteenth day of first Adar, and the second sale about fifteen days later, on the first day of second Adar. The next year, being regular, has only one Adar, so that the year of grace ends about two weeks earlier for the house that was purchased two weeks later, since the count runs from Adar to Adar.

7. ʿAr 9:4. This rule was enacted by Hillel, who noticed that the buyer would often hide on the last day in order to evade redemption and obtain permanent possession of the house.

8. B. 'Ar 31b; Sif Lev. *Bĕ-har* 4:8.

9. B. 'Ar 31b.

"the house is not restored"—see Lev. 25:30.

10. 'Ar 9:7; Sif Lev. *Bĕ-har* 4:6.

"hamlet"—literally "houses of courtyards," a settlement consisting of less than three courtyards, each containing at least two houses, and not likely to be surrounded by a wall; cf. below, Sec. 14.

"not surrounded by a proper wall"—for a definition of a proper wall see below, Sec. 15.

"governing . . . fields"—see above, xi, 4–5.

"and houses surrounded by a wall"—see above, Sec. 1.

11. 'Ar 9:5; Sif Lev. *Bĕ-har* 4:5.

12. B. Suk 3a; 'Ar 32b: BḲ 82b; 'Ar 9:5.

"less than four cubits square"—Scripture specifies *a dwelling house* (Lev. 25:29), which excludes such a 4 x 4 booth.

"in Jerusalem"—since the Holy City was not assigned to any individual Tribe when the Land was distributed, the law of walled cities cannot apply to it.

13. 'Ar 9:6; B. 'Ar 32a.

"gardens"—variant reading "roofs," i.e., the houses are so close together that they form a wall.

14. 'Ar 9:6; B. 'Ar 33b; Meḡ 3b.

15. 'Ar 9:6; B. 'Ar 32b; Ḥul 7a; cf. above, iii, i, 5.

16. This Section would seem to lend support to the hypothesis of Prof. Solomon Zeitlin that Maimonides believed in the imminent advent of the Messiah and the return of Israel to the Holy Land, and intended his Code to serve as the constitution for the restored State of Israel. Cf. S. Zeitlin, *Maimonides,* 2nd ed. (New York, 1955), pp. 83–86.

Chapter XIII

1. B. BB 122a; Mak 12b.

"was given no share in the Land"—see Deut. 18:1.

"had already been commanded"—Num. 35:2.

"open land"—see below, Sec. 2.

"cities of refuge," etc.—see Num. 35:6.

"When . . . additional cities of refuge will be designated"—see Deut. 19:8–9.

2. Soṭ 5:3; Tos Soṭ 5:13; Tos 'Ar 5:19.

3. B. Mak 12a.

4. 'Ar 9:8; Tos 'Ar 5:18. See the next Section. From the reasoning here it follows that neither may city space be converted into a field, or vice versa.

5. B. 'Ar 33b; 'Ar 9:8; Tos 'Ar 5:18.

6. Tos 'Ar 5:19.

"Similarily, in all the other cities of the Land of Israel"—this sentence is printed at the end of Sec. 5, but seems to belong to Sec. 6.

"nor plant the site of a ruined structure"—rather he should restore the structure. Generally it was more profitable to plant a piece of land than to use it as a building site.

7–9. ʿAr 9:8; B. ʿAr 33b.

7. "in the regular manner"—see above, xi, 4–5 and xii, 1.

"they may redeem them immediately"—and need not wait two years; see above, xi, 9–12.

"even after the lapse of several years"—unlike the houses of other people, where redemption is limited to one year; see above, xii, 1–4.

10. Sif Deut. 18:1; Sif Num. 18:20.

"it must be taken away from him"—no flogging is prescribed, since the piece of land can be repossessed intact; not so in the case of booty, where some or all of it may have been already consumed by the guilty Levite or priest.

11. "other lands"—cf. Num. 31:30–41, where the Tribe of Levi is stated to have been given a share in the Midianite booty.

12. Sif Num. *Maṭṭoṯ* 157; Num. Rabbah 1:12.

"his host"—usually translated *his substance.*

13. B. AZ 19b; Ber 32b.

GLOSSARY

'Am ha-'areṣ
see Unlearned person

Asherah
a tree used for idolatrous purposes

Associate scholar (*ḥaber*)
originally a member of a group that undertook to observe meticulously the rules governing heave offering and tithes, as well as the laws of cleanness and uncleanness. Later on the term was applied to scholarly persons who knew and observed all commandments, as distinct from unlearned persons (*'am ha-'areṣ*), who were regarded as unreliable in this respect

'Asufi
a foundling who knows neither his father nor his mother

Bastard (*mamzer*)
the issue of an adulterous or incestuous union

Benefit (*hănaʼah*)
indirect use of an article of food, such as use for sale or for fuel, as distinguished from the direct eating of it; or the deriving of any advantage or favor from another person

Bet kor
a piece of ground that may be sown with one kor of seed

Bet roba'
a piece of ground that may be sown with ¼ ḳab of seed

Bet sĕʼah
a piece of ground that may be sown with one sĕʼah of seed

Commandment, negative or positive
a commandment of the Torah stated in negative terms, "thou shalt not," or in positive terms, "thou shalt"

Denar (Latin *denarius*)
a silver or gold coin, the former worth 1/24 of the latter

Flogging for disobedience (*makkat mardut*, literally "lashes of rebellion")
flogging prescribed by Rabbinic law, the number of lashes being at the discretion of the court, in contradistinction to the Biblical lashes which are limited to no more than thirty-nine

Geonim (singular Gaon, "Excellency")
the title borne by the heads of the Babylonian academies, which flourished down to the 11th century

Geṭ
a bill of divorcement

Ḥaber see Associate scholar

Heave Offering, Great (*tĕrumah gĕdolah*)

a portion of the produce (about two percent on the average) which was given to the priests, who alone were permitted to consume it; cf. Num. 18:18; Lev. 22:10; Deut. 18:4

Heave offering of the tithe

out of the tithe which he received the Levite was obligated to give 1/10 to the priest; in other words, one percent of the original produce harvested by the Israelite; cf. Num. 18:25–32

Iron sheep property (*nikse ṣo'n barzel*)

property entrusted by one party to another (particularly by the wife to her husband), on condition that the trustee is to be responsible for any loss or damage thereto, regardless of the cause of such loss or damage

Israelite

a) a Jew, in contradistinction to a Gentile; b) a commoner who is neither priest nor Levite

'Issar (Latin *assarius,* later *as*)

a small Roman coin worth eight pĕruṭah, or four barleycorns of silver

Ḳab

a measure of capacity equal to four loḡ or 1/6 of a sĕ'ah

Ḳabbĕlan

a tenant who contracts to do some specific work in a field; or a surety who assumes a primary liability similar to that of the principal debtor

Kĕtubbah (literally "writ")

a) marriage contract specifying the mutual duties of husband and wife, as well as the husband's financial obligations toward the wife; b) the monetary amount of these obligations

Kor

a measure of capacity equal to thirty sĕ'ah

Koy

(a kind of bearded antelope) a creature concerning which the Rabbis were in doubt as to whether it was to be classified as a domestic animal or as a wild beast

Liṭra

measure of bulk equal to half a loḡ

Loḡ

a liquid measure equal to 1/4 of a ḳab, or the space occupied by six eggs

Lulab

the palm branch carried with the festive wreath during the Festival of Tabernacles (cf. Lev. 23:40); or more generally, the bouquet of palm branch, myrtle, and willow used with the ethrog on that festival

Ma'ămad ("station, place of standing")

the name given to groups of representatives (lay Israelites) from outlying districts, corresponding to the twenty-four divisions of the priests (*see Mišmar*). Some of them went up to the Temple as witnesses to the offering of the sacrifice (Ta 4:2); others came together in their own towns, where they held prayers at fixed times during the day, corresponding to the fixed times of sacrifice in the Temple

Mĕʿah

the smallest silver coin, equal in value to sixteen barleycorns of silver

Mĕlog̃ (literally "plucking") property.

property which belongs to the wife and of which the husband has only the usufruct, without any right to the principal or any responsibility for loss or deterioration

Mina (*maneh*)

a sum of money equal in value to 100 zuz or 100 denar

Minor

a girl up to the age of twelve years, and a boy up to the age of thirteen

Mišmar ("watch")

the priestly and Levitical families were divided into twenty-four watches, each being on duty for a week

Nathin (plural *Nethinim*)

a descendant of the Gibeonites who deceived Joshua and were condemned to be hewers of wood and drawers of water (Josh. 9:3–27)

Nazirite

a person who has uttered a vow to be a Nazirite was not permitted to drink wine or cut his hair until the expiration of his Nazirite term (Num. 6)

Nĕb̲elah

an animal slaughtered in any manner deviating, be it ever so slightly, from that prescribed by Jewish law.

ʿOmer

the sheaf of the first fruits of the harvest, also called the sheaf of waving, brought as an offering on the 16th of Nisan (Lev. 23:9–14); also a dry measure of capacity equal to 1/10 of an ephah.

ʿOrlah (literally "foreskin")

the fruit of newly planted trees in the first three years, which was forbidden to be eaten (Lev. 19:23–25)

Pĕruṭah

the smallest copper coin, equal to 1/8 of an ʾissar

Pondion

a large copper coin equal in value to two ʾissar

Priest (*Kohen*)

a descendant of Aaron, the first High Priest and the brother of Moses, in the male line, whose privileges and obligations as a minister are set forth in Lev. 21–22

Resident alien

semi-proselyte who renounces idolatry in order to acquire limited citizenship in the Land of Israel

Retroactive differentiation (or choice, *bĕrerah*)
the selection retrospectively of one object rather than another, as having been designated by a term equally applicable to both

Sages (or Scribes)
scholars of the post-Biblical period, the expounders and bearers of the Oral Law as recorded in the Talmud

Scribal law
law promulgated by the Scribes or Sages

Sĕ'ah
a measure of capacity equal to six ḳab̲

Sela'
a coin or weight equal to two common shekels or four *denar*

Symbolic act of barter (*ḳinyan*)
a formality of acquisition of an object, or of confirmation of an agreement, executed by the handing over of a kerchief or any other article by one of the parties to the other, or by the witnesses to the agreement, thus symbolizing that the object itself has been transferred or the obligation assumed

Ṭeb̲el
produce from which the priests' and Levites' dues have not yet been separated

Tĕrefah
an animal torn by a wild beast or suffering from a serious organic disease, whose flesh is forbidden even if it has been properly slaughtered according to Jewish law

Tithes (*ma'ăśer*)
were of three kinds: first tithe was given to the Levite in each of the first six years of the Sabbatical cycle; second tithe was set apart in the first, second, fourth, and fifth years of the cycle and was consumed by the owner in Jerusalem; poor man's tithe was given to the poor in the third and sixth years of the cycle

Ṭumṭum
a person whose sex cannot be determined with certainty

Unlearned person (*'am ha-'areṣ,* literally "people of the land")
an uninstructed person who is indifferent to the tithing of produce and to the observance of ritual cleanness and uncleanness

Zuz
a coin equal in value to a denar

BOTANICAL GLOSSARY

(Notes signed H. are by Professor G. Evelyn Hutchinson)

[The nomenclature follows the *Flora Palaestina* (Fl. Pal.), of which two volumes by Michael Zohary (Jerusalem, Israel Academy of Sciences and Humanities, 1966–) have so far appeared. For the families not yet treated, unfortunately including all the monocotyledons, the taxonomic treatment follows *Flora of Syria, Palestine, and Sinai,* by G. E. Post, 2nd edition, revised by J. E. Dinsmore (Beirut, American Press, 1933) (Post).—H.]

Acorns (?)	[*Quercus* spp.—H.]	קליסין, כליסין

[A number of species occur.—H.]

Almond	*Amygdalus communis* Linn.	שקד
Aloe	*Aquilaria agallocha* Roxb.	אהל

[Though not native to the region, is usually regarded as the correct identification. The English name could mean *Aloe vera* Linn.—H.]

Amaranth	*Amaranthus* sp.	ירבוז

[It is almost certain that there is a confusion here that I have not been able to resolve. Fl. Pal. gives ten species as established in the region. Of these eight are certainly of American origin, and cannot have occurred in antiquity. Two, *A. graecizans* Linn. and *A. gracile* Desf., are now of immensely wide distribution in the warmer parts of all continents. *A. gracile* is used occasionally as a vegetable. Their home is apparently unknown, but they may well be American in origin also, though there may also be a center of dispersal of the genus in Eastern Asia. I think there must be a mistake in translation here.—H.]

Amaranth, wild		ירבוז שוטה
Apple	*Pyrus malus* Linn.	תפוח
Artichoke	*Cynara scolymus* Linn.	קנרס, קנדס
Arum	*Arum dioscoridis* Sibth. & Sm. *Arum palaestinus* Boiss.	לוף

[Both yield edible roots. The first-named seems to be the more widespread.—H.]

Arum, wild	*Arum dioscoridis*	לוף שוטה
Asafetida	*Ferula assafoetida* Linn.	חלתית
Balsam	*Commiphora opobalsamum* (Linn.) Eng.	קטף

[The English name could mean *Balanites aegyptiaca* (L.) Del., for which Post gives the Arabic name *zakkūm,* and Fl. Pal. the Hebrew name צרי.—H.]

Barley	*Hordeum vulgare* Linn.	שעורה
Barley, two-rowed (foxtail)	*Hordeum distichon* Linn.	שבולת שועל
Bean (broad bean)	*Vicia faba* Linn.	פול
Bean, Egyptian, *see* Cowpea, Nile		
Bean, hyacinth (white bean)	*Dolichos lablab* Linn.	פול לבן
Boxthorn	*Lycium europaeum* Linn.	אטד
Cabbage, garden	*Brassica oleracea* var. *capitata* Linn.	תרובתור
Cabbage, kale	*B. oleracea var. acephala* DC.	כרוב
Caper berry	*Capparis*	אביונה, נצפה
Caper bush	*Capparis spinosa* Linn.	צלף
Caper flower		קפרים
Carob	*Ceratonia siliqua* Linn.	חרוב
Cedar	*Cedrus libani* Loud.	ארז
Celery	*Apium graveolens* Linn.	כרפס
Ceterach (scale fern)	*Ceterach officinarum* Lam. & DC.	דנדנה
Charlock	*Sinapis arvensis* Linn.	לפסן
Chick-pea	*Cicer arietinum* Linn.	אפון

[Not native.—H.]

Chicory	*Cichorium intybus* Linn.	עולשין
Chicory, wild	*C. pumilum* Jacq.	עולשין השדה
Citron (ethrog)	*Citrus medica* Linn.	אתרוג
Colocynth	*Citrullus colocynthis* (Linn.) Schrad.	פקועה
Coriander, garden	*Coriandrum sativum* Linn.	כסבר

Coriander, mountain	*Smyrnium connatum* Boiss.	כסבר הרים
Costus root	Costus	קושט

[There is a tropical genus of this name in the *Zinziberaceae;* I cannot find that its root was used. The name may have had another meaning in antiquity.—H.]

Cotton	*Gossypium arboreum* Linn.	צמר גפן
Cowpea	*Vigna sinensis* (Linn.) Savi.	פול מצרי

[Said by Post not to be cultivated in the area.—H.]

Cowpea, Nile (Egyptian pea)	*Vigna luteola* (Jacq.) Benth. (=*V. nilotica* Del.)	שעועית
Cucumber, *see* Melon, cucumber		
Cumin	*Cuminum cyminum* Linn.	כמון
Cypress; *see also* Juniper	*Cupressus sempervirens* Linn.	ברוש
Darnel (tare)	*Lolium temulentum* Linn.	זון

[The English word tares (Post) seems also to be used for *Vicia sativa* Linn.—H.]

Date berries		כפניות
Date palm	*Phoenix dactylifera* Linn.	תמר, דקל
Dates, stony		שיצין
Dill	*Anethum graveolens*	שבת
Durra	*Sorghum vulgare* Pers.	דוחן
Dyer's rocket, *see* Rocket, dyer's		
Emmer	*Triticum dicoccoides* (Koern.) A. Schult.	כוסמין
Fennel	*Foeniculum vulgare* Mill.	גופנין
Fennel flower	*Nigella sativa* Linn.	קצח

[I find no authority for the English name fennel flower. *Nigella sativa* has the same Hebrew name in Fl. Pal. as here, so it is certainly correct. Post gives nutmeg flower or fitches as the English name.—H.]

Fenugreek	*Trigonella foenum-graecum* Linn.	תלתן
Fig; *see also* Sycamore	*Ficus carica* Linn.	תאנה

[The fig is said to be "rather common in the wild state as well as extensively cultivated." There is in addition the wild false sycamore, *F. pseudo-sycomorus,* as well as the sycamore.—H.]

Figs, green or unripe		פגים
Figs, inferior		כליסים
Figs, white		בנות שוח
Figs, wild		שתין
Flax	*Linum usitatissimum* Linn.	פשתן
Foxtail, *see* Barley, two-rowed		
Garden cress	*Lepidium sativum* Linn.	שחלים
Garlic	*Allium sativum* Linn.	שום
Garlic, Baalbek		שום בעל בכי

[Not traceable; there are 49 species of *Allium* recorded by Post, none with this vernacular name.—H.]

Gourd, *see* Pumpkin		
Grape vine	*Vitis vinifera* Linn.	גפן
Grapes		ענבים
Grapes, unripe		בסר
Grass pea	*Lathyrus sativus* Linn.	טפח

[Apparently not native.—H.]

Grass pea, red	*Lathyrus cicera* Linn.	פורקדן
Hawthorn (berry)	*Crataegus azarolus*	עוזרד
Heliothrope	*Heliothropium europaeum*	ערקבנין
Hemp	*Cannabis sativa*	קנבס
Henna	*Lawsonia alba*	כופר
Horse-radish	*Sonchus oleraceus*	מרור
Hyacinth bean, *see* Bean, Hyacinth		
Hyssop	*Origanum syriacum* Linn. (*Hyssopus officinalis*)	אזוב
Iris	*Iris* spp.	אירוס

[About forty species are known from the area. *I. pallida* is not the commonest, and I would prefer merely *Iris* spp. unless there is clear contextual evidence of the species.—H.]

Ivy	*Hedera helix* Linn.	קיסום
Jujube	*Zizyphus jujuba* Mill.	שזיפין
Jujube, Damascene		דרמסקניות

[Probably *Z. lotus* (Linn.) Willd.—H.]

Jujube, wild	*Zizyphus spina-christi* (Linn.) Willd.	רים
Juniper; *see also* Cypress	*Juniperus phoenicia* Linn.	ברוש
Kale, *see* Cabbage, kale		
La(b,u)danum (rock rose)	*Cistus creticus* Linn.	לוט

[Or possibly *C. salviaefolius* Linn.; roots of the latter used in folk medicine as a styptic or in bronchitis.—H.]

Leek, field	*Allium ampeloprasum* Linn.	כרישי שדה
Leek, garden; *see also* Porret	*Allium porrum* Linn.	כרישה, חציר
Lentil	*Lens culinaris* Medit.	עדשים
Lettuce, garden	*Lactuca sativa* Linn.	חזרת
Lettuce, wild	*Lactuca scariola* Linn.	חזרת גלים
Lily, white	*Lilium candidum* Linn.	שושנה
Lupine	*Lupinus termis*	תורמוס

[*Lupinus album* Linn. seems to be the commonest cultivated species; others are known wild in the area.—H.]

Madder	*Rubia tinctorum* Linn.	פואה
Melon, chate	*Cucumis melo* var. *chate* Linn.	קישוא, קישות
Melon, cucumber	*Cucumis melo* Linn.	מלפפון
Millet	*Panicum miliaceum* Linn.	פרגין
Mulberry	*Morus nigra* Linn.	תות

[Almost certainly this species in Old Testament times; it is said that *M. alba* was imported from China for feeding silkworms, though both are now common.—H.]

Mushroom	*Boletus*	פטריה

[The ordinary mushroom, wild or cultivated in Europe and now in North America, is a species of *Agaricus,* usually *A. campestris,* but *Boletus* is also much eaten.—H.]

Mustard, black	*Brassica nigra* (Linn.) Koch	חרדל

Mustard, field, *see* Charlock		
Mustard, white	*Sinapis alba* Linn.	חרדל מצרי
Olive	*Olea europea* Linn.	זית
Onion	*Allium cepa* Linn.	בצל
Orach	*Atriplex halimus* Linn.	לעונין
Palm, *see* Date palm		
Palm bud, terminal		קור
Peach	*Amygdalus persica* Linn.	אפרסק
Pear	*Pyrus communis* Linn.	אגס
Pear, crustuminian	*Pyrus crustumina*	קרוסטמל

[I cannot find the crustuminian pear.—H.]
(cf. Jastrow, p. 1414 b)

Pear, Syrian	*Pyrus syriaca* Boiss.	חזרר
Pepper, black	*Piper nigrum*	פלפל
Pine, stone, cone of	*Pinus pinea*	אצטרבול
Pistachio	*Pistacia vera* Linn.	בטנה
Pomegranate	*Punica granatum* Linn.	רמון
Porret; *see also* Leek	*Allium porrum* Linn.	קפלוט
Pumpkin	*Cucurbita pepo* Linn.	דלעת

[And its varieties, notably *C. italica* ("zucchini"), widely cultivated. It is not clear that any are of ancient origin.—H.]

Purslane	*Portulaca oleracea* Linn.	חלגלוגת
Quince	*Cydonia oblonga* Mill.	פריש
Radish	*Raphanus sativus* Linn.	צנון
Rape	*Brassica napus* Linn.	נפום

[Rarely, if ever, cultivated with us (Post).—H.]

Reed	*Arundo donax* Linn.	קנה
Rice	*Oryza sativa* Linn.	אורז
Rocket, dyer's	*Reseda luteola* Linn.	רכפה
Rocket, garden	*Eruca sativa* Linn.	גרגר, גרגיר
Rose	*Rosa damascena* Mill.	ורד

[Much cultivated; it is probably a hybrid of *R. gallica* Linn., the ancestor of most garden roses, and the dog rose, *R. canina* As. *R. gallica* occurs in Asia Minor, and *R. canina* extends south into Palestine. *R. damascena* could have originated not far to the north. I suspect, however, that the name has no direct connection with Damascus, but rather an indirect one by way of damask. If the context is of cultivated roses, I would give *R. damascena;* if wild, probably *R. canina* is meant, though there are other species in the area. The taxonomy of the entire genus is difficult.—H.]

Rue	*Ruta graveolens* Linn.	פיגם

[*Ruta chalepensis* Linn. seems to be the local species; *R. graveolens* may have been cultivated; it was a very important medical plant in antiquity.—H.]

Safflower	*Carthamus tinctorius* Linn.	קוצה, חריע

[Post gives קוץ as the Hebrew name of *C. tenuis* (Boiss. & Bl.) Bornm, Several other species occur in the region.—H.]

Saffron	*Crocus sativus* Linn.	כרכום
Savory	*Satureja thymbra* Linn.	סיאה
Service tree	*Eriolobus trilobata* (Labill.) M. Roem.	זרד
Sesame	*Sesamum orientalis*	שומשום

[Not in Post.—H.]

Spelt	*Triticum spelta* Linn.	שיפון
Spice tips		ראשי בשמים
Spinach, beet	*Beta vulgaris* var. *cicla* Linn.	תרד

[Ordinarily known as Swiss chard.—H.]

Sumac	*Rhus coriaria* Linn.	אוג
Sycamore; *see also* Fig	*Ficus sycomorus* Linn.	שקמה
Sycamore fig		גמזית
Terebinth	*Pistacia atlantica* Desf.	אלה

[Perhaps also *P. palaestina* Boiss.—H.]

Thistle	*Centaurea* (?)	דרדר

[*Carduus, Cirsium,* and some other genera, perhaps including *Centaurea.*—H.]

Thistle, sow, *see* Horse-radish		
Thorns	*Notobasis syriaca* (Linn.) Cass.	קוץ, חוח
Thyme	*Thyma capitatus* (Linn.) Hoffm.	קורנית, קרנס
Truffle	*Tuber* spp.	כמהין

[Or allied species.—H.]

Turnip	*Brassica rapa* Linn.	לפת

Vetch, bitter	*Vicia ervilia* (Linn.) Willd.	כרשינה
Vetch, French	*Vicia narbonensis* Linn.	ספיר
Walnut	*Juglans regia* Linn.	אגוז
Watermelon	*Citrullus vulgaris* Schrad.	אבטיח
Wheat	*Triticum* spp.	חטה
Woad	*Isatis tinctoria* Linn.	אסטים

[Presumably this species when the context involves a dye. Post records it doubtfully from Gennesaret; Fl. Pal. omits this species.—H.]

ZOOLOGICAL GLOSSARY

(Notes signed H. are by Professor G. Evelyn Hutchinson)

[In general the nomenclature follows Ellerman's and Morrison-Scott's *Checklist of Palaearctic and Indian Mammals* and Peters' *Birds of the World*.—H.]

Addax	*Addax nasomaculatus* (Blainville)	יחמור
Ant	*Messor semirufus*	נמלה

[Presumably this species.—H.]

Ass	*Equus* (*Asinus*) *asinus* Linn.	חמור
Ass, wild	*Equus* (*Asinus*) *hemionus hemippus* I. Geoffr.	ערוד

[*E. h. onager* is the Persian subspecies. Both are supposedly mentioned in Job 39:5.—H.]

Carp	*Cyprinus carpio*	שיבוט (?)
Cattle, domestic	*Bos taurus* Linn.	שור (בקר)
Cattle, wild	*Bos primigenius*	שור הבר
Dog	*Canis familiaris* Linn.	כלב
Elephant	*Loxodonta africanum* (Blumenbach)	פיל
Fallow deer	*Dama dama* Linn.	יחמור

[Or *Dama mesopotamica* Brooke. Which species occurred in Biblical times, seems to be uncertain—probably *D. mesopotamica*.—H.]

Fox	*Vulpes vulpes aegyptiaca* (Sonnini)	שועל
Gazelle; *see also* Hind	*Gazella gazella* Pallas	צבי

[Or *G. dorcas Saudiya* Carruthers & Schwarz.—H.]

Goat	*Capra hircus* Linn.	עז, תיש
Goose, domestic	*Anser anser domesticus*	אוז
Goose, wild		אוז בר
Hare	*Lepus* spp.	ארנבת

[*Lepus capensis aegyptius* Desmarests, or *L. europaeus judeae* Gray—both are known in Palestine.—H.]

Hind; *see also* Gazelle		צביה
Horse	*Equus* (*equus*) *caballus* Linn.	סוס
House-fly	*Musca domestica* Linn.	זבוב
Ibex	*Capra ibex nubiana* F. Cuvier	יעל
Jackal	*Canis aureus syriacus* Hemprich & Ehrenberg	שועל
Koy, see Sheep, wild		כוי
Lion	*Panthera leo* Linn.	ארי
Locust	*Schistocerca gregaria* Forsk.	גובאי
Mule	*Equus asinus mulus*	פרד
Mule foaled by horse dam		רמך
Mullet	*Mugil cephalus* Linn.	שיבוט (?)
Onager, *see* Ass, wild		
Ox, *see* Cattle		
Sheep	*Ovis aries* Linn.	כבש, רחל
Sheep, wild	*Ovis orientalis* Gmelin	כוי (?)

[This is the ancestor of *O. aries;* a subspecies still exists in Iran. Doubtless the one caught in the thicket that saved Isaac's life belonged here.—H.]

Swine	*Sus scrofa domesticus* Brisson	חזיר
Wolf	*Canis lupus* Linn.	זאב

SCRIPTURAL REFERENCES

GENESIS

EXODUS

LEVITICUS

NUMBERS

DEUTERONOMY

JOSHUA

2 SAMUEL

1 KINGS

2 KINGS

ISAIAH

JEREMIAH

MALACHI

PSALMS

PROVERBS

JOB

RUTH

DANIEL

EZRA

NEHEMIAH

INDEX